Galileo's

Sidereus Nuncius

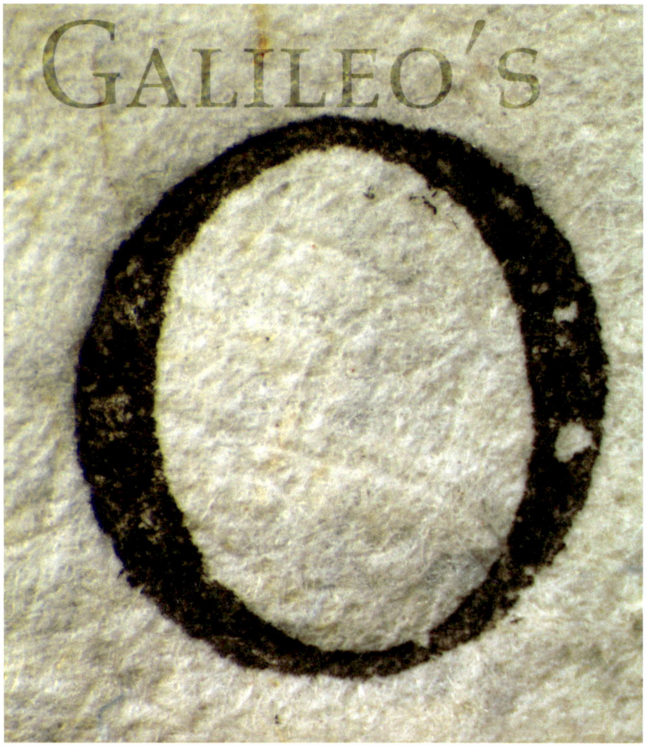

Edited by Horst Bredekamp Volume I

Galileo's
Sidereus Nuncius

A comparison of the proof copy (New York) with other paradigmatic copies

Edited by
Irene Brückle and Oliver Hahn

Akademie Verlag

Published with support by the Max-Planck-Gesellschaft,
the Alexander von Humboldt-Stiftung and the
BAM Bundesanstalt für Materialforschung und -prüfung

Bibliographical information of the German National Library
The German National Library lists this publication in the German national bibliography; bibliographic details are available on http://dnb.d-nb.de.

A scientific publisher of the Oldenbourg publisher group.

www.akademie-verlag.de

Organization of both volumes by Alexis Ruccius
Fine tuning and proofreading of Volume I by Theresa Smith

Graphic design and layout by Petra Florath, Berlin
Printing and binding: DZA Druckerei zu Altenburg GmbH, Altenburg

This paper is resistant to ageing (DIN/ISO 9706).

ISBN 978-3-05-005095-9

CONTENTS

IV THE PRINTING MATERIALS

V AFTERWORD

VI APPENDICES

PREFACE TO VOLUMES I AND II

CHRONOLOGY OF A DISCOVERY

JULY 2005–JANUARY 2006: THE CONFRONTATION

The *Sidereus Nuncius* made Galileo Galilei famous. This slim work, published in 1610 in Venice, captured Galileo's observations of the moon as well as the Milky Way, demonstrated that Jupiter had satellites, and created a new framework for questions of cosmology. Above all, with its five depictions of the moon and its numerous Jupiter-satellite configurations, this book created a standard: it made the sequence the rule for natural scientific observations and experiments. The five etchings (from four plates, one repeated) of the moon showed a surface whose rugged, pocked appearance contradicted the teaching that the heavenly bodies were smooth, ethereal spheres. In brilliant descriptions that provided the foundation of his reputation as a high stylist, Galileo was also able to relay his observations in a most elaborate language. With the uneven moon, the earth lost its exceptional position in the universe, and the newly discovered satellites of Jupiter put an end to the uniqueness of the known earth-moon system. Galileo was instantly famous.

In July 2005, I received an e-mail from the New York antiquarian booksellers Martayan Lan, which left me hardly less perplexed than Galileo must have been when he first directed his gaze through the telescope at the moon. Richard Lan laconically asked whether I could comment upon the enclosed attachment, page 8r of the copy of the *Sidereus Nuncius* (the *New York copy*) that was in his possession.[1] When the attachment began loading on my monitor, what appeared on this page was not the expected etching (Fig. 1), but a drawing of the moon (Fig. 2). Apparently, the area for the print of the moon was taken up by a wash drawing. Since nothing comparable in Galileiana has appeared in more than a hundred years, I was simultaneously electrified and skeptical. My first impression left every possibility open, from it being an authentic document to a contemporaneous or later copy, to a counterfeit.

When I conveyed to the owners the possible interpretations, including my skepticism, Richard Lan sent me high-resolution digital photographs of the remaining four wash drawings in the book and the title page, under whose last line was displayed the signature: "Io Galileo Galilei f." In September 2005, I compared the handwriting with many signed Galileo letters in the *Biblioteca Nazionale Centrale di Firenze* (BNCF, Central National Library of Florence). Through these investigations and through discussions with Isabella Trucci, a specialist in the handwriting of Galileo, I came to the conclusion that the signature is definitely authentic. This inscription alone makes the book a singular discovery.

1 Galileo Galilei, *Sidereus Nuncius*, Venice 1610, in the possession of Richard Lan and Seyla Martayan, of the antiquarian book and map firm Martayan Lan, in New York.

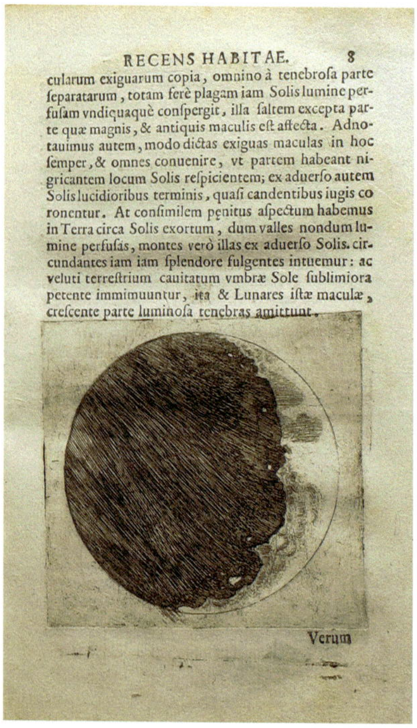
RECENS HABITAE. 8

culärum exiguarum copia, omnino à tenebroſa parte ſeparatarum, totam ferè plagam iam Solis lumine perfuſam vndiquaquè conſpergit, illa ſaltem excepta parte quæ magnis, & antiquis maculis eſt affecta. Adnotauimus autem, modo dictas exiguas maculas in hoc ſemper, & omnes conuenire, vt partem habeant nigricantem locum Solis reſpicientem; ex aduerſo autem Solis lucidioribus terminis, quaſi candentibus iugis coronentur. At conſimilem penitus aſpectum habemus in Terra circa Solis exortum, dum valles nondum lumine perfuſas, montes verò illas ex aduerſo Solis circundantes iam iam ſplendore fulgentes intuemur: ac veluti terreſtrium cauitatum vmbræ Sole ſublimiora petente imminuuntur, ita & Lunares iſtæ maculæ, creſcente parte luminoſa tenebras amittunt.

Verum

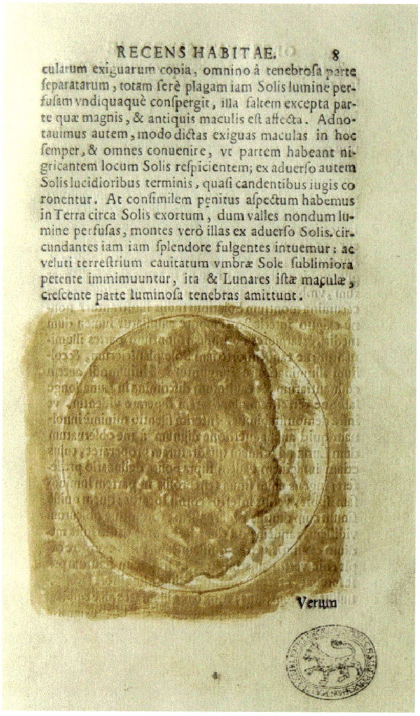
RECENS HABITAE. 8

culärum exiguarum copia, omnino à tenebroſa parte ſeparatarum, totam ferè plagam iam Solis lumine perfuſam vndiquaquè conſpergit, illa ſaltem excepta parte quæ magnis, & antiquis maculis eſt affecta. Adnotauimus autem, modo dictas exiguas maculas in hoc ſemper, & omnes conuenire, vt partem habeant nigricantem locum Solis reſpicientem; ex aduerſo autem Solis lucidioribus terminis, quaſi candentibus iugis coronentur. At conſimilem penitus aſpectum habemus in Terra circa Solis exortum, dum valles nondum lumine perfuſas, montes verò illas ex aduerſo Solis circundantes iam iam ſplendore fulgentes intuemur: ac veluti terreſtrium cauitatum vmbræ Sole ſublimiora petente imminuuntur, ita & Lunares iſtæ maculæ, creſcente parte luminoſa tenebras amittunt.

Verum

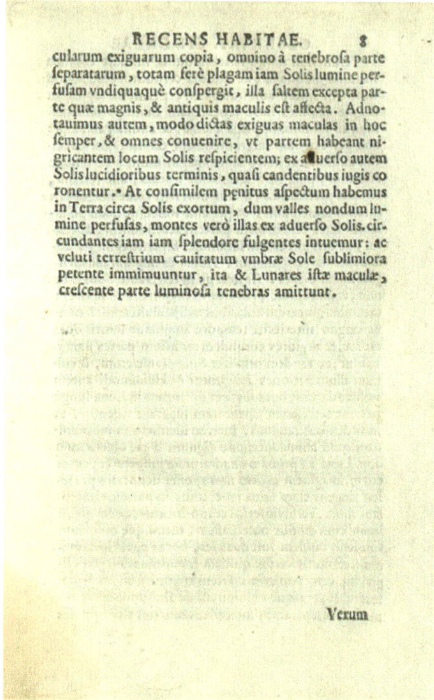
RECENS HABITAE. 8

culärum exiguarum copia, omnino à tenebroſa parte ſeparatarum, totam ferè plagam iam Solis lumine perfuſam vndiquaquè conſpergit, illa ſaltem excepta parte quæ magnis, & antiquis maculis eſt affecta. Adnotauimus autem, modo dictas exiguas maculas in hoc ſemper, & omnes conuenire, vt partem habeant nigricantem locum Solis reſpicientem; ex aduerſo autem Solis lucidioribus terminis, quaſi candentibus iugis coronentur. At conſimilem penitus aſpectum habemus in Terra circa Solis exortum, dum valles nondum lumine perfuſas, montes verò illas ex aduerſo Solis circundantes iam iam ſplendore fulgentes intuemur: ac veluti terreſtrium cauitatum vmbræ Sole ſublimiora petente imminuuntur, ita & Lunares iſtæ maculæ, creſcente parte luminoſa tenebras amittunt.

Verum

Fig. 1: Galileo Galilei, Sidereus Nuncius, Venice 1610, p. 8r, Munster copy.

Fig. 2: Galileo Galilei, Sidereus Nuncius, Venice 1610, p. 8r, New York copy.

Fig. 3: Galileo Galilei, Sidereus Nuncius, Venice 1610, p. 8r, Zurich copy.

These first results and the following investigations, carried out over a period of two years with the methods of forensics and comparison, and which have been part of my broader investigations of Galileo's draftsman-minded hand, have gone into my book *Galilei der Künstler* (*Galileo the Artist*).[2] It might be helpful to indicate the chronological stages of the investigation.

First, I ordered reproductions of the *Sidereus Nuncius* from libraries at which I could find it listed. Among the photographic and digital copies that came to me in the following months were nine in which the etchings were indeed lacking (Fig. 3).[3] Galileo himself spoke of thirty copies that did not contain etchings.[4]

Suspecting that I was confronted with one of these copies, and armed with detailed photos of the Florentine drawings, in November 2005 I inspected the New York *Sidereus Nuncius* for the first time. A mere glance sufficed to ascertain that the pictures of the moon were not carried out as miniatures, but as sketches whose mixture of fidgetiness and precision was impressive. Until now, seven drawings of the moon were known from Galileo's Florentine legacy; six of them are collected on a bifolium at the BNCF (see p. 51). I began a

2 Bredekamp 2007, pp. 131–160.

3 Staatsbibliothek zu Berlin, Stiftung Preußischer Kulturbesitz (Mat F30/Mat F591), Liège, Bibliothèque de l'Université (R. 190c; Brussels, facsimile, printed 1967); Durham, Palace Green Library (SB 0255); Cologne, Staats- und Universitätsbibliothek (N4/612); *New York copy*, private ownership (Seyla Martayan and Richard Lan); United States, private ownership (former Gudesche Bibliothek Rendsburg, Q 282); Paris, Bibliothèque Nationale (Rés.p.V.743); Paris, Observatoire (20174 (1)); St. Andrews University, Department of Special Collections (QB41.G2SA Rare Books); Zurich: ETH-Bibliothek (Rar 4342).

4 Letter to Belisario Vinta, March 19, 1610, *Opere*, vol. X, p. 300, lines 94–97.

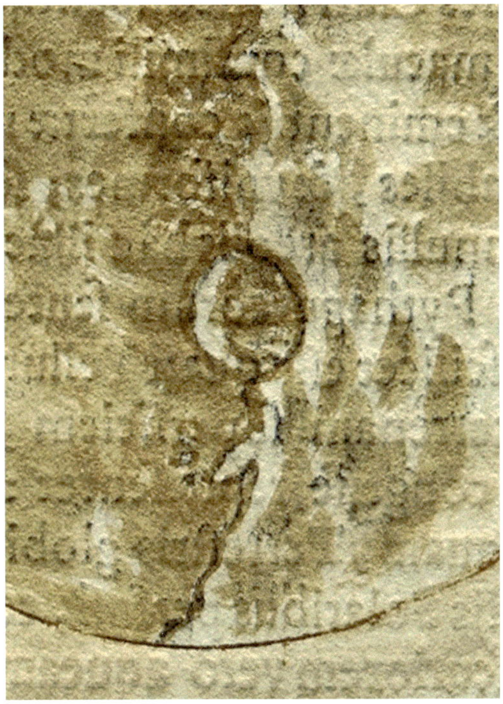

Fig. 4: Galileo Galilei, Sidereus Nuncius, Venice 1610, p. 9v, detail, New York copy.

renewed investigation of these well-known drawings. It became clear to me that their deviations from the etchings are so marked that they cannot possibly be the immediate models,[5] but as authentic documents of one of Galileo's greatest discoveries, they possess extraordinary value. Following the preliminary work of Samuel Egerton, in the 1990s I analyzed their style, moon by moon.[6] Now I was all the more impressed by both the similarities and the differences between the New York and Florentine drawings. It turned out that some of the differences were caused by the paper of the *New York copy* (see Brückle et al. p. 127ff.), whereas the stylistic similarities left no doubt that the same hand was at work.

Equally important results came out of a comparison between the *New York copy* drawings and the etchings. I saw that the drawings carry more information than the printed pictures of the moon, thus they must have come first. This becomes all the more clear through a detail visible on closer inspection. Only when I studied the *New York copy* a second time in New York, looking through a magnifying glass at page 9v (Fig. 4), for example, did I notice the systematic use of a black material over the drawings in significant areas. Since this black material lies over the colored drawing, it cannot have served as a preliminary sketch; on the other hand, because the intentionally highlighted features respond to the primary lines of the moon, it was apparently used as a medium of transfer.[7]

5 Bredekamp 2007, pp. 146–148.
6 Bredekamp 2000; Edgerton 1984; idem 1991.
7 Bredekamp 2007, pp. 198–208.

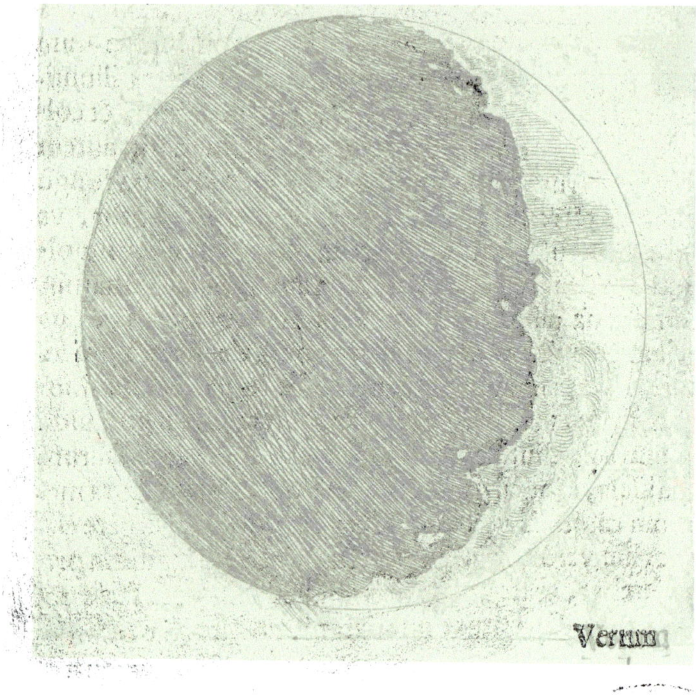

Fig. 5: Galileo Galilei, Sidereus Nuncius, Venice 1610, p. 8r, isolated black material of the New York copy overlaid on the etching of the Florence copy.

FEBRUARY 2006: THE FIRST BERLIN INVESTIGATION

This phenomenon, difficult to explain, led me to suggest additional laboratory tests to the owners. They were willing to send their book to Berlin for a week. Within a few days all the relevant institutions of the *Stiftung Preußischer Kulturbesitz* (Prussian Cultural Heritage Foundation) were able to develop a common strategy. Hein-Th. Schulze Altcappenberg agreed to keep the book in the *Kupferstichkabinett Berlin* (Museum of Prints and Drawings) for the duration of the analysis in February 2006 and also allowed the tests to be carried out there; Holm Bevers organized the particulars, while Irene Brückle, since 2008 Professor at the *Staatliche Akademie der Bildenden Künste Stuttgart* (State Academy of Art and Design Stuttgart), and Theresa Smith carried out the investigations of the paper, the media including the drawing technique, and took on the overall organization of the collaboration.

Oliver Hahn from the *BAM Bundesanstalt für Materialforschung und -prüfung* (Federal Institute for Materials Research and Testing) and Timo Wolff from the *Technische Universität Berlin* (Technical University Berlin) carried out additional materials-science investigations with the aid of X-ray fluorescence analysis. This was carried out in collaboration with the *Rathgen-Forschungslabor* (Rathgen Research Laboratory) under the supervision of Stefan Simon, who devoted his attention in particular to the black material overlaying the drawings. Using digital image analysis, Sonja Krug was able to isolate the pattern of this layer of material (Fig. 5), overlay it to scale on digital photographs of the etchings, and conduct a systematic analysis.[8]

8 Bredekamp 2007, pp. 198–205.

This collaboration between the specialists of various institutions, exemplary in its unbureaucratic and precise interplay, was not only able to differentiate this black material from the printing ink of text and etchings, it also clarified a number of details, for example how the drawings were transferred to the printing plates. All these elements went into the book *Galilei der Künstler*, though they could not be explained in full detail there. For this reason, from the beginning we considered presenting the results in a separate publication to document not only this unique object, but also the model of collaboration.

APRIL AND MAY 2008: COMPARISON OF THE *GRAZ*, *NEW YORK* AND *WASHINGTON COPIES*

During the preparation of this book, the desire arose to repeat a number of investigations because the evaluation of the first results presented new questions and possibilities. Above all, it seemed indispensable to check the data by means of comparison with other examples of the *Sidereus Nuncius* and to confirm their interpretation. In April and May 2008, the *New York copy* was again brought to Berlin to be analyzed at the *Kupferstichkabinett Berlin*, but now with more time available: an entire month. The examination that had already been conducted was expanded. Now it also included the *Fraunhofer-Institut für Umwelt-, Sicherheits- und Energietechnik UMSICHT* (Fraunhofer-Institute for Environmental, Safety and Energy Technology UMSICHT) in Oberhausen which took confocal photomicrographs using digital microscopy.

The generous cooperation of the *Karl-Franzens Universität* of Graz (Karl-Franzens University), Austria and of a book dealer in Paris made it possible to bring two of the most telling copies of the *Sidereus Nuncius* to Berlin during the period of testing. A collection of copies was thus brought together whose number had almost certainly never been in one place since their printing in March 1610, after which they were sold in all directions. The *Library of Congress* in Washington bought the former Paris copy in November 2008. Because of this shift we name it in the following the *Washington copy*.

The copies from Graz and Washington differ in that the Graz title page is covered by the initials and marks of changing owners (Fig. 6), while the Washington title page has remained free of notation (Fig. 7). The reason is that the book from Washington was never permanently bound, and thus preserves the form in which the printer delivered it.

The dimensions of the paper changed our impression of the book. The copy from Graz is the usual trimmed, smooth-edged, bound book. The *Washington copy*, by contrast, creates a downright majestic impression: with a broad skirt of paper, the text is situated in a generous environment. This is definitely more than an aesthetic element. Most of the surviving copies of the *Sidereus Nuncius* have been trimmed along the edges; the double page of the belt and sword of Orion and the constellation of the Pleiades, for example, demonstrate this (Fig. 8). The width of the pages is typically about sixteen centimeters, and the swaths of stars are hard on the edge of the page. This has been regarded as an aesthetic weakness, but no one has previously imagined that the depiction itself could have suffered, and that is precisely the case.

In the *Washington copy*, this star group shows that the *Sidereus Nuncius*, when uncut by later bookbinders, did full justice to the depiction (Fig. 9). The swath of this stellar struc-

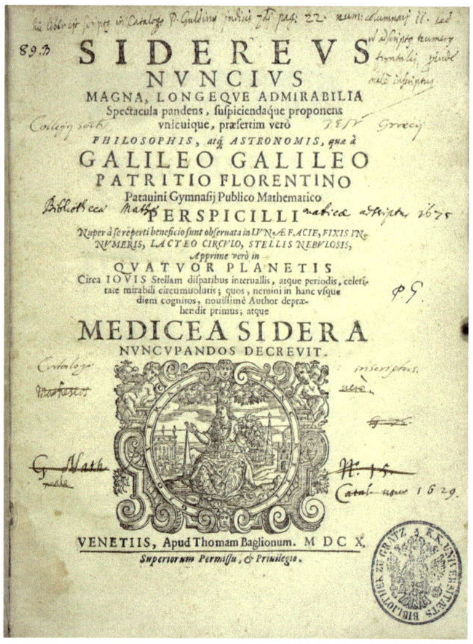

Fig. 6: Galileo Galilei, Sidereus Nuncius, Venice 1610, title page, Graz copy.

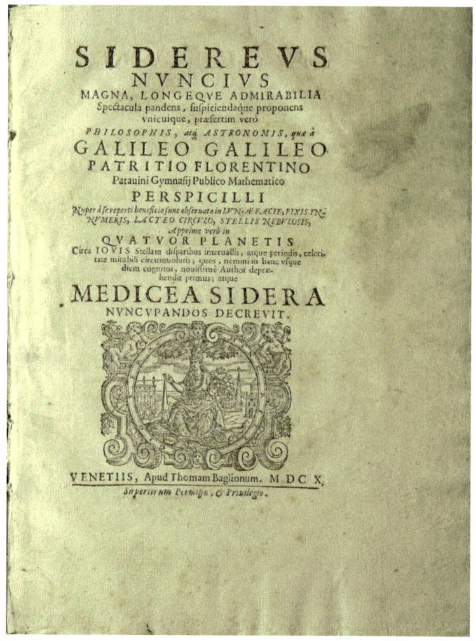

Fig. 7: Galileo Galilei, Sidereus Nuncius, Venice 1610, title page, Washington copy.

ture takes on another character, seeming to swing out over the surface as in a celestial dance. Trimming the paper edges destroys not only this impression, but also the information itself. In most copies, the outermost stars on the left are cut off. The bookbinders followed their rules of book production rigidly, at the cost of a not unimportant star.

The *Graz copy* played an important role in the book *Galilei der Künstler* because the printing errors in both demonstrate the *New York copy's* status as a proof copy.[9] Now, in the space of one month, all three books – the *Graz, New York,* and *Washington copies* – were compared with regard to their paper, watermarks, and the printing of their letters and illustrations.

Especially revealing was a bibliographic method that relied on identifying printing errors and tracing the history of their removal, which was carried out by Paul Needham of the *Scheide Library (Princeton),* who had joined the research group in early 2008. As a specialist in the history of the early modern book, he pursued the error analysis which I began in *Galilei der Künstler,* but from a much broader base and with more systematic experience. On the basis of more than 80 copies, his analysis confirms with unquestionable evidence that the *New York copy* is the proof copy. In this way it was possible to bring the other copies into a chronology of the production process over the weeks preceding completion of the edition on 13 March 1610. Finally, Ruth Tesmar of the *Humboldt-Universität* (Humboldt University) in Berlin, a specialist in artistic techniques, recreated the process of transferring the drawings from the *New York copy* to the printing plates and, with Irene Brückle, modeled the sequence of the procedure.

9 Bredekamp 2007, pp. 187–189.

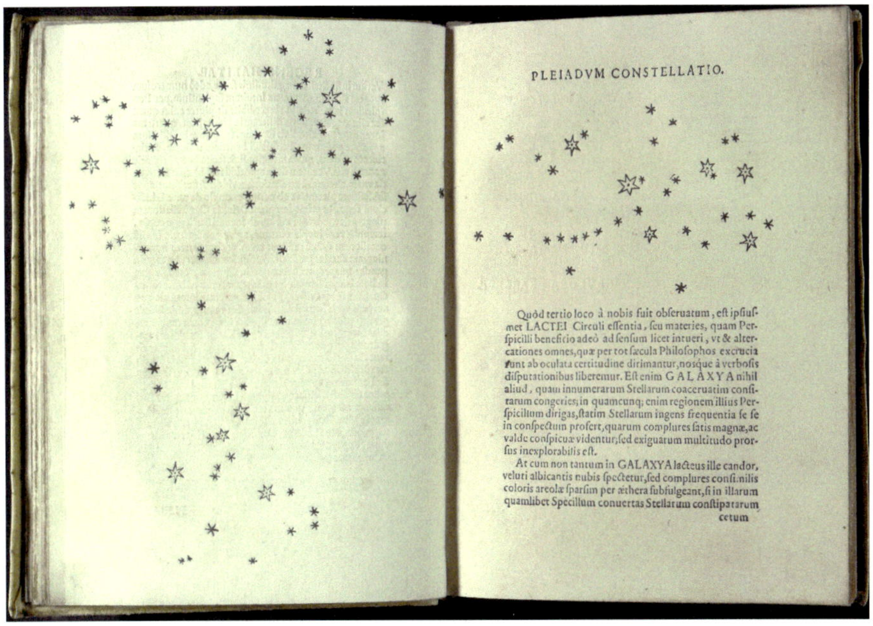

PLEIADVM CONSTELLATIO.

Quòd tertio loco à nobis fuit obſeruatum, eſt ipſiuſmet LACTEI Circuli eſſentia, ſeu materies, quam Perſpicilli beneficio adeò ad ſenſum licet intueri, vt & altercationes omnes, quæ per tot ſæcula Philoſophos excruciarunt ab oculata certitudine dirimantur, noſque à verboſis diſputationibus liberemur. Eſt enim GALAXYA nihil aliud, quam innumerarum Stellarum coaceruatim conſitarum congeries; in quamcunq; enim regionem illius Perſpicillum dirigas, ſtatim Stellarum ingens frequentia ſe ſe in conſpectum profert, quarum complures ſatis magnæ, ac valde conſpicuæ videntur; ſed exiguarum multitudo prorſus inexplorabilis eſt.

At cum non tantum in GALAXYA lacteus ille candor, veluti albicantis nubis ſpectetur, ſed complures conſimilis coloris areolæ ſparſim per æthera ſubfulgeant, ſi in illarum quamlibet Specillum conuertas Stellarum conſtipatarum

cetum

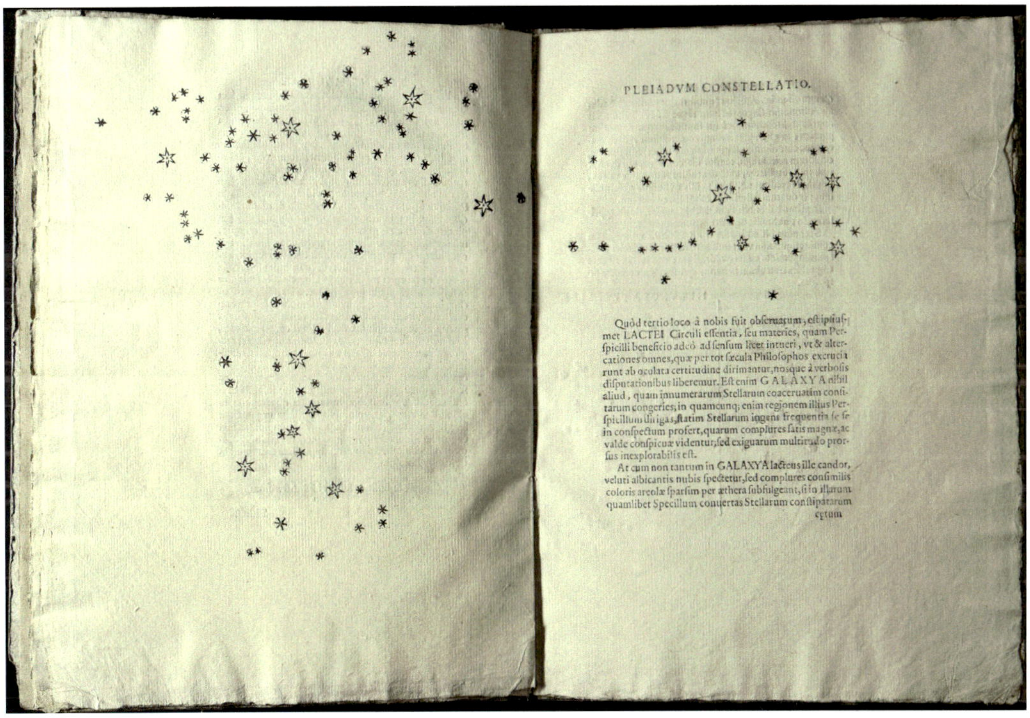

PLEIADVM CONSTELLATIO.

Quòd tertio loco à nobis fuit obſeruatum, eſt ipſiuſmet LACTEI Circuli eſſentia, ſeu materies, quam Perſpicilli beneficio adeò ad ſenſum licet intueri, vt & altercationes omnes, quæ per tot ſæcula Philoſophos excruciarunt ab oculata certitudine dirimantur, noſque à verboſis diſputationibus liberemur. Eſt enim GALAXYA nihil aliud, quam innumerarum Stellarum coaceruatim conſitarum congeries; in quamcunq; enim regionem illius Perſpicillum dirigas, ſtatim Stellarum ingens frequentia ſe ſe in conſpectum profert, quarum complures ſatis magnæ, ac valde conſpicuæ videntur; ſed exiguarum multitudo prorſus inexplorabilis eſt.

At cum non tantum in GALAXYA lacteus ille candor, veluti albicantis nubis ſpectetur, ſed complures conſimilis coloris areolæ ſparſim per æthera ſubfulgeant, ſi in illarum quamlibet Specillum conuertas Stellarum conſtipatarum

cetum

Fig. 8: Galileo Galilei, Sidereus Nuncius, Venice 1610, unnumbered pages showing the Pleiades, New York copy.

Fig. 9: Galileo Galilei, Sidereus Nuncius, Venice 1610, unnumbered pages showing the Pleiades, Washington copy.

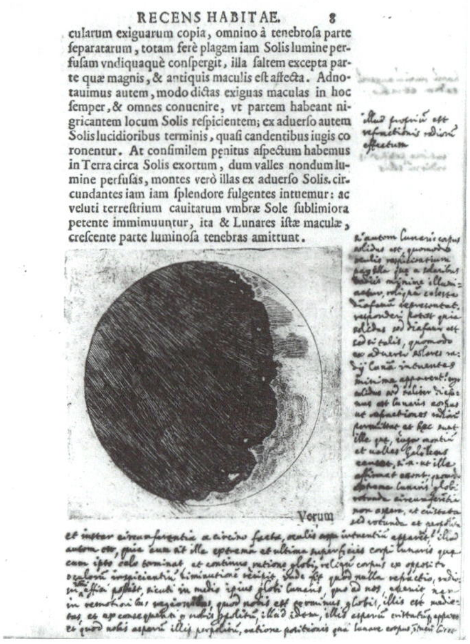

RECENS HABITAE. 8

cularum exiguarum copia, omnino à tenebroſa parte ſeparatarum, totam ferè plagam iam Solis lumine perfuſam vndiquaquè conſpergit, illa ſaltem excepta parte quæ magnis, & antiquis maculis eſt affecta. Adnotauimus autem, modo dictas exiguas maculas in hoc ſemper, & omnes conuenire, vt partem habeant nigricantem locum Solis reſpicientem; ex aduerſo autem Solis lucidioribus terminis, quaſi candentibus iugis coronentur. At conſimilem pęnitus aſpectum habemus in Terra circa Solis exortum, dum valles nondum lumine perfuſas, montes verò illas ex aduerſo Solis circundantes iam iam ſplendore fulgentes intuemur: ac veluti terreſtrium cauitatum vmbræ Sole ſublimiora petente immimuuntur, ita & Lunares iſtæ maculæ, creſcente parte luminoſa tenebras amittunt.

Verum

Fig. 10: Galileo Galilei, Sidereus Nuncius, Venice 1610, p. 8r, Florence copy.

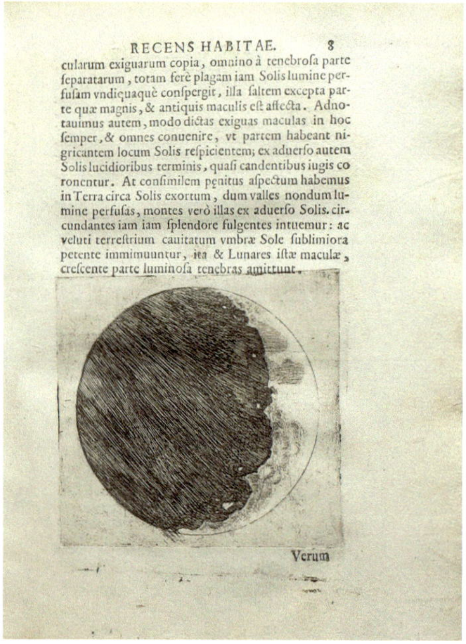

RECENS HABITAE. 8

cularum exiguarum copia, omnino à tenebroſa parte ſeparatarum, totam ferè plagam iam Solis lumine perfuſam vndiquaquè conſpergit, illa ſaltem excepta parte quæ magnis, & antiquis maculis eſt affecta. Adnotauimus autem, modo dictas exiguas maculas in hoc ſemper, & omnes conuenire, vt partem habeant nigricantem locum Solis reſpicientem; ex aduerſo autem Solis lucidioribus terminis, quaſi candentibus iugis coronentur. At conſimilem pęnitus aſpectum habemus in Terra circa Solis exortum, dum valles nondum lumine perfuſas, montes verò illas ex aduerſo Solis circundantes iam iam ſplendore fulgentes intuemur: ac veluti terreſtrium cauitatum vmbræ Sole ſublimiora petente immimuuntur, ita & Lunares iſtæ maculæ, creſcente parte luminoſa tenebras amittunt.

Verum

Fig. 11: Galileo Galilei, Sidereus Nuncius, Venice 1610, p. 8r, Munster copy.

JULY–OCTOBER 2008: *FLORENCE, PISA* AND *MUNSTER COPIES*

In May 2008, William S. Shea from *Università degli Studi di Padova* (Padua University) visited Berlin to see the *New York copy*. After discussing the latest list of print errors, he and his students analyzed several copies of the *Sidereus Nuncius* that neither Needham nor I had been able to see. They were able to confirm and enlarge upon our results.[10]

Stimulated by the discussions with Needham and Shea, I continued to study other copies in Pisa, Florence and Munster. Especially telling is the copy of the BNCF, famous for its wide margins. It is also well known because it consists of brighter and thicker paper than most known copies. A comparison with the copy in the *Universitäts- und Landesbibliothek* (University and Regional Library) at Munster[11] shows that though the BNCF copy is unique, as has always been thought, at Munster it has a "sister" (Figs. 10 and 11). The quality of paper, watermarks and printing errors are also the same as in the *Florence copy*. This demonstrates again that the *Sidereus Nuncius* was deliberately printed on different papers: two distinct paper qualities in the *New York copy*, the "regular" paper of most other known copies, and the fine, presumably more costly paper of the *Florence* and *Munster copies* (see Brückle et al. p. 134; Needham, p. 21f.).

10 Shea 2009. These experiences motivated Shea to reject Owen Gingerich's doubts about the authenticity of the *New York copy* (Gingerich 2009, pp. 141–165; Shea, 2010, pp. 97–110).

11 Signature: 4° M4 278 So. Mag.

MAY 2008–SEPTEMBER 2010: FINAL WORKS

Paul Needham's book-historical research produced surprising results month after month, shifting his perspective from an analysis of the status of the *New York copy* to a fundamental reconstruction of the production of the *Sidereus Nuncius*. In October 2009, when we jointly studied the watermarks in the papers of Galileo's estate in the Biblioteca Nazionale, we discussed the idea of not only publishing the investigations in a book, but also of summarizing Needham's analysis in a second volume. In this way, a history of the *Sidereus Nuncius* emerged that reveals the day-by-day production process in an unprecedented manner. The starting point of the investigations of the *New York copy* expanded to become a comprehensive analysis of the book. In this expansion of perspective, the status of the *New York copy* as a proof copy has been confirmed even more convincingly.

Enrica Schettini Piazza, Curator in the *Accademia Nazionale dei Lincei* (National Academy of Linceans) in Rome, identified the stamp in the *New York copy* as the one that was used for the private library of the head of this institution, Frederico Cesi. This makes it clear that the book was given to Cesi by Galileo, probably on the occasion of his stay in Rome 1612. The *New York copy* was never removed from its seventeenth-century binding; it consists of proof-copy paper; it is the proof copy of the printed book; it has the stamp of Federico Cesi's personal library on its title page and elsewhere in the text;[12] the title page carries the signature of Galileo; its writing and drawing materials are typical of the early seventeenth century; the black material on top of the drawings indicates a transfer process to the copperplate, and the style of drawing is the same as that of Galileo's sunspots and the Florentine drawings.

Through the comparison of the *New York copy* with the two other paradigmatic copies from Graz and Washington and numerous other copies, a level of knowledge has been gained previously achieved only for the Gutenberg Bible. The participants believe that this alone justified the research whose results are recorded here. These data and their interpretation make this book comprehensible in its physical and aesthetic valences. The present book was intended as a contribution to the 400th birthday of the *Sidereus Nuncius*. We did not meet the very year, but we are happy at least to have met the following: 2011.

ACKNOWLEDGEMENTS

It may seem inappropriate for a researcher who was directly involved to thank the participating institutions and other researchers, because such thanks imply a hierarchy that was not present in this way. Rather, this was a rare form of collegial collaboration with the editors Irene Brückle, Oliver Hahn and Paul Needham as well as with all of the authors. I am grateful for that, above all.

The work could not have been performed without the generosity and assistance of Seyla Martayan and Richard Lan, who lent their copy of the *Sidereus Nuncius* twice for alto-

12 Enrica Schettini Piazza corrected me in this; I identified it as the stamp of the Accademia dei Lincei (Bredekamp 2007, p. 149).

gether five weeks in Berlin. Throughout the long period of research, they acted as true scholars in the most generous sense.

I would like to single out Irene Brückle for organizing and overseeing the investigations at the *Kupferstichkabinett Berlin* with prudence and precision. The same goes for Alexis Ruccius, to whom special thanks are due for editorial organization and supervision. Managing the illustrations alone was a tremendous effort that could be accomplished only with his conscientious attention. Ulrike Feist is to be thanked for her editorial help, Mitch Cohen for the careful reading and thorough translation of those texts that were written originally in German. Theresa Smith undertook the immense work of editing the final English version, especially regarding the technical terms and the homogeneity of texts stemming from different authors. Gratitude also goes to David Freedberg, Jürgen Renn, and William S. Shea, Moritz Wullen for discussions, valuable remarks, corrections and further support. Barbara Herrenkind has invested all of her experience as a photographer. Petra Florath has designed the two books with utmost care and engagement.

Without the support of the *Max-Planck-Gesellschaft* (Max Planck Society) and the *Alexander von Humboldt-Stiftung* (Alexander von Humboldt-Foundation) in the framework of the *Max Planck Research Award of 2006*, as well as the *BAM Bundesanstalt für Materialforschung und -prüfung* in Berlin, neither book could have appeared in the present shape. Final thanks go the *Akademie Verlag*, which has supported the project from the beginning.

Horst Bredekamp, June 2011.

INTRODUCTION TO VOLUME I

In 2006, a single, small, outwardly unassuming, parchment-bound volume of astronomical writings became the focus of an increasingly intense and soon multidisciplinary study. The book was first brought to the attention of Horst Bredekamp at *Humboldt-Universität* (Humboldt University) in Berlin by its New York owners, Richard Lan and Seyla Martayan. It contains a compilation of published works by Galileo Galilei, observations of the moon and other cosmological considerations, prefaced by the short treatise *Sidereus Nuncius*, which set a new scientific stage for astronomy. This seminal publication has been well-researched, but the New York *Sidereus Nuncius* was a new discovery because it featured, instead of the five customary etchings of the textured moon surfaces, their images as hand-drawn illustrations. Until then, several wash drawings at Florence were considered the closest relation to the published etchings. The discovery of this singular copy raised several questions. Where did these newly discovered drawings fit in the overall picture of Galileo's publication? Was this exemplar a modern fake, was it connected with Galileo's time, or even with the publishing process of *Sidereus Nuncius*, and therefore with Galileo himself? A wide range of possible answers had to be considered. If it could be assumed that the drawings were indeed authentic documents of the time, then several possible explanations of their origin would still have to be considered. They could be copies of Galileo's work, most likely taken from a printed exemplar of *Sidereus Nuncius*, or from other lost preparatory drawings. In this case they could have been drawn by someone other than Galileo, perhaps a studio assistant, as a presentation copy for a supporter of his work. If the drawings were by Galileo, then it was also possible they might be records of the final stages of his research just before translation into printing.

Starting from the questions that immediately arose during the initial examination of the New York *Sidereus Nuncius*, a study was launched with the hope of shedding light on the nature of the drawings and elucidating their function and origin. Aside from the historical, astronomical and stylistic study of the drawings in the context of Galileo's manuscript preparation, it was decided that the study of the materials present in the book would support the overall investigation. It would narrow the range of possible answers to the initial question, namely, how closely was the New York *Sidereus Nuncius* related to Galileo Galilei and his immediate working environment during the crucial winter of 1609–1610 leading up to the March 1610 printing? Consequently, the study focused on the paper, the writing and drawing inks, the inks used for the printing of the letterpress text and the etchings and, crucially, the unusual black material that appears on selected areas of the wash drawings. To find out as much as possible about their material composition and their relation to the fully printed *Sidereus Nuncius* required the cooperation of specialists in the allied disciplines of conservation and conservation science, as well as scientists in other fields, working in communication with the art historians and testing hypotheses in cooperation with an artist. Reaching across the institutional boundaries of *Humboldt-Universität*, the *Kupferstichkabinett Berlin* (Museum of Prints and Drawings), the *Rathgen Forschungslabor* (Rathgen Research Laboratory), the latter two institutions part of the *Stiftung Preußischer Kulturbesitz* (Prussian Cultural Heritage Foundation), the *BAM Bundesanstalt für Materialforschung und -prüfung* (Federal Institute for Materials Research and Testing), also in Berlin, and the *Staatliche Akademie der Bildenden Künste Stuttgart* (State Academy of Art and Design Stuttgart), the partners in this project were united in a multidisciplinary dialogue. Discussions evolved between specialists at distant locations in Berlin, Graz, Florence, Boston, Stuttgart and Princeton.

Our current understanding of the New York *Sidereus Nuncius* did not arrive in a single stroke of insight, but developed gradually as we drew together our different perspectives, looking repeatedly at the book while juxtaposing, reworking, and narrowing the focus of our investigations. Our observations made along the course of research required detours and sometimes return trips to previous questions. At the end of any investigation probing a historical object in a particular physical state – produced from a combination of different materials, used and aged over time, and preserved in this condition for our inspection – some uncertainties may linger that cannot be swept aside even by the most sophisticated course of study. We were the more conscious that our inquiry would have to be rigorous and self-critical to avoid misinterpretation.

The preface by Horst Bredekamp that precedes our introduction recounts the origin and evolution of the project. An account of Federico Cesi's library, which originally housed the New York *Sidereus Nuncius*, sets the stage for the following contributions which assemble the results of the interdisciplinary work conducted by the research team. At the center of the technical examination lies the visual conservation inspection, which throughout enabled the interpretation of the analytical results. This book is divided into six parts, the first dealing with the New York *Sidereus Nuncius*, the second presenting photographs of the individual pages of this *New York copy*, the third discussing the function of the drawings in relation to the printed etchings, followed, in the fourth part, by a discussion of pertinent features in the fully printed standard copies. While not every question can be answered with certainty – for example, it must remain open to what extent Galileo was involved in the preparation of the etchings – many pertinent details concerning the physical evolution of the book come to light. The fifth part, the afterword by Horst Bredekamp, embeds the foregoing discussions in an encompassing review that highlights once more the significance of this extraordinary find. Finally, the appendices provide selected information.

The investigation, especially as concerns the material technology of the *Sidereus Nuncius*, centered around two viewings, February 2006 and April–May 2008, when the unique *New York copy* was generously made available by its New York owners and hosted at the *Kupferstichkabinett Berlin* for direct examination. Several fully printed *Sidereus Nuncius* copies were made available for comparison, among them, notably, the copy at the *Karl-Franzens Universität* (Karl-Franzens University) at Graz, another copy then owned by a dealer in Paris, now at the *Library of Congress* in Washington, and the copy at the *Universitäts- und Landesbibliothek* (University and Regional Library) at Munster (in the following, these copies are distinguished as the *Graz copy*, the *Washington copy*, and the *Munster copy*). The copies at the *Biblioteca Marciana* in Venice (the *Venice copy*) and the Postillata 110 in Florence (the *Florence copy*) were viewed on site.

On the basis of our analyses, our research proves some crucial points: the New York *Sidereus Nuncius* is not a modern forgery – we did not find any evidence of materials that were not genuine to the time. We show that even inconspicuous technical details can shed light on the pertinent physical processes of drawing the images, etching the printing plates, and producing the book. Considered separately, many of the individual craft steps that went into this production were surely not uncommon at the time. Seen together, they reveal ingenious idiosyncrasies that are unique to the *Sidereus Nuncius* and that can be glimpsed only on close inspection. We also demonstrate that the drawings may very well have been made by Galileo Galilei himself in preparation of the printing of *Sidereus Nuncius*, a plausible explanation in light of the facts uncovered.

ACKNOWLEDGEMENTS

Seconding Horst Bredekamp's words of acknowledgement, we, in our turn, would like to thank several people who were particularly important to accomplishing the project in our immediate environment at the *Kupferstichkabinett Berlin*. It was Hein-Th. Schulze Altcappenberg, Director of the *Kupferstichkabinett Berlin*, who immediately saw the significance of this project and supported the museum conservation department as the center of the investigation. We would also like to acknowledge the insightful contributions of Theresa Smith, especially during the crucial initial phases of the project when major research directions were set, while she was an advanced graduate fellow from the *Art Conservation Department* at *Buffalo State College* working at the *Kupferstichkabinett Berlin*. Alexis Ruccius, senior student at *Humboldt-Universität* and assistant to Horst Bredekamp on the project, most effectively managed the task of coordinating the publication. Barbara Herrenkind, photographer at *Humboldt-Universität*, expertly supported the photographic documentation of the project at the *Kupferstichkabinett Berlin*. Carmen Alonso, senior student at *Humboldt-Universität* graciously assisted work at the *Kupferstichkabinett Berlin*. Creative discussions concerning the transfer method were held with Ruth Tesmar from *Humboldt-Universität*. Nancy Ash from the *Philadelphia Museum of Art* provided helpful information on watermark questions. Christopher Schmidt at the *Gemäldegalerie, Stiftung Preußischer Kulturbesitz*, created infrared reflectography images. The designers *Kraft plus Wiechmann*, in Berlin, are thanked for preparing the diagrams. And, finally, we would like to extend our gratitude to Petra Florath who patiently and expertly transformd the manuscript into a book, and to the *Akademie Verlag*, which supported the book production process throughout.

Irene Brückle (Stuttgart, formerly Berlin) / Oliver Hahn (Berlin)

In vol. I, page numbers of the Sidereus Nuncius are according to the book's pagination. In vol. II, the bibliographic nomenclature is used. A concordance can be found in vol. I, pp. 150–151.

I

THE NEW YORK *SIDEREUS NUNCIUS*

ENRICA SCHETTINI PIAZZA

THE *SIDEREUS NUNCIUS* IN FEDERICO CESI'S LIBRARY

Some years ago I undertook the project of reconstructing the library of Federico Cesi, Roman prince of Sant'Angelo and San Polo, Duke of Acquasparta, Marquis of Monticelli, who founded the Academy of Linceans in Rome on August 17, 1603.[1] The first Academy was short lived, and its final breath coincided more or less with the death of Cesi in August 1630. While Cesi was alive, this innovative project in European cultural history prospered despite the difficulties and opposition it encountered along the way. But with his demise, the motive-force and political acumen that had sustained it died with him.

It was Cesi's intention that the Academy be the repository of his "paper museum" [*museum chartaceum*] as well as his library of around 3000 volumes. As he died without a will, however, his library was lumped together with the rest of his estate. The cost of supporting the Academy had already reduced the family fortune, and soon after Cesi's death his widow, Isabella Salviati, guardian of their young daughters Teresa and Olimpia, was forced to inventory everything of value and put some of it up for sale. Meanwhile, there was the brief hope of a successor to Cesi in the person of cardinal-nephew Franceso Barberini, strategically selected to membership among the Linceans on the eve of the Papal election of his uncle Maffeo Barberini (Urban VIII). But the cardinal withdrew his support and the Academy was dissolved.[2]

It is not an easy task to follow the fate of Cesi's manuscripts and books across the centuries – documents so fundamental to our understanding of the interests and reading habits of the first members of the Academy. Three years after Cesi's death, fellow Lincean Francesco Stelluti came to the family with an offer to purchase part of the library on behalf of Academy member Cassiano dal Pozzo (1588–1657). Patron, erudite, collector, great connoisseur of art and a diplomat (having served in missions to France and Spain in 1625–26) – dal Pozzo was then in the service of Cardinal Francesco Barberini.

In a collection of seventeenth-century documents, Archivio Linceo 13,[3] on deposit in the Library of the National Academy of Lincei and Corsiniana in Rome, next to letters from Stelluti and dal Pozzo and among various receipts and financial documents, are three inventories of the sale to dal Pozzo entitled *Indici della Libreria di Federico Cesi*. They are in the form of three alphabetical lists totaling approximately 2000 printed items, each distin-

1 Schettini Piazza 2005, pp. 129–154.

2 Schettini Piazza 2007, pp. 117–126.

3 This collection, formerly Albani 286, with the stamp of the Albani Library (p. 5r n.n.) was sold in 1857–58 by the heirs of the Albani family, together with other Lincean collections, to the Prince Baldassare Boncompagni, who in turn gave them to the Academy of Linceans.

guished by format, and each priced by the German bookseller and publisher Hermann Scheus.

On January 16, 1633, came the authorization from Father Niccolò Riccardi, General of the Dominicans, to sell the books to dal Pozzo. Nicknamed "Padre Mostro" (literally, 'Father Monster') for his large girth and formidable erudition, Father Riccardi became Master of the Sacred Palace in 1629.[4] His order reads, *"Bibliotheca et libri Exc. mi q Principis Caesi S. Angeli possunt vendi impune Ill.mo D Cassiano De Puteo quod attinet ad nos"* (Arch. Linc. 13, pp. 124r–125r). In another part of the same codex (p. 59r), however, there is reference to a second document whose contents are today unknown. It lists those of Cesi's books blocked from being sold by Father Riccardi, who had second thoughts, reversing the previous authorization, *"Vendantur impune exceptis qui in separato folio denotantur."*

Finally, on January 23, 1633, the volumes sold by Cesi's widow for 758 scudi were incorporated into dal Pozzo's library in his residence on via dei Chiavari, near Sant'Andrea della Valle.

Cesi's library reflects his scholarly interests as well as his varied intellectual curiosities. He not only studied natural history, traditionally emphasized in histories of the Academy, but was also taken with astronomy, literature, history, philosophy, medicine, antiquarianism, alchemy, etc. which, together, reveal a multivalent personality imbued with the encyclopedism typical of his era. Naturally, of course, the library was full of the publications of his fellow Linceans, as well as manuscripts of works yet to be printed under the Academy's auspices.

On the shelves of his library was also a copy of the first edition of the *Sidereus Nuncius,* with 5 drawings of the surface of the moon in Galileo's hand, discovered by Horst Bredekamp and documented in his fascinating work on Galileo.[5] The connection to the Academy is clear from the frontispiece, which bears the ex libris stamp of Federico Cesi.

4 Eszer 1983, pp. 428–461.
5 Bredekamp 2007.

THE STAMP

We observe an oval stamp printed in black, with a spotted lynx in the center facing left, surrounded by a legend in a double frame:

EX BIBLIOTH.[ECA] LYNCÆA FEDERICI CÆSII L.[YNCEORUM] P.[RINCIPIS] MARCH.[IONIS] MONT.[IS] CÆL.[II] II[6]

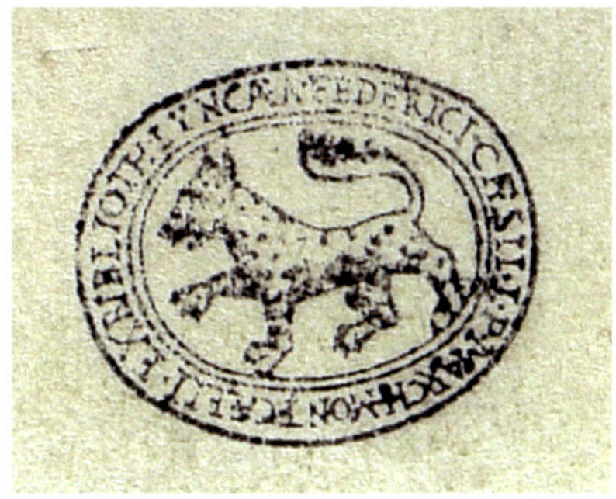

Fig. 1: Galileo Galilei, Sidereus Nuncius, Venice 1610, p. 8r, New York copy, detail.

This stamp, in addition to appearing on the title page, typically appears again on another randomly-chosen page of the same book: thus in the *Sidereus Nuncius*, we find this stamp once more on p. 8r (Fig. 1).

It is important to emphasize that this stamp is *not* the stamp of the Library of the Academy, which did not exist as such in the early years of the Linceans: in its first ten years (1603–1612), in fact, the library was synonymous with the personal library of Cesi, who made it available to fellow members at his residence in Rome on via della Maschera d'Oro, before it was subsequently transferred to the family seat in Acquasparta. It was this collegial library, full of manuscripts and printed items and identified by the proprietary character of the Cesi stamp, that was entrusted to Lincean Angelo de Filiis di Terni, brother of Anastasio de Filiis, one of its four founding members. De Filiis' duties as librarian were recorded as such in the minutes of the meeting of the Academy on June 7, 1612, where we read that he was responsible for looking after the publications of the Linceans as well as taking care of their distribution. It was, in fact, Angelo de Filiis who, at the meeting of the Linceo on November 9, 1612, would propose the Academy's sponsorhip of Galileo's *Macchie Solari*, and who signed the dedication letter to the Lincean Filippo Salviati and the notice to the reader.[7]

The emblem the Academy used to show its sponsorship of works is completely different from Cesi's ex libris. As already noted, the emblem of the Academy of Linceans is the lynx, its namesake, framed by laurel and surmounted by the marquis's crown of Federico Cesi (Fig. 3). Chosen for its symbolic value, the lynx connects the idea of penetrating vision with a belief in the intelligibility of the world. This, however, was not an invention of the Linceans. Cesi found the image of the lynx in two works by Giambattista Della Porta,

6 Federico Cesi was the Marquis of Monticelli. Only after January 28, 1613, when Pope Paul V (Camillo Borghese) raised to principality the marquesates of Sant'Angelo (today Sant'Angelo Romano) and of San Polo (today San Polo dei Cavalieri), was he titled Prince as the firstborn of the family. He became the Duke of Acquasparta for a few months in 1630, after the death of his father.

7 Gabrieli 1989, pp. 497–550.

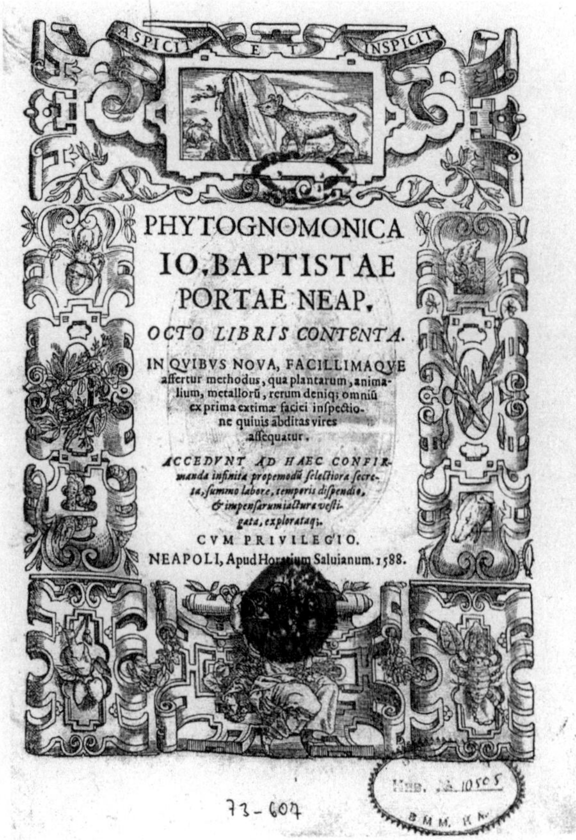

ASPICIT ET INSPICIT

PHYTOGNOMONICA
IO. BAPTISTAE
PORTAE NEAP.
OCTO LIBRIS CONTENTA.
IN QVIBVS NOVA, FACILLIMAQVE affertur methodus, qua plantarum, animalium, metallorũ, rerum deniq; omniũ ex prima extimæ faciei inspectione quiuis abditas vires assequatur.
ACCEDVNT AD HAEC CONFIRmanda infinita propemodũ selectiora secreta, summo labore, temporis dispendio, & impensarum iactura vestigata, explorataq;.
CVM PRIVILEGIO.
NEAPOLI, Apud Horatium Saluianum. 1588.

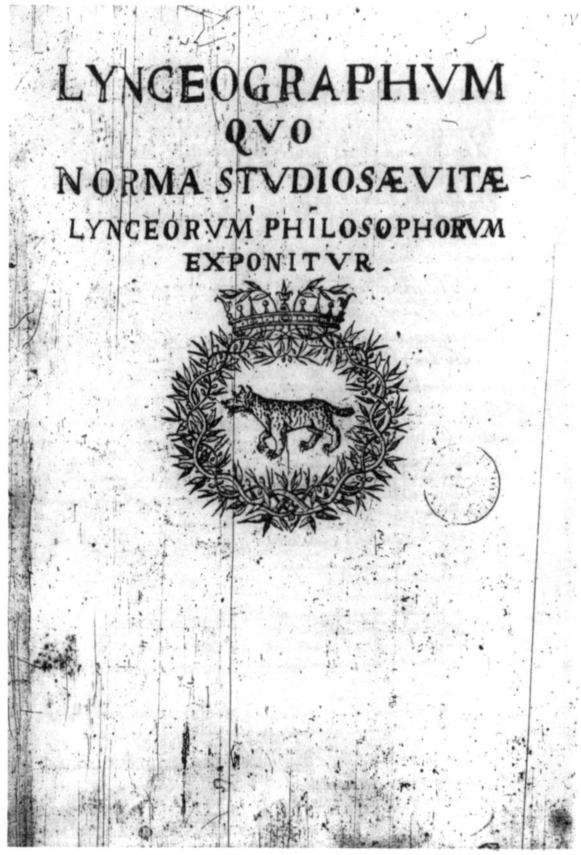

LYNCEOGRAPHVM
QVO
NORMA STVDIOSÆ VITÆ
LYNCEORVM PHILOSOPHORVM
EXPONITVR.

Fig. 2: Giambattista Della Porta, Phytognomonica, Naples 1588, title page.

Fig. 3: Lynceographum quo norma studiosae vitae, title page.

elected to membership in the Academy on July 6, 1610: *Phytognomonica octo libris contenta* (Naples, apud Horatium Salvianum, 1588) and *Magiae naturalis libri XX* (Naples, apud Horatium Salvianum, 1589) – among numerous titles by Della Porta in Cesi's library. Here, on the Della Porta frontispiece, one sees in a rectilinear frame a lynx facing left before a wall of cliffs. Above, one reads *Aspicit et inspicit,* which for the Academy will become *Sagacius ista* (Fig. 2).

The lynx as a heraldic device for the Academy appears for the first time in the *Lynceographum,* its founding charter, which was initially written in 1605, revised in its last years but never completed, and remained unpublished for centuries.[8] This document, fundamental to the story of the Academy, can be found at the beginning of the codex miscellany Archivio Linceo 4, Albani 318, entitled *Varia Principis Caesii,* on deposit in the Library of the National Academy. There on the title page one finds a drawing of the lynx in profile facing left which, except for brief periods of time, would be used thereafter by the Academy, in the same way Cesi's ex libris was used before (Fig. 3). On page 244r of the same codex we find another drawing, colored this time, of the lynx, profile facing right, with dark brown spots against a white background, within two intertwining laurel branches in the form of a wreath with a thin red ribbon below and a marquis's crown above (Fig. 4).

8 Nicoló 2001.

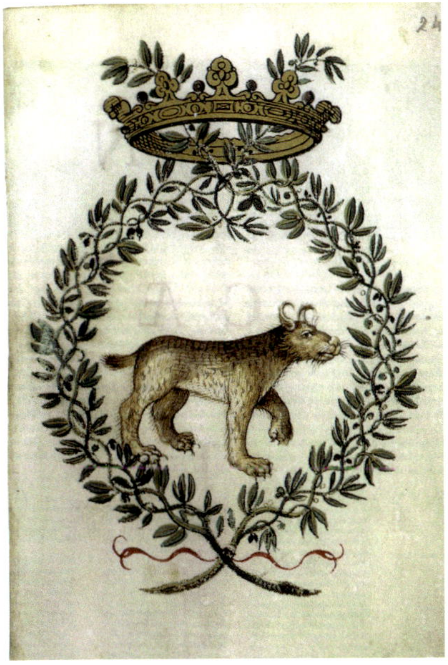

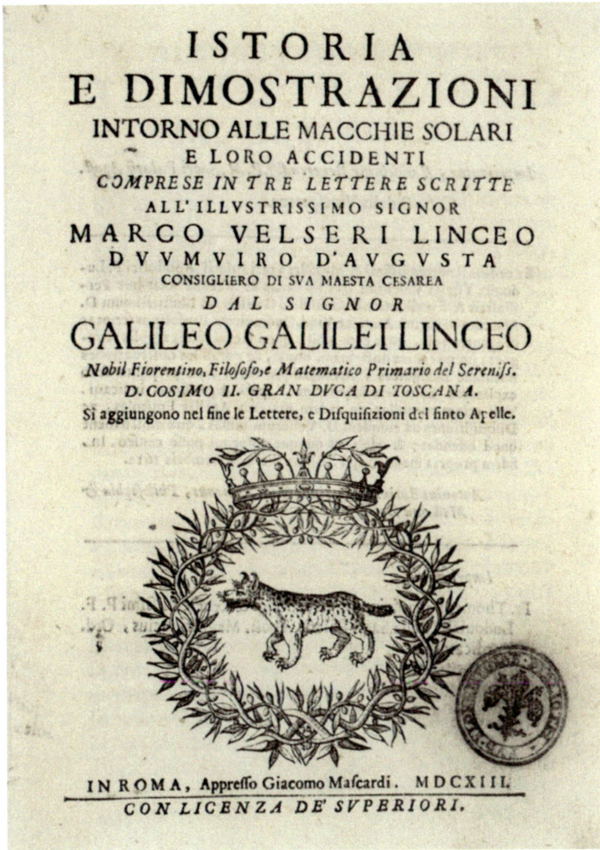

ISTORIA
E DIMOSTRAZIONI
INTORNO ALLE MACCHIE SOLARI
E LORO ACCIDENTI
COMPRESE IN TRE LETTERE SCRITTE
ALL' ILLVSTRISSIMO SIGNOR
MARCO VELSERI LINCEO
DVVMVIRO D'AVGVSTA
CONSIGLIERO DI SVA MAESTA CESAREA
DAL SIGNOR
GALILEO GALILEI LINCEO
Nobil Fiorentino, Filosofo, e Matematico Primario del Sereniss.
D. COSIMO II. GRAN DVCA DI TOSCANA.
Si aggiungono nel fine le Lettere, e Disquisizioni del finto Apelle.

IN ROMA, Appresso Giacomo Mascardi. MDCXIII.
CON LICENZA DE' SVPERIORI.

Fig. 4: Adam Elsheimer, p. 244r, drawing of lynx.

Fig. 5: Galileo Galilei, Istoria e dimostrazioni intorno alle macchie solari, Rome 1613, title page.

To find the precise date of the first appearance of this emblem in print, we must go to 1613, to the title page of a work by Galileo, his *Istoria e dimostrazioni intorno alle macchie solari,* which is particularly noteworthy for being the first Lincean publication (Fig. 5). Strictly speaking, two other works by Della Porta, dedicated to Prince Cesi and owned by him (*De aeris transmutationibus,* Romae apud Zannettum, 1610 and *Elementorum curvilineorum libri tres,* Romae, Zannettum, 1610) both bear the Cesi family arms with a rampant lynx on their title pages (Figs. 6 and 7). This emblem is not the heraldic device of the Academy, however, but a personal tribute, or dedication, from one scientist to the founder of the Linceans.[9]

The *Sidereus Nuncius* was printed at Venice in 550 copies, which were then put into circulation. Certainly it was the interest in the discoveries it proclaimed and the "talk" it aroused which induced Cesi to offer Galileo membership among the Linceans, made official on April 25, 1611, with the addition of his signature in the sixth place in the album of signatories of the Academy.

Of the impact of the *Sidereus Nuncius* on the community of the Academy, we find testimony in a letter written by Francesco Stelluti from Rome to his brother Giovanbattista

9 The emblem of the lynx is also maintained today on the frontispiece of works in which the National Academy is editor.

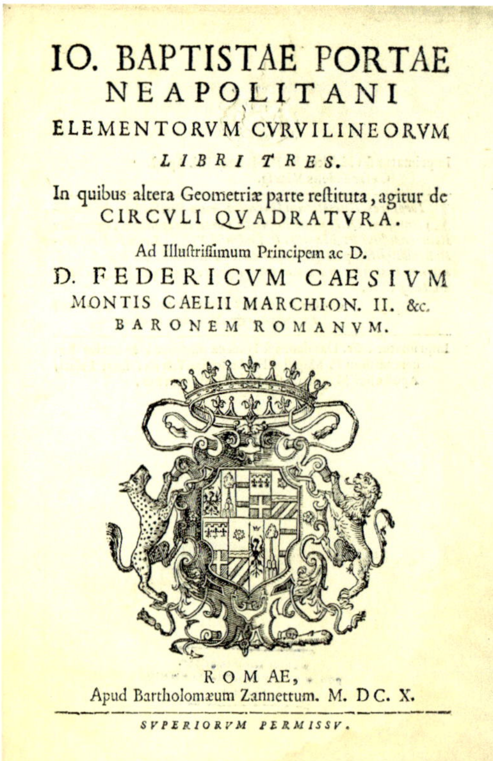
IO. BAPTISTAE PORTAE
NEAPOLITANI
ELEMENTORVM CVRVILINEORVM
LIBRI TRES.
In quibus altera Geometriæ parte reſtituta, agitur de
CIRCVLI QVADRATVRA.
Ad Illuſtriſſimum Principem ac D.
D. FEDERICVM CAESIVM
MONTIS CAELII MARCHION. II. &c.
BARONEM ROMANVM.
ROMAE,
Apud Bartholomæum Zannettum. M. DC. X.
SVPERIORVM PERMISSV.

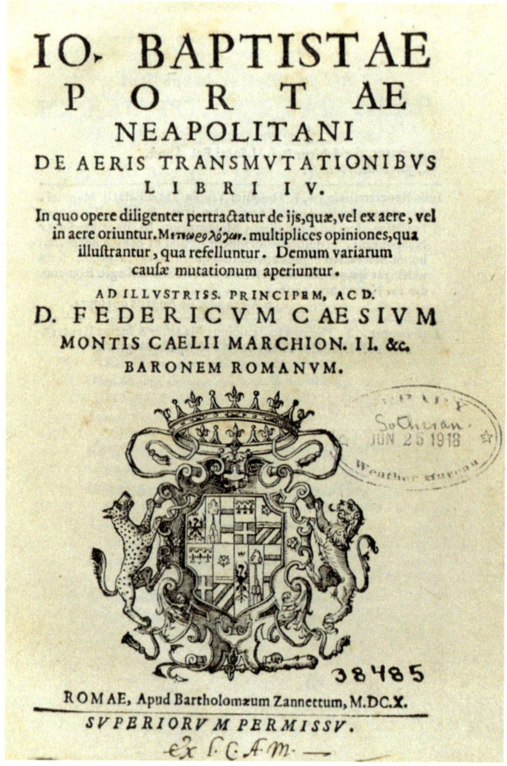
IO. BAPTISTAE
PORTAE
NEAPOLITANI
DE AERIS TRANSMVTATIONIBVS
LIBRI IV.
In quo opere diligenter pertractatur de ijs, quæ, vel ex aere, vel in aere oriuntur. Μετεωρολόγων. multiplices opiniones, qua illuſtrantur, qua refelluntur. Demum variarum cauſæ mutationum aperiuntur.
AD ILLVSTRISS. PRINCIPEM, AC D.
D. FEDERICVM CAESIVM
MONTIS CAELII MARCHION. II. &c.
BARONEM ROMANVM.
38485
ROMAE, Apud Bartholomæum Zannettum, M.DC.X.
SVPERIORVM PERMISSV.

Fig. 6: Giambattista Della Porta, Elementorum curvilineorum, Rome 1610, title page.

Fig. 7: Giambattista Della Porta, De aeris transmutationibus, Rome 1610, title page.

Stelluti in Fabriano on September 15, 1610, "We have just seen the Galileo, that is, the Sidereus Nuncius, and the great things which it says. But, now, [Johannes] Kepler, student of Tico [Tycho Brahe], has already written against him and already from Venice a copy of the book has come into the hands of Padre [Christophorus] Clavius."[10]

Cesi's copy of the *Sidereus Nuncius* containing the lunar drawings in the astronomer's hand is evidently a *unicum*; and one could well imagine that before coming to Rome Galileo had sent Cesi this very special copy of his book. But it is also possible that Galileo gave it to him when he arrived on the 29th of March. For, on the evening of April 14, at the Vigna Malvasia outside Porta San Pancrazio, on the Janiculum, the Linceans organized a meeting of fellow scientists, a meeting that famously continued until midnight. The purpose of the meeting was to provide Galileo the opportunity to demonstrate his new scientific discoveries with the *telescopio* – a term first employed by the Linceans, probably by Cesi himself, instead of *occhiale* or *ispecillo celeste*. Likewise, the term *microscopio* instead of *occhialino* was also coined by a Lincean – the German Johann Faber of Bamberg, who was chancellor of the Academy.[11]

In a letter dated April 30 of that year, Cesi wrote to Stelluti, who was not present at that historic moment, regretting his absence, "If ever there was a moment for you to be in Rome,

10 Gabrieli 1996, pp. 152–154.

11 Johann Faber's interesting note on the inventions of the telescope and microscope, and on the choice of new words, may be read in the manuscript Archivio Linceo 2, p. 51r, ed. in: Academia nationale dei Lincei 1965, pp. 198–199.

it is now; and if I could wish it, so much more would I wish it now: the reason itself you partly know already; but let me tell you what else: Every night, when it is calm, we observe new things in the sky, things verified by our members; Jupiter with its 4 [satellites] and their revolutions; the moon -- mountainous, cavernous, sinuous and aqueous. There is horned Venus and the three rings of Saturn that in the early morning one can see. Of the fixed stars, I shall say no more ... Nor is it a small problem that the earth is the center of the orbits ..."[12]

It was also Federico Cesi who was the catalyst of the meeting between Galileo and the Jesuit astronomers of the Collegio Romano – a meeting that took place on March 13 and was attended by cardinals, bishops, priests, nobles, scientists etc., and publicized in all the Roman *Avvisi* of the day. And on August 20, writing to Galileo, who had already returned to Florence, Cesi exhorts, "I must, out of respect to VS, urge you to publish as soon as you can the supplement to your Sidereus Nuncius. You have to write something on the horns of Venus and the triple rings of Saturn. Do it, please, before some brazen prelate declares 'your children' as his own..."[13]

From the catalogues of the sale of the library to dal Pozzo, we know that Cesi also owned the following works by Galilei: *Discorso... intorno alle cose, che stanno in su l'acqua* (Florence, Giunti, 1612), the *Istoria e dimostrazioni intorno alle macchie solari* (Rome, Mascardi, 1613) and *Il Saggiatore* (Rome, Mascardi, 1623) published under the auspices of the Linceans, with its dedicatory letter to Urban VIII, who had just ascended to the papacy, and undersigned by the Academy.

The *Sidereus Nuncius* is not among the books listed in these sales catalogues, limited as they are to a partial view of the original contents of the Cesi library. Also missing are a large part of Cesi's manuscripts. For the rest, the books and manuscripts of the Linceans were, over the course of centuries, subjected to numerous and significant dispersions.

After the deaths of Cassiano dal Pozzo and his brother Carlo Antonio junior (1606–1689), who had continued to build the family collection, the library was passed on to the great-grandson Cosimo Antonio dal Pozzo, who then sold the books to the Vatican Library on August 8, 1703. The asking price of 4000 *scudi* was paid in advance by the Vatican's Apostolic Camera, and the collection was added to the personal library of the Pope at the Palazzo del Quirinale at Monte Cavallo. But at the wish of Pope Clemente XI (Giovanni Francesco Albani), the dal Pozzo collection was sold again, this time for 4500 *scudi*, on January 8, 1714 to his nephew, the cardinal Alessandro Albani (1691–1779), patron, bibliophile, and art lover. After the death of Clemente XI, the books were incorporated into the magnificent papal library left to the family. The library was thereafter protected under the *fidecommisso istituito* of the Pope and had to remain in the new Palazzo Albani alle Quattro Fontane. But in 1798, during the French occupation of Rome and the consequent installment of the Roman Jacobin Republic, revolutionaries confiscated the property of the Albani and removed most of their art collection and books to France. During the Napoleonic era

12 Letter to Francesco Stelluti 1611, Gabrieli 1996, pp. 158–160; the original manuscript is in the Biblioteca Apostolica, vat. Lat. 9684.

13 Letter to Galileo Galilei 1611, Gabrieli 1996, pp. 171–172.

and after the "Restoration," the Albani family managed to regain a substantial portion of the books they had lost, reconstructing their collection.

With the absence of any male heir to the patrimony, there followed a second dispersion of the Albani library in 1856. The female heirs of the Albani, princesses who married into the Castelbarco of Milan and the Guidi di Bagno-Chigi of Modena, decided to sell the printed books. But the manuscripts and miscellanea, forming a larger group of material, were sold to the Royal Library of Berlin through the mediation of Theodor Mommsen, with the assistance of Michelangelo Caetani, Duke of Sermoneta, who acted as guarantor. The sale, concluded on March 15, 1862, for the total price of 14,500 francs in gold and silver, was followed by a letter from Mommsen dated March 19 to the Prussian Ministry of Culture and Education.[14] The shipment of the volumes, which was uninsured, was initiated on August 11, 1863 through Fiumicino by ship and should have arrived at Hamburg at the Consulate General for forwarding to Berlin. However, having arrived at Gibraltar with a leak, which was evidently poorly repaired, the ship was subsequently lost in the Atlantic and never arrived at its destination.

What remains of the library, then, are those books that were sold or donated by the Albani heirs to institutions or private parties.

Missing from the New York *Sidereus Nuncius* is the stamp of the Albani library, present in other Lincean books and manuscripts that shared this fate, next to the Cesi ex libris. The *New York copy*, however, traveled other unseen and less documentable paths – in fact, not just paths, but sea routes! Its provenance, however, is certain: it goes back to the library of Federico Cesi and its reappearance in New York four hundred years later.

14 Mommsen's unpublished report is preserved in the Staatsbibliothek zu Berlin, Stiftung Preußischer Kulturbesitz (Slg. Darmst. 2 f 1850 (7): Mommsen, Theodor, Mp. 31, Bl. 7–9).

HORST BREDEKAMP

THE SIGNATURE INSCRIPTION

THE DEDICATION

Fig. 1: Galileo Galilei, Sidereus Nuncius, Venice 1610, title page, New York copy, detail of Galileo's signature.

The signature "Io Galileo Galilei f." (Fig. 1), written at the bottom of the title page, is the first unique feature of the *New York copy*. Since the concluding "f" is lower case, it cannot stand for "F[lorentinus]", as the title page might seem to suggest: "GALILEO GALILEO / PATRITIO FLORENTINO". Rather, it should be interpreted as "f[eci]": "I, Galileo Galilei, made this."

Galileo's practice of using the space at the bottom of the title page for dedications was reconstructed in an exhibition of the *Biblioteca Nazionale Centrale Firenze* in 2008. The first documented dedication of this kind stems from Galileo's *Difesa* against Baldessare Capra, published 1607 in Venice (Fig. 2). It is directed to Silvio Piccolomini, *Signore di Sticciano,* soldier and educator of the coming *Granduca* Cosimo II de'Medici. Galileo does not sign with his name, but with the neutral remark: "l' aut[o]^{re}".[1] Unknown until now, there is another copy of the *Difesa* which Galileo dedicated in a similar manner (Fig. 3). In this case the book was presented to Angelo Contarini, his pupil in Paduan times. Here, too, Galileo signed with "l' autore".[2]

Also to be added to the Florentine catalogue is a dedication with which Galileo presented the *Sidereus Nuncius* to the poet Gabriello Chiabrera. This copy is owned by the University of Oklahoma (Fig. 4).[3] It might be a sign of growing pride that Galileo here used his name for the presentation inscription, "Galileo Galilei", and it might also be due to this psychological shift that on the occasion of presenting it to Frederico Cesi during his stay in

1 Galilei 1607, BNCF; Benucci 2008, p. 80.

2 Galilei 1607, University of St. Andrews, Special Collections Department, QB41.G2D5.

3 Galilei, *Sidereus Nuncius,* 1610. University of Oklahoma, History of Science Collections, Class 520, Book G 133s.

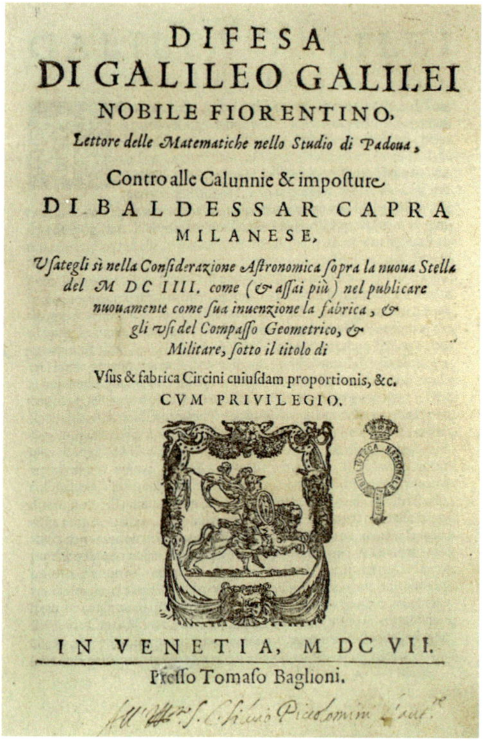

DIFESA
DI GALILEO GALILEI
NOBILE FIORENTINO.
Lettore delle Matematiche nello Studio di Padoua,
Contro alle Calunnie & imposture
DI BALDESSAR CAPRA
MILANESE,
Vsategli si nella Considerazione Astronomica sopra la nuoua Stella del M DC IIII. come (& assai più) nel publicare nuouamente come sua inuenzione la fabrica, & gli vsi del Compasso Geometrico, & Militare, sotto il titolo di
Vsus & fabrica Circini cuiusdam proportionis, &c.
CVM PRIVILEGIO.
IN VENETIA, M DC VII.
Presso Tomaso Baglioni.

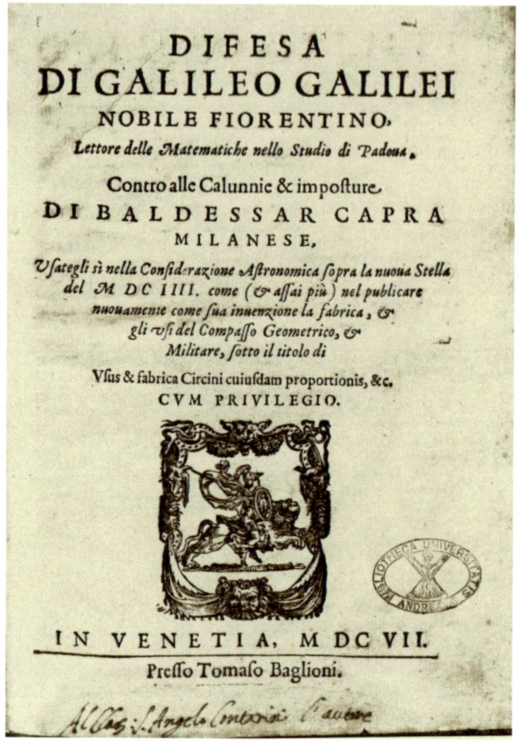

DIFESA
DI GALILEO GALILEI
NOBILE FIORENTINO.
Lettore delle Matematiche nello Studio di Padoua,
Contro alle Calunnie & imposture
DI BALDESSAR CAPRA
MILANESE,
Vsategli si nella Considerazione Astronomica sopra la nuoua Stella del M DC IIII. come (& assai più) nel publicare nuouamente come sua inuenzione la fabrica, & gli vsi del Compasso Geometrico, & Militare, sotto il titolo di
Vsus & fabrica Circini cuiusdam proportionis, &c.
CVM PRIVILEGIO.
IN VENETIA, M DC VII.
Presso Tomaso Baglioni.

Fig. 2: Galileo Galilei, Difesa, Venice 1607, title page, Florence copy.

Fig. 3: Galileo Galilei, Difesa, Venice 1607, title page, St. Andrews copy.

Rome in 1612 (see Schettini Piazza p. 28), Galileo signed the *New York copy* with the declamatory gesture of an artist, "I, Galileo Galilei, made this" (Figs. 1 and 5).

In his dedication of the second edition of his book on the behaviour of things in water, however, Galileo gave no hint as to the author at all (Fig. 6).[4] The reason for this new shift may have been the rank of the person to whom the book was gifted: prince Giovanni de'Medici. Modesty was a tool to uplift persons approaching members of the ruling family. The last of this kind of dedication stems from the presentation of the *Dialogo* to the medical doctor of the Medici, Giovanni Ronconi, in 1632 (Fig. 7). In this case, Galileo goes back to the formula he used in 1607 when presenting the *Difesa*, "l'Autore".[5]

THE NAME

These dedications demonstrate that Galileo signed only the *Sidereus Nuncius* with his full name and these inscriptions are closely connected. The special quality of Galileo's name written in the *New York copy*, the change between the softly developed tail of the "G" in "Galileo" and the sharp-angled turn of the tail of the "G" in "Galilei" (Fig. 8) can be similarly observed in the dedication of the Oklahoma copy (Fig. 9). The "G" in "Galileo" of the *New York copy* comes close to the "G" of "Gabriel" in the upper line of the dedication to Chiabrera, whereas the sharp-angled tail of the "G" in "Galilei" of the *New York copy*

4 Galilei 1612; Benucci 2008, p. 81.
5 Galilei 1632; Benucci 2008, p. 87.

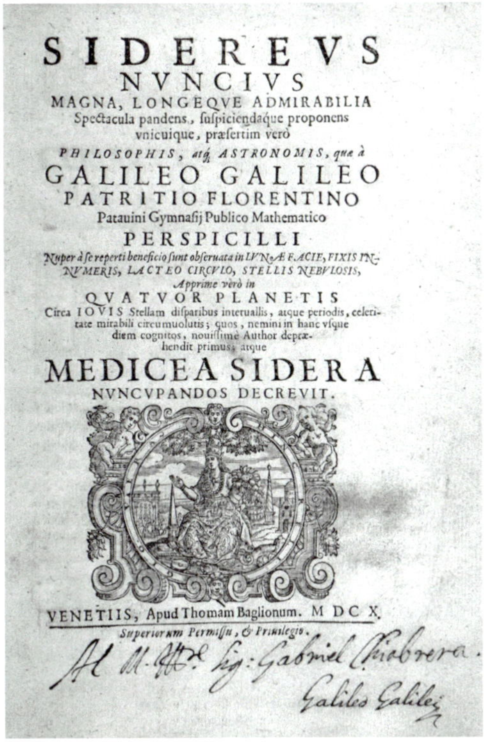

SIDEREVS
NVNCIVS
MAGNA, LONGEQVE ADMIRABILIA
Spectacula pandens, suspiciendaque proponens
vnicuique, præsertim verò
PHILOSOPHIS, atq; ASTRONOMIS, quæ à
GALILEO GALILEO
PATRITIO FLORENTINO
Patauini Gymnasij Publico Mathematico
PERSPICILLI
Nuper à se reperti beneficio sunt obseruata in LVNÆ FACIE, FIXIS IN-NVMERIS, LACTEO CIRCVLO, STELLIS NEBVLOSIS,
Apprime verò in
QVATVOR PLANETIS
Circa IOVIS Stellam disparibus interuallis, atque periodis, celeritate mirabili circumuolutis; quos, nemini in hanc vsque diem cognitos, nouissimè Author depræhendit primus; atque
MEDICEA SIDERA
NVNCVPANDOS DECREVIT.
VENETIIS, Apud Thomam Baglionum. M DC X.
Superiorum Permissu, & Priuilegio.
Al M. Ill.re Sig: Gabriel Chiabrera
Galileo Galilei

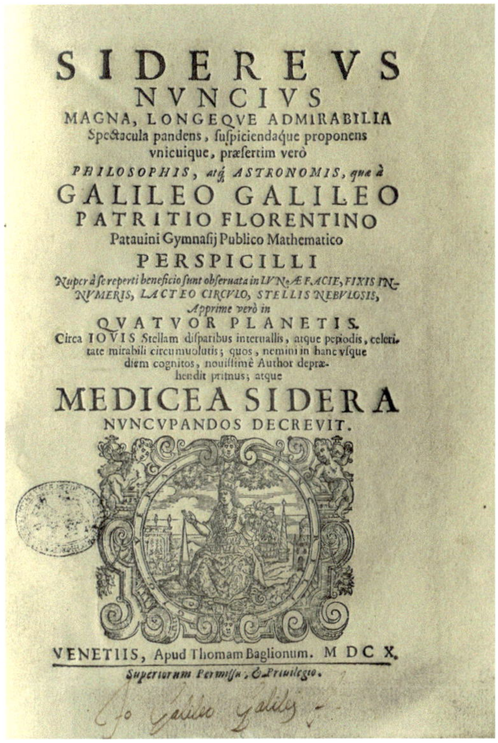

SIDEREVS
NVNCIVS
MAGNA, LONGEQVE ADMIRABILIA
Spectacula pandens, suspiciendaque proponens
vnicuique, præsertim verò
PHILOSOPHIS, atq; ASTRONOMIS, quæ à
GALILEO GALILEO
PATRITIO FLORENTINO
Patauini Gymnasij Publico Mathematico
PERSPICILLI
Nuper à se reperti beneficio sunt obseruata in LVNÆ FACIE, FIXIS IN-NVMERIS, LACTEO CIRCVLO, STELLIS NEBVLOSIS,
Apprime verò in
QVATVOR PLANETIS
Circa IOVIS Stellam disparibus interuallis, atque periodis, celeritate mirabili circumuolutis; quos, nemini in hanc vsque diem cognitos, nouissimè Author depræhendit primus; atque
MEDICEA SIDERA
NVNCVPANDOS DECREVIT.
VENETIIS, Apud Thomam Baglionum. M DC X.
Superiorum Permissu, & Priuilegio.
Jo Galileo Galilei

Fig. 4: Galileo Galilei, Sidereus Nuncius, Venice 1610, title page, Oklahoma copy.

Fig. 5: Galileo Galilei, Sidereus Nuncius, Venice 1610, title page, New York copy.

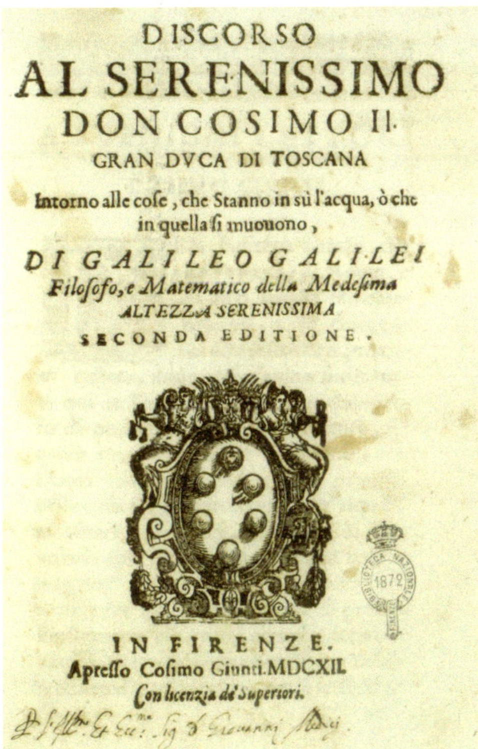

DISCORSO
AL SERENISSIMO
DON COSIMO II.
GRAN DVCA DI TOSCANA
Intorno alle cose, che Stanno in sù l'acqua, ò che in quella si muouono,
DI GALILEO GALILEI
Filosofo, e Matematico della Medesima
ALTEZZA SERENISSIMA
SECONDA EDITIONE.
IN FIRENZE.
Appresso Cosimo Giunti. MDCXII.
Con licenzia de' Superiori.

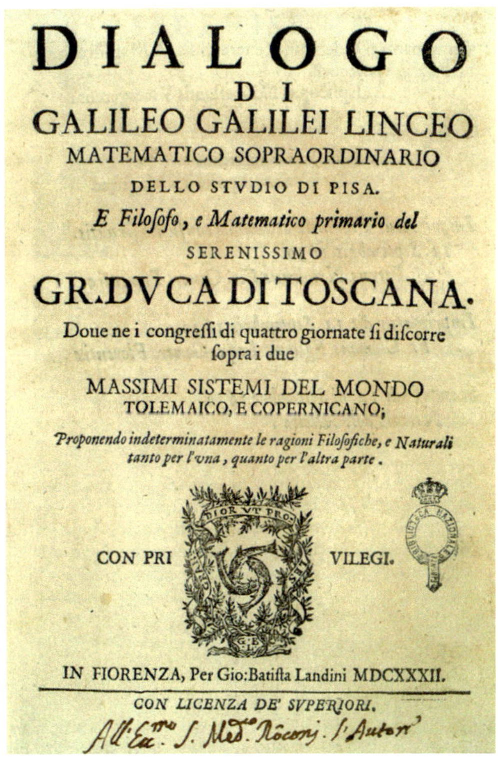

DIALOGO
DI
GALILEO GALILEI LINCEO
MATEMATICO SOPRAORDINARIO
DELLO STVDIO DI PISA.
E Filosofo, e Matematico primario del
SERENISSIMO
GR. DVCA DI TOSCANA.
Doue ne i congressi di quattro giornate si discorre
sopra i due
MASSIMI SISTEMI DEL MONDO
TOLEMAICO, E COPERNICANO;
Proponendo indeterminatamente le ragioni Filosofiche, e Naturali
tanto per l'vna, quanto per l'altra parte.
CON PRI VILEGI.
IN FIORENZA, Per Gio: Batista Landini MDCXXXII.
CON LICENZA DE' SVPERIORI.

Fig. 6: Galileo Galilei, Discorso, Florence 1612, title page.

Fig. 7: Galileo Galilei, Dialogo, Florence 1632, title page.

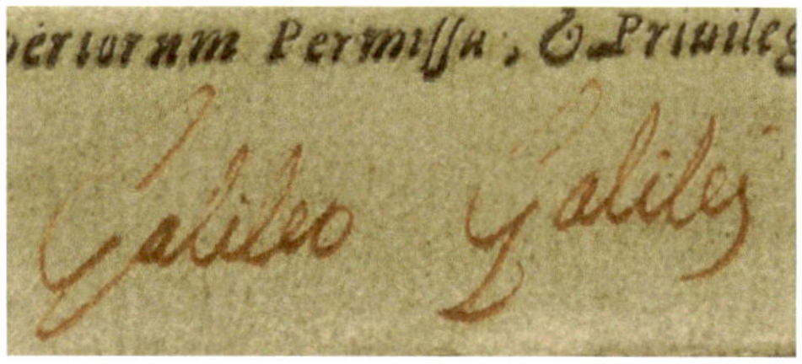

Fig. 8: Galileo Galilei, Sidereus Nuncius, Venice 1610, title page, New York copy, detail of Galileo's signature.

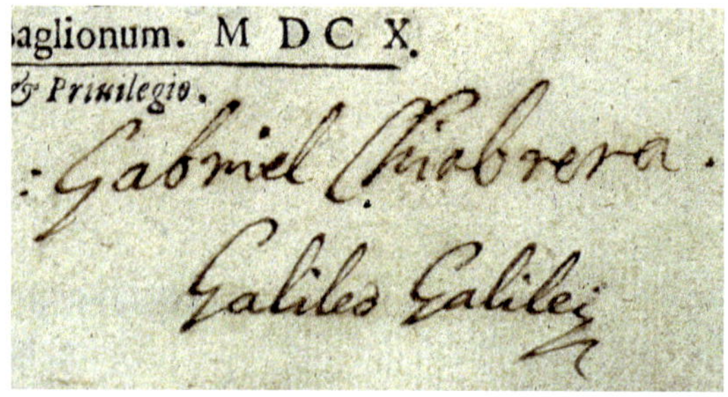

Fig. 9: Galileo Galilei, Sidereus Nuncius, Venice 1610, title page, Oklahoma copy, detail of the dedication with Galileo's signature.

resembles more closely both "G"s of Galileo's name in the Oklahoma copy. The distance between the "G" and the "a" in "Galilei" of the *New York copy* is present again in "Gabriel" of the Oklahoma dedication. All the other letters of the signatures resemble each other closely.

The technical analysis showed that the composition of the ink is commensurate with inks used in manuscripts of Galileo's time. The stylistic closeness of the signatures also leaves no doubt about the genuineness of the *New York copy* inscription.[6]

THE SHARP-ANGLED TAIL OF THE "G"

With the first and the last letters of "Galilei" in the *New York copy*, though, problems remain. The "G" shows a curve unrolling like an arabesque from the bottom to the upper right. None of the known dedications or property marks bear this form. However, on an envelope of a letter which Galileo wrote to his brother Michelangelo Galilei in May 1606 the same shape appears, only slightly less exposed (Fig. 10).[7] Not connected to his name and not only seen in the "G", but also with "p" and "q", this shape is drawn in numerous examples in Galileo's letters. In a letter written to Cristina di Lorena, ca. September 1608, three of them parade in a line (Fig. 11).[8]

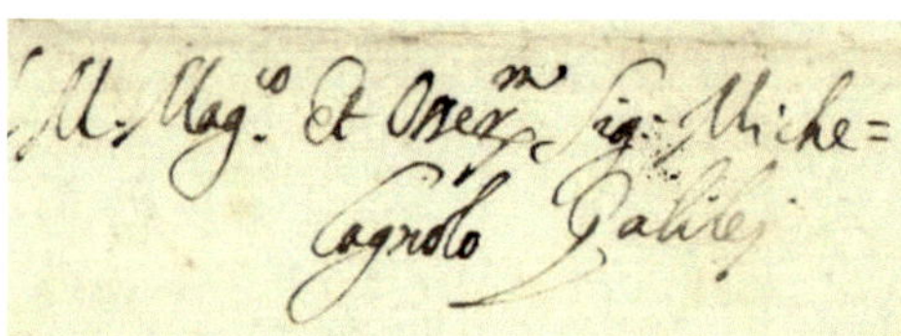

Fig. 10: Galileo Galilei, letter to his brother Michelangelo, May 1606, BNCF, Mss. Galileani, no. 14, fol. 18v, detail of the envelope with Galileo's signature.

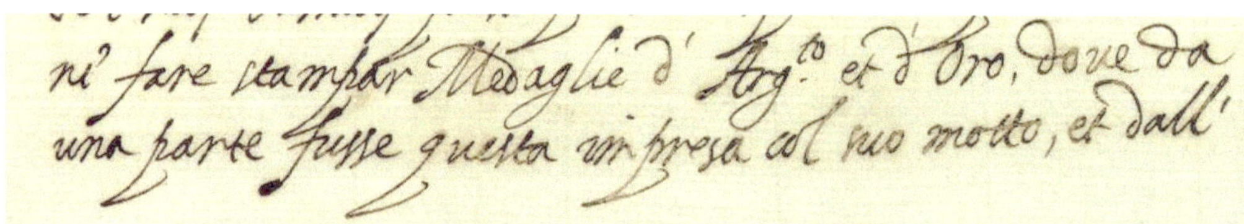

Fig. 11: Galileo Galilei, letter to Cristina di Lorena, September 1608, BNCF, Mss. Galileani, no. 14, fol. 30r, detail.

6 Bredekamp 2007, pp. 151–153, with examples from his letters.

7 BNCF, Mss. Galileani, no. 14, fol. 18v; *Opere*, vol. X, p. 158, l. 29.

8 BNCF, Mss. Galileani, no. 14, fol. 30r; *Opere*, vol. X, p. 223, l. 55.

THE CURVED "I"

The second stroke to note is associated with the final "i" in "Galilei" of the *New York copy* signature (Fig. 8). The tail of the letter swings to the left in a curve. Similar strokes can be observed in other examples of Galileo's signature. Among the ownership marks Galileo wrote in the books of his property, there are entries which come close to the shape of this "i". One of these signatures was written in the year of the *Sidereus Nuncius*, 1610; here the final "i" angles down to the left (Fig. 12).[9] This is also apparent in another mark of possession probably from the same year (Fig. 13).[10]

Until his last days, Galileo's signature is rather consistent, with an "i" which comes straight down. But curves to the right or left occur as well, as the signature of the *New York copy* shows. One of the most similar was written in a letter to Geri Bocchineri in May 1634 (Fig. 14).[11] Thus the final "i" in the *New York copy* inscription is not an exception.

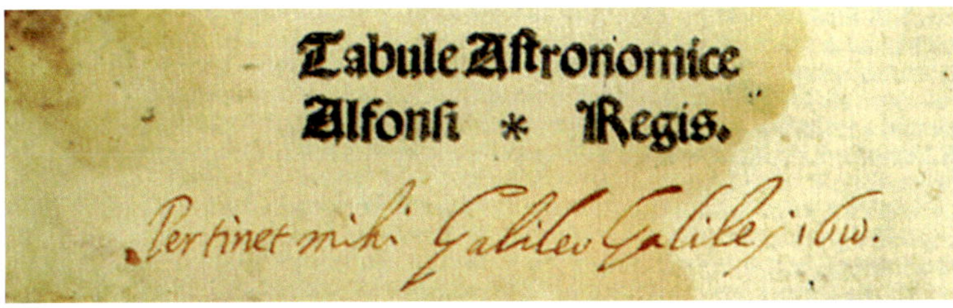

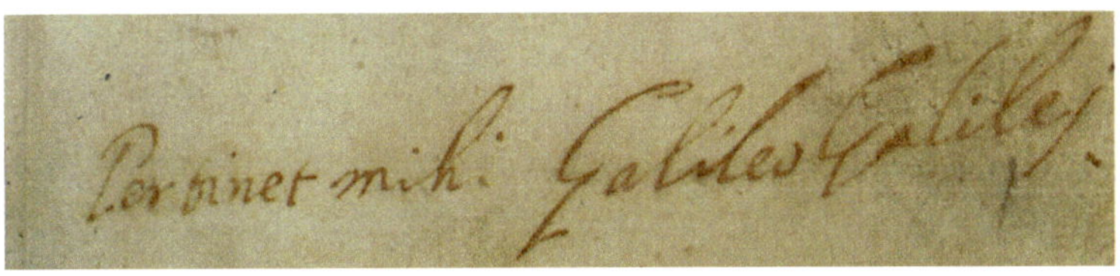

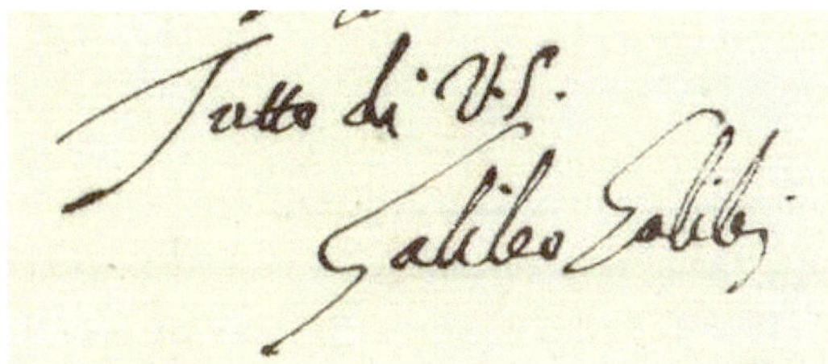

Fig. 12: Tabulae astronomicae [Alphonsine Tables], Venice 1492, title page, detail of Galileo's signature.

Fig. 13: Horatius Flaccus, Quintus, Poemata, Venice 1509, title page, detail of Galileo's signature.

Fig. 14: Galileo Galilei, letter to Geri Bocchineri, May 1634, BNCF, Mss. Galileani, no. 14, fol. 98r, detail of Galileo's signature.

THE "IO"

The first and last words of the *New York copy* inscription, "Io" ("I") and the abbreviated "f[eci]" ("made this") are also characteristic. The "Io" is characterized by a curved ascending stroke to the top of the letter and then a downward counter stroke, slightly bowed and unrolling horizontally to the left at the bottom (Fig. 15). In the middle, there is a small horizontal stroke more typical of an "F". The "o", rather small in comparison to the initial "I", is apart a small distance, so that the letters are not connected. This form can be observed in numerous letters in which his sentences start with the Italian "Io" ("I"), so that the first letter is capitalized. The extremes of Galileo's design of this word are designated by the most theatrical form in a letter written to Cosimo de'Medici in December 1605 (Fig. 17)[12] and by the more ascetic and dense variation in a letter from May 1608, directed to Belisario

9 Tabulae astronomicae [Alphonsine Tables] 1492; Benucci 2008, p. 102.

10 Horatius Flaccus, Venice 1509; Benucci 2008, p. 120. The words *"Pertinet mihi Galileo Galilei"* come very close to the dated inscription of 1610.

11 BNCF, Mss. Galileani, no. 14, fol. 98r; *Opere*, vol. XVI, p. 98, l. 9.

12 BNCF, Mss. Galileani, no. 14, fol. 15r; *Opere*, vol. X, p. 153, l. 2.

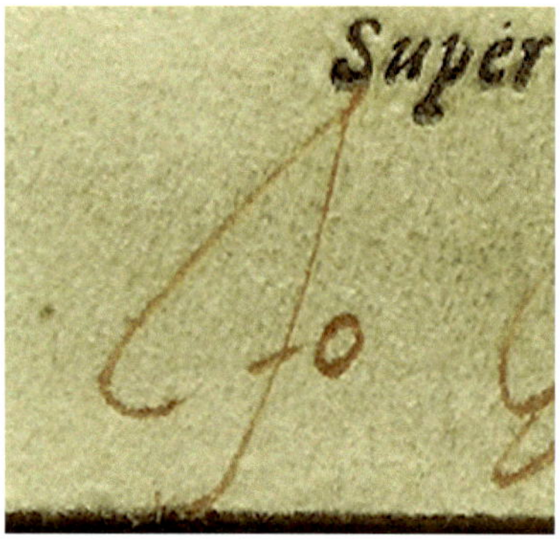

Fig. 15: Galileo Galilei, Sidereus Nuncius, Venice 1610, title page, New York copy, detail of Galileo's inscription.

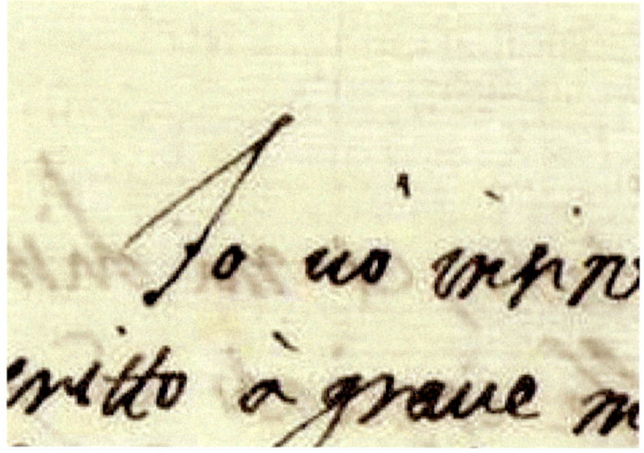

Fig. 16: Galileo Galilei, letter to Belisario Vinta, May 1608, BNCF, Mss. Galileani, no. 14, fol. 26r, detail.

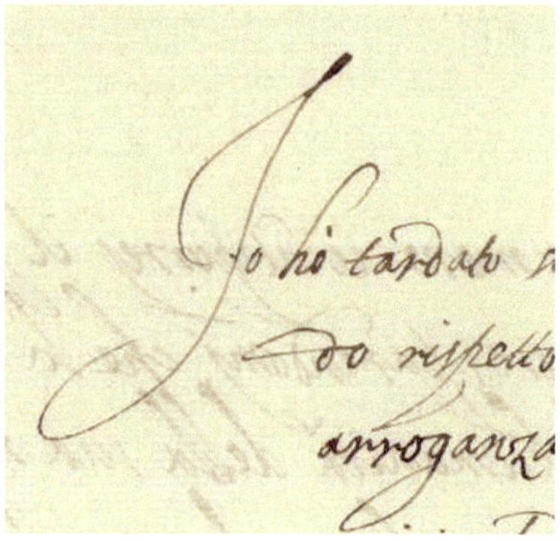

Fig. 17: Galileo Galilei, letter to Cosimo de'Medici, December 1605, BNCF, Mss. Galileani, no. 14, fol. 15r, detail.

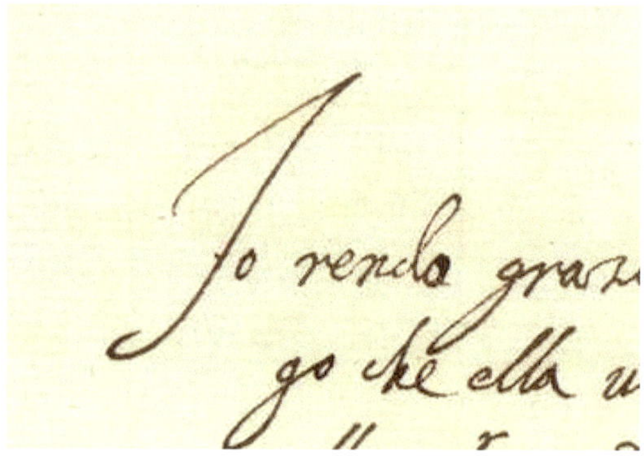

Fig. 18: Galileo Galilei, letter to Curzio Picchena, December 1615, BNCF, Mss. Galileani, no. 14, fol. 57r, detail.

Vinta, secretary of the Granduce (Fig. 16).[13] A median form, as represented by the *New York copy* inscription, is also present in a letter to the secretary of the Granduca, Curzio Picchena, December 1615 (Fig. 18).[14]

THE FINAL "F."

A special problem is the bottom of the final "f." where the elegantly curving form of the upper part of the letter breaks off into an open rectangular shape (Fig. 19). The angle to the right at the bottom of the letter can be observed regularly, as in the letter to Belisario Vinta from May 1608 (Fig. 20).[15] But this does not explain why the curve is more of a triangle.

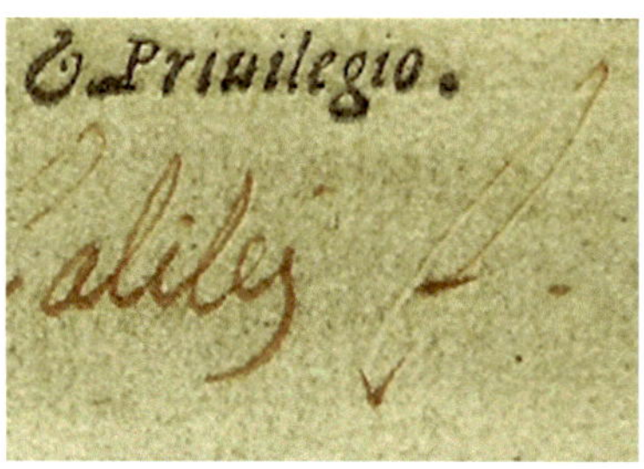

Fig. 19: Galileo Galilei, Sidereus Nuncius, Venice 1610, title page, New York copy, detail of Galileo's signature.

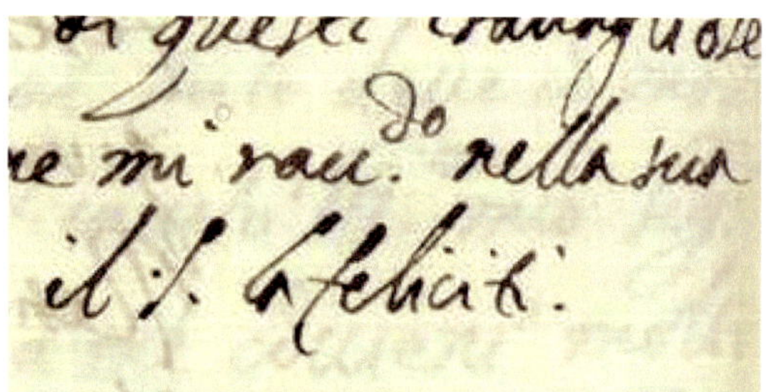

Fig. 20: Galileo Galilei, letter to Belisario Vinta, May 1608, BNCF, Mss. Galileani, no. 14, fol. 26v, detail.

13 BNCF, Mss. Galileani, no. 14, fol. 26r; *Opere*, vol. X, p. 212, l. 61.
14 BNCF, Mss. Galileani, no. 14, fol. 57r; *Opere*, vol. XII, p. 211, l. 2.
15 BNCF, Mss. Galileani, no. 14, fol. 26v; *Opere*, vol. X, p. 213, l. 102.

The strange phenomenon can be explained by the peculiarities of writing with a quill pen that requires regular reloading with ink. With each stroke, the ink in the quill pen is gradually depleted until the nib runs dry. Ideally, the pen is dipped into the ink well again before this point, but a stroke may be completed with the emptied pen, leaving, at best, an indentation caused by the pressure of the tool. Just such an indentation can be seen on this particularly soft paper, showing the sinuous curve of the "f". This is a tell-tale sign that Galileo set down his signature in one uninterrupted motion as the ink ran out while he was writing. The pen was dry once the down-stroke was reached but the "f" was completed in one sweep-

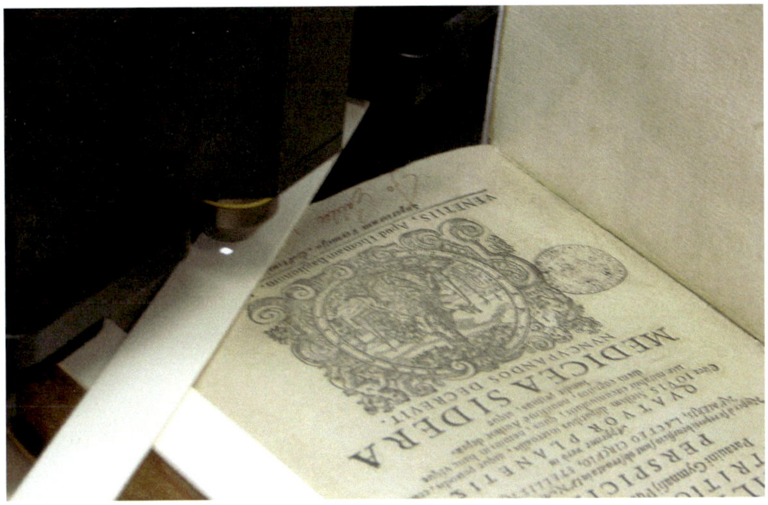

Fig. 21: Set-up for 3D confocal microscopy with white light microscope to analyze the title page of the New York copy.

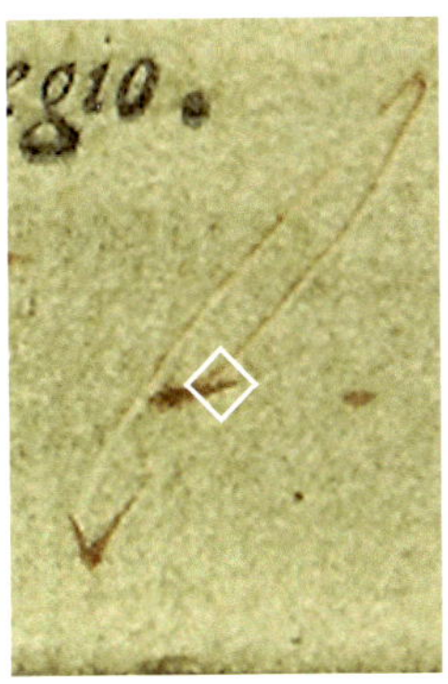

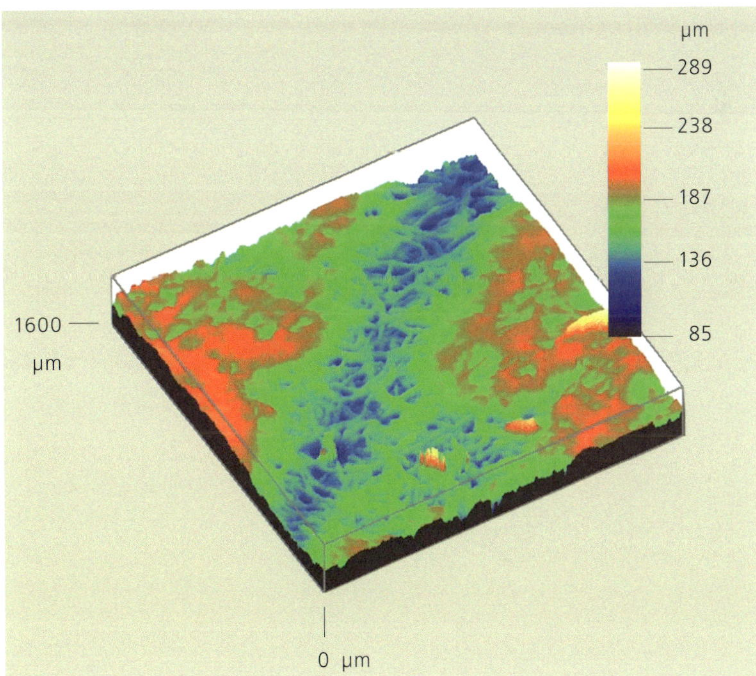

Fig. 22: Galileo Galilei, Sidereus Nuncius, Venice 1610, title page, New York copy, detail of Galileo's inscription; square marks the area of interest.

Fig. 23: Topographic profile obtained from the 3D confocal microscope.

ing motion, regardless. The result was an indented line, quasi blind embossed into the paper, as can be verified by the depth of the impressed area (Fig. 22) which lies about 35 µm below the paper surface (Fig. 23). When checking the freshly applied inscription, Galileo was confronted with the void and added the angled quill stroke at the bottom to close the form, as well as another horizontal line in the middle of the letter.

Each element of the inscription in the *New York copy*, indeed its whole ambience, is so inseparably connected to Galileo's style of writing that its authenticity is beyond doubt. This is all the more clear when looking at the psychology of this sculpture-like style of writing, as demonstrated in the upper curve of the second "G" which likewise lacks ink. The palpable pride with which Galileo evoked the grandeur of an artist's signature is manifest in the pressure of the writing. No counterfeiter could have imagined it. The relief-like character of this script is not only another confirmation of its genuineness, it is a strong hint at the importance of the copy which Galileo proclaimed so proudly to be of his own making.

OLIVER HAHN / TIMO WOLFF /
IRENE BRÜCKLE / THERESA SMITH

THE SIGNATURE AND MOON OUTLINE INKS

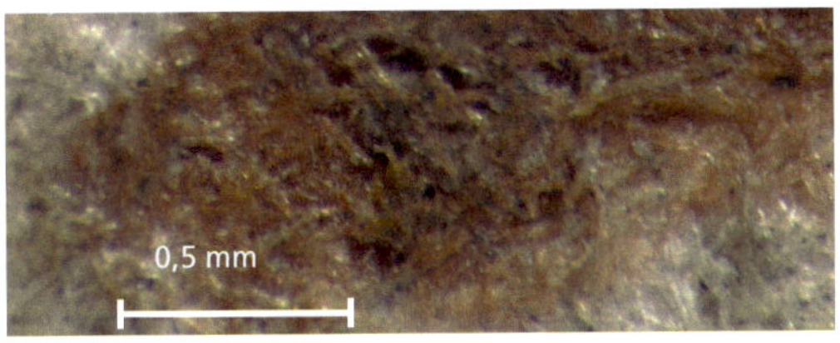

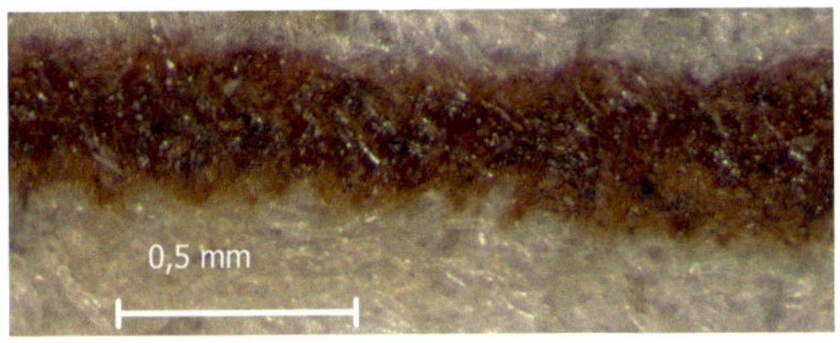

Fig. 1: Galileo Galilei, Sidereus Nuncius, Venice 1610, title page, New York copy, detail of the letter "f" in the signature inscription, reddish-brown ink.

Fig. 2: Galileo Galilei, Sidereus Nuncius, Venice 1610, p. 9v, New York copy, detail of the moon outline, reddish-brown ink.

A reddish-brown ink was used for the signature on the title page (Fig. 1). The moon drawings were executed in two different liquid drawing media, one a bistre ink (see Smith and Brückle pp. 41–45), the other a reddish-brown ink used to draw the outlines of the moons (Fig. 2). While the inks are similar in color, they show differences in their absorption of UV radiation that indicate they are not the same. Under near ultraviolet radiation, the ink of the circles absorbs more strongly and appears darker than that of the signature.

With micro X-ray fluorescence (XRF) analysis, only small mass fractions of inorganic compounds could be identified in both inks. It is therefore assumed that in both cases, though very small amounts of iron were present, the inks were made from plant dyes and are not iron gall ink. The use of plant-based inks in the beginning of the seventeenth century is

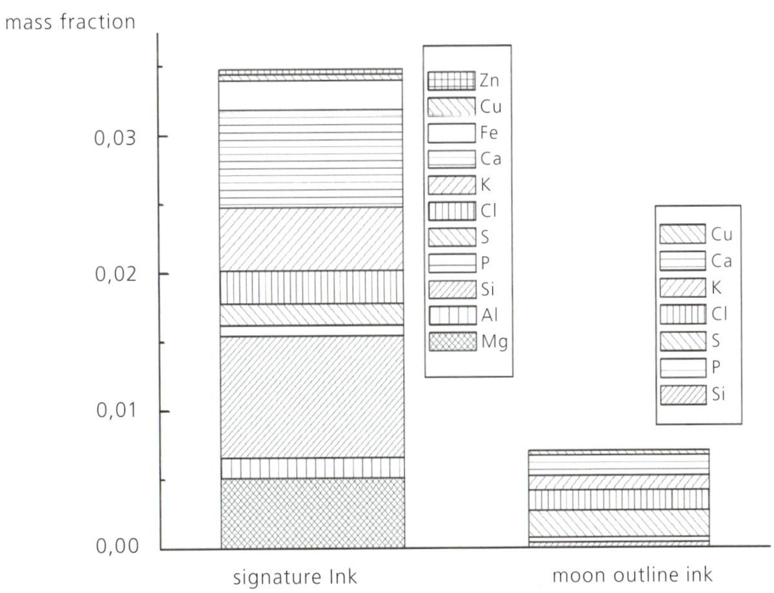

Fig. 3: Quantitative results of the micro XRF analysis.

rather exceptional given that iron gall ink was the most prevalent writing ink since the Middle Ages in Europe. The total amounts of the inorganic compounds detected with micro XRF and the ratios of these elements in the two samples differ significantly. Some elements like iron (Fe) and zinc (Zn) were only detected in the signature ink. The quantitative results are shown in Fig. 3. These findings prove that the inks of the signature and the moon outlines were made from similar, though not the same materials.

THERESA SMITH / IRENE BRÜCKLE

BISTRE IN THE DRAWING PROCESS

The five drawings were created by a confident hand working rapidly, and are more complex in the technical details of their execution than the Florentine drawings. Close inspection allows us to describe these aspects of the drawing process because almost every stroke in brush and ink left a distinctive and separately distinguishable mark on the paper. Specific characteristics common to all of the drawings are pointed out with reference to two of the three drawings on pages 10r and 10v, the two that lie back to back on the opposite sides of the same folio.

The drawing medium was determined to be bistre based on its characteristic appearance and the lack of morphological and visual features indicative of other available liquid media, namely iron gall ink, watercolor made from earth pigments, or plant ink as in the reddish moon outline. Infrared reflectography and micro X-ray fluorescence analysis conducted during this study corroborate these conclusions.

Bistre was a widely used brown-colored liquid drawing medium apparently first documented in the fifteenth century. In his art-theoretical treatise published in Milan in 1584, Lomazzo wrote of bistre as "caligine, which is much used for laying washes on paper".[1] Bistre was produced by the incomplete combustion of wood in a fireplace, which typically released tarry substances that collected on the inner walls of the chimney, while the fully carbonized parts escaped as smoke. The glossy, brittle granules were collected, heated in water or an alkaline solution to dissolve them, filtered to separate carbonized residues (sometimes also used for drawing) and dried until a gummy consistency was reached. This product reportedly did not require the addition of a binder because the tar content ensured a good working consistency.[2] Bistre was easily worked with a brush to produce washes on paper. The color of the drawing fluid varied depending on the type of wood used as raw material, producing tonal shades that ranged from "brownish saffron yellow to dark blackish brown".[3]

The most striking feature of the *New York copy* drawings is that invariably, the ink has deeply sunk into the paper matrix instead of remaining on its surface. This is, in part, explained by the highly absorbent quality of the paper, but the effect was likely further enhanced by the natural tendency of bistre to sink into paper to which it was applied. Bistre could be more easily absorbed into paper than other drawing media such as watercolor, because its coloring matter was dominated by dye-like substances and not pigment particles.

1 Meder 1978, vol. 1, p. 45; p. 56, footnote 115.

2 Several recipes for bistre production, the earliest dating from 1549, are discussed in Schwahn 1985, pp. 20–26.

3 Meder 1978, vol. 1, p. 46.

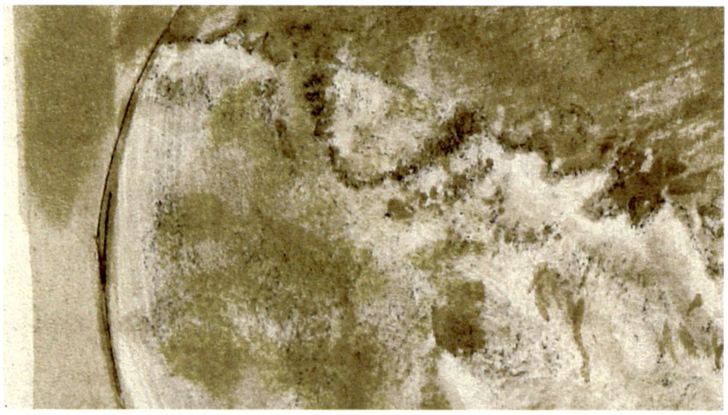

Fig. 1: Galileo Galilei, Sidereus Nuncius, Venice 1610, p. 10v, New York copy, detail of upper moon.

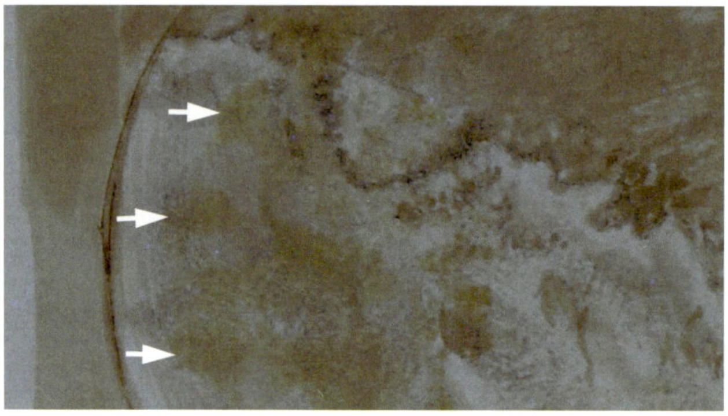

Fig. 2: Galileo Galilei, Sidereus Nuncius, Venice 1610, p. 10v, New York copy, detail of upper moon, UVA. Arrows show bistre ink migration from p. 10r.

The slightly muted tonality of the drawings consequently results from the complex interplay of the colorant itself and the porous, textured paper surface into which it penetrated. Due to this effect, bistre, in extreme cases, could produce quite "deadening results, the warm brown color being transformed into an innocuous gray or grayish-brown stain".[4] The *New York copy* drawings were evidently made with bistre that preserved its rich brown tonality after application, and were protected effectively from any potential light-induced color changes by remaining enclosed inside the book.

Another interesting phenomenon is the yellowish staining around heavy ink applications, most evident where the ink has bled to the verso of the paper, for example, on p. 10v (Fig. 1). There, amid the light brown ink washes of the moon, yellow stains can be distinguished that stem from the drawing on p. 10r. This is explained as follows: when the ink on p. 10r was freshly applied, it sank into the paper, but not all of its components traveled an equal distance. Yellow-colored components of the ink migrated further than the brown components, causing them to be visible on the verso of the sheet. In other words, the paper acted as a selective sieve that allowed the yellow components to pass through its matrix more easily than the brown ones. Viewed under long wave ultraviolet radiation (UVA),[5] the yellow stain shows a slight fluorescence (Fig. 2), which is an identifying mark of bistre ink.

4 Watrous 1957, p. 78.
5 Source: Sylvania Black Light Blue F 8WBLB.

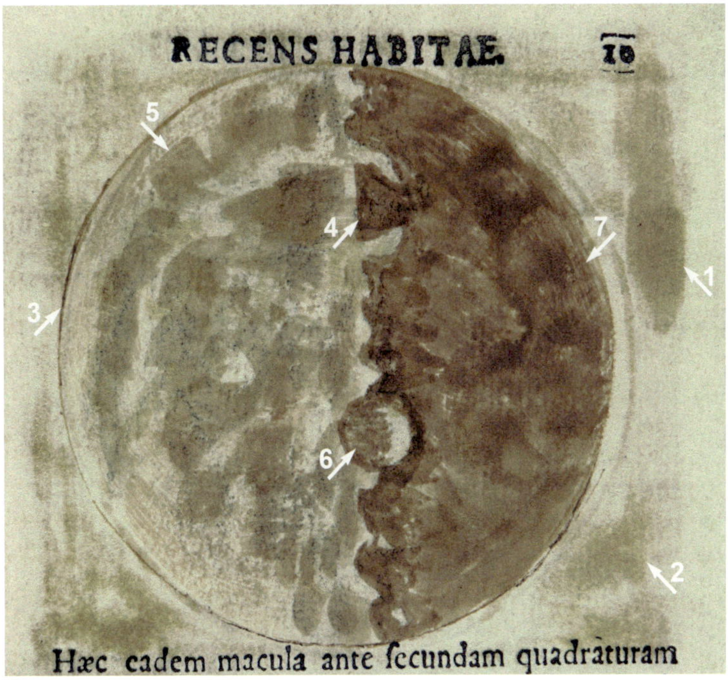

Fig. 3: Galileo Galilei, Sidereus Nuncius, Venice 1610, p. 10r, New York copy.

Arrow 1: bistre ink that bled through from p. 10v; arrow 2: yellowish halo from bistre ink that bled through from p. 10v; arrow 3: mismatch of the outlines; arrow 4: narrow day-night dividing line; arrow 5: short, wide brush strokes; arrow 6: short, narrow brush strokes; arrow 7: feathery ink application showing use of a "dry" brush.

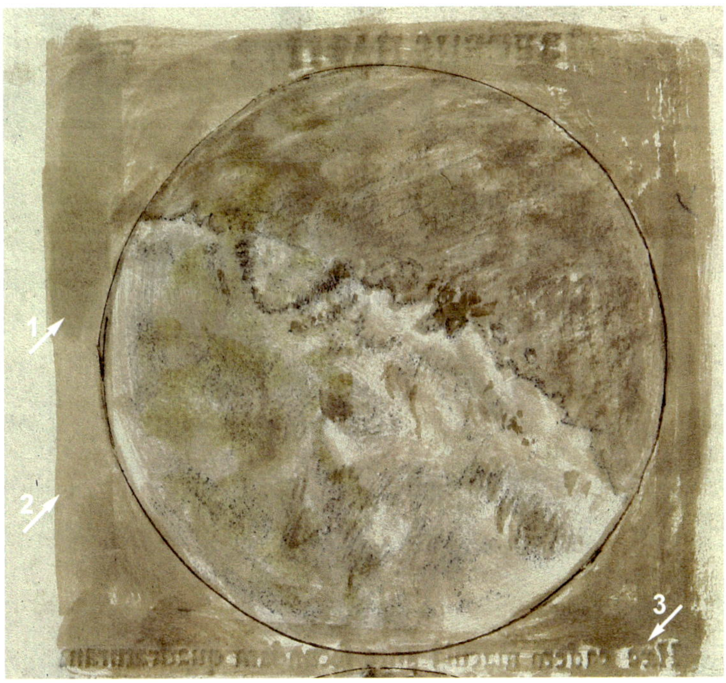

Fig. 4: Galileo Galilei, Sidereus Nuncius, Venice 1610, p. 10v, New York copy.

Arrow 1: heavy bistre ink that bled through to p. 10r; arrow 2: wide vertical brush stroke from top to bottom; arrow 3: unevenly inked areas where the brush skipped across the texture of the printed paper.

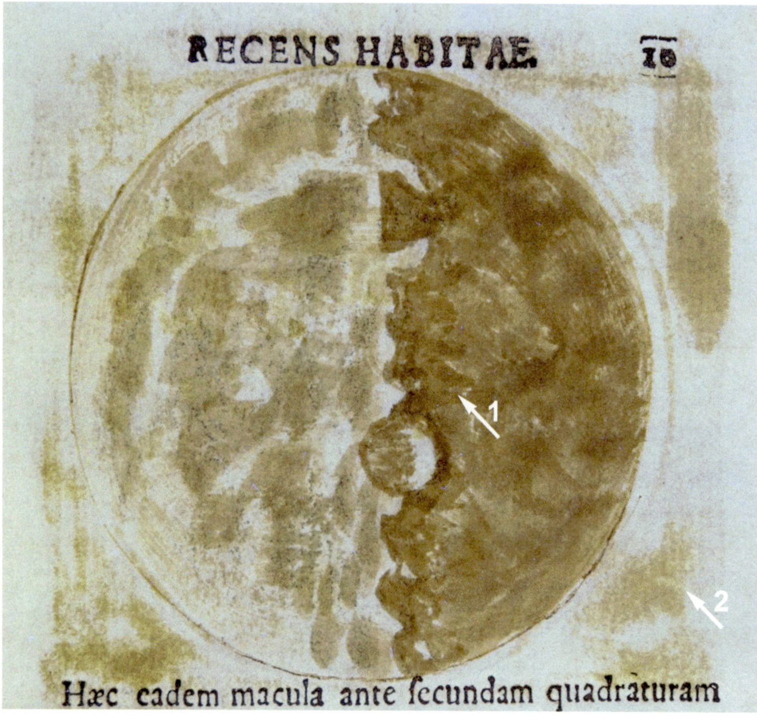

Fig. 5: Galileo Galilei, Sidereus Nuncius, Venice 1610, p. 10r, New York copy, UVA.

Arrow 1: brush strokes in concentrated ink appear dark; arrow 2: fluorescence at the edges of the bistre ink application.

The inked areas show a brownish-purple color, which is also characteristic of bistre in that it is neither completely absorbing nor distinctly fluorescing (see also Fig. 5).[6]

These characteristic features of the bistre drawings can be seen on pages 10r and 10v (Figs. 3 and 4). Where two brush strokes were applied on top of each other (Fig. 4, arrow 1), the ink bled through the paper and is visible on the other side (Fig. 3, arrow 1). This indicates a rapid working method: had the first ink stroke been allowed to dry, the second stroke on top would have been unlikely to bleed through the paper so profusely. The yellowish haloing effect associated with selective migration along the periphery of inked areas is not always as easily discerned under regular daylight illumination as under UVA radiation. For example, compare the lower right corner of the drawing on p. 10r in Fig. 3, arrow 2, with Fig. 5, arrow 2.

From the execution of the drawings, it is clear that the tendency of the ink to sink into the paper was evident to the artist while working. This likely inspired the decision to leave two of the three moons on these pages without the shaded background that indicates outer space in the other drawings (see part II, pp. 9v, 10r, 10v and 11r). The bleed-through effect of the ink background would have likely muddied and confused the back-to-back images. The selective use of the dark background also serves to visually separate the two moons on p. 10v. The etchings, by comparison, realize something of an intermediate stage between the shaded and non-shaded drawings in that they do not bother with added etched design

6 Colbourne 2001; Baker 1985.

elements to indicate the space surrounding the moons, but rely effectively on an ample residue of plate tone bordered by the plate mark to indicate space and separate the images (see Figs. 7 and 8, p. 107).

In constructing the drawings, it appears that initially, only a tentative indication of the moon outline was made and later intensified by the circle in reddish ink. Whether it was drawn with a stencil or compass cannot be conclusively said. There is no center hole to indicate the use of a compass, and the textured page could have caused irregularities and misalignment of the ends of lines in either case (Fig. 3, arrow 3; see also Fig. 8, p. 101).[7] The line that marks the all-important day-and-night divide would have logically occurred early in the drawing process (Fig. 3, arrow 4). The artist may then have turned his attention to the pertinent features of the moon landscape, filling in crescents, craters and other textural features. Shading was laid with a broad, probably flat brush of almost one centimeter width, as can be seen where a single short brush stroke stands clear on the page (Fig. 3, arrow 5). Furthermore, it took several consecutive ink applications to intensify the drawing in its key features, especially along the day-and-night divide, for example, in the crater (Fig. 3, arrow 6; see also Fig. 6, p. 93). There, the brush strokes are short and narrow, and may stem either from the tip of the flat brush held at an angle, or from a smaller round brush. The whole area also received final emphasis through brush work with a more concentrated, and therefore darker ink, which stands out more distinctly under UVA radiation (Fig. 5, arrow 1). Long strokes with a loaded flat brush mark some shaded areas of the night side of the moons (for example, see Fig. 1, p. 50) and the dark spaces around the moons. Carrying dilute ink, long strokes fill the spandrels and occasionally continue the length of the rectangle that frames the moon (Fig. 4, arrow 2). Strokes like this were meant to cover a substantial paper area swiftly, in some cases setting several strokes next to each other in a moment's time (see also the upper edge of the drawing on p. 9v, part II). These strokes are difficult to place accurately and evidence a precision that only comes with assured draughtsmanship.

The artist was quite clearly aware that he was working on a less than ideal surface, an unsized paper which behaved much like a blotting paper, eagerly absorbing the ink intended to be spread across the paper surface. However, he coped with this disadvantage by altering the brush motion to regulate the flow of the ink. At times he used a "dry" brush, one that was not fully loaded with ink and would therefore deposit a small amount of bistre on the paper, creating feathery lines (Fig. 3, arrow 7). At other times, he moved the fully loaded brush quickly across the absorbent paper to prevent the ink from draining too much into any one area of the paper. This explains the swift quality and length of some of the brush strokes, as well as their uneven skipping effect wherever they ran across the raised, relief printed text (Fig. 4, arrow 3).

Overall, the drawings suggest a nimble use of the available tools. They were executed quickly, but deftly with the artist's skill and knowledge of drawing techniques overcoming some disadvantages imposed by the absorbent and highly textured paper.

7 In the case of the Florentine drawings at the Biblioteca Nazionale Centrale Firenze, Gal. 48, the use of a compass is clearly indicated by the hole in the center of each circle (see Fig. 4, p. 51).

OLIVER HAHN / TIMO WOLFF / IRENE BRÜCKLE /
THERESA SMITH / ERICH JELEN

THE BLACK DEPOSIT

Upon close inspection with the naked eye, one can see that important features of the moon drawings were faintly marked with an unidentified black material. The material was found to contain no relevant inorganic compounds in the micro XRF analysis and is assumed to contain a carbon-based pigment in an organic binder. Such pigments, generated by burning wood or other organic materials in closed containers, were commonly used in paints, drawing materials and printing inks. The black deposit has a slightly shiny appearance when viewed under magnification in specular light. There is no powdery deposit present as would be the case if dry drawing media, such as graphite, charcoal or black chalk had been used (see Brückle and Tesmar p. 98ff.). The presence of a binding agent is likely responsible for the slight gloss observed. These black deposits are isolated areas smaller than a square mil-

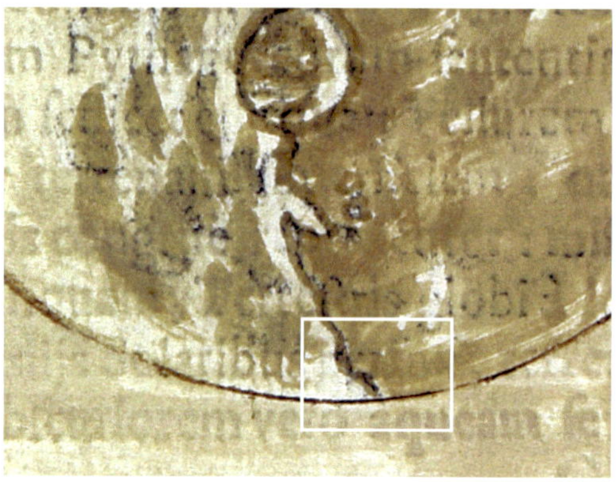

Fig. 1: Galileo Galilei, Sidereus Nuncius, Venice 1610, p. 9v, New York copy, detail; rectangle marks the area of interest.

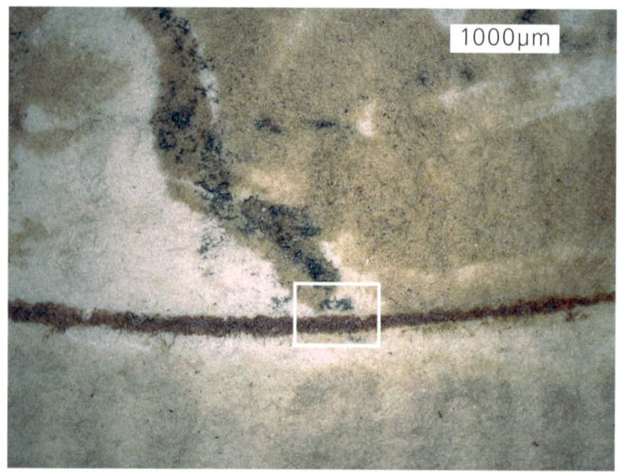

Fig. 2: Galileo Galilei, Sidereus Nuncius, Venice 1610, p. 9v, New York copy, detail of Fig. 1.

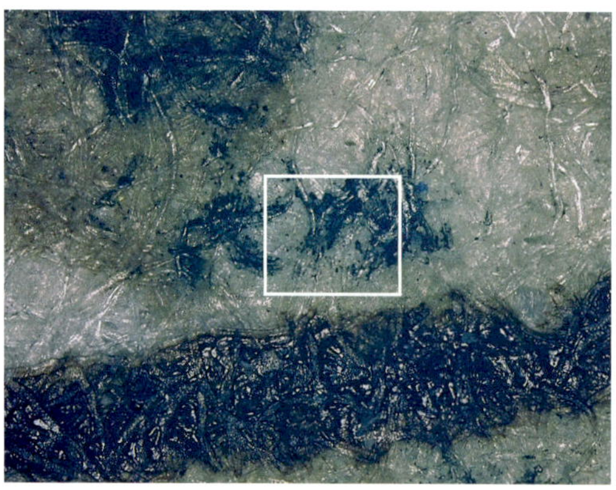

Fig. 3: Galileo Galilei, Sidereus Nuncius, Venice 1610, p. 9v, New York copy, detail of Fig. 2 at higher magnification.

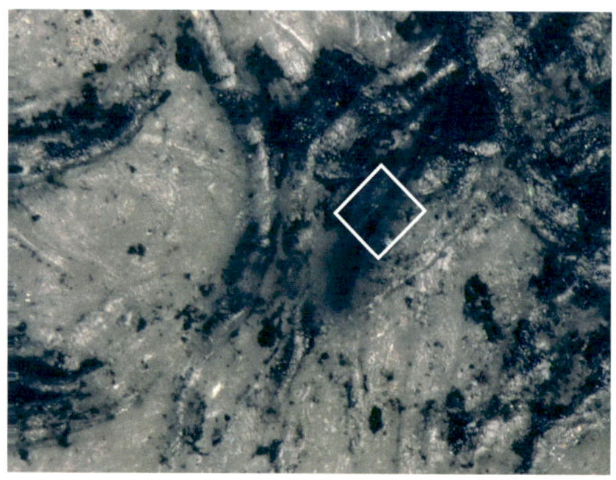

Fig. 4: Galileo Galilei, Sidereus Nuncius, Venice 1610, p. 9v, New York copy, detail of Fig. 3 at higher magnification.

limeter. The marks are located predominantly in areas of bistre ink application that mark the main features of the drawing. This is most obvious in the demarcation line separating the light and dark sides of the moon. This feature, central to the drawings of the moon phases, received particular attention when the black material was applied, as indicated by the fact that it is deposited in a line following the edge of the shaded area.

Digital microscopy and 3D confocal microscopy were used to try to locate the exact placement of the black deposit. Of special interest was the area where two lines cross at the south pole (Fig. 1). To look closer at the surface texture (Fig. 2), photomicrographs at high magnification were taken through the digital microscope (Figs. 3 and 4). These images confirm the original visual conclusions about the black deposit: it lies on top of the wash drawing on the paper. Confocal microscopy was used to make a precise measurement of the material's location (Fig. 6). The height data from the confocal measurement are shown in (Fig. 5). As can be clearly seen, the black material lies on top of the drawing on the paper surface. The particle detected at the bottom of the image was analyzed by measuring its profile. The particle was about 560 µm in width and 96 µm in height. Several facts were ascertained by examining the drawings under magnification.

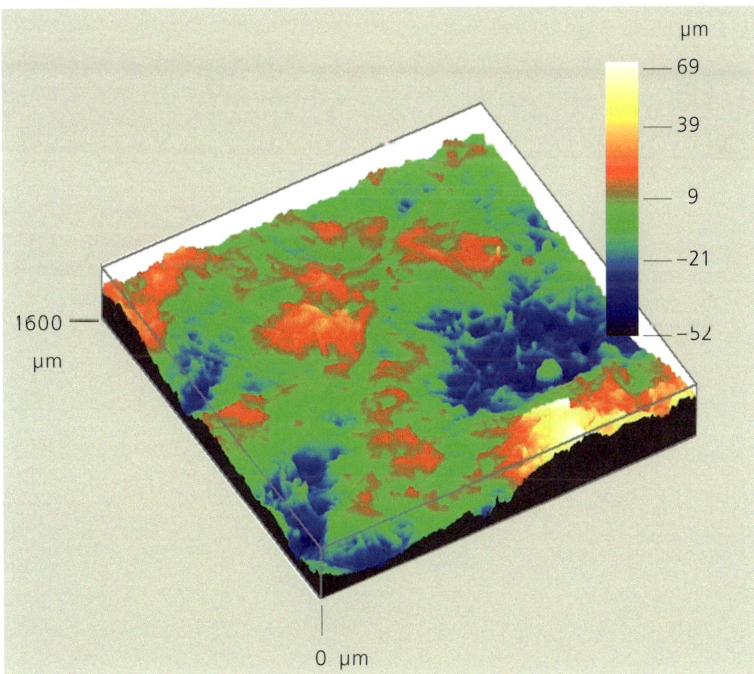

Fig. 5: Topographic profile obtained from 3D confocal microscopy.

1. The printed text was present before the black material was applied.
2. The drawings were executed before the black material was applied.
3. The black material is deposited in two patterns that are assumed to have been created simultaneously in a single working process:
   - a linear pattern loosely follows the demarcation line in the center of the moon;
   - a diffuse pattern follows the shaded areas of the moon.
4. Both patterns are interrupted by the texture of the paper and the relief created by the letterpress printing. The black material was preferentially deposited on the high points of the paper.
5. The black material does not appear to be applied directly to the page by a tool such as a brush, pen, or a stylus.

Comparative image analysis (see Krug and Simon p. 94ff.) demonstrates that the black material bears a direct relationship to the printed copperplate engravings of *Sidereus Nuncius*.

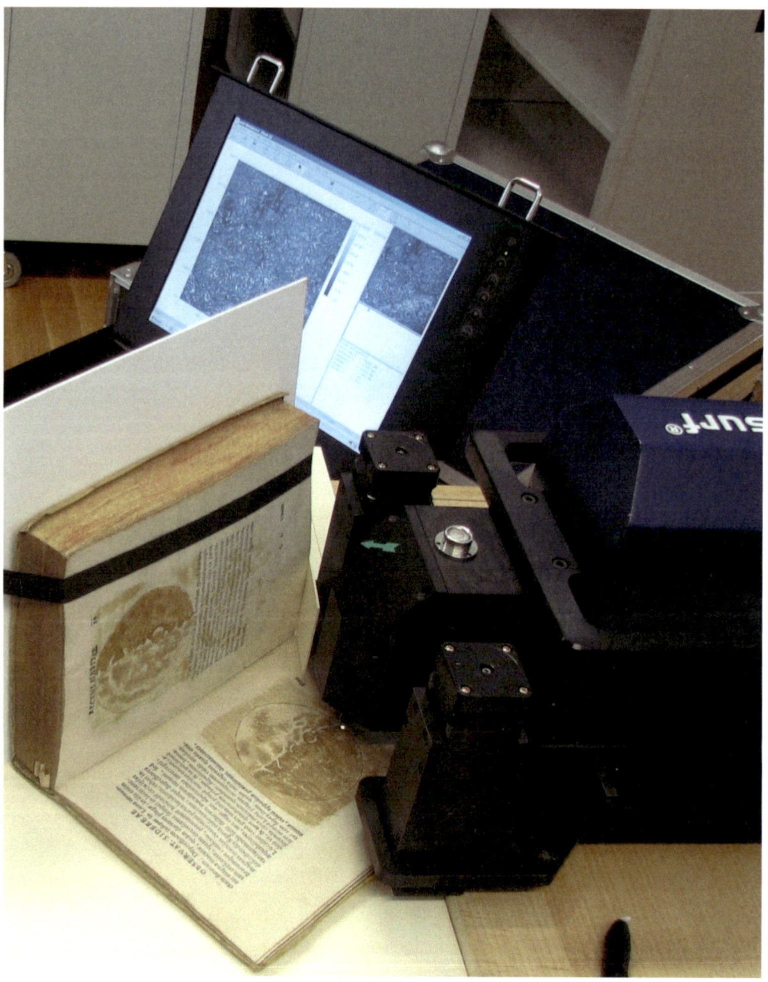

Fig. 6: Set-up of the 3D confocal microscope to analyze p. 9v of the New York copy.

HORST BREDEKAMP

THE DRAWINGS: THE EXAMPLE OF PAGE 8R

The first ink drawing in the *New York copy* appears on page 8r (Fig. 1). It is painted on a brown square whose lower right corner surrounds the printed word "Verum". The strokes marking outer space are executed broadly and rapidly; in the corners they follow the curve of the moon's circle except at the outermost edge, where they parallel the boundary of the square. Darkened areas are overlaid on this basic pattern, as if outer space were not a homogeneous region but consisted of a mass of energy unfolding in a rhythm of darkening and brightening. The strokes are impressive in the rapid, almost nervous manner of execution, which is sure enough to achieve the desired effect without considering details.

The artist filled the corners directly adjoining the circular moon outline on the left-hand side in short, careful strokes that were probably first laid parallel to the curve of the circle and then emphasized the corners. Then he followed the same procedure in a slightly weakened form on the opposite, right-hand side. The margins of the printed space on the obverse side, visible through the paper, are covered and exceeded on the left-hand side with a broad vertical brushstroke that continues in one sweep to the lower margin. Over this lies a broad, diagonal brushstroke that is picked up and set down several times on its way to the bottom margin.

In the interior of the circle, the boundary between light and dark was probably drawn first. The line of demarcation is carried out with a marked stroke whose emphasis offers a defining orientation for the entire form. It stands out as a distinct line in the lower area, but in the upper field it is softly mediated toward the dark zone. This night region is characterized by brown tones applied in slightly curved strokes. Gleaming bright spots testify to painterly finesse in many passages that were not painted, but simply left blank (Fig. 2). They allow the bright paper to shine as if from within. In the lower area, broad brushstrokes from the upper left to the lower right are conspicuous; these echo the parallel strokes of the outer space area on the left. Within the crescent, another mode of drawing is also noticeable in the form of diagonal strokes carried out with a dry brush only partly loaded with ink, which breaks up the broad strokes into thin lines (Fig. 3). They define a first layer of shading upon which darker brown tones are laid with a narrower brush strokes.

The overall painting gesture of this moon is ambivalent. Its abbreviated fidgetiness disturbs, but at the same time it also has a convincing, generously self-confident side. In particular, the curved craters of the illuminated crescent and the small zones of white paper that catch the eye in the night zone allow the viewer to almost physically feel the sureness with which these passages were executed.

A comparison with Galileo's known drawings of the moon in the BNCF helps to understand these peculiarities. The seven watercolor drawings that are bound in the original

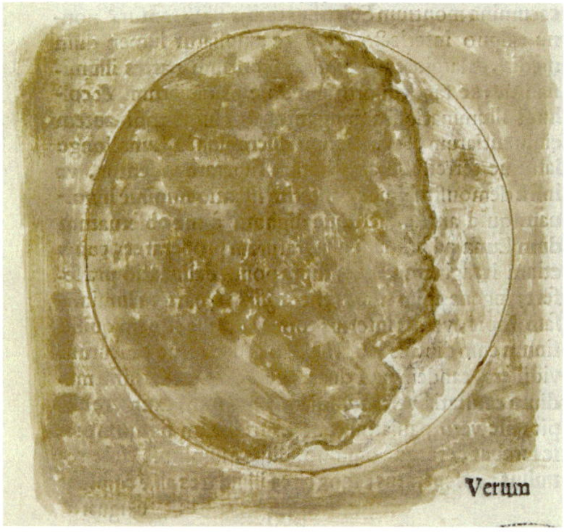

Fig. 1: Galileo Galilei, Sidereus Nuncius, Venice 1610, p. 8r, New York copy.

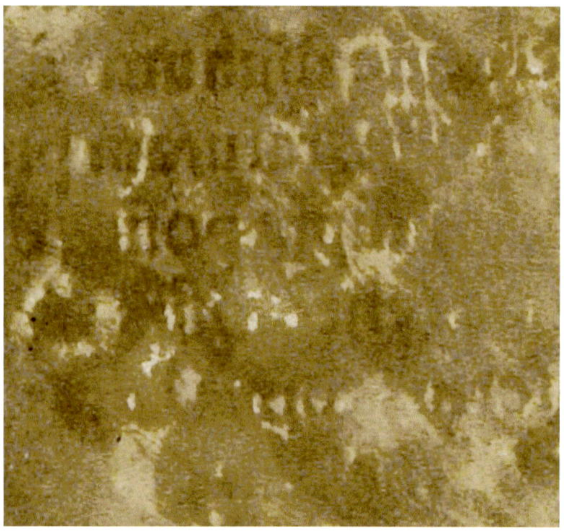

Fig. 2: Galileo Galilei, Sidereus Nuncius, Venice 1610, p. 8r, New York copy, detail of Fig. 1.

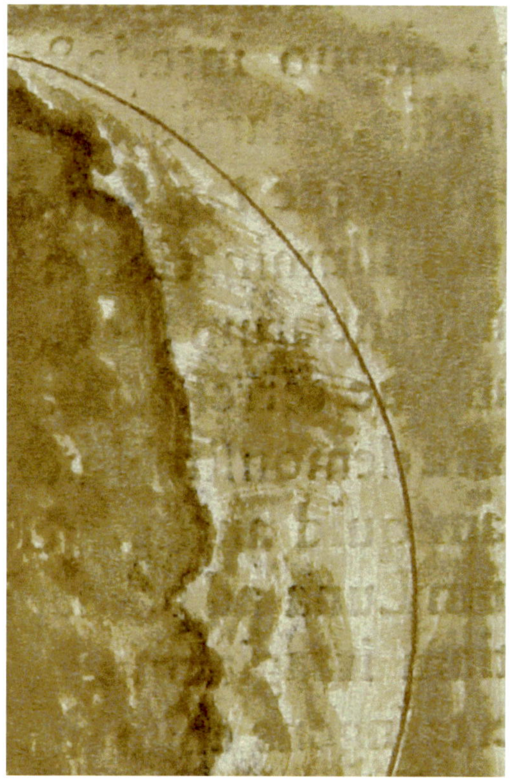

Fig. 3: Galileo Galilei, Sidereus Nuncius, Venice 1610, p. 8r, New York copy, detail of Fig. 1 showing variations in the color application.

manuscript stored in Florence are entered on the double page of folia 28 and 29, which, as paper for watercolor drawings, is somewhat smaller and heavier than the text pages.[1] On page 28r, two moons are painted in each of three rows (Fig. 4). The seventh moon appears in isolation on the reverse of folio 29 (Needham, p. 93).

Galileo's drawings of the moon were long known only through their reproduction in the *Opere*.[2] As meritorious as this image was, its almost muted appearance could not reproduce the quality of the watercolor drawings; recent research has for the first time perceived the latter's artistic rank.[3] With their relatively narrow dimensions of 57 to 59 mm, the drawings are impressive in their technique of depicting the plastic surface of the moon using a brush.

1 BNCF, Gal. 48, fol. 28 and 29. At 25.8 × 18.3 cm, it is slightly smaller than the other pages, each of which is 30 × 20 cm.
2 Galilei, *Opere*, vol. III/1, p. 48.
3 First: Edgerton 1984.

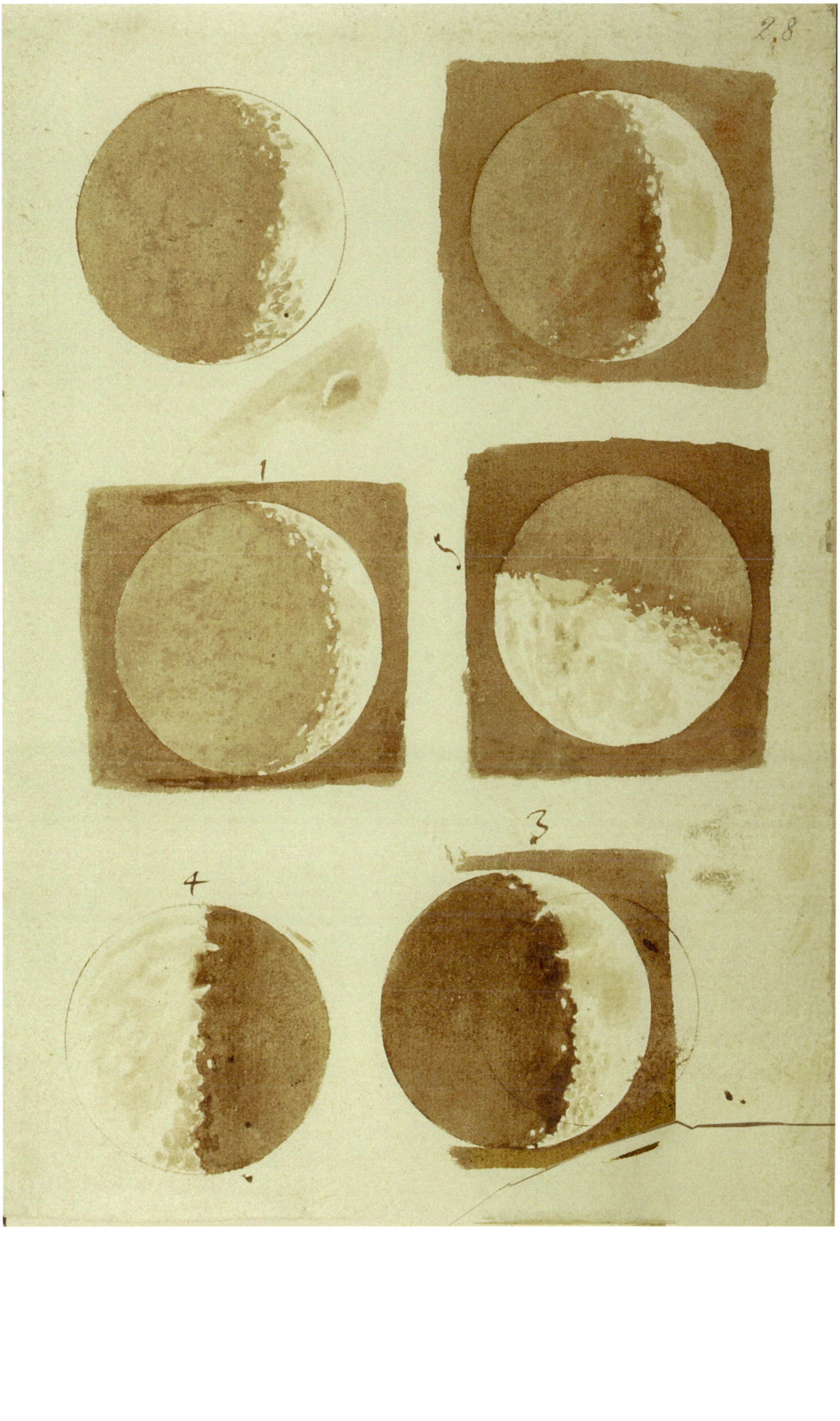

Fig. 4: Galileo Galilei, moon phases, watercolor, 1610, BNCF, Gal. 48, fol. 28r.

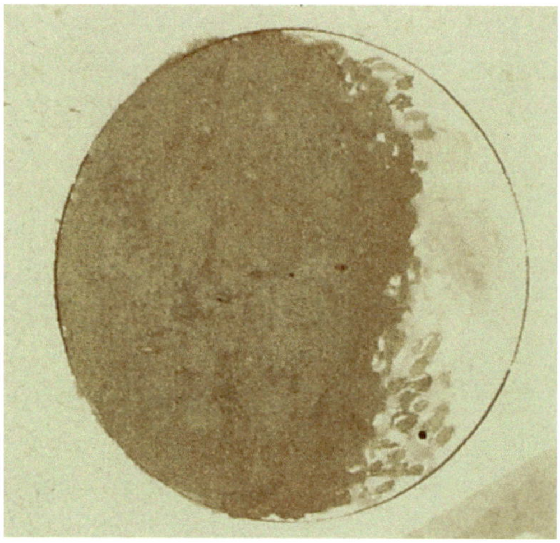

Fig. 5: Galileo Galilei, moon phase, watercolor, 1610, BNCF, Gal. 48, fol. 28r.

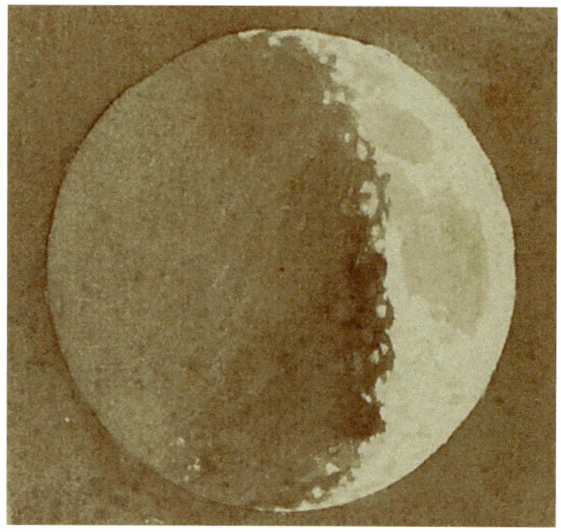

Fig. 6: Galileo Galilei, moon phase, watercolor, 1610, BNCF, Gal. 48, fol. 28r.

The particular quality of the drawings is owing to the use of brown watercolor washes applied in various concentrations that made it possible to modulate from a deep, shadowy tone to beige, lightening almost to white. As the top left depiction of the waxing moon already shows, Galileo used the matte cream of the paper as a color of its own; it defines moonlight in full reflection (Fig. 5). All the other tones are in relation to this standard.

The background – the night of outer space – offers the second axis of the color coordinate system (Fig. 6). To indicate this as a completed unit, Galileo placed a dark square around the circle of the moon. Visible at the upper and lower margin are his attempts to adequately illustrate the darkness of night, in that he made more darkening brushstrokes here. The result is that the deepest darkness is used at the periphery, on the lower and upper margin of the picture, as if the satellite suspended in outer space had received a kind of color axis as a visual anchor. In the interior of the moon the phenomenon of "second light" is applied with remarkable finesse. On the line where light and shadow meet, a browning, a nuance more intense than the abutting, ever-brightening sphere follows to the left. At the margin this blends into a bright arc of a circle, answered in turn by the depth of the nocturnal sky. These brightenings in the dark side of the moon display the most delicate structures and gradations of darkening.

A comparison makes it clear that the bistre ink drawing on page 8r of the *New York copy* is atmospherically related to the first three of the Florentine depictions (Fig. 4, see the drawing at top left, and those numbered "1" and "3"), but does not correspond in detail with any of them. Thus the ink drawing is definitely not a copy of these detailed watercolor drawings, but part of a series of its own.

Conspicuous are both differences from and correspondences with the first Florentine drawing (Fig. 5, Fig. 4 top left). For example, the gesture of the application of ink and watercolor, respectively, is different. While the nuances of the nocturnal sky in the Florentine drawing blur and produce fluid modulations both in outer space and on the night side of the moon, in the drawing in the *New York copy* on page 8r (Fig. 1), the ink layers abut each other energetically.[4] In addition, the broadly, almost sketchily brushed execution of the darkenings in the lower night zone in the *New York copy* drawing do not appear in the Florentine drawing; the same is true for the line separating day and night, which is intensely emphasized, while it is weak in the Florentine drawing.

No less obvious, however, are the correspondences. For example, the overall design is identical. The circle of the moon, sharply drawn in the Florentine drawings, is surrounded by a square marking outer space as background, and already in this area subtle resemblances are visible despite all the differences. The strokes beginning above the lower margin of the drawing on page 8r (Fig. 1) and extending across the south pole to the left correspond with the Florentine drawing numbered "1" (Fig. 6), which is drawn with the same generosity in ample watercolor from the lower right corner to the left. The same is true of the zone above the north pole in this Florentine drawing, where lines parallel the edge and correspond with the details of the gesture of this passage in the nocturnal sky on page 8r of the *New York copy*. Finally, this impression is underscored by the upper left corner of the square, where in both cases the brushstrokes are straight at the edge, but run in curves paralleling the horizon of the moon in its neighborhood.

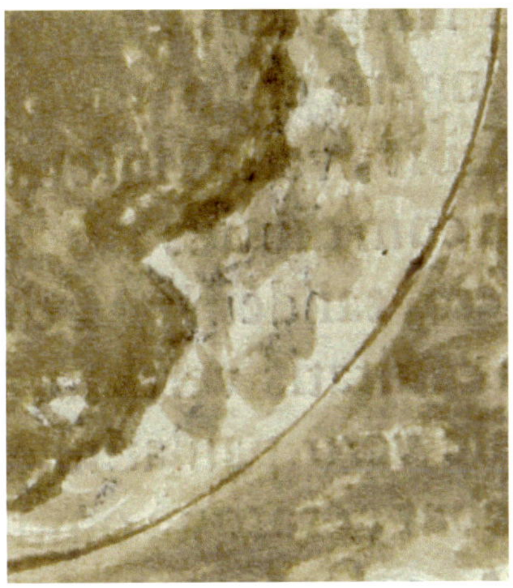

Fig. 7: Galileo Galilei, Sidereus Nuncius, Venice 1610, p. 8r, New York copy, detail of Fig. 1.

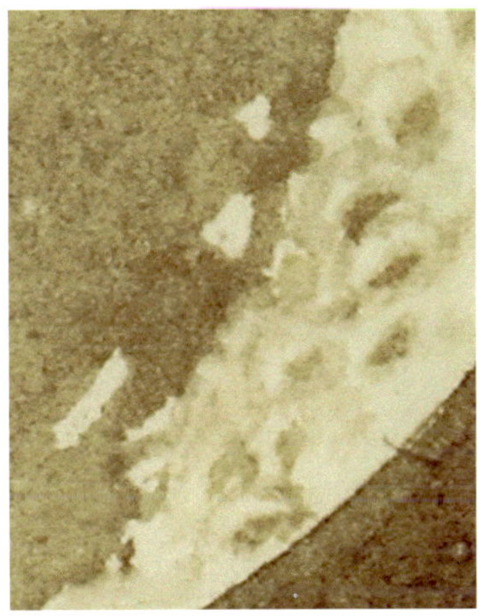

Fig. 8: Galileo Galilei, moon phase, watercolor, 1610, BNCF, Gal. 48, fol. 28r, detail of Fig. 6.

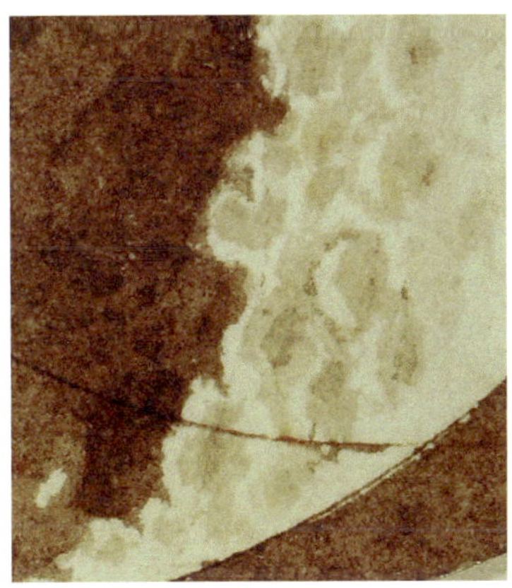

Fig. 9: Galileo Galilei, moon phase, watercolor, 1610, BNCF, Gal. 48, fol. 28r, detail of moon numbered "3" (Fig. 4).

4 It should be pointed out that some of the most subtle ink markings in the drawings result from uneven settlement of the ink on the paper.

Above all, the spots of the arc-like craters, in part isolated and in part connected with the dark zone, carried out in curves in the lower right with short hand motions, have almost identical formations (Fig. 7). The corresponding craters in the Florentine drawing (Fig. 6, numbered "1" in Fig. 4) show similarly turned-in, thick dabs in the lower half of the illuminated crescent (Fig. 8). The relationship is even more prominent in the upper left Florentine drawing (Fig. 5), in which the craters are characterized with abbreviated dabs similar to those in the drawing on page 8r. However, the character of the brushstrokes is nowhere as similar as in the field of craters in the Florentine moon, numbered "3" in Fig. 4, which is among Galileo's most impressive achievements as a painter. The small moons presented beside the deep shading of the night zone show the same curved crescents of shadow and light (Fig. 9) as is found, somewhat more coarsely, in the drawings of the *New York copy*.

Also impressive is the identical way the spots of light were surrounded and left standing in the nocturnal zone. Another draftsman might have depicted the bright peaks emerging from the dark of the lunar night by applying white paint to the dark background, but here, no white paint is applied anywhere. Rather, the spots of light in the Florentine drawings are formed by being left out and surrounded by watercolor applied with great sureness, despite the detailed technique of painting. On page 8r, the ability of the seemingly careless hand to repeatedly and constructively stop the act of painting to leave highlights in the shaded areas reveals a quality developed through practice – a quality also found in the Florentine drawings.

The differences in the painterly gesture have a material explanation in the differing consistencies of the paper. The pages of the *New York copy* are unsized or only weakly sized paper which was appropriate and customary for printing, but which is hardly suitable for ink or watercolor drawings because it is too absorbent (see Brückle et al. p. 140f.). This paper quality necessitated the high speed of execution. The style, which at first sight seems careless in comparison with the Florentine drawings, is a result of the material of the paper. Whereas in the Florentine watercolor drawings, Galileo was able to capture his observations of the moon on a paper suitable to this purpose without any time pressure, it was necessary to execute the bistre drawings swiftly in the *New York copy*.

The initial comparison between the New York and the Florentine drawings produces a clear result. The resemblances are so many and so subtle that they are evidence that the two series are depictions from Galileo's own hand. Neither a pupil, nor an unmentioned assistant, nor an imitator could have assimilated these characteristics of the motion of the brush that appear within the narrowest space, sometimes a fraction of a millimeter.[5]

5 A comprehensive analysis of all drawings: Bredekamp 2007, pp. 155–174.

II

TURNING THE PAGES OF THE NEW YORK *SIDEREUS NUNCIUS*

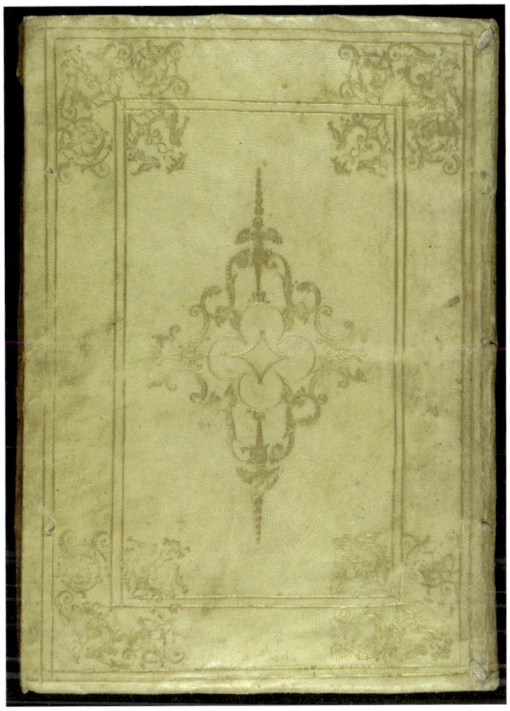

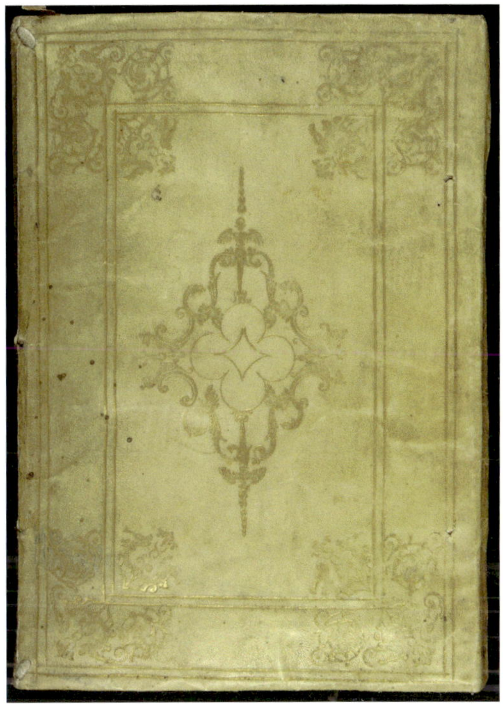

BARBARA HERRENKIND

THE PHOTOGRAPHS

The goal of the photography was to capture the texture of the paper and the dimensional relief of the letterpress printing. To this end, the book was supported on padded blocks covered with velvet and the blocks were adjusted so that each set of facing pages was level and in focus. The book was lit with diffuse raking light to accentuate the texture while making the text and images legible.

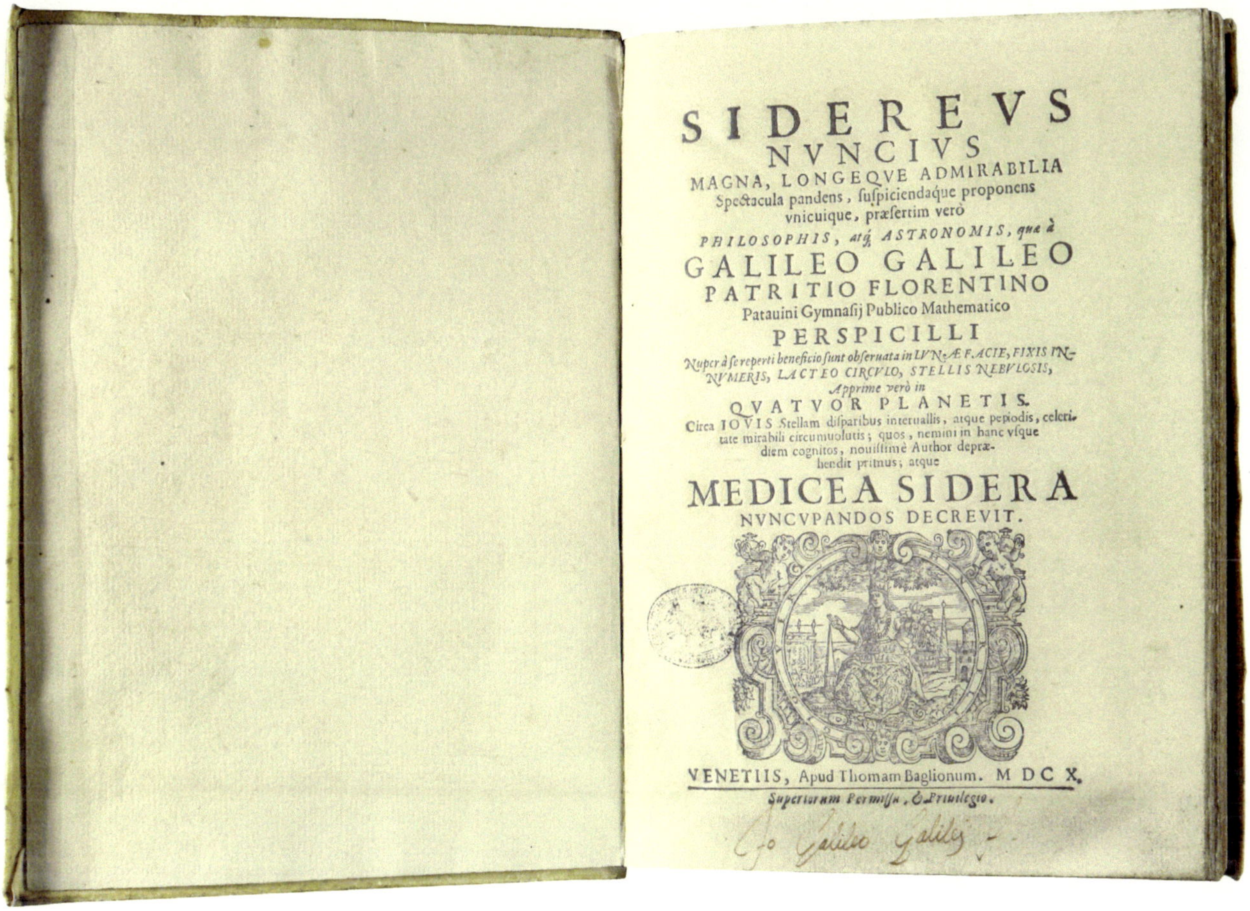
SIDEREVS
NVNCIVS
MAGNA, LONGEQVE ADMIRABILIA
Spectacula pandens, suspiciendaque proponens
vnicuique, præsertim verò
PHILOSOPHIS, atq; ASTRONOMIS, quæ à
GALILEO GALILEO
PATRITIO FLORENTINO
Patauini Gymnasij Publico Mathematico
PERSPICILLI
Nuper à se reperti beneficio sunt obseruata in LVNÆ FACIE, FIXIS INNVMERIS, LACTEO CIRCVLO, STELLIS NEBVLOSIS,
Apprime verò in
QVATVOR PLANETIS
Circa IOVIS Stellam disparibus interuallis, atque periodis, celeritate mirabili circumuolutis; quos, nemini in hanc vsque diem cognitos, nouissimè Author depræhendit primus; atque
MEDICEA SIDERA
NVNCVPANDOS DECREVIT.
VENETIIS, Apud Thomam Baglionum. M DC X.
Superiorum Permissu, & Priuilegio.

Galileo Galilei, Sidereus Nuncius, Venice 1610, p. 1r, New York copy.

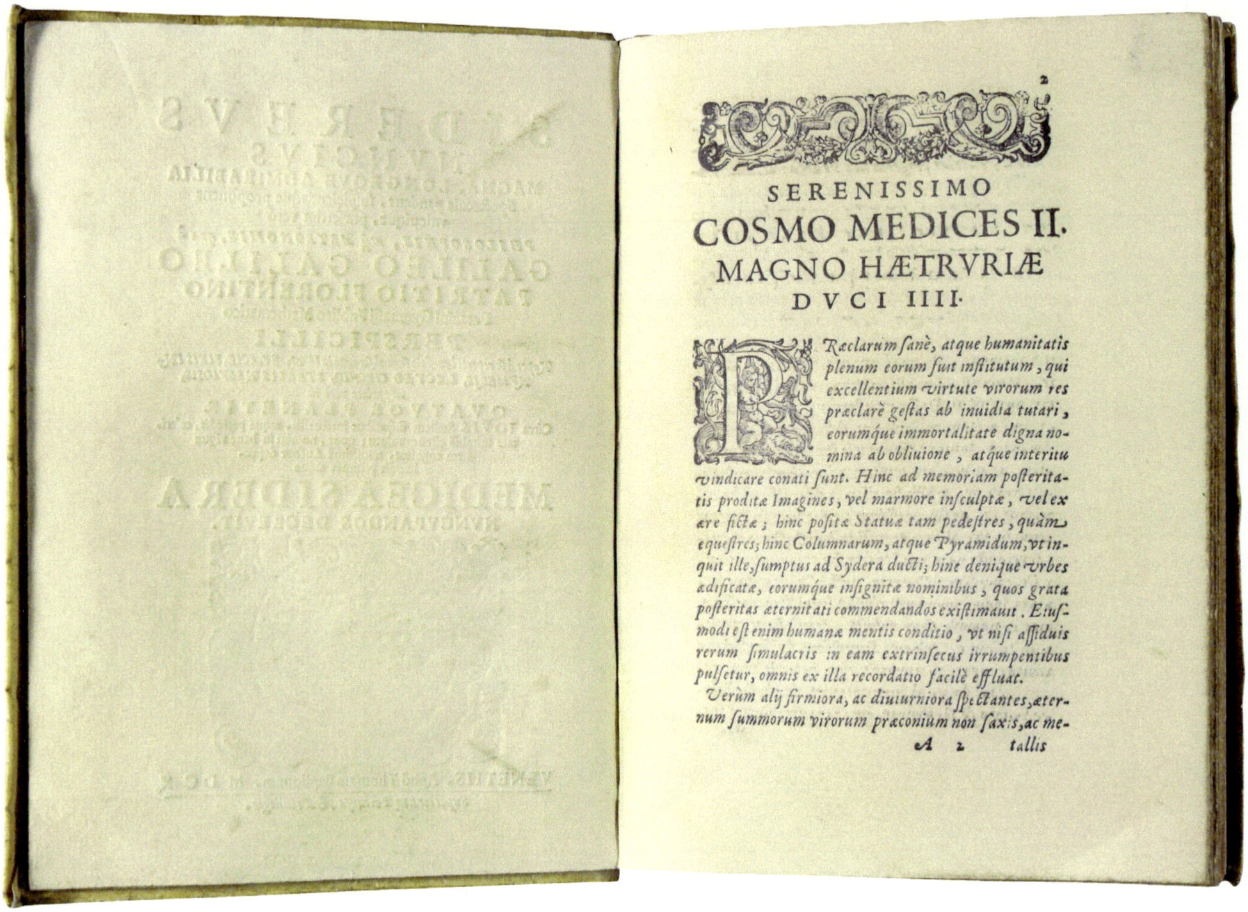

2

SERENISSIMO
COSMO MEDICES II.
MAGNO HÆTRVRIÆ
DVCI IIII.

PRæclarum sanè, atque humanitatis plenum eorum fuit institutum, qui excellentium virtute virorum res præclarè gestas ab inuidia tutari, eorumque immortalitate digna nomina ab obliuione, atque interitu vindicare conati sunt. Hinc ad memoriam posteritatis prodita Imagines, vel marmore insculptæ, vel ex ære fictæ; hinc positæ Statuæ tam pedestres, quàm equestres; hinc Columnarum, atque Pyramidum, vt inquit ille, sumptus ad Sydera ducti; hinc denique vrbes ædificatæ, eorumque insignitæ nominibus, quos grata posteritas æternitati commendandos existimauit. Eiusmodi est enim humanæ mentis conditio, vt nisi assiduis rerum simulacris in eam extrinsecus irrumpentibus pulsetur, omnis ex illa recordatio facilè effluat.

Verùm alij firmiora, ac diuturniora spectantes, æternum summorum virorum præconium non saxis, ac metallis

A 2 tallis

Galileo Galilei, Sidereus Nuncius, Venice 1610, pp. 1v and 2r, New York copy.

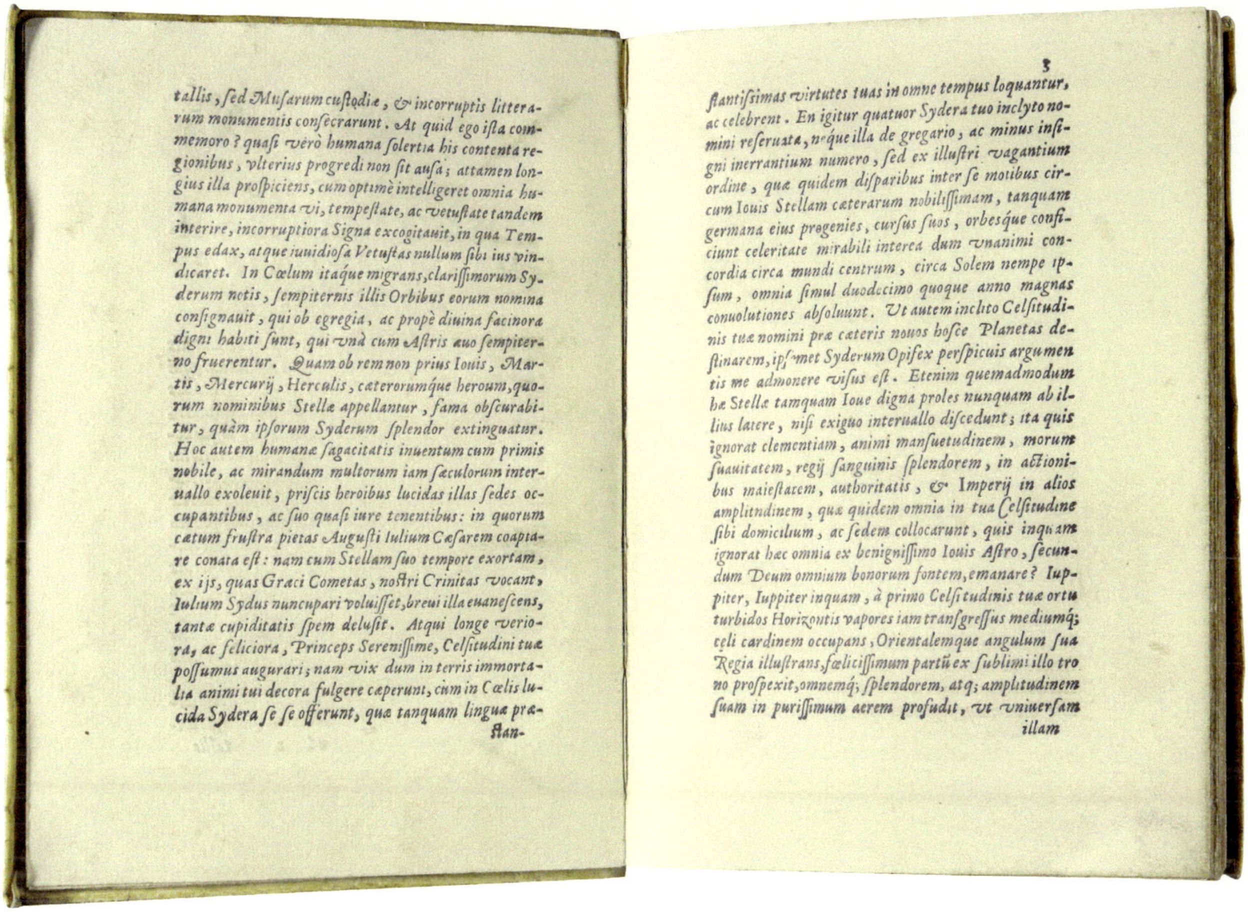
tallis, ſed Muſarum cuſtodiæ, & incorruptis littera-
rum monumentis conſecrarunt. At quid ego iſta com-
memoro? quaſi verò humana ſolertia his contenta re-
gionibus, vlterius progredi non ſit auſa; attamen lon-
gius illa proſpiciens, cum optimè intelligeret omnia hu-
mana monumenta vi, tempeſtate, ac vetuſtate tandem
interire, incorruptiora Signa excogitauit, in qua Tem-
pus edax, atque inuidioſa Vetuſtas nullum ſibi ius vin-
dicaret. In Cœlum itaque migrans, clariſſimorum Sy-
derum notis, ſempiternis illis Orbibus eorum nomina
conſignauit, qui ob egregia, ac propè diuina facinora
digni habiti ſunt, qui vnà cum Aſtris æuo ſempiter-
no fruerentur. Quam ob rem non prius Iouis, Mar-
tis, Mercurij, Herculis, cæterorumque heroum, quo-
rum nominibus Stellæ appellantur, fama obſcurabi-
tur, quàm ipſorum Syderum ſplendor extinguatur.
Hoc autem humanæ ſagacitatis inuentum cum primis
nobile, ac mirandum multorum iam ſæculorum inter-
uallo exoleuit, priſcis heroibus lucidas illas ſedes oc-
cupantibus, ac ſuo quaſi iure tenentibus: in quorum
cætum fruſtra pietas Auguſti Iulium Cæſarem coapta-
re conata eſt: nam cum Stellam ſuo tempore exortam,
ex ijs, quas Græci Cometas, noſtri Crinitas vocant,
Iulium Sydus nuncupari voluiſſet, breui illa euaneſcens,
tantæ cupiditatis ſpem deluſit. Atqui longe verio-
ra, ac feliciora, Princeps Sereniſſime, Celſitudini tuæ
poſſumus augurari; nam vix dum in terris immorta-
lia animi tui decora fulgere cœperunt, cùm in Cœlis lu-
cida Sydera ſe ſe offerunt, quæ tanquam linguæ præ-
ſtan-

3

ſtantiſſimas virtutes tuas in omne tempus loquantur,
ac celebrent. En igitur quatuor Sydera tuo inclyto no-
mini reſeruata, neque illa de gregario, ac minus inſi-
gni inerrantium numero, ſed ex illuſtri vagantium
ordine, quæ quidem diſparibus inter ſe motibus cir-
cum Iouis Stellam cæterarum nobiliſſimam, tanquam
germana eius progenies, curſus ſuos, orbesque confi-
ciunt celeritate mirabili interea dum vnanimi con-
cordia circa mundi centrum, circa Solem nempe ip-
ſum, omnia ſimul duodecimo quoque anno magnas
conuolutiones abſoluunt. Vt autem inclito Celſitudi-
nis tuæ nomini præ cæteris nouos hoſce Planetas de-
ſtinarem, ipſemet Syderum Opifex perſpicuis argumen
tis me admonere viſus eſt. Etenim quemadmodum
hæ Stellæ tamquam Ioue digna proles nunquam ab il-
lius latere, niſi exiguo interuallo diſcedunt; ita quis
ignorat clementiam, animi manſuetudinem, morum
ſuauitatem, regij ſanguinis ſplendorem, in actioni-
bus maieſtatem, authoritatis, & Imperij in alios
amplitudinem, quæ quidem omnia in tua Celſitudine
ſibi domicilium, ac ſedem collocarunt, quis inquam
ignorat hæc omnia ex benigniſſimo Iouis Aſtro, ſecun-
dum Deum omnium bonorum fontem, emanare? Iup-
piter, Iuppiter inquam, à primo Celſitudinis tuæ ortu
turbidos Horizontis vapores iam tranſgreſſus mediumq;
cęli cardinem occupans, Orientalemque angulum ſua
Regia illuſtrans, fœliciſſimum partũ ex ſublimi illo tro
no proſpexit, omnemq; ſplendorem, atq; amplitudinem
ſuam in puriſſimum aerem profudit, vt vniuerſam
illam

Galileo Galilei, Sidereus Nuncius, Venice 1610, pp. 2v and 3r, New York copy.

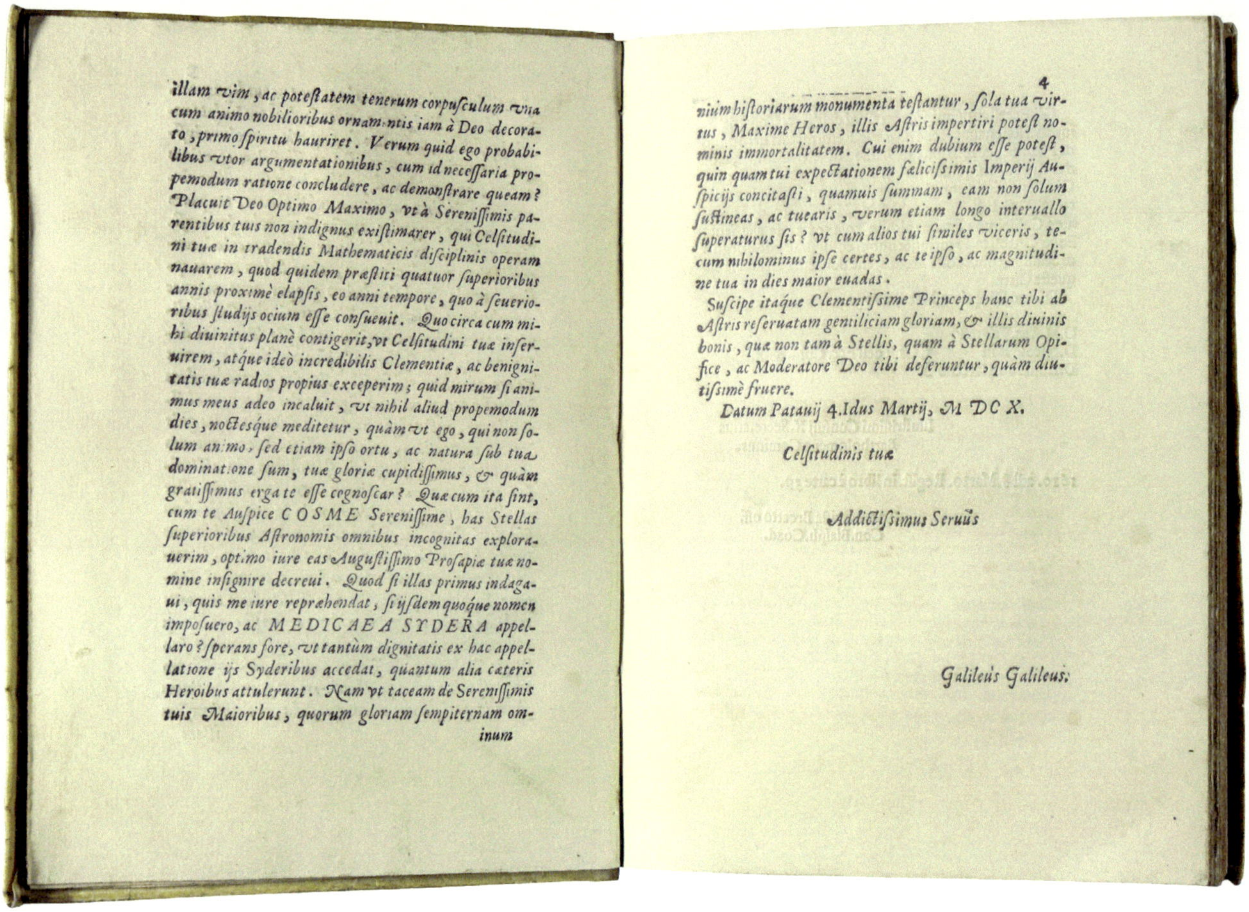
illam vim, ac potestatem tenerum corpusculum vna cum animo nobilioribus ornam:ntis iam à Deo decorato, primo spiritu hauriret. Verum quid ego probabilibus vtor argumentationibus, cum id necessaria propemodum ratione concludere, ac demonstrare queam? Placuit Deo Optimo Maximo, vt à Serenissimis parentibus tuis non indignus existimarer, qui Celsitudini tuæ in tradendis Mathematicis disciplinis operam nauarem, quod quidem præstiti quatuor superioribus annis proximè elapsis, eo anni tempore, quo à seuerioribus studijs ocium esse consueuit. Quo circa cum mihi diuinitus planè contigerit, vt Celsitudini tuæ inseruirem, atque ideò incredibilis Clementiæ, ac benignitatis tuæ radios propius exceperim; quid mirum si animus meus adeo incaluit, vt nihil aliud propemodum dies, noctesque meditetur, quàm vt ego, qui non solum animo, sed etiam ipso ortu, ac natura sub tua dominatione sum, tuæ gloriæ cupidissimus, & quàm gratissimus erga te esse cognoscar? Quæ cum ita sint, cum te Auspice COSME Serenissime, has Stellas superioribus Astronomis omnibus incognitas explorauerim, optimo iure eas Augustissimo Prosapiæ tuæ nomine insignire decreui. Quod si illas primus indagaui, quis me iure reprehendat, si ijsdem quoque nomen imposuero, ac MEDICAEA SYDERA appellaro? sperans fore, vt tantùm dignitatis ex hac appellatione ijs Syderibus accedat, quantum alia cæteris Heroibus attulerunt. Nam vt taceam de Serenissimis tuis Maioribus, quorum gloriam sempiternam om-

inum

4

nium historiarum monumenta testantur, sola tua virtus, Maxime Heros, illis Astris impertiri potest nominis immortalitatem. Cui enim dubium esse potest, quin quam tui expectationem fælicissimis Imperij Auspicijs concitasti, quamuis summam, eam non solum sustineas, ac tuearis, verum etiam longo interuallo superaturus sis? vt cum alios tui similes viceris, tecum nihilominus ipse certes, ac te ipso, ac magnitudine tua in dies maior euadas.

Suscipe itaque Clementissime Princeps hanc tibi ab Astris reseruatam gentiliciam gloriam, & illis diuinis bonis, quæ non tam à Stellis, quam à Stellarum Opifice, ac Moderatore Deo tibi deferuntur, quàm diutissimè fruere.

Datum Patauij 4.Idus Martij, M DC X.

Celsitudinis tuæ

Addictissimus Seruus

Galileus Galileus.

Galileo Galilei, Sidereus Nuncius, Venice 1610, pp. 3v and 4r, New York copy.

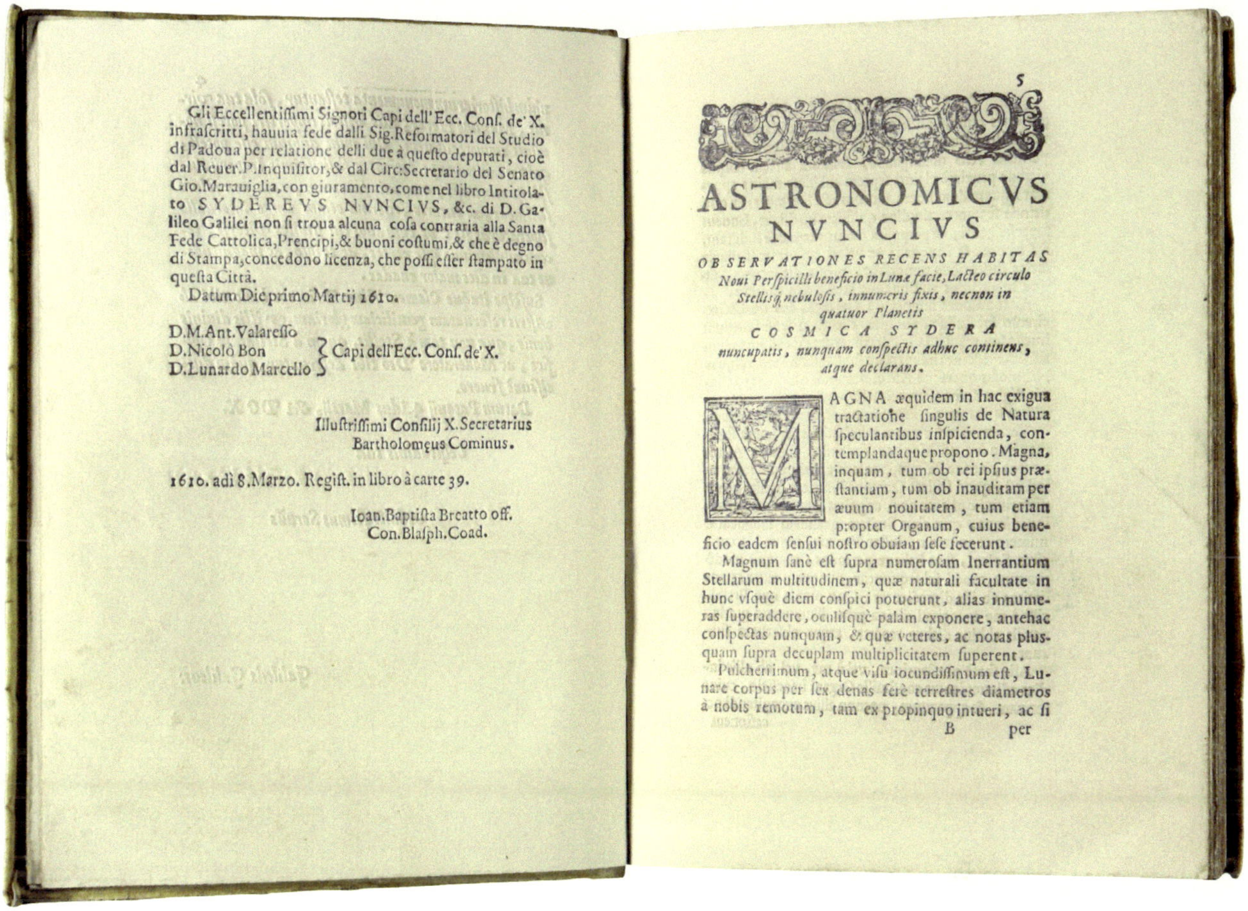

Gli Eccellentiſſimi Signori Capi dell'Ecc. Conſ. de' X. infraſcritti, hauuta fede dalli Sig. Reformatori del Studio di Padoua per relatione delli due à queſto deputati, cioè dal Reuer. P. Inquiſitor, & dal Circ: Secretario del Senato Gio. Marauiglia, con giuramento, come nel libro Intitolato SYDEREVS NVNCIVS, &c. di D. Galileo Galilei non ſi troua alcuna coſa contraria alla Santa Fede Cattolica, Prencipi, & buoni coſtumi, & che è degno di Stampa, concedono licenza, che poſſi eſſer ſtampato in queſta Città.

Datum Die primo Martij 1610.

D.M.Ant.Valareſſo
D.Nicolò Bon } Capi dell'Ecc. Conſ. de' X.
D.Lunardo Marcello

Illuſtriſſimi Conſilij X. Secretarius
Bartholomæus Cominus.

1610. adì 8. Marzo. Regiſt. in libro à carte 39.

Ioan. Baptiſta Breatto off.
Con. Blaſph. Coad.

5

ASTRONOMICVS
NVNCIVS

OBSERVATIONES RECENS HABITAS Noui Perſpicilli beneficio in Lunæ facie, Lacteo circulo Stellisq; nebuloſis, innumeris fixis, necnon in quatuor Planetis
COSMICA SYDERA
nuncupatis, nunquam conſpectis adhuc continens, atque declarans.

MAGNA æquidem in hac exigua tractatione ſingulis de Natura ſpeculantibus inſpicienda, contemplandaque propono. Magna, inquam, tum ob rei ipſius præſtantiam, tum ob inauditam per æuum nouitatem, tum etiam propter Organum, cuius beneficio eadem ſenſui noſtro obuiam ſeſe fecerunt.

Magnum ſanè eſt ſupra numeroſam Inerrantium Stellarum multitudinem, quæ naturali facultate in hunc vſquè diem conſpici potuerunt, alias innumeras ſuperaddere, oculiſquè palàm exponere, antehac conſpectas nunquam, & quæ veteres, ac notas plusquam ſupra decuplam multiplicitatem ſuperent.

Pulcherrimum, atque viſu iocundiſſimum eſt, Lunare corpus per ſex denas ferè terreſtres diametros à nobis remotum, tam ex propinquo intueri, ac ſi

B per

Galileo Galilei, Sidereus Nuncius, Venice 1610, pp. 4v and 5r, New York copy.

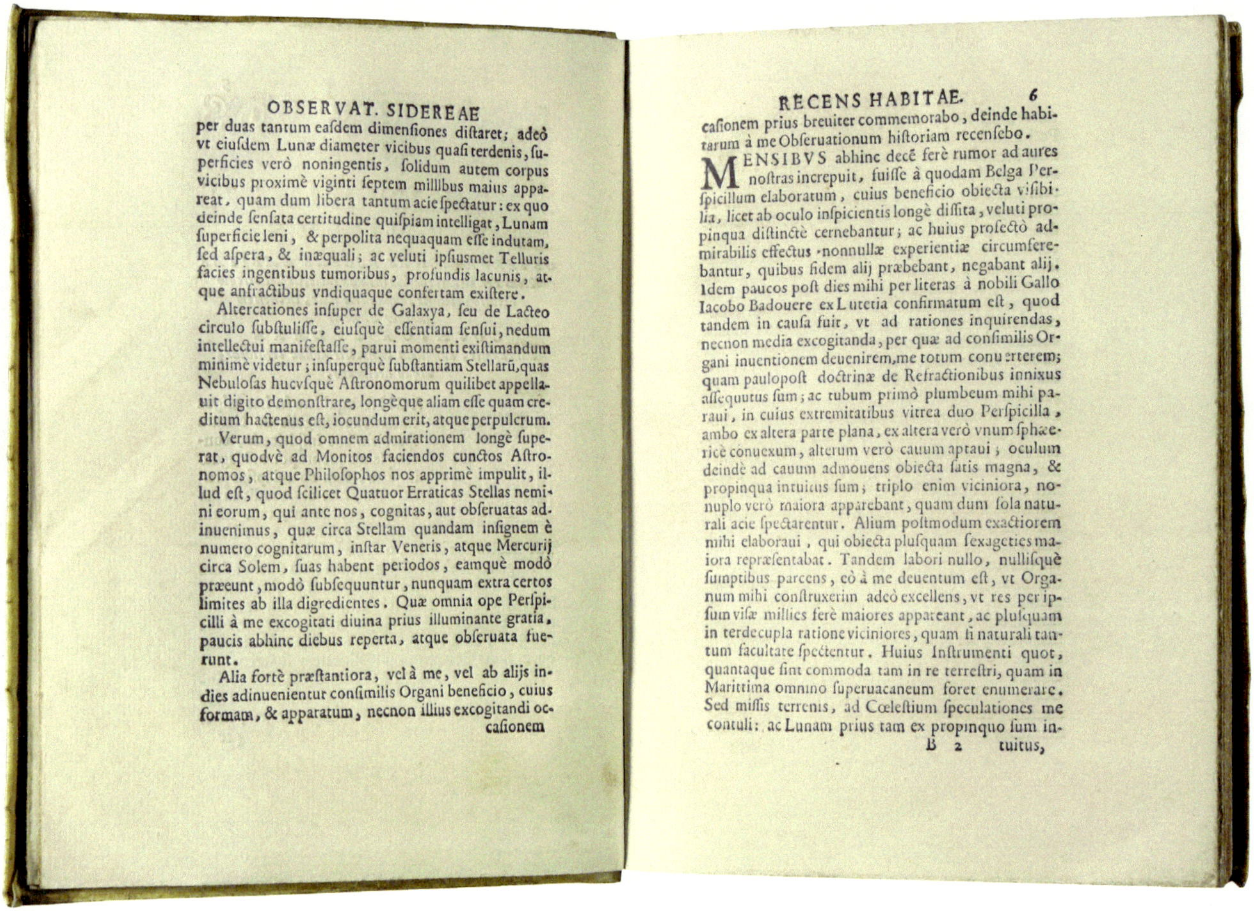

OBSERVAT. SIDEREAE

per duas tantum easdem dimensiones distaret; adeò vt eiusdem Lunæ diameter vicibus quasi terdenis, superficies verò noningentis, solidum autem corpus vicibus proximè viginti septem millibus maius appareat, quam dum libera tantum acie spectatur: ex quo deinde sensata certitudine quispiam intelligat, Lunam superficie leni, & perpolita nequaquam esse indutam, sed aspera, & inæquali; ac veluti ipsiusmet Telluris facies ingentibus tumoribus, profundis lacunis, atque anfractibus vndiquaque confertam existere.

Altercationes insuper de Galaxya, seu de Lacteo circulo substulisse, eiusquè essentiam sensui, nedum intellectui manifestasse, parui momenti existimandum minimè videtur; insuperquè substantiam Stellarũ, quas Nebulosas hucvsquè Astronomorum quilibet appellauit digito demonstrare, longèque aliam esse quam creditum hactenus est, iocundum erit, atque perpulcrum.

Verum, quod omnem admirationem longè superat, quodvè ad Monitos faciendos cunctos Astronomos, atque Philosophos nos apprimè impulit, illud est, quod scilicet Quatuor Erraticas Stellas nemini eorum, qui ante nos, cognitas, aut obseruatas adinuenimus, quæ circa Stellam quandam insignem è numero cognitarum, instar Veneris, atque Mercurij circa Solem, suas habent periodos, eamquè modò præeunt, modò subsequuntur, nunquam extra certos limites ab illa digredientes. Quæ omnia ope Perspicilli à me excogitati diuina prius illuminante gratia, paucis abhinc diebus reperta, atque obseruata fuerunt.

Alia fortè præstantiora, vel à me, vel ab alijs indies adinuenientur consimilis Organi beneficio, cuius formam, & apparatum, necnon illius excogitandi occasionem

RECENS HABITAE. 6

casionem prius breuiter commemorabo, deinde habitarum à me Obseruationum historiam recensebo.

MENSIBVS abhinc decẽ ferè rumor ad aures nostras increpuit, fuisse à quodam Belga Perspicillum elaboratum, cuius beneficio obiecta visibilia, licet ab oculo inspicientis longè dissita, veluti propinqua distinctè cernebantur; ac huius profectò admirabilis effectus nonnullæ experientiæ circumferebantur, quibus fidem alij præbebant, negabant alij. Idem paucos post dies mihi per literas à nobili Gallo Iacobo Badouere ex Lutetia confirmatum est, quod tandem in causa fuit, vt ad rationes inquirendas, necnon media excogitanda, per quæ ad consimilis Organi inuentionem deuenirem, me totum conuerterem; quam paulopost doctrinæ de Refractionibus innixus assequutus sum; ac tubum primò plumbeum mihi paraui, in cuius extremitatibus vitrea duo Perspicilla, ambo ex altera parte plana, ex altera verò vnum sphæricè conuexum, alterum verò cauum aptaui; oculum deindè ad cauum admouens obiecta satis magna, & propinqua intuitus sum; triplo enim viciniora, nonuplo verò maiora apparebant, quam dum sola naturali acie spectarentur. Alium postmodum exactiorem mihi elaboraui, qui obiecta plusquam sexagecies maiora repræsentabat. Tandem labori nullo, nullisquè sumptibus parcens, eò à me deuentum est, vt Organum mihi construxerim adeò excellens, vt res per ipsum visæ millies ferè maiores appareant, ac plusquam in terdecupla ratione viciniores, quam si naturali tantum facultate spectentur. Huius Instrumenti quot, quantaque sint commoda tam in re terrestri, quam in Marittima omnino superuacaneum foret enumerare. Sed missis terrenis, ad Cœlestium speculationes me contuli: ac Lunam prius tam ex propinquo sum in-

B 2 tuitus,

Galileo Galilei, Sidereus Nuncius, Venice 1610, pp. 5v and 6r, New York copy.

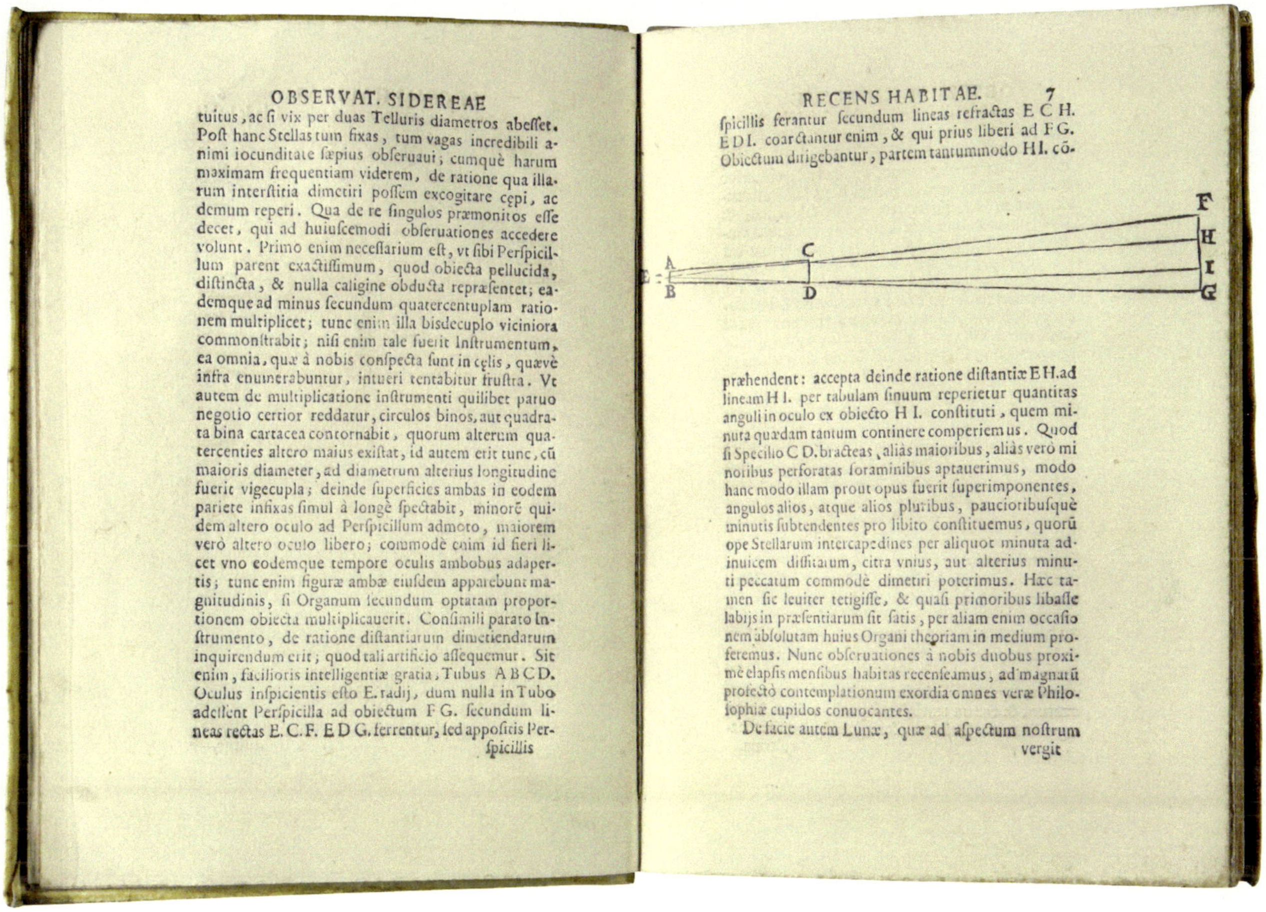

OBSERVAT. SIDEREAE

tuitus, ac ſi vix per duas Telluris diametros abeſſet. Poſt hanc Stellas tum fixas, tum vagas incredibili animi iocunditate ſæpius obſeruaui; cumquè harum maximam frequentiam viderem, de ratione qua illarum interſtitia dimetiri poſſem excogitare cępi, ac demum reperi. Qua de re ſingulos præmonitos eſſe decet, qui ad huiuſcemodi obſeruationes accedere volunt. Primo enim neceſſarium eſt, vt ſibi Perſpicillum parent exactiſſimum, quod obiecta pellucida, diſtincta, & nulla caligine obducta repræſentet; eademque ad minus ſecundum quatercentuplam rationem multiplicet; tunc enim illa bisdecuplo viciniora commonſtrabit; niſi enim tale fuerit Inſtrumentum, ea omnia, quæ à nobis conſpecta ſunt in cęlis, quævè infra enumerabuntur, intueri tentabitur fruſtra. Vt autem de multiplicatione inſtrumenti quilibet paruo negotio certior reddatur, circulos binos, aut quadrata bina cartacea contornabit, quorum alterum quatercenties altero maius exiſtat, id autem erit tunc, cũ maioris diameter, ad diametrum alterius longitudine fuerit vigecupla; deinde ſuperficies ambas in eodem pariete infixas ſimul à longè ſpectabit, minorẽ quidem altero oculo ad Perſpicillum admoto, maiorem verò altero oculo libero; commodè enim id fieri licet vno eodemque tempore oculis ambobus adapertis; tunc enim figuræ ambæ eiuſdem apparebunt magnitudinis, ſi Organum ſecundum optatam proportionem obiecta multiplicauerit. Conſimili parato Inſtrumento, de ratione diſtantiarum dimetiendarum inquirendum erit; quod tali artificio aſſequemur. Sit enim, faciliotis intelligentiæ gratia, Tubus A B C D. Oculus inſpicientis eſto E. radij, dum nulla in Tubo adeſſent Perſpicilla ad obiectum F G. ſecundum lineas rectas E. C. F. E D G. ferrentur, ſed appoſitis Per-

ſpicillis

RECENS HABITAE. 7

ſpicillis ferantur ſecundum lineas refractas E C H. E D I. coarctantur enim, & qui prius liberi ad F G. Obiectum dirigebantur, partem tantummodo H I. cõ-

A, B, C, D, E, F, G, H, I

præhendent: accepta deinde ratione diſtantiæ E H. ad lineam H I. per tabulam ſinuum reperietur quantitas anguli in oculo ex obiecto H I. conſtituti, quem minuta quædam tantum continere comperiemus. Quod ſi Specilio C D. bracteas, aliàs maioribus, aliàs verò minoribus perforatas foraminibus aptauerimus, modo hanc modo illam prout opus fuerit ſuperimponentes, angulos alios, atque alios pluribus, paucioribuſquè minutis ſubtendentes pro libito conſtituemus, quorũ ope Stellarum intercapedines per aliquot minuta adinuicem diſſitarum, citra vnius, aut alterius minuti peccatum commodè dimetiri poterimus. Hæc tamen ſic leuiter tetigiſſe, & quaſi primoribus libaſſe labijs in præſentiarum ſit ſatis, per aliam enim occaſionem abſolutam huius Organi theoriam in medium proferemus. Nunc obſeruationes à nobis duobus proximè elapſis menſibus habitas recenſeamus, ad magnarũ profectò contemplationum exordia omnes veræ Philoſophiæ cupidos conuocantes.

De facie autem Lunæ, quæ ad aſpectum noſtrum

vergit

Galileo Galilei, Sidereus Nuncius, Venice 1610, pp. 6v and 7r, New York copy.

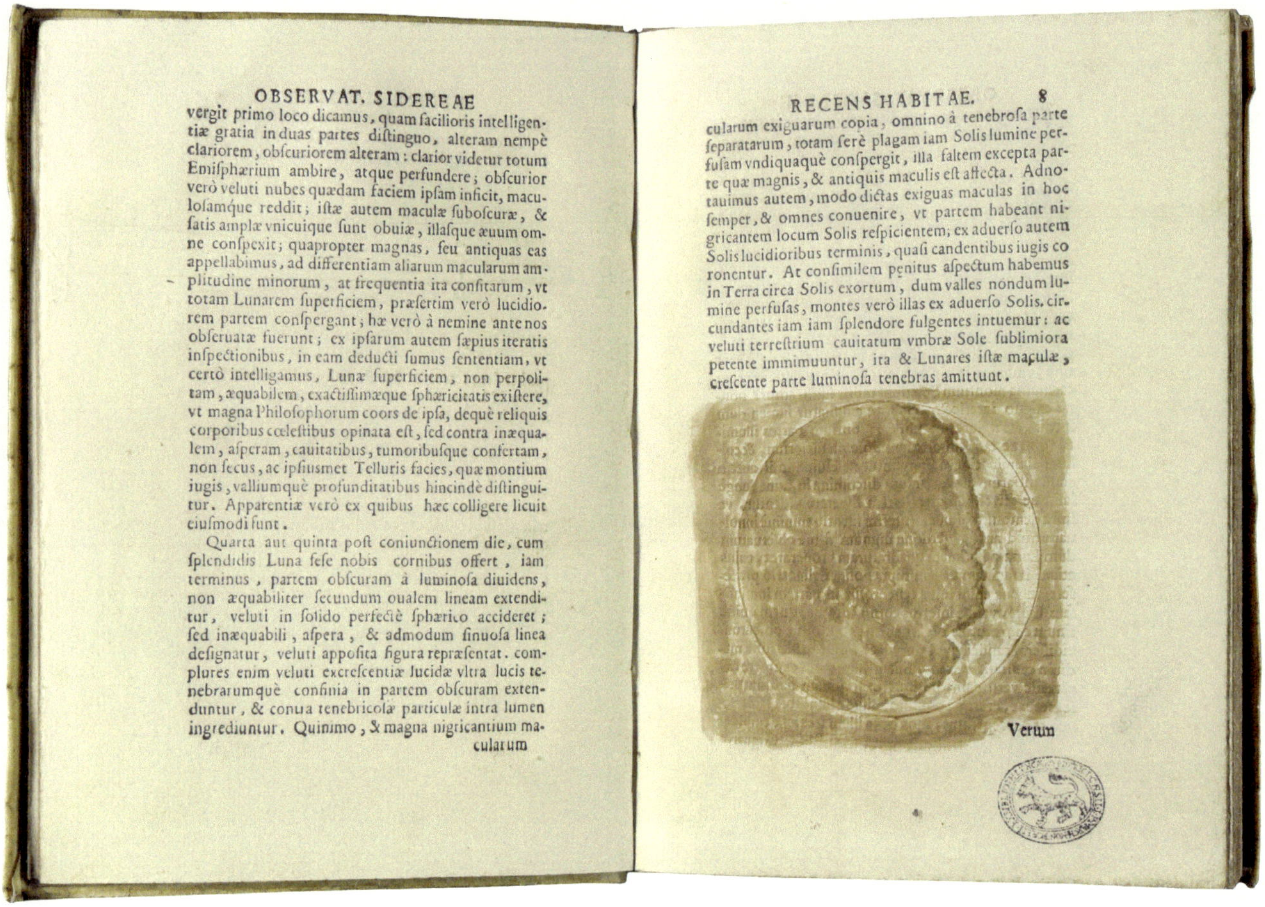

OBSERVAT. SIDEREAE
vergit primo loco dicamus, quam facilioris intelligen-
tiæ gratia in duas partes distinguo, alteram nempè
clariorem, obscuriorem alteram: clarior videtur totum
Emisphærium ambire, atque perfundere; obscurior
verò veluti nubes quædam faciem ipsam inficit, macu-
losamque reddit; istæ autem maculæ subobscuræ, &
satis amplæ vnicuique sunt obuiæ, illasque æuum om-
ne conspexit; quapropter magnas, seu antiquas eas
appellabimus, ad differentiam aliarum macularum am-
plitudine minorum, at frequentia ita consitarum, vt
totam Lunarem superficiem, præsertim verò lucidio-
rem partem conspergant; hæ verò à nemine ante nos
obseruatæ fuerunt; ex ipsarum autem sæpius iteratis
inspectionibus, in eam deducti sumus sententiam, vt
certò intelligamus, Lunæ superficiem, non perpoli-
tam, æquabilem, exactissimæque sphæricitatis existere,
vt magna Philosophorum coors de ipsa, dequè reliquis
corporibus cœlestibus opinata est, sed contra inæqua-
lem, asperam, cauitatibus, tumoribusque confertam,
non secus, ac ipsiusmet Telluris facies, quæ montium
jugis, valliumquè profunditatibus hincindè distingui-
tur. Apparentiæ verò ex quibus hæc colligere licuit
eiusmodi sunt.
Quarta aut quinta post coniunctionem die, cum
splendidis Luna sese nobis cornibus offert, iam
terminus, partem obscuram à luminosa diuidens,
non æquabiliter secundum oualem lineam extendi-
tur, veluti in solido perfectè sphærico accideret;
sed inæquabili, aspera, & admodum sinuosa linea
designatur, veluti apposita figura repræsentat. com-
plures enim veluti excrescentiæ lucidæ vltra lucis te-
nebrarumquè confinia in partem obscuram exten-
duntur, & contra tenebricolæ particulæ intra lumen
ingrediuntur. Quinimo, & magna nigricantium ma-
cularum

RECENS HABITAE. 8
cularum exiguarum copia, omnino à tenebrosa parte
separatarum, totam ferè plagam iam Solis lumine per-
fusam vndiquaquè conspergit, illa saltem excepta par-
te quæ magnis, & antiquis maculis est affecta. Adno-
tauimus autem, modo dictas exiguas maculas in hoc
semper, & omnes conuenire, vt partem habeant ni-
gricantem locum Solis respicientem; ex aduerso autem
Solis lucidioribus terminis, quasi candentibus iugis co
ronentur. At consimilem penitus aspectum habemus
in Terra circa Solis exortum, dum valles nondum lu-
mine perfusas, montes verò illas ex aduerso Solis. cir-
cundantes iam iam splendore fulgentes intuemur: ac
veluti terrestrium cauitatum vmbræ Sole sublimiora
petente imminuuntur, ita & Lunares istæ maculæ,
crescente parte luminosa tenebras amittunt.
Verum

Galileo Galilei, Sidereus Nuncius, Venice 1610, pp. 7v and 8r, New York copy.

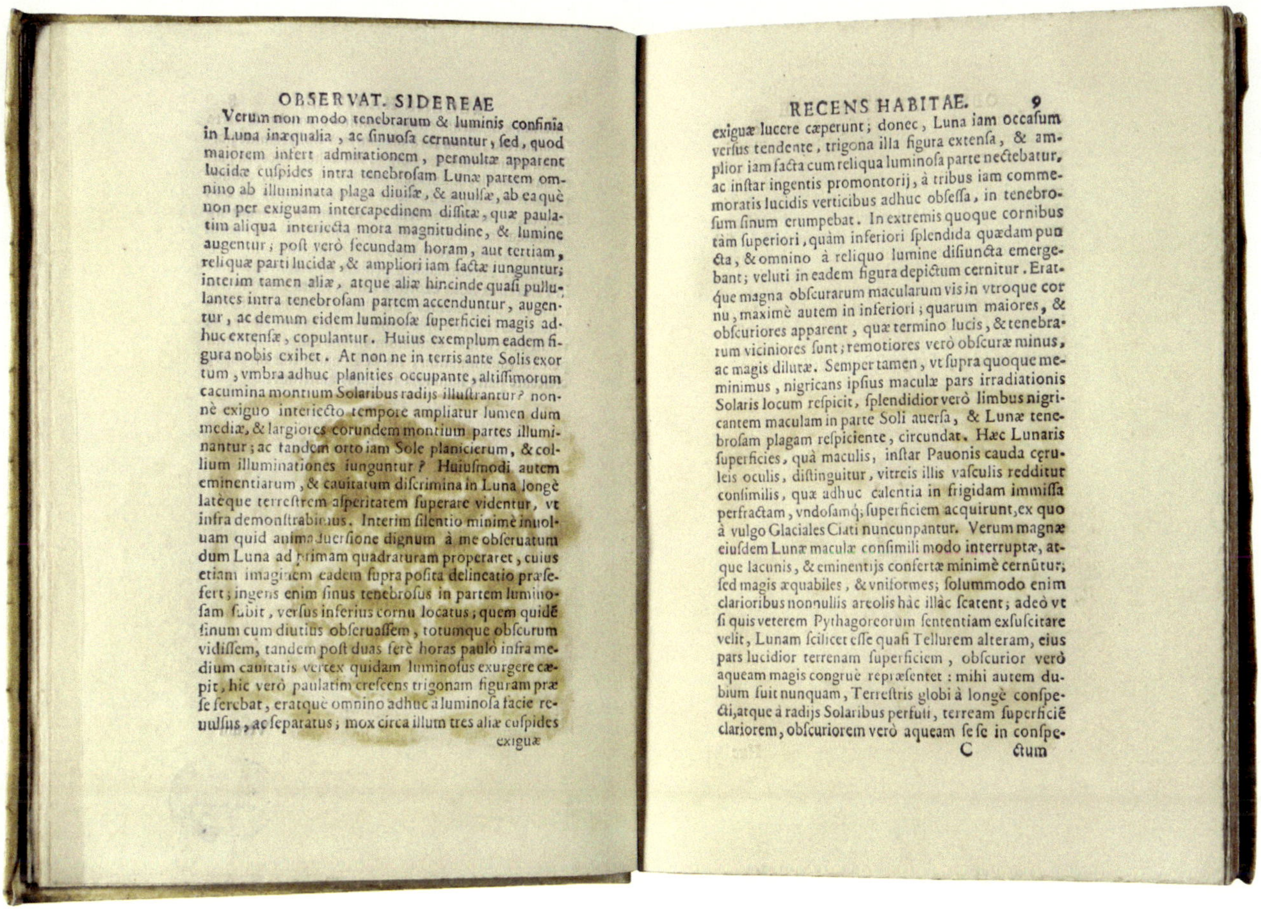
OBSERVAT. SIDEREAE

Verum non modo tenebrarum & luminis confinia
in Luna inæqualia, ac ſinuoſa cernuntur, ſed, quod
maiorem infert admirationem, permultæ apparent
lucidæ cuſpides intra tenebroſam Lunæ partem om-
nino ab illuminata plaga diuiſæ, & auulſæ, ab ea què
non per exiguam intercapedinem diſſitæ, quæ paula-
tim aliqua intericcta mora magnitudine, & lumine
augentur; poſt verò ſecundam horam, aut tertiam,
reliquæ parti lucidæ, & ampliori iam factæ iunguntur;
interim tamen aliæ, atque aliæ hincinde quaſi pullu-
lantes intra tenebroſam partem accenduntur, augen-
tur, ac demum eidem luminoſæ ſuperficiei magis ad-
huc extenſæ, copulantur. Huius exemplum eadem fi-
gura nobis exibet. At non ne in terris ante Solis exor
tum, vmbra adhuc planities occupante, altiſſimorum
cacumina montium Solaribus radijs illuſtrantur? non-
nè exiguo interiecto tempore ampliatur lumen dum
mediæ, & largiores eorundem montium partes illumi-
nantur; ac tandem orto iam Sole planicierum, & col-
lium illuminationes iunguntur? Huiuſmodi autem
eminentiarum, & cauitatum diſcrimina in Luna longè
latèque terreſtrem aſperitatem ſuperare videntur, vt
infra demonſtrabimus. Interim ſilentio minimè inuol-
uam quid animaduerſione dignum à me obſeruatum
dum Luna ad primam quadraturam properaret, cuius
etiam imaginem eadem ſupra poſita delineatio præſe-
fert; ingens enim ſinus tenebroſus in partem lumino-
ſam ſubit, verſus inferius cornu locatus; quem quidē
ſinum cum diutius obſeruaſſem, totumque obſcurum
vidiſſem, tandem poſt duas ferè horas paulò infra me-
dium cauitatis vertex quidam luminoſus exurgere cæ-
pit, hic verò paulatim creſcens trigonam figuram præ
ſe ferebat, eratquè omnino adhuc à luminoſa facie re-
uulſus, ac ſeparatus; mox circa illum tres aliæ cuſpides
exiguæ

RECENS HABITAE. 9

exiguæ lucere cæperunt; donec, Luna iam occaſum
verſus tendente, trigona illa figura extenſa, & am-
plior iam facta cum reliqua luminoſa parte nectebatur,
ac inſtar ingentis promontorij, à tribus iam comme-
moratis lucidis verticibus adhuc obſeſſa, in tenebro-
ſum ſinum erumpebat. In extremis quoque cornibus
tàm ſuperiori, quàm inferiori ſplendida quædam pun
cta, & omnino à reliquo lumine diſiuncta emerge-
bant; veluti in eadem figura depictum cernitur. Erat-
que magna obſcurarum macularum vis in vtroque cor
nu, maximè autem in inferiori; quarum maiores, &
obſcuriores apparent, quæ termino lucis, & tenebra-
rum viciniores ſunt; remotiores verò obſcuræ minus,
ac magis dilutæ. Semper tamen, vt ſupra quoque me-
minimus, nigricans ipſius maculæ pars irradiationis
Solaris locum reſpicit, ſplendidior verò limbus nigri-
cantem maculam in parte Soli auerſa, & Lunæ tene-
broſam plagam reſpiciente, circundat. Hæc Lunaris
ſuperficies, quà maculis, inſtar Pauonis cauda cęru-
leis oculis, diſtinguitur, vitreis illis vaſculis redditur
conſimilis, quæ adhuc calentia in frigidam immiſſa
perfractam, vndoſamq; ſuperficiem acquirunt, ex quo
à vulgo Glaciales Ciati nuncupantur. Verum magnæ
eiuſdem Lunæ maculæ conſimili modo interruptæ, at-
que lacunis, & eminentijs confertæ minimè cernūtur;
ſed magis æquabiles, & vniformes; ſolummodo enim
clarioribus nonnullis areolis hàc illàc ſcatent; adeò vt
ſi quis veterem Pythagoreorum ſententiam exſuſcitare
velit, Lunam ſcilicet eſſe quaſi Tellurem alteram, eius
pars lucidior terrenam ſuperficiem, obſcurior verò
aqueam magis congruè repræſentet: mihi autem du-
bium fuit nunquam, Terreſtris globi à longè conſpe-
cti, atque à radijs Solaribus perfuſi, terream ſuperficiē
clariorem, obſcuriorem verò aqueam ſeſe in conſpe-
C ctum

Galileo Galilei, Sidereus Nuncius, Venice 1610, pp. 8v and 9r, New York copy.

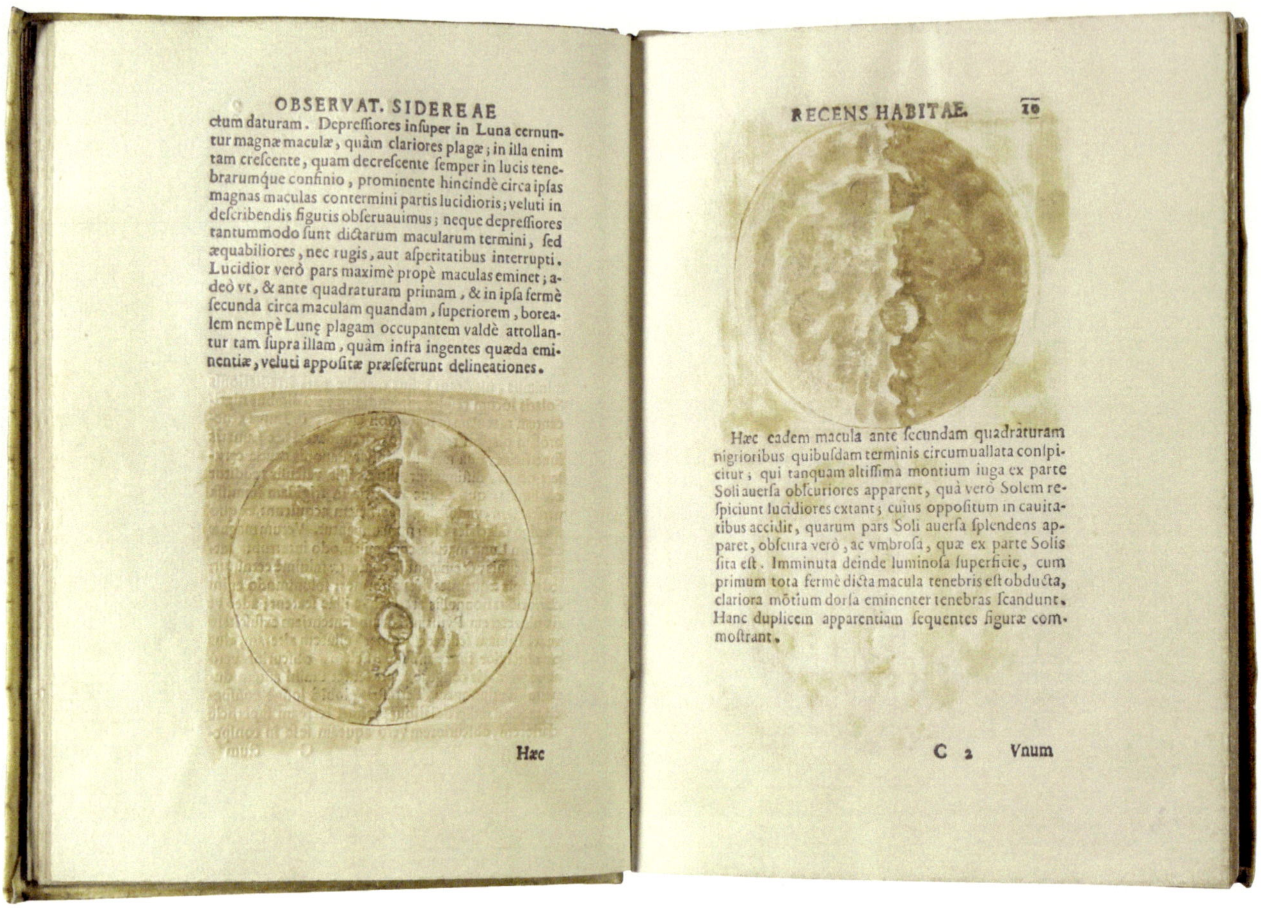
OBSERVAT. SIDEREAE

ctum daturam. Depreſſiores inſuper in Luna cernuntur magnæ maculæ, quàm clariores plagæ; in illa enim tam creſcente, quam decreſcente ſemper in lucis tenebrarumque confinio, prominente hincindè circa ipſas magnas maculas contermini partis lucidioris; veluti in deſcribendis figuris obſeruauimus; neque depreſſiores tantummodo ſunt dictarum macularum termini, ſed æquabiliores, nec rugis, aut aſperitatibus interrupti. Lucidior verò pars maximè propè maculas eminet; adeò vt, & ante quadraturam primam, & in ipſa fermè ſecunda circa maculam quandam, ſuperiorem, borealem nempè Lunę plagam occupantem valdè attollantur tam ſupra illam, quàm infra ingentes quæda eminentiæ, veluti appoſitæ præſeferunt delineationes.

Hæc

RECENS HABITAE. 10

Hæc eadem macula ante ſecundam quadràturam nigrioribus quibuſdam terminis circumuallata conſpicitur; qui tanquam altiſſima montium iuga ex parte Soli auerſa obſcuriores apparent, quà verò Solem reſpiciunt lucidiores extant; cuius oppoſitum in cauitatibus accidit, quarum pars Soli auerſa ſplendens apparet, obſcura verò, ac vmbroſa, quæ ex parte Solis ſita eſt. Imminuta deinde luminoſa ſuperficie, cum primum tota fermè dicta macula tenebris eſt obducta, clariora mõtium dorſa eminenter tenebras ſcandunt. Hanc duplicem apparentiam ſequentes figuræ commoſtrant.

C 2 Vnum

Galileo Galilei, Sidereus Nuncius, Venice 1610, pp. 9v and 10r, New York copy.

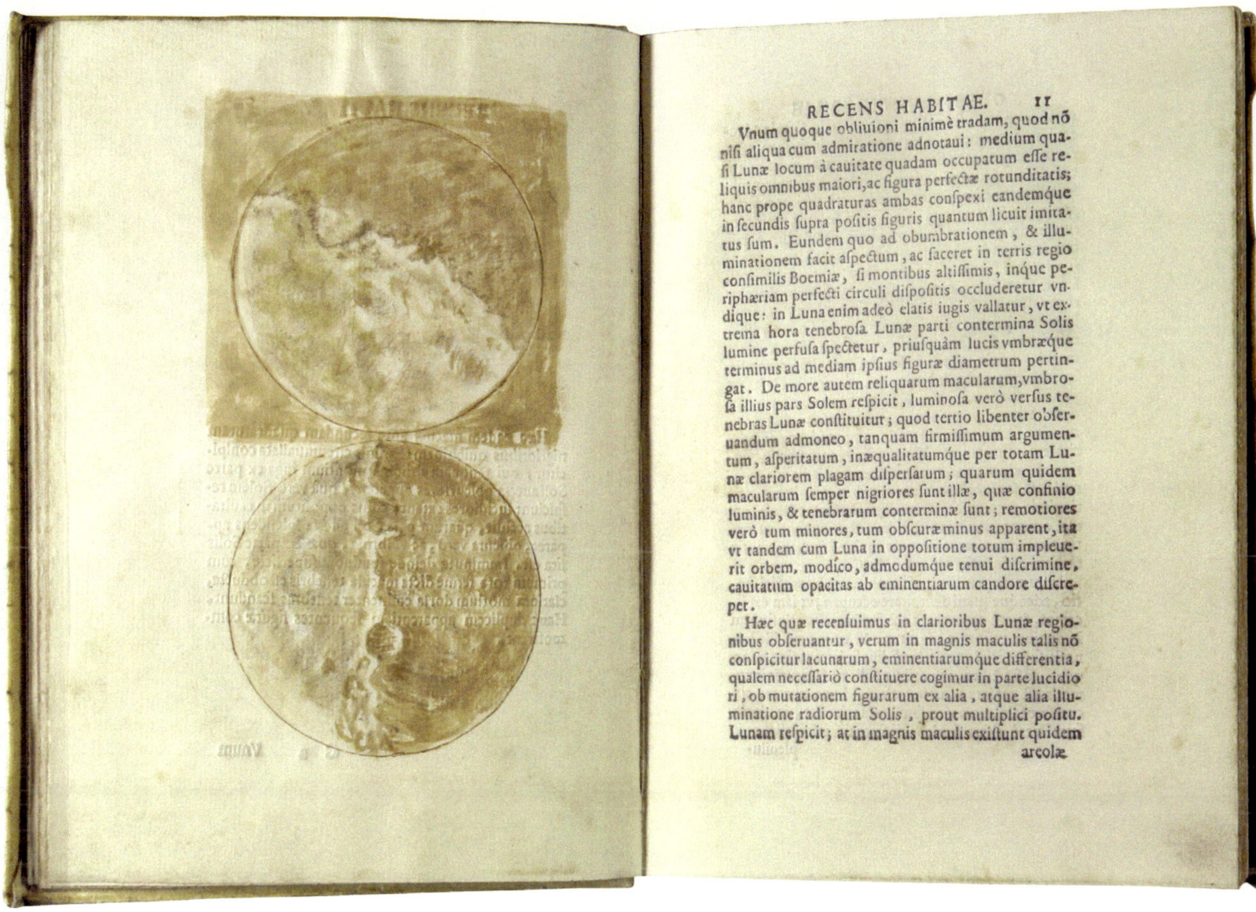

RECENS HABITAE. 11

Vnum quoque obliuioni minimè tradam, quod nõ
nisi aliqua cum admiratione adnotaui: medium qua-
si Lunæ locum à cauitate quadam occupatum esse re-
liquis omnibus maiori, ac figura perfectæ rotunditatis;
hanc prope quadraturas ambas conspexi eandemque
in secundis supra positis figuris quantum licuit imita-
tus sum. Eundem quo ad obumbrationem, & illu-
minationem facit aspectum, ac faceret in terris regio
consimilis Boemiæ, si montibus altissimis, inque pe-
riphæriam perfecti circuli dispositis occluderetur vn-
dique: in Luna enim adeò elatis iugis vallatur, vt ex-
trema hora tenebrosa Lunæ parti contermina Solis
lumine perfusa spectetur, priusquàm lucis vmbræque
terminus ad mediam ipsius figuræ diametrum pertin-
gat. De more autem reliquarum macularum, vmbro-
sa illius pars Solem respicit, luminosa verò versus te-
nebras Lunæ constituitur; quod tertio libenter obser-
uandum admoneo, tanquam firmissimum argumen-
tum, asperitatum, inæqualitatumque per totam Lu-
næ clariorem plagam dispersarum; quarum quidem
macularum semper nigriores sunt illæ, quæ confinio
luminis, & tenebrarum conterminæ sunt; remotiores
verò tum minores, tum obscuræ minus apparent, ita
vt tandem cum Luna in oppositione totum impleue-
rit orbem, modico, admodumque tenui discrimine,
cauitatum opacitas ab eminentiarum candore discre-
pet.

Hæc quæ recensuimus in clarioribus Lunæ regio-
nibus obseruantur, verum in magnis maculis talis nõ
conspicitur lacunarum, eminentiarumque differentia,
qualem necessariò constituere cogimur in parte lucidio-
ri, ob mutationem figurarum ex alia, atque alia illu-
minatione radiorum Solis, prout multiplici positu
Lunam respicit; at in magnis maculis existunt quidem
areolæ

Galileo Galilei, Sidereus Nuncius, Venice 1610, pp. 10v and 11r, New York copy.

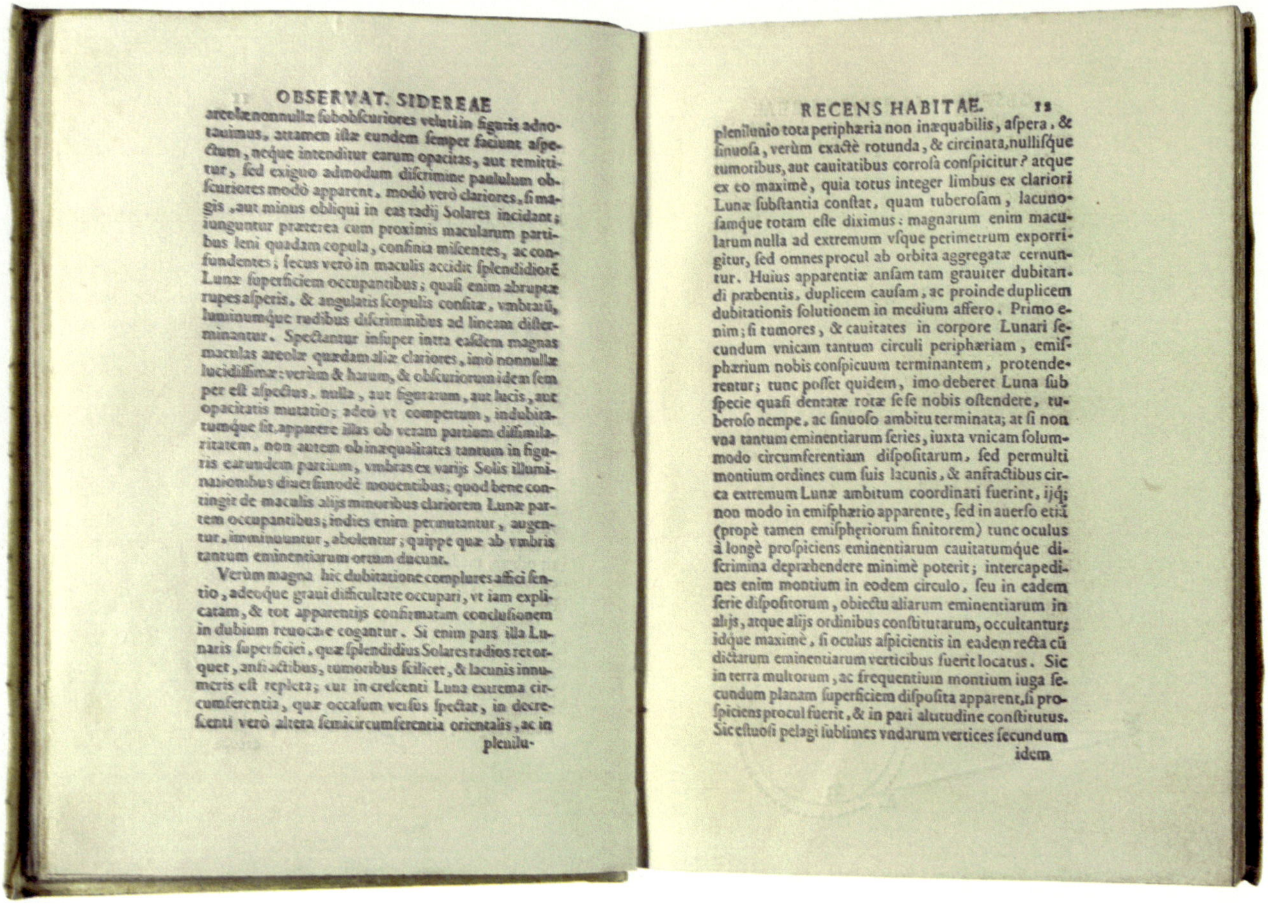

OBSERVAT. SIDEREAE

areolæ nonnullæ subobscuriores veluti in figuris adnotauimus, attamen istæ eundem semper faciunt aspectum, neque intenditur earum opacitas, aut remittitur, sed exiguo admodum discrimine paululum obscuriores modò apparent, modò verò clariores, si magis, aut minus obliqui in eas radij Solares incidant; iunguntur præterea cum proximis macularum partibus leni quadam copula, confinia miscentes, ac confundentes; secus verò in maculis accidit splendidiorē Lunæ superficiem occupantibus; quasi enim abruptæ rupes asperis, & angulatis scopulis consitæ, vmbrarū, luminumque radibus discriminibus ad lineam disterminantur. Spectantur insuper intra easdem magnas maculas areolæ quædam aliæ clariores, imò nonnullæ lucidissimæ: verùm & harum, & obscuriorum idem semper est aspectus, nulla, aut figurarum, aut lucis, aut opacitatis mutatio; adeò vt compertum, indubitatumque sit, apparere illas ob veram partium dissimilaritatem, non autem ob inæqualitates tantum in figuris earundem partium, vmbras ex varijs Solis illuminationibus diuersimodè mouentibus; quod bene contingit de maculis alijs minoribus clariorem Lunæ partem occupantibus; indies enim permutantur, augentur, imminuuntur, abolentur; quippe quæ ab vmbris tantum eminentiarum ortum ducunt.

Verùm magna hic dubitatione complures affici sentio, adeoque graui difficultate occupari, vt iam explicatam, & tot apparentijs confirmatam conclusionem in dubium reuocare cogantur. Si enim pars illa Lunaris superficiei, quæ splendidius Solares radios retorquet, anfractibus, tumoribus scilicet, & lacunis innumeris est repleta; cur in crescenti Luna extrema circumferentia, quæ occasum versus spectat, in decrescenti verò altera semicircumferentia orientalis, ac in plenilu-

RECENS HABITAE. 12

plenilunio tota periphæria non inæquabilis, aspera, & sinuosa, verùm exactè rotunda, & circinata, nullisque tumoribus, aut cauitatibus corrosa conspicitur? atque ex eo maximè, quia totus integer limbus ex clariori Lunæ substantia constat, quam tuberosam, lacunosamque totam esse diximus: magnarum enim macularum nulla ad extremum vsque perimetrum exporrigitur, sed omnes procul ab orbita aggregatæ cernuntur. Huius apparentiæ ansam tam grauiter dubitandi præbentis, duplicem causam, ac proinde duplicem dubitationis solutionem in medium affero. Primo enim; si tumores, & cauitates in corpore Lunari secundum vnicam tantum circuli periphæriam, emisphærium nobis conspicuum terminantem, protenderentur; tunc posset quidem, imo deberet Luna sub specie quasi dentatæ rotæ sese nobis ostendere, tuberoso nempe, ac sinuoso ambitu terminata; at si non vna tantum eminentiarum series, iuxta vnicam solummodo circumferentiam dispositarum, sed permulti montium ordines cum suis lacunis, & anfractibus circa extremum Lunæ ambitum coordinati fuerint, ijq; non modo in emisphærio apparente, sed in auerso etiā (propè tamen emispheriorum finitorem) tunc oculus à longè prospiciens eminentiarum cauitatumque discrimina depræhendere minimè poterit; intercapedines enim montium in eodem circulo, seu in eadem serie dispositorum, obiectu aliarum eminentiarum in alijs, atque alijs ordinibus constitutarum, occultantur; idque maximè, si oculus aspicientis in eadem recta cū dictarum eminentiarum verticibus fuerit locatus. Sic in terra multorum, ac frequentium montium iuga secundum planam superficiem disposita apparent, si prospiciens procul fuerit, & in pari altitudine constitutus. Sic estuosi pelagi sublimes vndarum vertices secundum idem

Galileo Galilei, Sidereus Nuncius, Venice 1610, pp. 11v and 12r, New York copy.

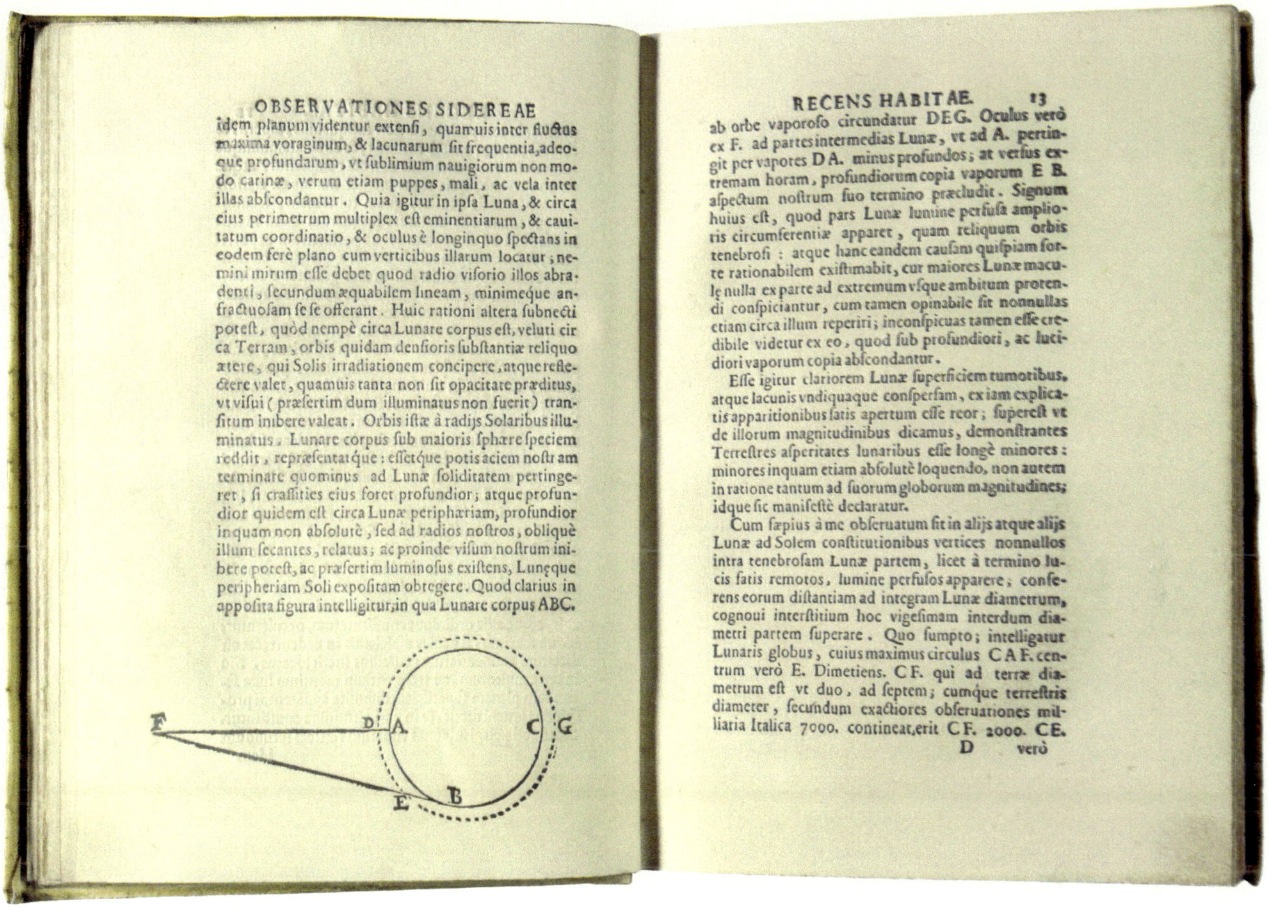

OBSERVATIONES SIDEREAE

idem planum videntur extenſi, quamuis inter fluctus maxima voraginum, & lacunarum ſit frequentia, adeoque profundarum, vt ſublimium nauigiorum non modo carinæ, verum etiam puppes, mali, ac vela inter illas abſcondantur. Quia igitur in ipſa Luna, & circa eius perimetrum multiplex eſt eminentiarum, & cauitatum coordinatio, & oculus è longinquo ſpectans in eodem ferè plano cum verticibus illarum locatur; nemini mirum eſſe debet quod radio viſorio illos abradenti, ſecundum æquabilem lineam, minimeque anfractuoſam ſe ſe offerant. Huic rationi altera ſubnecti poteſt, quòd nempè circa Lunare corpus eſt, veluti circa Terram, orbis quidam denſioris ſubſtantiæ reliquo ætere, qui Solis irradiationem concipere, atque reflectere valet, quamuis tanta non ſit opacitate præditus, vt viſui (præſertim dum illuminatus non fuerit) tranſitum inibere valeat. Orbis iſtæ à radijs Solaribus illuminatus, Lunare corpus ſub maioris ſphære ſpeciem reddit, repræſentatque: eſſetque potis aciem noſtram terminare quominus ad Lunæ ſoliditatem pertingeret, ſi craſſities eius foret profundior; atque profundior quidem eſt circa Lunæ periphæriam, profundior inquam non abſolutè, ſed ad radios noſtros, obliquè illum ſecantes, relatus; ac proinde viſum noſtrum inibere poteſt, ac præſertim luminoſus exiſtens, Lunęque peripheriam Soli expoſitam obtegere. Quod clarius in appoſita figura intelligitur, in qua Lunare corpus ABC.

RECENS HABITAE. 13

ab orbe vaporoſo circundatur DEG. Oculus verò ex F. ad partes intermedias Lunæ, vt ad A. pertingit per vapores DA. minus profundos; at verſus extremam horam, profundiorum copia vaporum EB. aſpectum noſtrum ſuo termino præcludit. Signum huius eſt, quod pars Lunæ lumine perfuſa amplioris circumferentiæ apparet, quam reliquum orbis tenebroſi: atque hanc eandem cauſam quiſpiam forte rationabilem exiſtimabit, cur maiores Lunæ maculę nulla ex parte ad extremum vſque ambitum protendi conſpiciantur, cum tamen opinabile ſit nonnullas etiam circa illum reperiri; inconſpicuas tamen eſſe credibile videtur ex eo, quod ſub profundiori, ac lucidiori vaporum copia abſcondantur.

Eſſe igitur clariorem Lunæ ſuperficiem tumoribus, atque lacunis vndiquaque conſperſam, ex iam explicatis apparitionibus ſatis apertum eſſe reor; ſupereſt vt de illorum magnitudinibus dicamus, demonſtrantes Terreſtres aſperitates lunaribus eſſe longè minores: minores inquam etiam abſolutè loquendo, non autem in ratione tantum ad ſuorum globorum magnitudines; idque ſic manifeſtè declaratur.

Cum ſæpius à me obſeruatum ſit in alijs atque alijs Lunæ ad Solem conſtitutionibus vertices nonnullos intra tenebroſam Lunæ partem, licet à termino lucis ſatis remotos, lumine perfuſos apparere; conferens eorum diſtantiam ad integram Lunæ diametrum, cognoui interſtitium hoc vigeſimam interdum diametri partem ſuperare. Quo ſumpto; intelligatur Lunaris globus, cuius maximus circulus CAF. centrum verò E. Dimetiens. CF. qui ad terræ diametrum eſt vt duo, ad ſeptem; cumque terreſtris diameter, ſecundum exactiores obſeruationes milliaria Italica 7000. contineat, erit CF. 2000. CE.

D verò

Galileo Galilei, Sidereus Nuncius, Venice 1610, pp. 12v and 13r, New York copy.

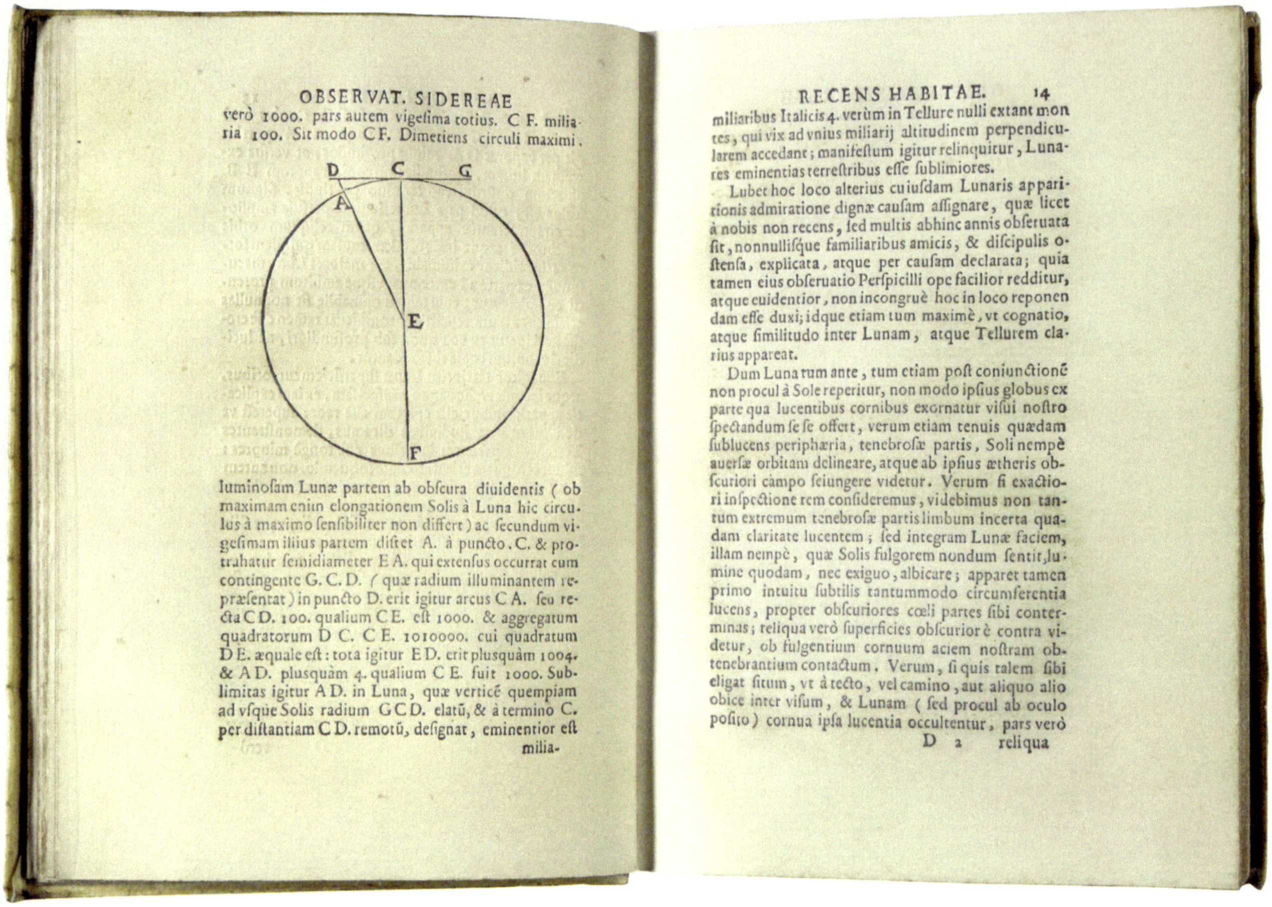

OBSERVAT. SIDEREAE

verò 1000. pars autem vigeſima totius. C F. milia-
ria 100. Sit modo CF. Dimetiens circuli maximi.

luminoſam Lunæ partem ab obſcura diuidentis (ob
maximam enim elongationem Solis à Luna hic circu-
lus à maximo ſenſibiliter non differt) ac ſecundum vi-
geſimam illius partem diſtet A. à puncto . C. & pro-
trahatur ſemidiameter E A. qui extenſus occurrat cum
contingente G. C. D. (quæ radium illuminantem re-
præſentat) in puncto D. erit igitur arcus C A. ſeu re-
cta C D. 100. qualium C E. eſt 1000. & aggregatum
quadratorum D C. C E. 1010000. cui quadratum
D E. æquale eſt : tota igitur E D. erit plusquàm 1004.
& A D. plusquàm 4. qualium C E. fuit 1000. Sub-
limitas igitur A D. in Luna, quæ verticē quempiam
ad vſque Solis radium G C D. elatū, & à termino C.
per diſtantiam C D. remotū, deſignat, eminentior eſt
milia-

RECENS HABITAE. 14

miliaribus Italicis 4. verùm in Tellure nulli extant mon-
tes, qui vix ad vnius miliarij altitudinem perpendicu-
larem accedant; manifeſtum igitur relinquitur, Luna-
res eminentias terreſtribus eſſe ſublimiores.

Lubet hoc loco alterius cuiuſdam Lunaris appari-
tionis admiratione dignæ cauſam aſſignare, quæ licet
à nobis non recens, ſed multis abhinc annis obſeruata
ſit, nonnulliſque familiaribus amicis, & diſcipulis o-
ſtenſa, explicata, atque per cauſam declarata; quia
tamen eius obſeruatio Perſpicilli ope facilior redditur,
atque euidentior, non incongruè hoc in loco reponen
dam eſſe duxi; idque etiam tum maximè, vt cognatio,
atque ſimilitudo inter Lunam, atque Tellurem cla-
rius appareat.

Dum Luna tum ante, tum etiam poſt coniunctionē
non procul à Sole reperitur, non modo ipſius globus ex
parte qua lucentibus cornibus exornatur viſui noſtro
ſpectandum ſe ſe offert, verum etiam tenuis quædam
ſublucens periphæria, tenebroſæ partis, Soli nempè
auerſæ orbitam delineare, atque ab ipſius ætheris ob-
ſcuriori campo ſeiungere videtur. Verum ſi exactio-
ri inſpectione rem conſideremus, videbimus non tan-
tum extremum tenebroſæ partis limbum incerta qua-
dam claritate lucentem; ſed integram Lunæ faciem,
illam nempè, quæ Solis fulgorem nondum ſentit, lu-
mine quodam, nec exiguo, albicare; apparet tamen
primo intuitu ſubtilis tantummodo circumferentia
lucens, propter obſcuriores cœli partes ſibi conter-
minas; reliqua verò ſuperficies obſcurior è contra vi-
detur, ob fulgentium cornuum aciem noſtram ob-
tenebrantium contactum. Verum, ſi quis talem ſibi
eligat ſitum, vt à tecto, vel camino, aut aliquo alio
obice inter viſum, & Lunam (ſed procul ab oculo
poſito) cornua ipſa lucentia occultentur, pars verò
D 2 reliqua

Galileo Galilei, Sidereus Nuncius, Venice 1610, pp. 13v and 14r, New York copy.

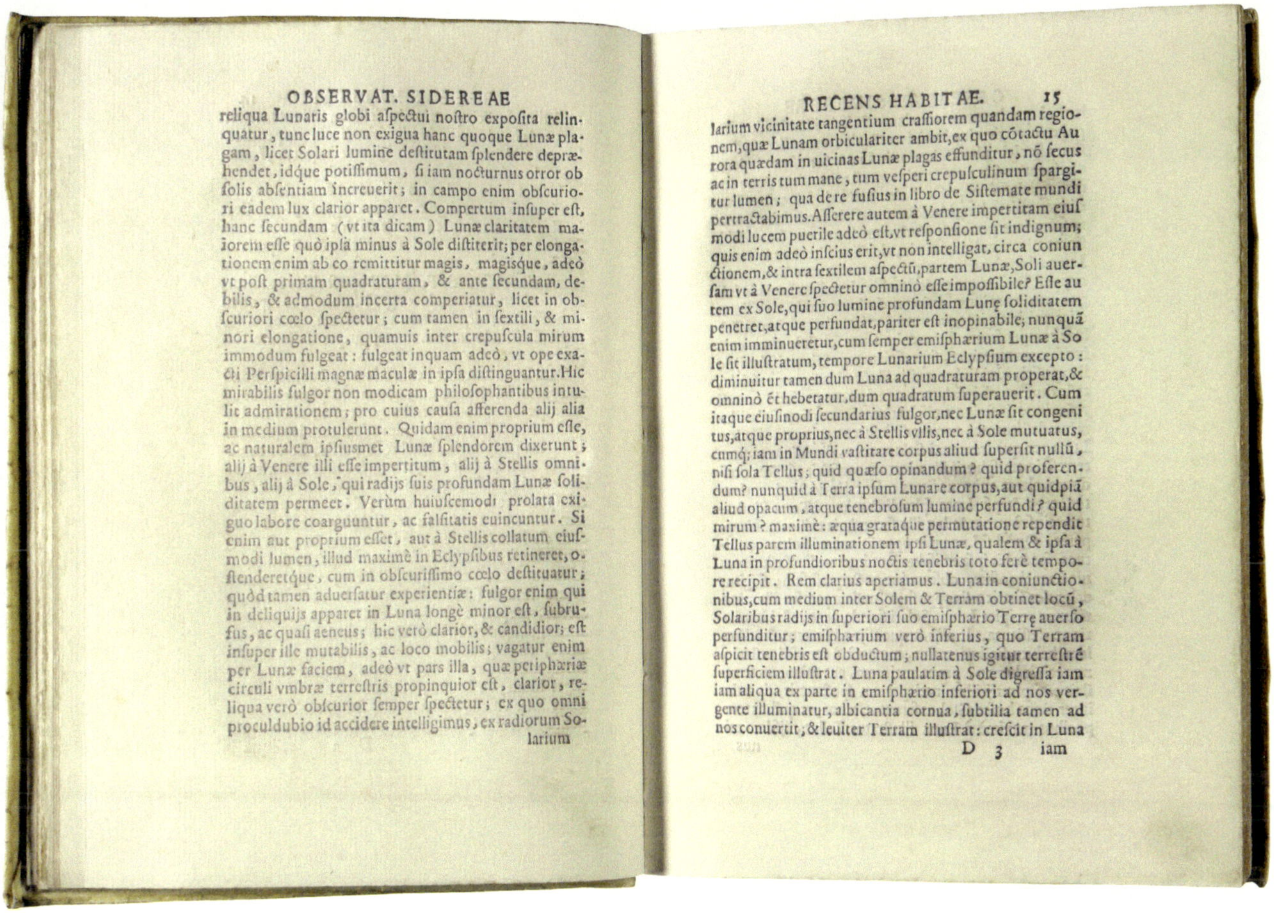

OBSERVAT. SIDEREAE
reliqua Lunaris globi aſpectui noſtro expoſita relin-
quatur, tunc luce non exigua hanc quoque Lunæ pla-
gam, licet Solari lumine deſtitutam ſplendere depræ-
hendet, idq́ue potiſſimum, ſi iam nocturnus orror ob
ſolis abſentiam increuerit; in campo enim obſcurio-
ri eadem lux clarior apparet. Compertum inſuper eſt,
hanc ſecundam (vt ita dicam) Lunæ claritatem ma-
iorem eſſe quò ipſa minus à Sole diſtiterit; per elonga-
tionem enim ab eo remittitur magis, magisq́ue, adeò
vt poſt primam quadraturam, & ante ſecundam, de-
bilis, & admodum incerta comperiatur, licet in ob-
ſcuriori cœlo ſpectetur; cum tamen in ſextili, & mi-
nori elongatione, quamuis inter crepuſcula mirum
immodum fulgeat: fulgeat inquam adeò, vt ope exa-
cti Perſpicilli magnæ maculæ in ipſa diſtinguantur. Hic
mirabilis fulgor non modicam philoſophantibus intu-
lit admirationem; pro cuius cauſa afferenda alij alia
in medium protulerunt. Quidam enim proprium eſſe,
ac naturalem ipſiusmet Lunæ ſplendorem dixerunt;
alij à Venere illi eſſe impertitum, alij à Stellis omni-
bus, alij à Sole, qui radijs ſuis profundam Lunæ ſoli-
ditatem permeet. Verùm huiuſcemodi prolata exi-
guo labore coarguuntur, ac falſitatis euincuntur. Si
enim aut proprium eſſet, aut à Stellis collatum eiuſ-
modi lumen, illud maximè in Eclypſibus retineret, o-
ſtenderetq́ue, cum in obſcuriſſimo cœlo deſtituatur;
quòd tamen aduerſatur experientiæ: fulgor enim qui
in deliquijs apparet in Luna longè minor eſt, ſubru-
fus, ac quaſi aeneus; hic verò clarior, & candidior; eſt
inſuper ille mutabilis, ac loco mobilis; vagatur enim
per Lunæ faciem, adeò vt pars illa, quæ peripheriæ
circuli vmbræ terreſtris propinquior eſt, clarior, re-
liqua verò obſcurior ſemper ſpectetur; ex quo omni
proculdubio id accidere intelligimus, ex radiorum So-
larium

RECENS HABITAE. 15
larium vicinitate tangentium craſſiorem quandam regio-
nem, quæ Lunam orbiculariter ambit, ex quo cōtactu Au
rora quædam in uicinas Lunæ plagas effunditur, nō ſecus
ac in terris tum mane, tum veſperi crepuſculinum ſpargi-
tur lumen; qua de re fuſius in libro de Siſtemate mundi
pertractabimus. Aſſerere autem à Venere impertitam eiuſ
modi lucem puerile adeò eſt, vt reſponſione ſit indignum;
quis enim adeò inſcius erit, vt non intelligat, circa coniun
ctionem, & intra ſextilem aſpectū, partem Lunæ, Soli auer-
ſam vt à Venere ſpectetur omninò eſſe impoſſibile? Eſſe au
tem ex Sole, qui ſuo lumine profundam Lunę ſoliditatem
penetret, atque perfundat, pariter eſt inopinabile; nunquā
enim imminueretur, cum ſemper emiſphærium Lunæ à So
le ſit illuſtratum, tempore Lunarium Eclypſium excepto:
diminuitur tamen dum Luna ad quadraturam properat, &
omninò ēt hebetatur, dum quadratum ſuperauerit. Cum
itaque eiuſinodi ſecundarius fulgor, nec Lunæ ſit congeni
tus, atque proprius, nec à Stellis vllis, nec à Sole mutuatus,
cumq́; iam in Mundi vaſtitate corpus aliud ſuperſit nullū,
niſi ſola Tellus; quid quæſo opinandum? quid proferen-
dum? nunquid à Terra ipſum Lunare corpus, aut quidpiā
aliud opacum, atque tenebroſum lumine perfundi? quid
mirum? maximè: æqua grataq́ue permutatione rependit
Tellus parem illuminationem ipſi Lunæ, qualem & ipſa à
Luna in profundioribus noctis tenebris toto ferè tempo-
re recipit. Rem clarius aperiamus. Luna in coniunctio-
nibus, cum medium inter Solem & Terram obtinet locū,
Solaribus radijs in ſuperiori ſuo emiſphærio Terrę auerſo
perfunditur; emiſphærium verò inferius, quo Terram
aſpicit tenebris eſt obductum; nullatenus igitur terreſtrē
ſuperficiem illuſtrat. Luna paulatim à Sole digreſſa iam
iam aliqua ex parte in emiſphærio inferiori ad nos ver-
gente illuminatur, albicantia cornua, ſubtilia tamen ad
nos conuertit; & leuiter Terram illuſtrat: creſcit in Luna
D 3 iam

Galileo Galilei, Sidereus Nuncius, Venice 1610, pp. 14v and 15r, New York copy.

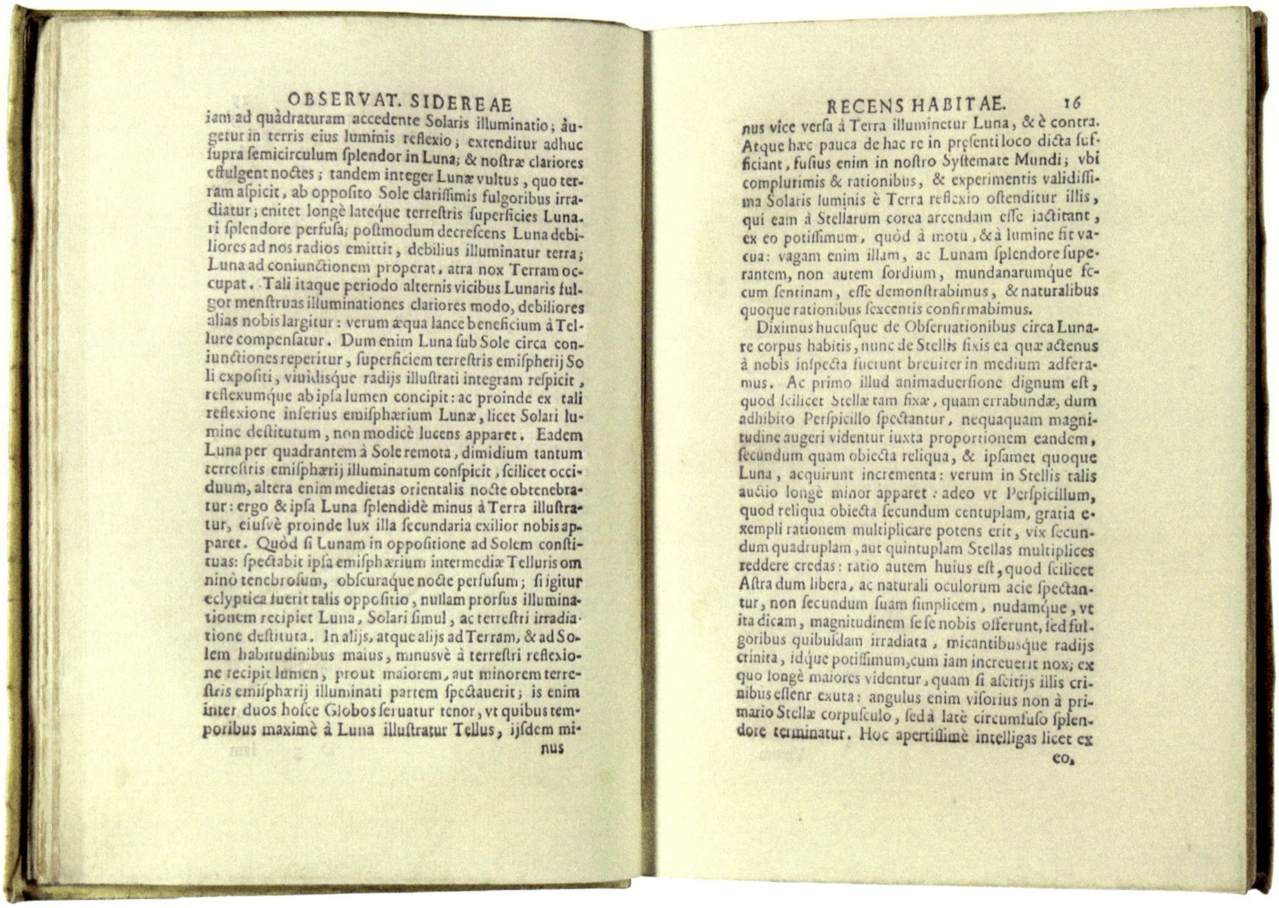

OBSERVAT. SIDEREAE

iam ad quàdraturam accedente Solaris illuminatio; àu-
getur in terris eius luminis reflexio; extenditur adhuc
ſupra ſemicirculum ſplendor in Luna; & noſtræ clariores
effulgent noctes; tandem integer Lunæ vultus, quo ter-
ram aſpicit, ab oppoſito Sole clariſſimis fulgoribus irra-
diatur; enitet longè lateque terreſtris ſuperficies Luna-
ri ſplendore perfuſa; poſtmodum decreſcens Luna debi-
liores ad nos radios emittit, debilius illuminatur terra;
Luna ad coniunctionem properat, atra nox Terram oc-
cupat. Tali itaque periodo alternis vicibus Lunaris ful-
gor menſtruas illuminationes clariores modo, debiliores
alias nobis largitur: verum æqua lance beneficium à Tel-
lure compenſatur. Dum enim Luna ſub Sole circa con-
iunctiones reperitur, ſuperficiem terreſtris emiſpherij So
li expoſiti, viuidiſque radijs illuſtrati integram reſpicit,
reflexumque ab ipſa lumen concipit: ac proinde ex tali
reflexione inferius emiſphærium Lunæ, licet Solari lu-
mine deſtitutum, non modicè lucens apparet. Eadem
Luna per quadrantem à Sole remota, dimidium tantum
terreſtris emiſphærij illuminatum conſpicit, ſcilicet occi-
duum, altera enim medietas orientalis nocte obtenebra-
tur: ergo & ipſa Luna ſplendidè minus à Terra illuſtra-
tur, eiuſvè proinde lux illa ſecundaria exilior nobis ap-
paret. Quòd ſi Lunam in oppoſitione ad Solem conſti-
tuas: ſpectabit ipſa emiſphærium intermediæ Telluris om
ninò tenebroſum, obſcuraque nocte perfuſum; ſi igitur
eclyptica fuerit talis oppoſitio, nullam prorſus illumina-
tionem recipiet Luna, Solari ſimul, ac terreſtri irradia-
tione deſtituta. In alijs, atque alijs ad Terram, & ad So-
lem habitudinibus maius, minusvè à terreſtri reflexio-
ne recipit lumen, prout maiorem, aut minorem terre-
ſtris emiſphærij illuminati partem ſpectauerit; is enim
inter duos hoſce Globos ſeruatur tenor, vt quibus tem-
poribus maximè à Luna illuſtratur Tellus, ijſdem mi-
nus

RECENS HABITAE. 16

nus vice verſa à Terra illuminetur Luna, & è contra.
Atque hæc pauca de hac re in præſenti loco dicta ſuf-
ficiant, fuſius enim in noſtro Syſtemate Mundi; vbi
complurimis & rationibus, & experimentis validiſſi-
ma Solaris luminis è Terra reflexio oſtenditur illis,
qui eam à Stellarum corea arcendam eſſe iactitant,
ex eo potiſſimum, quòd à motu, & à lumine fit va-
cua: vagam enim illam, ac Lunam ſplendore ſupe-
rantem, non autem ſordium, mundanarumque fe-
cum ſentinam, eſſe demonſtrabimus, & naturalibus
quoque rationibus ſexcentis confirmabimus.

Diximus hucuſque de Obſeruationibus circa Luna-
re corpus habitis, nunc de Stellis fixis ea quæ actenus
à nobis inſpecta fuerunt breuiter in medium adfera-
mus. Ac primo illud animaduerſione dignum eſt,
quod ſcilicet Stellæ tam fixæ, quam errabundæ, dum
adhibito Perſpicillo ſpectantur, nequaquam magni-
tudine augeri videntur iuxta proportionem eandem,
ſecundum quam obiecta reliqua, & ipſamet quoque
Luna, acquirunt incrementa: verum in Stellis talis
auctio longè minor apparet: adeo vt Perſpicillum,
quod reliqua obiecta ſecundum centuplam, gratia e-
xempli rationem multiplicare potens erit, vix ſecun-
dum quadruplam, aut quintuplam Stellas multiplices
reddere credas: ratio autem huius eſt, quod ſcilicet
Aſtra dum libera, ac naturali oculorum acie ſpectan-
tur, non ſecundum ſuam ſimplicem, nudamque, vt
ita dicam, magnitudinem ſeſe nobis offerunt, ſed ful-
goribus quibuſdam irradiata, micantibusque radijs
crinita, idque potiſſimum, cum iam increuerit nox; ex
quo longè maiores videntur, quam ſi aſcitijs illis cri-
nibus eſſent exuta: angulus enim viſorius non à pri-
mario Stellæ corpuſculo, ſed à latè circumfuſo ſplen-
dore terminatur. Hoc apertiſſimè intelligas licet ex
eo,

Galileo Galilei, Sidereus Nuncius, Venice 1610, pp. 15v and 16r, New York copy.

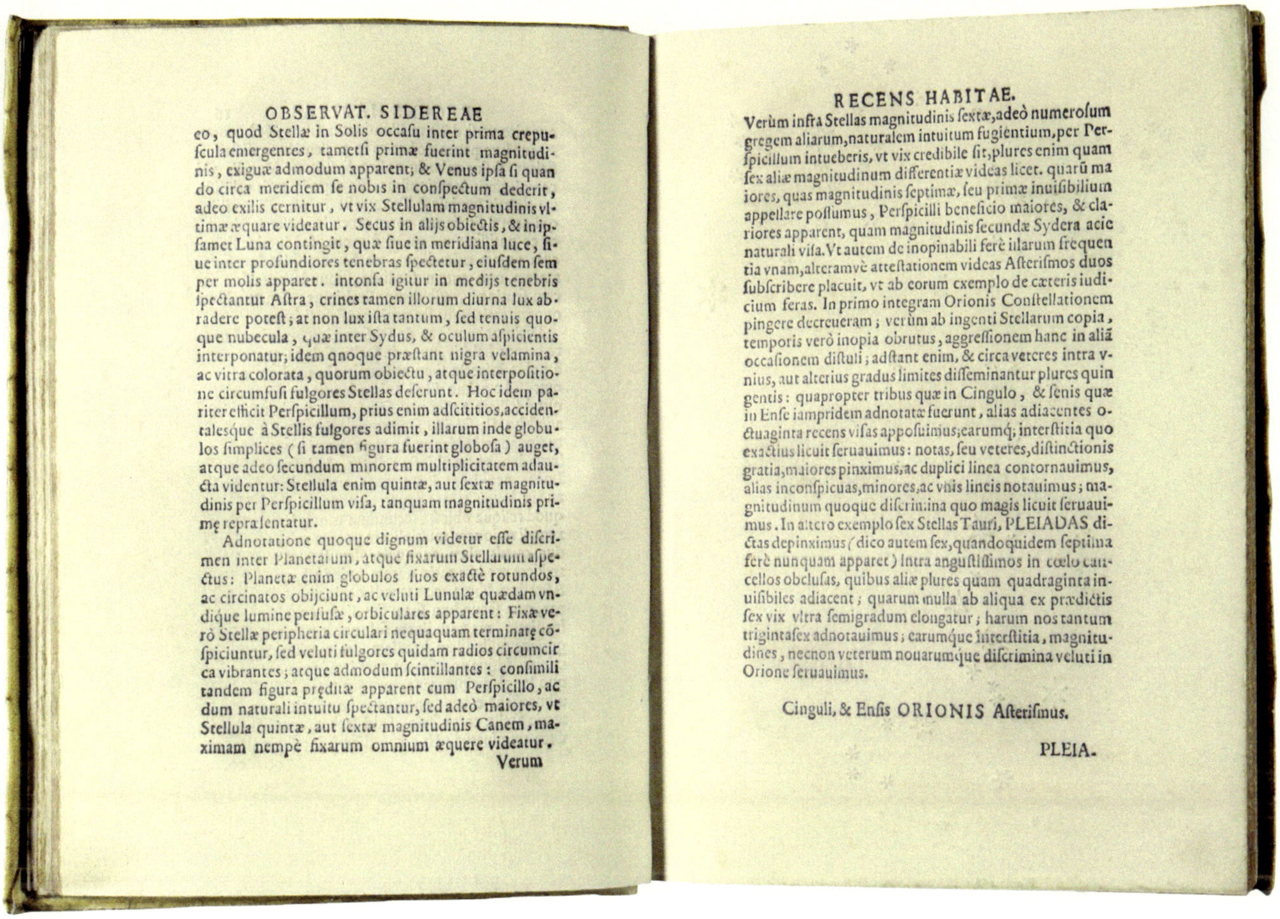

OBSERVAT. SIDEREAE

eo, quod Stellæ in Solis occaſu inter prima crepu-
ſcula emergentes, tametſi primæ fuerint magnitudi-
nis, exiguæ admodum apparent; & Venus ipſa ſi quan
do circa meridiem ſe nobis in conſpectum dederit,
adeo exilis cernitur, vt vix Stellulam magnitudinis vl-
timæ æquare videatur. Secus in alijs obiectis, & in ip-
ſamet Luna contingit, quæ ſiue in meridiana luce, ſi-
ue inter profundiores tenebras ſpectetur, eiuſdem ſem
per molis apparet. intonſa igitur in medijs tenebris
ſpectantur Aſtra, crines tamen illorum diurna lux ab-
radere poteſt; at non lux iſta tantum, ſed tenuis quo-
que nubecula, quæ inter Sydus, & oculum aſpicientis
interponatur; idem qnoque præſtant nigra velamina,
ac vitra colorata, quorum obiectu, atque interpoſitio-
ne circumfuſi fulgores Stellas deſerunt. Hoc idem pa-
riter efficit Perſpicillum, prius enim adſcititios, acciden-
taleſque à Stellis fulgores adimit, illarum inde globu-
los ſimplices (ſi tamen figura fuerint globoſa) auget,
atque adeo ſecundum minorem multiplicitatem adau-
cta videntur: Stellula enim quintæ, aut ſextæ magnitu-
dinis per Perſpicillum viſa, tanquam magnitudinis pri-
mę repræſentatur.

Adnotatione quoque dignum videtur eſſe diſcri-
men inter Planetarum, atque fixarum Stellarum aſpe-
ctus: Planetæ enim globulos ſuos exactè rotundos,
ac circinatos objiciunt, ac veluti Lunulæ quædam vn-
dique lumine perfuſæ, orbiculares apparent: Fixæ ve-
rò Stellæ peripheria circulari nequaquam terminatę cō-
ſpiciuntur, ſed veluti fulgores quidam radios circumcir
ca vibrantes; atque admodum ſcintillantes: conſimili
tandem figura pręditæ apparent cum Perſpicillo, ac
dum naturali intuitu ſpectantur, ſed adeò maiores, vt
Stellula quintæ, aut ſextæ magnitudinis Canem, ma-
ximam nempè fixarum omnium æquere videatur.

Verum

RECENS HABITAE.

Verùm infra Stellas magnitudinis ſextæ, adeò numeroſum
gregem aliarum, naturalem intuitum fugientium, per Per-
ſpicillum intueberis, vt vix credibile ſit, plures enim quam
ſex aliæ magnitudinum differentiæ videas licet. quarū ma
iores, quas magnitudinis ſeptimæ, ſeu primæ inuiſibilium
appellare poſſumus, Perſpicilli beneficio maiores, & cla-
riores apparent, quam magnitudinis ſecundæ Sydera acie
naturali viſa. Vt autem de inopinabili ferè illarum frequen
tia vnam, alteramvè atteſtationem videas Aſteriſmos duos
ſubſcribere placuit, vt ab eorum exemplo de cæteris iudi-
cium feras. In primo integram Orionis Conſtellationem
pingere decreueram; verùm ab ingenti Stellarum copia,
temporis verò inopia obrutus, aggreſſionem hanc in aliā
occaſionem diſtuli; adſtant enim, & circa veteres intra v-
nius, aut alterius gradus limites diſſeminantur plures quin
gentis: quapropter tribus quæ in Cingulo, & ſenis quæ
in Enſe iampridem adnotatæ fuerunt, alias adiacentes o-
ctuaginta recens viſas appoſuimus; earumq; interſtitia quo
exactius licuit ſeruauimus: notas, ſeu veteres, diſtinctionis
gratia, maiores pinximus, ac duplici linea contornauimus,
alias inconſpicuas, minores, ac vnis lineis notauimus; ma-
gnitudinum quoque diſcrimina quo magis licuit ſeruaui-
mus. In altero exemplo ſex Stellas Tauri, PLEIADAS di-
ctas depinximus (dico autem ſex, quandoquidem ſeptima
ferè nunquam apparet) intra anguſtiſſimos in cœlo can-
cellos occluſas, quibus aliæ plures quam quadraginta in-
uiſibiles adiacent; quarum nulla ab aliqua ex prædictis
ſex vix vltra ſemigradum elongatur; harum nos tantum
trigintaſex adnotauimus; earumque interſtitia, magnitu-
dines, necnon veterum nouarumque diſcrimina veluti in
Orione ſeruauimus.

Cinguli, & Enſis ORIONIS Aſteriſmus.

PLEIA-

Galileo Galilei, Sidereus Nuncius, Venice 1610, p. 16v and recto of first unnumbered page, New York copy.

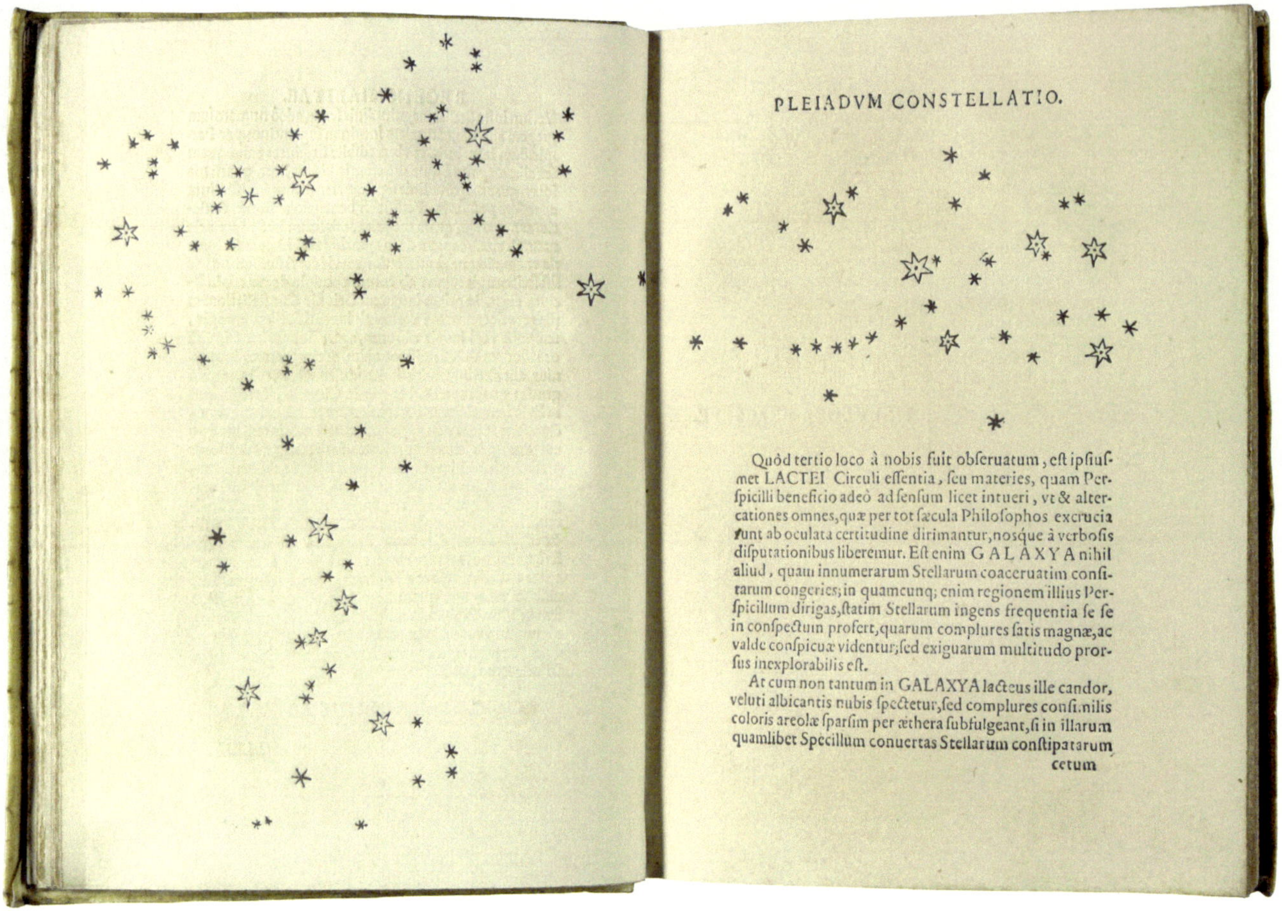

Galileo Galilei, Sidereus Nuncius, Venice 1610, verso of first and recto of second unnumbered pages, New York copy.

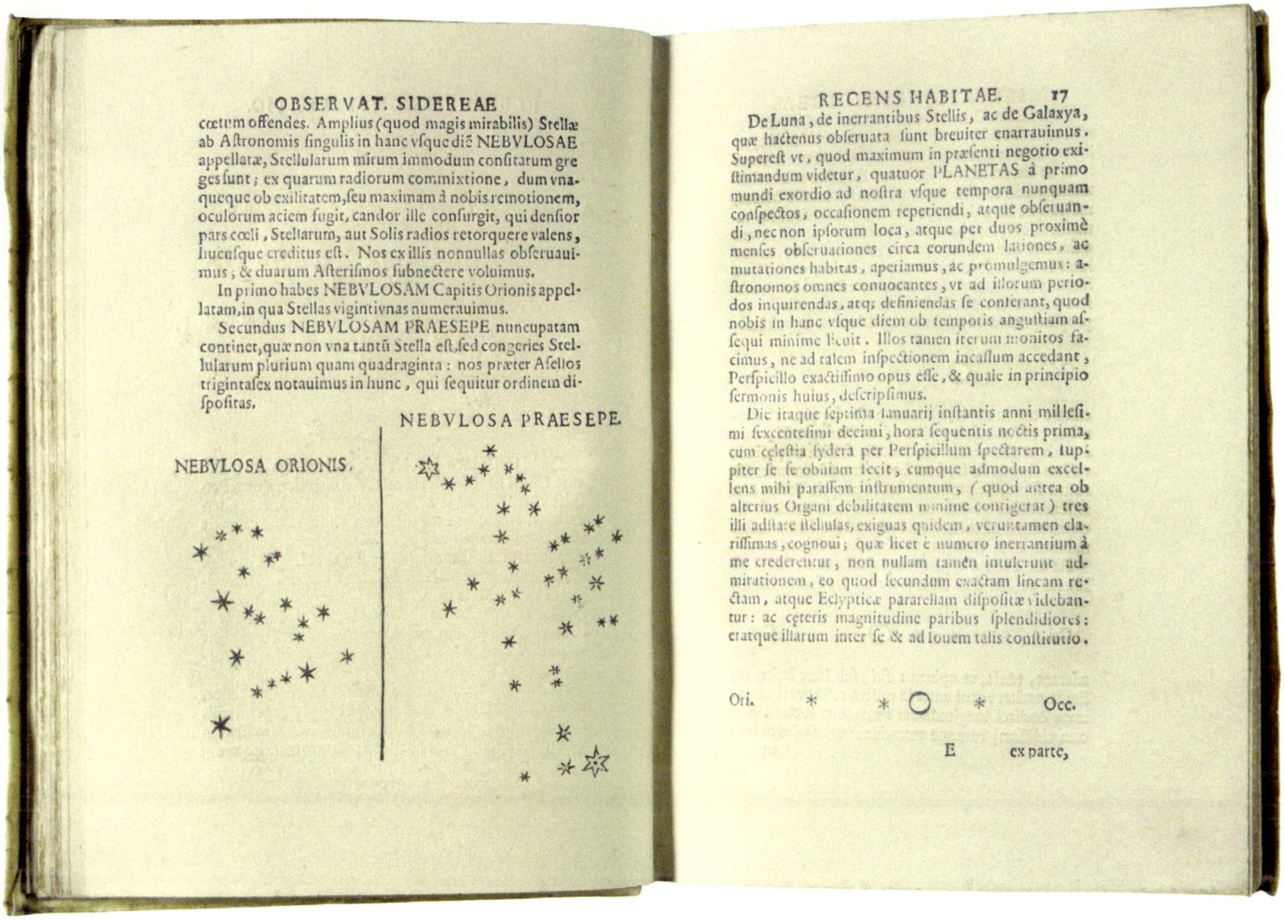

OBSERVAT. SIDEREAE

cœtum offendes. Amplius (quod magis mirabilis) Stellæ ab Aſtronomis ſingulis in hanc vſque diē NEBVLOSAE appellatæ, Stellularum mirum immodum conſitarum greges ſunt; ex quarum radiorum commixtione, dum vnaquæque ob exilitatem, ſeu maximam à nobis remotionem, oculorum aciem fugit, candor ille conſurgit, qui denſior pars cœli, Stellarum, aut Solis radios retorquere valens, hucuſque creditus eſt. Nos ex illis nonnullas obſeruauimus; & duarum Aſteriſmos ſubnectere voluimus.

In primo habes NEBVLOSAM Capitis Orionis appellatam, in qua Stellas vigintivnas numerauimus.

Secundus NEBVLOSAM PRAESEPE nuncupatam continet, quæ non vna tantū Stella eſt, ſed congeries Stellularum plurium quam quadraginta: nos præter Aſellos trigintaſex notauimus in hunc, qui ſequitur ordinem diſpoſitas.

RECENS HABITAE. 17

DeLuna, de inerrantibus Stellis, ac de Galaxya, quæ hactenus obſeruata ſunt breuiter enarrauimus. Supereſt vt, quod maximum in præſenti negotio exiſtimandum videtur, quatuor PLANETAS à primo mundi exordio ad noſtra vſque tempora nunquam conſpectos, occaſionem reperiendi, atque obſeruandi, nec non ipſorum loca, atque per duos proximè menſes obſeruationes circa eorundem lationes, ac mutationes habitas, aperiamus, ac promulgemus: aſtronomos omnes conuocantes, vt ad illorum periodos inquirendas, atq; definiendas ſe conferant, quod nobis in hanc vſque diem ob temporis anguſtiam aſſequi minime licuit. Illos tamen iterum monitos facimus, ne ad talem inſpectionem incaſſum accedant, Perſpicillo exactiſſimo opus eſſe, & quale in principio ſermonis huius, deſcripſimus.

Die itaque ſeptima Ianuarij inſtantis anni milleſimi ſexcenteſimi decimi, hora ſequentis noctis prima, cum cęleſtia ſydera per Perſpicillum ſpectarem, Iupiter ſe ſe obuiam fecit, cumque admodum excellens mihi paraſſem inſtrumentum, (quod antea ob alterius Organi debilitatem minime contigerat) tres illi adſtare ſtellulas, exiguas quidem, veruntamen clariſſimas, cognoui; quæ licet è numero inerrantium à me crederentur, non nullam tamen intulerunt admirationem, eo quod ſecundum exactam lineam rectam, atque Eclypticæ pararellam diſpoſitæ videbantur: ac cęteris magnitudine paribus ſplendidiores: eratque illarum inter ſe & ad Iouem talis conſtitutio.

Ori. * * ○ * Occ.

E ex parte,

Galileo Galilei, Sidereus Nuncius, Venice 1610, verso of second unnumbered page and p. 17r, New York copy.

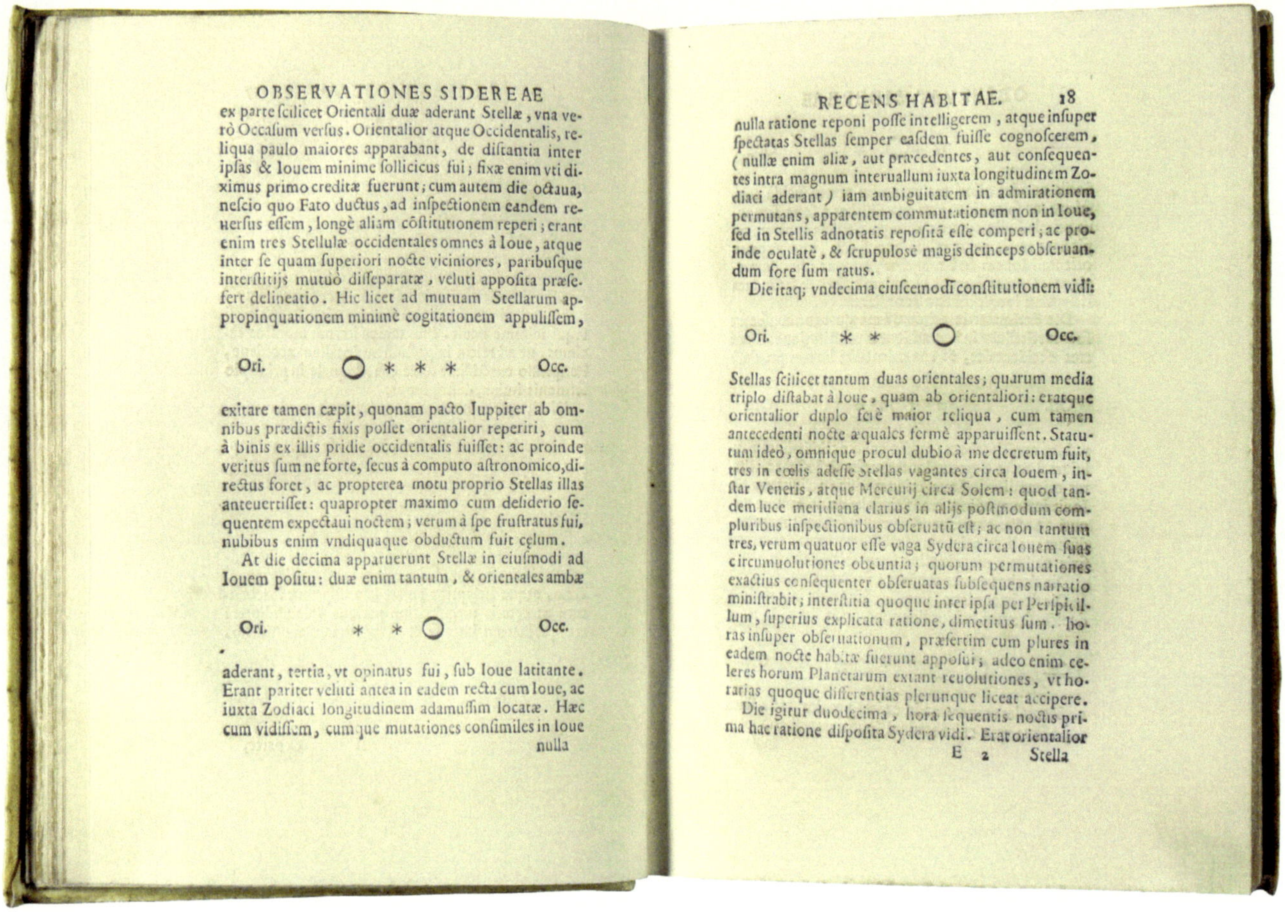
OBSERVATIONES SIDEREAE

ex parte ſcilicet Orientali duæ aderant Stellæ, vna verò Occaſum verſus. Orientalior atque Occidentalis, reliqua paulo maiores apparabant, de diſtantia inter ipſas & Iouem minime ſollicicus fui; fixæ enim vti diximus primo creditæ fuerunt; cum autem die octaua, neſcio quo Fato ductus, ad inſpectionem eandem reuerſus eſſem, longè aliam cõſtitutionem reperi; erant enim tres Stellulæ occidentales omnes à Ioue, atque inter ſe quam ſuperiori nocte viciniores, paribuſque interſtitijs mutuò diſſeparatæ, veluti appoſita præſefert delineatio. Hic licet ad mutuam Stellarum appropinquationem minimè cogitationem appuliſſem,

Ori. ○ * * * Occ.

exitare tamen cæpit, quonam pacto Iuppiter ab omnibus prædictis fixis poſſet orientalior reperiri, cum à binis ex illis pridie occidentalis fuiſſet: ac proinde veritus ſum ne forte, ſecus à computo aſtronomico, directus foret, ac propterea motu proprio Stellas illas anteuertiſſet: quapropter maximo cum deſiderio ſequentem expectaui noctem; verum à ſpe fruſtratus fui, nubibus enim vndiquaque obductum fuit cęlum.

At die decima apparuerunt Stellæ in eiuſmodi ad Iouem poſitu: duæ enim tantum, & orientales ambæ

Ori. * * ○ Occ.

aderant, tertia, vt opinatus fui, ſub Ioue latitante. Erant pariter veluti antea in eadem recta cum Ioue, ac iuxta Zodiaci longitudinem adamuſſim locatæ. Hæc cum vidiſſem, cumque mutationes conſimiles in Ioue

nulla

RECENS HABITAE. 18

nulla ratione reponi poſſe intelligerem, atque inſuper ſpectatas Stellas ſemper eaſdem fuiſſe cognoſcerem, (nullæ enim aliæ, aut præcedentes, aut conſequentes intra magnum interuallum iuxta longitudinem Zodiaci aderant) iam ambiguitatem in admirationem permutans, apparentem commutationem non in Ioue, ſed in Stellis adnotatis repoſitã eſſe comperi; ac proinde oculatè, & ſcrupuloſè magis deinceps obſeruandum fore ſum ratus.

Die itaq; vndecima eiuſcemodi conſtitutionem vidi:

Ori. * * ○ Occ.

Stellas ſcilicet tantum duas orientales; quarum media triplo diſtabat à Ioue, quam ab orientaliori: eratque orientalior duplo ferè maior reliqua, cum tamen antecedenti nocte æquales fermè apparuiſſent. Statutum ideò, omnique procul dubio à me decretum fuit, tres in cœlis adeſſe Stellas vagantes circa Iouem, inſtar Veneris, atque Mercurij circa Solem: quod tandem luce meridiana clarius in alijs poſtmodum compluribus inſpectionibus obſeruatũ eſt; ac non tantum tres, verum quatuor eſſe vaga Sydera circa Iouem ſuas circumuolutiones obeuntia; quorum permutationes exactius conſequenter obſeruatas ſubſequens narratio miniſtrabit; interſtitia quoque inter ipſa per Perſpicillum, ſuperius explicata ratione, dimetitus ſum. horas inſuper obſeruationum, præſertim cum plures in eadem nocte habitæ fuerunt appoſui; adeo enim celeres horum Planetarum extant reuolutiones, vt horarias quoque differentias plerunque liceat accipere.

Die igitur duodecima, hora ſequentis noctis prima hac ratione diſpoſita Sydera vidi. Erat orientalior

E 2 Stella

Galileo Galilei, Sidereus Nuncius, Venice 1610, pp. 17v and 18r, New York copy.

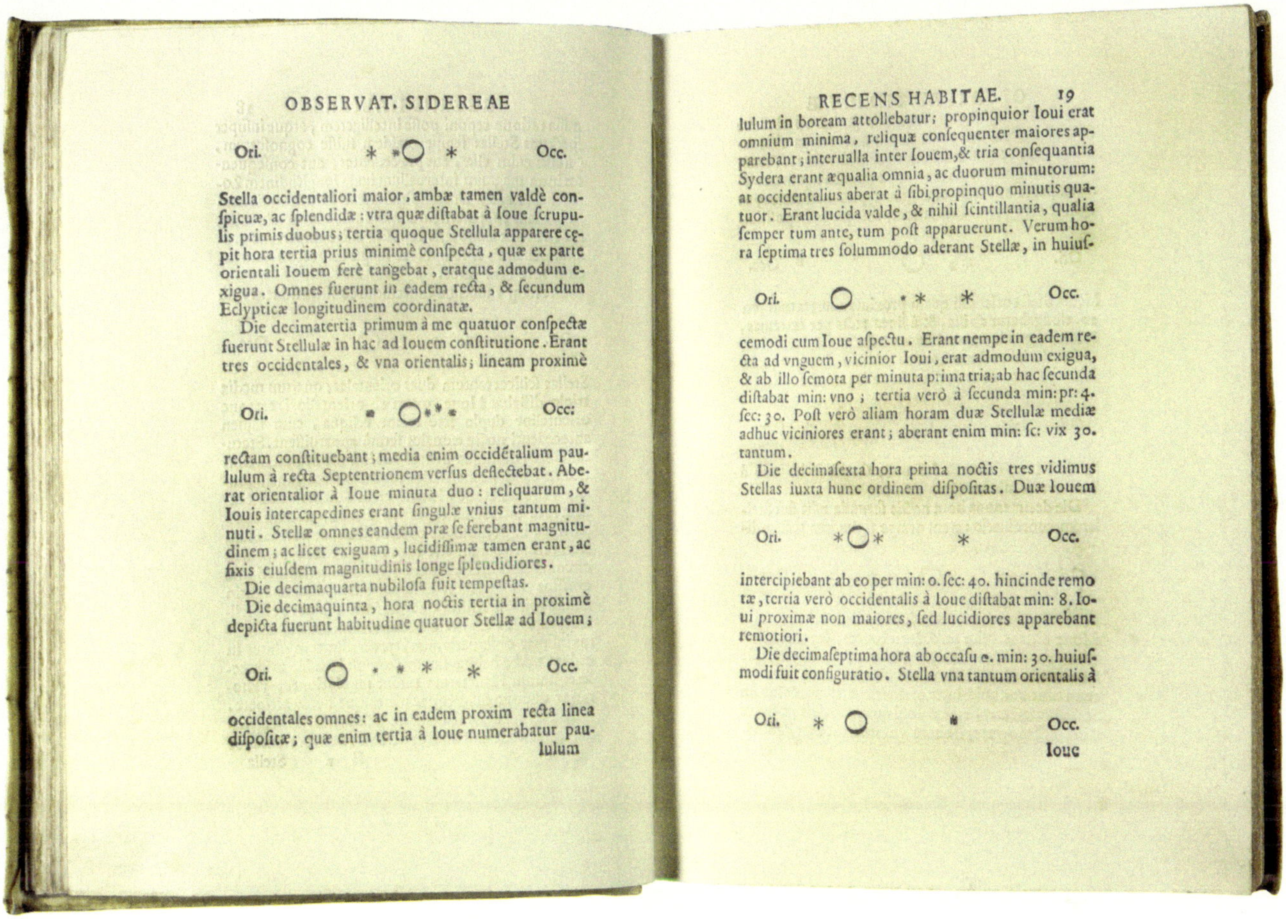
OBSERVAT. SIDEREAE

Ori. Occ.

Stella occidentaliori maior, ambæ tamen valdè conſpicuæ, ac ſplendidæ: vtra quæ diſtabat à Ioue ſcrupulis primis duobus; tertia quoque Stellula apparere cępit hora tertia prius minimè conſpecta, quæ ex parte orientali Iouem ferè tangebat, eratque admodum exigua. Omnes fuerunt in eadem recta, & ſecundum Eclypticæ longitudinem coordinatæ.

Die decimatertia primum à me quatuor conſpectæ fuerunt Stellulæ in hac ad Iouem conſtitutione. Erant tres occidentales, & vna orientalis; lineam proximè

Ori. Occ:

rectam conſtituebant; media enim occidētalium paululum à recta Septentrionem verſus deflectebat. Aberat orientalior à Ioue minuta duo: reliquarum, & Iouis intercapedines erant ſingulæ vnius tantum minuti. Stellæ omnes eandem præ ſe ferebant magnitudinem; ac licet exiguam, lucidiſſimæ tamen erant, ac fixis eiuſdem magnitudinis longe ſplendidiores.

Die decimaquarta nubiloſa fuit tempeſtas.

Die decimaquinta, hora noctis tertia in proximè depicta fuerunt habitudine quatuor Stellæ ad Iouem;

Ori. Occ.

occidentales omnes: ac in eadem proxim recta linea diſpoſitæ; quæ enim tertia à Ioue numerabatur paululum

RECENS HABITAE. 19

lulum in boream attollebatur; propinquior Ioui erat omnium minima, reliquæ conſequenter maiores apparebant; interualla inter Iouem, & tria conſequantia Sydera erant æqualia omnia, ac duorum minutorum: at occidentalius aberat à ſibi propinquo minutis quatuor. Erant lucida valde, & nihil ſcintillantia, qualia ſemper tum ante, tum poſt apparuerunt. Verum hora ſeptima tres ſolummodo aderant Stellæ, in huiuſcemodi cum Ioue aſpectu.

Ori. Occ.

Erant nempe in eadem recta ad vnguem, vicinior Ioui, erat admodum exigua, & ab illo ſemota per minuta prima tria; ab hac ſecunda diſtabat min: vno; tertia verò à ſecunda min: pr: 4. ſec: 30. Poſt verò aliam horam duæ Stellulæ mediæ adhuc viciniores erant; aberant enim min: ſc: vix 30. tantum.

Die decimaſexta hora prima noctis tres vidimus Stellas iuxta hunc ordinem diſpoſitas. Duæ Iouem

Ori. Occ.

intercipiebant ab eo per min: 0. ſec: 40. hincinde remotæ, tertia verò occidentalis à Ioue diſtabat min: 8. Ioui proximæ non maiores, ſed lucidiores apparebant remotiori.

Die decimaſeptima hora ab occaſu 0. min: 30. huiuſmodi fuit configuratio. Stella vna tantum orientalis à

Ori. Occ.

Ioue

Galileo Galilei, Sidereus Nuncius, Venice 1610, pp. 18v and 19r, New York copy.

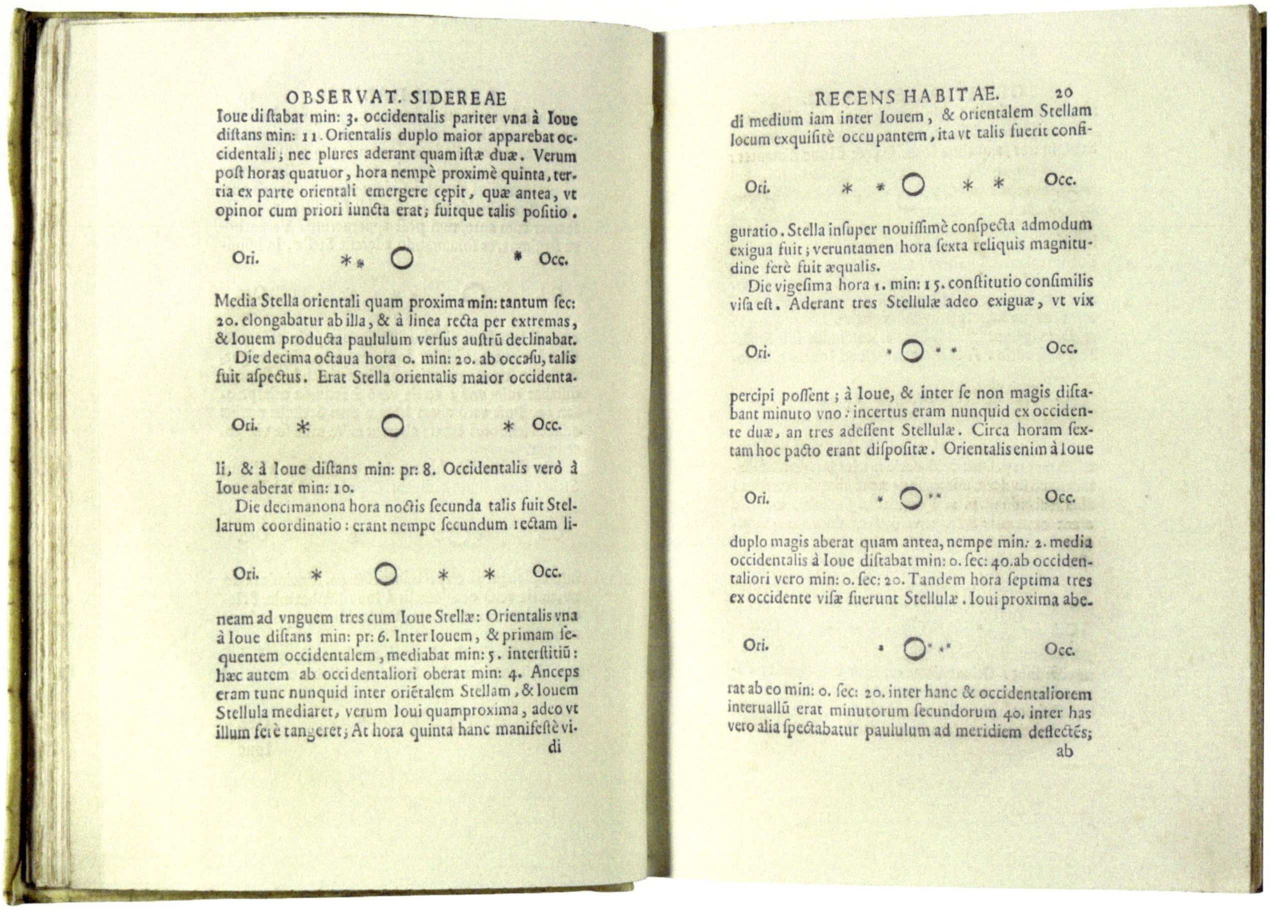

OBSERVAT. SIDEREAE

Ioue distabat min: 3. occidentalis pariter vna à Ioue distans min: 11. Orientalis duplo maior apparebat occidentali; nec plures aderant quam istæ duæ. Verum post horas quatuor, hora nempè proximè quinta, tertia ex parte orientali emergere cępit, quæ antea, vt opinor cum priori iuncta erat; fuitque talis positio.

Ori. * * ○ * Occ.

Media Stella orientali quam proxima min: tantum sec: 20. elongabatur ab illa, & à linea recta per extremas, & Iouem producta paululum versus austrū declinabat.

Die decima octaua hora 0. min: 20. ab occasu, talis fuit aspectus. Erat Stella orientalis maior occidenta-

Ori. * ○ * Occ.

li, & à Ioue distans min: pr: 8. Occidentalis verò à Ioue aberat min: 10.

Die decimanona hora noctis secunda talis fuit Stellarum coordinatio: erant nempe secundum rectam li-

Ori. * ○ * * Occ.

neam ad vnguem tres cum Ioue Stellæ: Orientalis vna à Ioue distans min: pr: 6. Inter Iouem, & primam sequentem occidentalem, mediabat min: 5. interstitiū: hæc autem ab occidentaliori oberat min: 4. Anceps eram tunc nunquid inter oriētalem Stellam, & Iouem Stellula mediaret, verum Ioui quamproxima, adeo vt illum ferè tangeret; At hora quinta hanc manifestè vi-
di

RECENS HABITAE. 20

di medium iam inter Iouem, & orientalem Stellam locum exquisitè occupantem, ita vt talis fuerit confi-

Ori. * * ○ * * Occ.

guratio. Stella insuper nouissimè conspecta admodum exigua fuit; veruntamen hora sexta reliquis magnitudine ferè fuit æqualis.

Die vigesima hora 1. min: 15. constitutio consimilis visa est. Aderant tres Stellulæ adeo exiguæ, vt vix

Ori. * ○ * * Occ.

percipi possent; à Ioue, & inter se non magis distabant minuto vno: incertus eram nunquid ex occidente duæ, an tres adessent Stellulæ. Circa horam sextam hoc pacto erant dispositæ. Orientalis enim à Ioue

Ori. * ○ * * Occ.

duplo magis aberat quam antea, nempe min: 2. media occidentalis à Ioue distabat min: 0. sec: 40. ab occidentaliori vero min: 0. sec: 20. Tandem hora septima tres ex occidente visæ fuerunt Stellulæ. Ioui proxima aberat ab eo min: 0. sec: 20. inter hanc & occidentaliorem

Ori. * ○ * * * Occ.

interuallū erat minutorum secundorum 40. inter has vero alia spectabatur paululum ad meridiem deflectēs;
ab

Galileo Galilei, Sidereus Nuncius, Venice 1610, pp. 19v and 20r, New York copy.

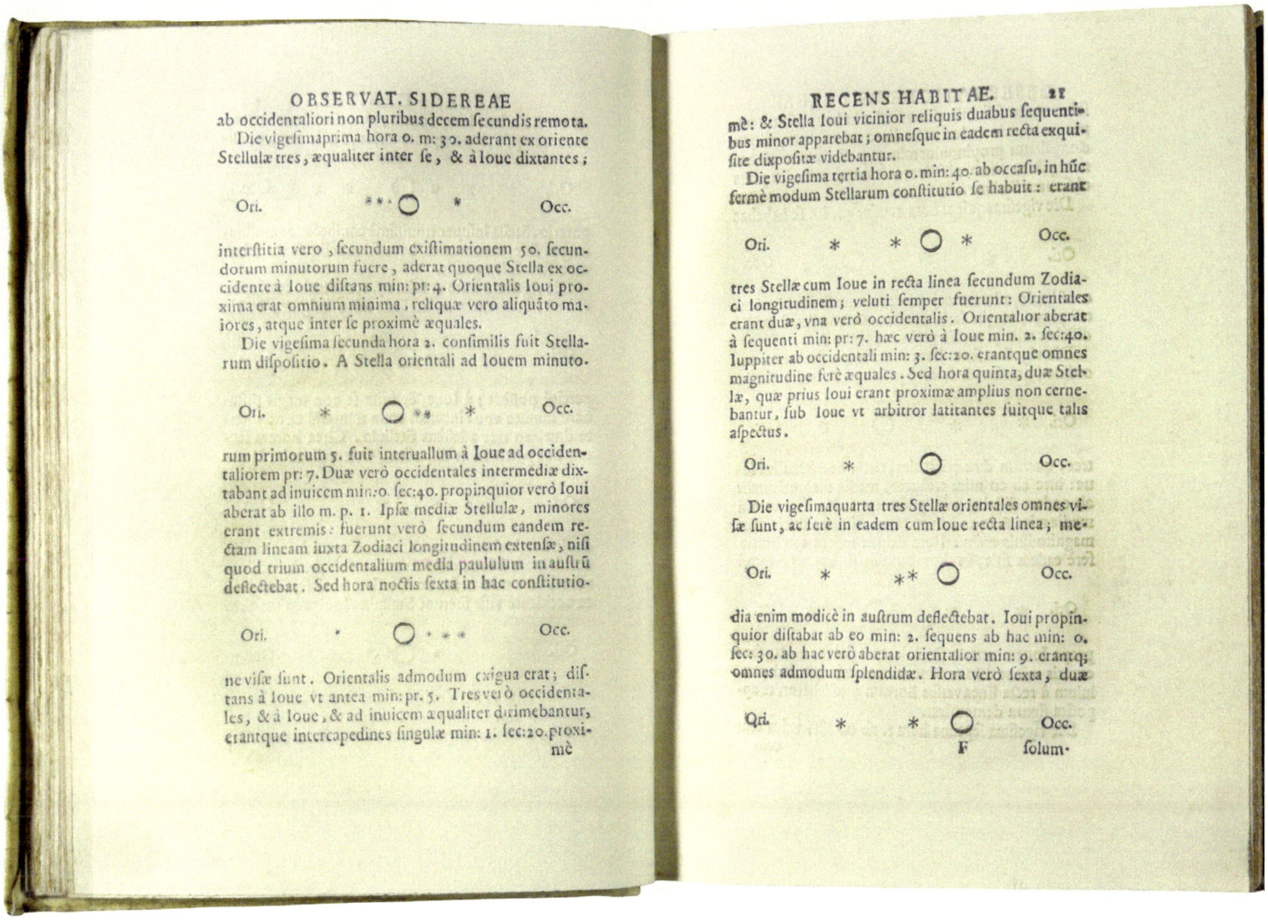
OBSERVAT. SIDEREAE

ab occidentaliori non pluribus decem ſecundis remota.

Die vigeſimaprima hora o. m: 30. aderant ex oriente Stellulæ tres, æqualiter inter ſe, & à Ioue dixtantes;

Ori. Occ.

interſtitia vero, ſecundum exiſtimationem 50. ſecundorum minutorum fuere, aderat quoque Stella ex occidente à Ioue diſtans min: pr: 4. Orientalis Ioui proxima erat omnium minima, reliquæ vero aliquãto maiores, atque inter ſe proximè æquales.

Die vigeſima ſecunda hora 2. conſimilis fuit Stellarum diſpoſitio. A Stella orientali ad Iouem minuto-

Ori. Occ.

rum primorum 5. fuit interuallum à Ioue ad occidentaliorem pr: 7. Duæ verò occidentales intermediæ dixtabant ad inuicem min: o. ſec: 40. propinquior verò Ioui aberat ab illo m. p. 1. Ipſæ mediæ Stellulæ, minores erant extremis: fuerunt verò ſecundum eandem rectam lineam iuxta Zodiaci longitudinem extenſæ, niſi quod trium occidentalium media paululum in auſtrũ deflectebat. Sed hora noctis ſexta in hac conſtitutio-

Ori. Occ.

ne viſæ ſunt. Orientalis admodum exigua erat; diſtans à Ioue vt antea min: pr. 5. Tres verò occidentales, & à Ioue, & ad inuicem æqualiter dirimebantur, erantque intercapedines ſingulæ min: 1. ſec: 20. proximè

RECENS HABITAE. 21

mè: & Stella Ioui vicinior reliquis duabus ſequentibus minor apparebat; omneſque in eadem recta exquiſite dixpoſitæ videbantur.

Die vigeſima tertia hora o. min: 40. ab occaſu, in hũc fermè modum Stellarum conſtitutio ſe habuit: erant

Ori. Occ.

tres Stellæ cum Ioue in recta linea ſecundum Zodiaci longitudinem; veluti ſemper fuerunt: Orientales erant duæ, vna verò occidentalis. Orientalior aberat à ſequenti min: pr: 7. hæc verò à Ioue min. 2. ſec: 40. Iuppiter ab occidentali min: 3. ſec: 20. erantque omnes magnitudine ferè æquales. Sed hora quinta, duæ Stellæ, quæ prius Ioui erant proximæ amplius non cernebantur, ſub Ioue vt arbitror latitantes fuitque talis aſpectus.

Ori. Occ.

Die vigeſimaquarta tres Stellæ orientales omnes viſæ ſunt, ac ferè in eadem cum Ioue recta linea; me-

Ori. Occ.

dia enim modicè in auſtrum deflectebat. Ioui propinquior diſtabat ab eo min: 2. ſequens ab hac min: o. ſec: 30. ab hac verò aberat orientalior min: 9. erantq; omnes admodum ſplendidæ. Hora verò ſexta, duæ

Ori. Occ.

F ſolum-

Galileo Galilei, Sidereus Nuncius, Venice 1610, pp. 20v and 21r, New York copy.

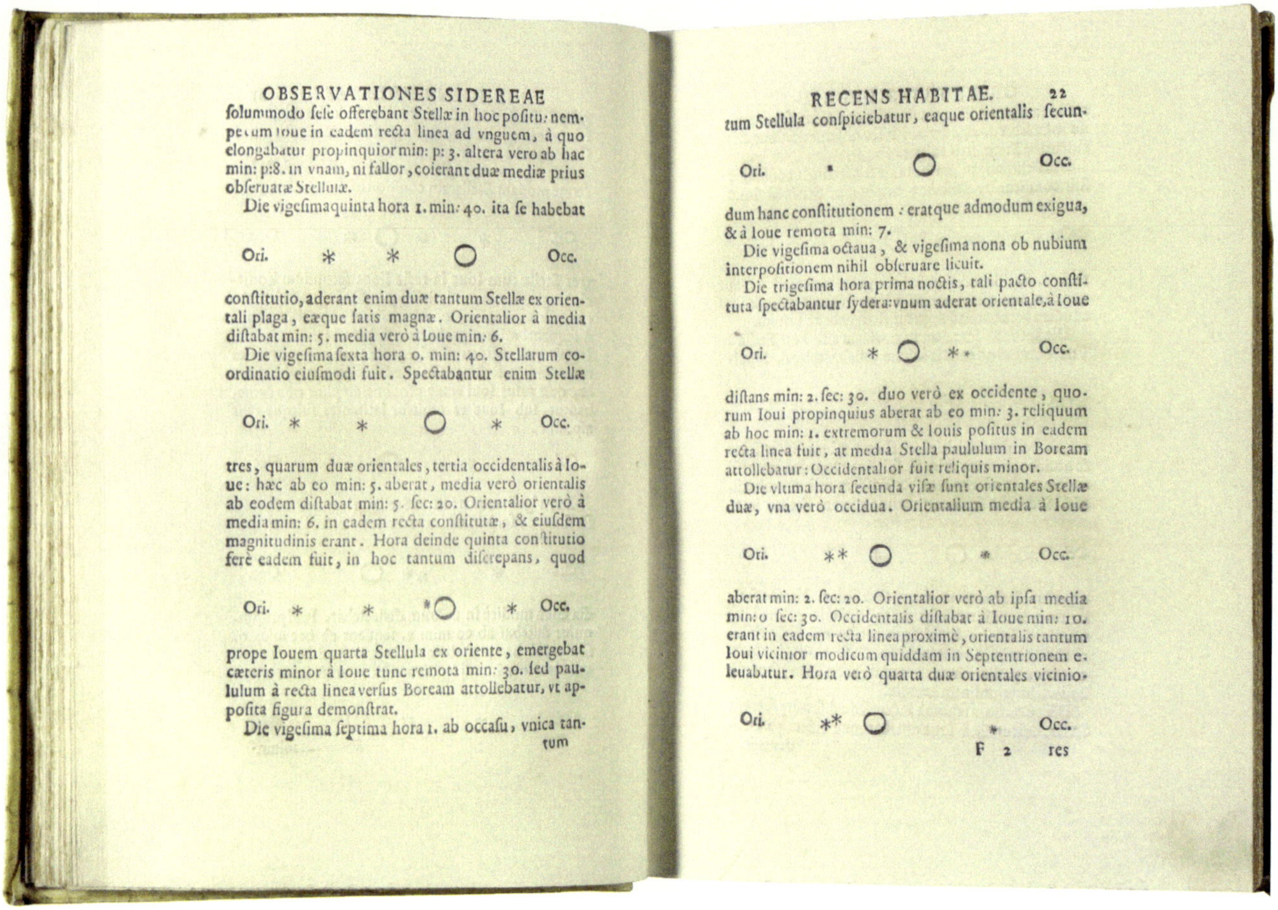
OBSERVATIONES SIDEREAE

ſolummodo ſeſe offerebant Stellæ in hoc poſitu: nempe cum Ioue in eadem recta linea ad vnguem, à quo elongabatur propinquior min: p: 3. altera vero ab hac min: p:8. in vnam, ni fallor, coierant duæ mediæ prius obſeruatæ Stellulæ.

Die vigeſimaquinta hora 1. min: 40. ita ſe habebat

Ori. * * ○ Occ.

conſtitutio, aderant enim duæ tantum Stellæ ex orientali plaga, eæque ſatis magnæ. Orientalior à media diſtabat min: 5. media verò à Ioue min: 6.

Die vigeſima ſexta hora 0. min: 40. Stellarum coordinatio eiuſmodi fuit. Spectabantur enim Stellæ

Ori. * * ○ * Occ.

tres, quarum duæ orientales, tertia occidentalis à Ioue: hæc ab eo min: 5. aberat, media verò orientalis ab eodem diſtabat min: 5. ſec: 20. Orientalior verò à media min: 6. in eadem recta conſtitutæ, & eiuſdem magnitudinis erant. Hora deinde quinta conſtitutio ferè eadem fuit, in hoc tantum diſcrepans, quod

Ori. * * *○ * Occ.

prope Iouem quarta Stellula ex oriente, emergebat cæteris minor à Ioue tunc remota min: 30. ſed paululum à recta linea verſus Boream attollebatur, vt appoſita figura demonſtrat.

Die vigeſima ſeptima hora 1. ab occaſu, vnica tantum

RECENS HABITAE. 22

tum Stellula conſpiciebatur, eaque orientalis ſecun-

Ori. · ○ Occ.

dum hanc conſtitutionem: eratque admodum exigua, & à Ioue remota min: 7.

Die vigeſima octaua, & vigeſima nona ob nubium interpoſitionem nihil obſeruare licuit.

Die trigeſima hora prima noctis, tali pacto conſtituta ſpectabantur ſydera: vnum aderat orientale, à Ioue

Ori. * ○ *· Occ.

diſtans min: 2. ſec: 30. duo verò ex occidente, quorum Ioui propinquius aberat ab eo min: 3. reliquum ab hoc min: 1. extremorum & Iouis poſitus in eadem recta linea fuit, at media Stella paululum in Boream attollebatur: Occidentalior fuit reliquis minor.

Die vltima hora ſecunda viſæ ſunt orientales Stellæ duæ, vna verò occidua. Orientalium media à Ioue

Ori. ** ○ * Occ.

aberat min: 2. ſec: 20. Orientalior verò ab ipſa media min: 0 ſec: 30. Occidentalis diſtabat à Ioue min: 10. erant in eadem recta linea proximè, orientalis tantum Ioui vicinior modicum quiddam in Septentrionem eleuabatur. Hora verò quarta duæ orientales vicinio-

Ori. ** ○ * Occ.

F 2 res

Galileo Galilei, Sidereus Nuncius, Venice 1610, pp. 21v and 22r, New York copy.

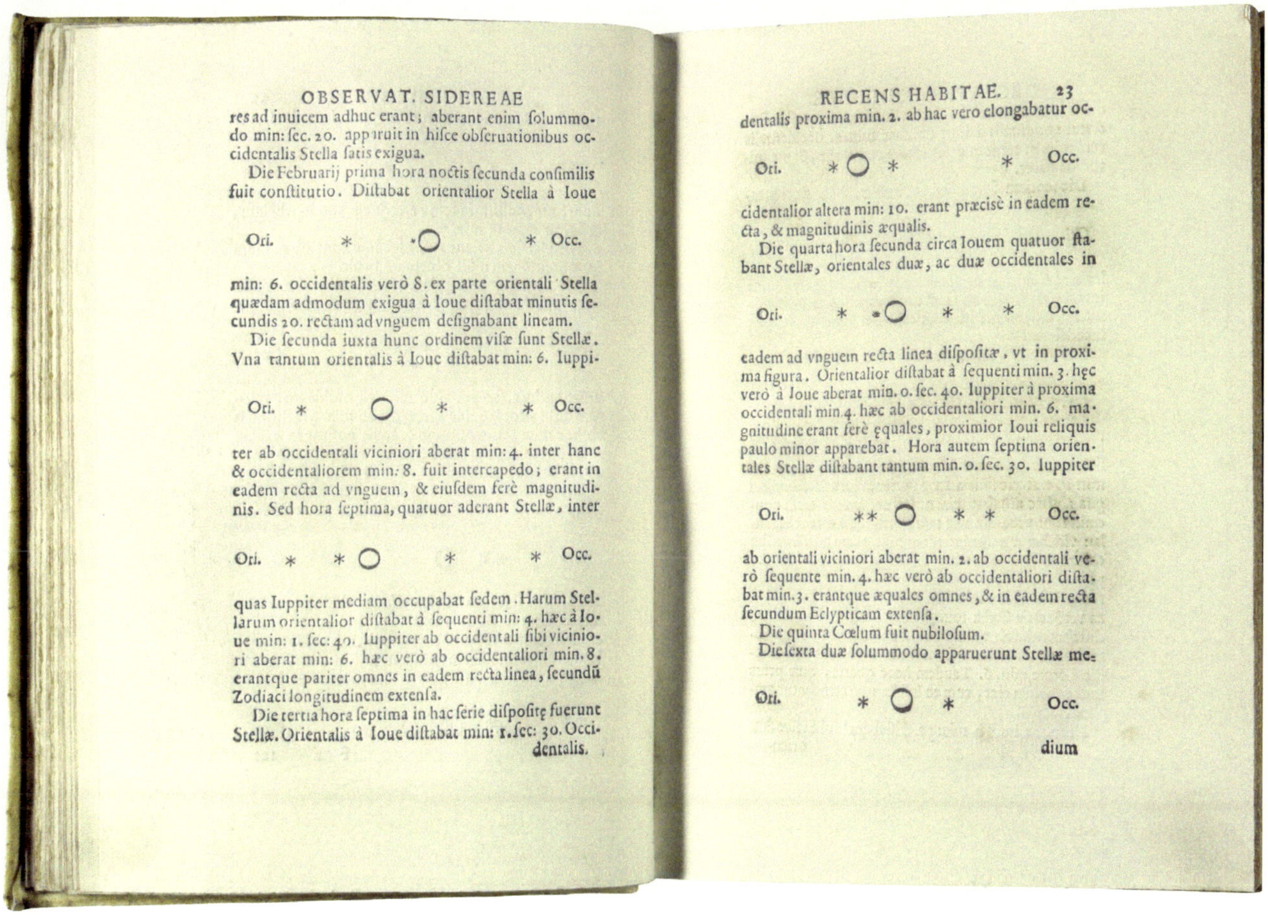

OBSERVAT. SIDEREAE

res ad inuicem adhuc erant; aberant enim ſolummodo min: ſec. 20. apparuit in hiſce obſeruationibus occidentalis Stella ſatis exigua.

Die Februarij prima hora noctis ſecunda conſimilis fuit conſtitutio. Diſtabat orientalior Stella à Ioue

Ori. * ·○ * Occ.

min: 6. occidentalis verò 8. ex parte orientali Stella quædam admodum exigua à Ioue diſtabat minutis ſecundis 20. rectam ad vnguem deſignabant lineam.

Die ſecunda iuxta hunc ordinem viſæ ſunt Stellæ. Vna tantum orientalis à Ioue diſtabat min: 6. Iuppi-

Ori. * ○ * * Occ.

ter ab occidentali viciniori aberat min: 4. inter hanc & occidentaliorem min: 8. fuit intercapedo; erant in eadem recta ad vnguem, & eiuſdem ferè magnitudinis. Sed hora ſeptima, quatuor aderant Stellæ, inter

Ori. * * ○ * * Occ.

quas Iuppiter mediam occupabat ſedem. Harum Stellarum orientalior diſtabat à ſequenti min: 4. hæc à Ioue min: 1. ſec: 40. Iuppiter ab occidentali ſibi viciniori aberat min: 6. hæc verò ab occidentaliori min. 8. erantque pariter omnes in eadem recta linea, ſecundũ Zodiaci longitudinem extenſa.

Die tertia hora ſeptima in hac ſerie diſpoſitę fuerunt Stellæ. Orientalis à Ioue diſtabat min: 1. ſec: 30. Occidentalis.

RECENS HABITAE. 23

dentalis proxima min. 2. ab hac vero elongabatur oc-

Ori. * ○ * * Occ.

cidentalior altera min: 10. erant præciſè in eadem recta, & magnitudinis æqualis.

Die quarta hora ſecunda circa Iouem quatuor ſtabant Stellæ, orientales duæ, ac duæ occidentales in

Ori. * ·○ * * Occ.

eadem ad vnguem recta linea diſpoſitæ, vt in proxima figura. Orientalior diſtabat à ſequenti min. 3. hęc verò à Ioue aberat min. o. ſec. 40. Iuppiter à proxima occidentali min. 4. hæc ab occidentaliori min. 6. magnitudine erant ferè ęquales, proximior Ioui reliquis paulo minor apparebat. Hora autem ſeptima orientales Stellæ diſtabant tantum min. o. ſec. 30. Iuppiter

Ori. ** ○ * * Occ.

ab orientali viciniori aberat min. 2. ab occidentali verò ſequente min. 4. hæc verò ab occidentaliori diſtabat min. 3. erantque æquales omnes, & in eadem recta ſecundum Eclypticam extenſa.

Die quinta Cœlum fuit nubiloſum.

Die ſexta duæ ſolummodo apparuerunt Stellæ me-

Ori. * ○ * Occ.

dium

Galileo Galilei, Sidereus Nuncius, Venice 1610, pp. 22v and 23r, New York copy.

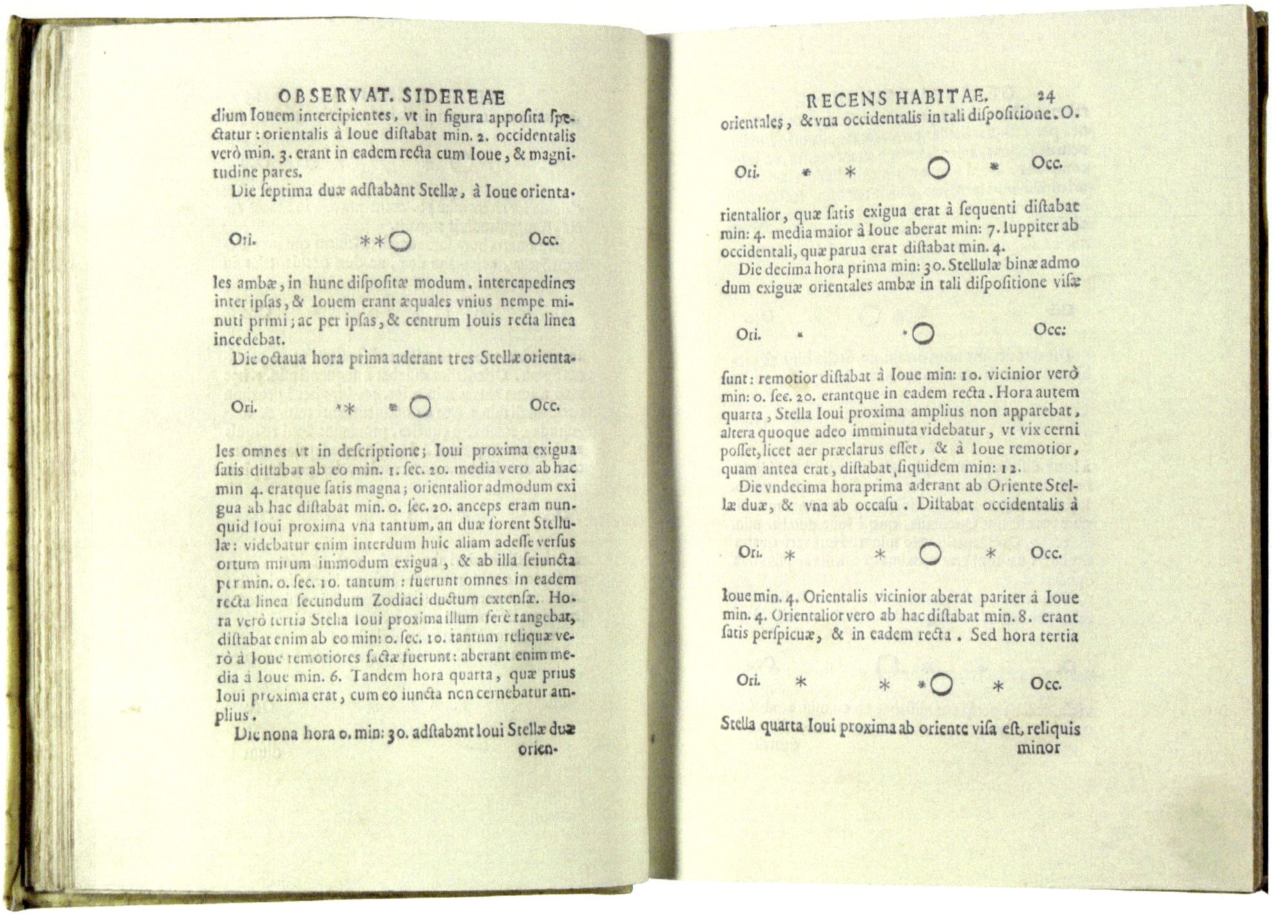

OBSERVAT. SIDEREAE

dium Iouem intercipientes, vt in figura apposita spectatur: orientalis à Ioue distabat min. 2. occidentalis verò min. 3. erant in eadem recta cum Ioue, & magnitudine pares.

Die septima duæ adstabant Stellæ, à Ioue orienta-

Ori. * * O Occ.

les ambæ, in hunc dispositæ modum. intercapedines inter ipsas, & Iouem erant æquales vnius nempe minuti primi; ac per ipsas, & centrum Iouis recta linea incedebat.

Die octaua hora prima aderant tres Stellæ orienta-

Ori. * * * O Occ.

les omnes vt in descriptione; Ioui proxima exigua satis distabat ab eo min. 1. sec. 20. media vero ab hac min 4. eratque satis magna; orientalior admodum exigua ab hac distabat min. o. sec. 20. anceps eram nunquid Ioui proxima vna tantum, an duæ forent Stellulæ: videbatur enim interdum huic aliam adesse versus ortum mirum immodum exigua, & ab illa seiuncta per min. o. sec. 10. tantum: fuerunt omnes in eadem recta linea secundum Zodiaci ductum extensæ. Hora verò tertia Stella Ioui proxima illum ferè tangebat, distabat enim ab eo min: o. sec. 10. tantum reliquæ verò à Ioue remotiores factæ fuerunt: aberant enim media à Ioue min. 6. Tandem hora quarta, quæ prius Ioui proxima erat, cum eo iuncta non cernebatur amplius.

Die nona hora o. min: 30. adstabant Ioui Stellæ duæ orien-

RECENS HABITAE. 24

orientales, & vna occidentalis in tali dispositione. O-

Ori. * * O * Occ.

rientalior, quæ satis exigua erat à sequenti distabat min: 4. media maior à Ioue aberat min: 7. Iuppiter ab occidentali, quæ parua erat distabat min. 4.

Die decima hora prima min: 30. Stellulæ binæ admodum exiguæ orientales ambæ in tali dispositione visæ

Ori. * *O Occ.

sunt: remotior distabat à Ioue min: 10. vicinior verò min: o. sec. 20. erantque in eadem recta. Hora autem quarta, Stella Ioui proxima amplius non apparebat, altera quoque adeo imminuta videbatur, vt vix cerni posset, licet aer præclarus esset, & à Ioue remotior, quam antea erat, distabat, siquidem min: 12.

Die vndecima hora prima aderant ab Oriente Stellæ duæ, & vna ab occasu. Distabat occidentalis à

Ori. * * O * Occ.

Ioue min. 4. Orientalis vicinior aberat pariter à Ioue min. 4. Orientalior vero ab hac distabat min. 8. erant satis perspicuæ, & in eadem recta. Sed hora tertia

Ori. * * *O * Occ.

Stella quarta Ioui proxima ab oriente visa est, reliquis minor

Galileo Galilei, Sidereus Nuncius, Venice 1610, pp. 23v and 24r, New York copy.

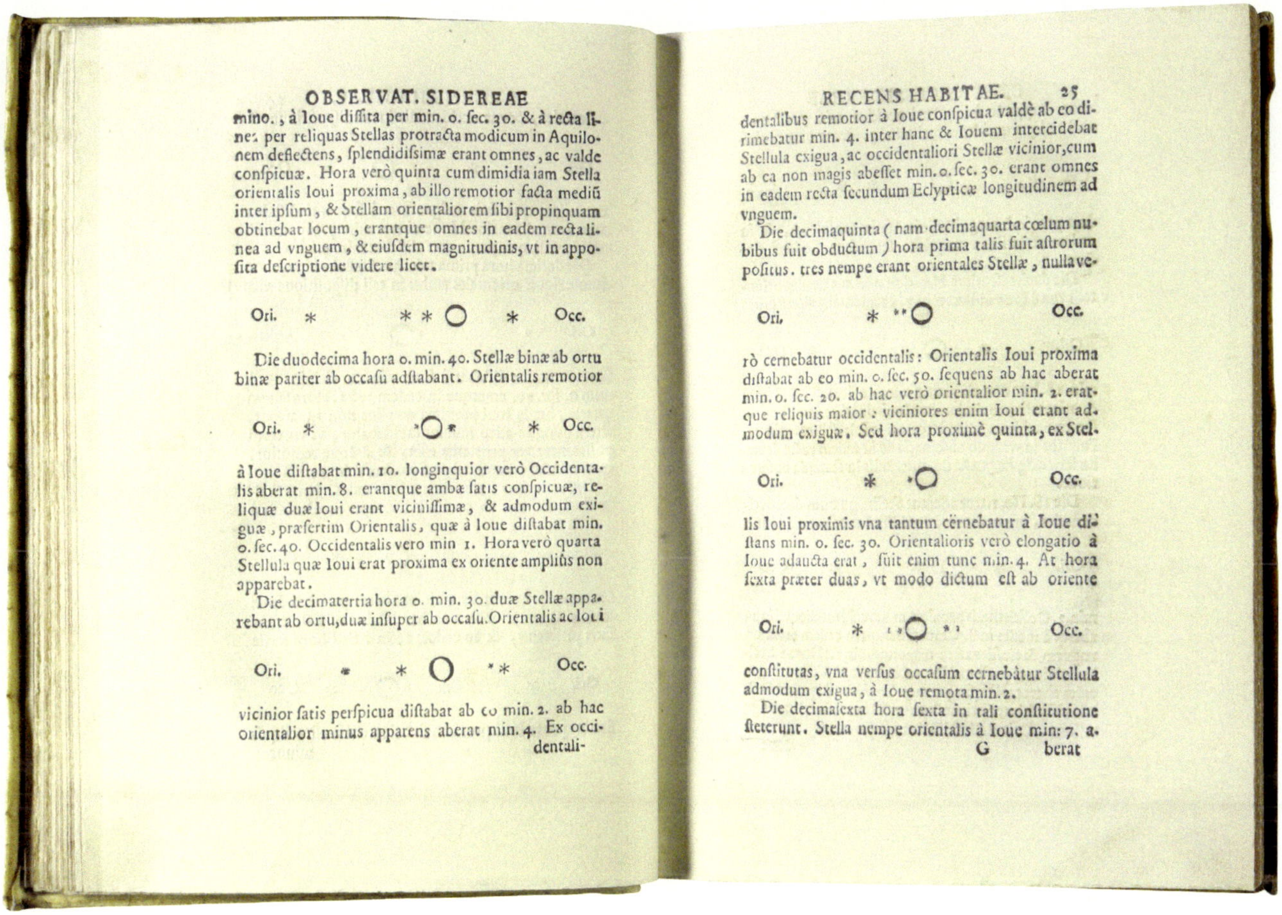

OBSERVAT. SIDEREAE

mino., à Ioue diſſita per min. o. ſec. 30. & à recta line: per reliquas Stellas protracta modicum in Aquilonem deflectens, ſplendidiſsimæ erant omnes, ac valde conſpicuæ. Hora verò quinta cum dimidia iam Stella orientalis Ioui proxima, ab illo remotior facta mediũ inter ipſum, & Stellam orientaliorem ſibi propinquam obtinebat locum, erantque omnes in eadem recta linea ad vnguem, & eiuſdem magnitudinis, vt in appoſita deſcriptione videre licet.

Ori. * * * O * Occ.

Die duodecima hora o. min. 40. Stellæ binæ ab ortu binæ pariter ab occaſu adſtabant. Orientalis remotior

Ori. * ·O* * Occ.

à Ioue diſtabat min. 10. longinquior verò Occidentalis aberat min. 8. erantque ambæ ſatis conſpicuæ, reliquæ duæ Ioui erant viciniſſimæ, & admodum exiguæ, præſertim Orientalis, quæ à Ioue diſtabat min. o. ſec. 40. Occidentalis vero min 1. Hora verò quarta Stellula quæ Ioui erat proxima ex oriente ampliùs non apparebat.

Die decimatertia hora o. min. 30. duæ Stellæ apparebant ab ortu, duæ inſuper ab occaſu. Orientalis ac Ioui

Ori. * * O ·* Occ.

vicinior ſatis perſpicua diſtabat ab eo min. 2. ab hac orientalior minus apparens aberat min. 4. Ex occi-

dentali-

RECENS HABITAE. 25

dentalibus remotior à Ioue conſpicua valdè ab eo dirimebatur min. 4. inter hanc & Iouem intercidebat Stellula exigua, ac occidentaliori Stellæ vicinior, cum ab ea non magis abeſſet min. o. ſec. 30. erant omnes in eadem recta ſecundum Eclypticæ longitudinem ad vnguem.

Die decimaquinta (nam decimaquarta cœlum nubibus fuit obductum) hora prima talis fuit aſtrorum poſitus. tres nempe erant orientales Stellæ, nulla ve-

Ori. * ··O Occ.

rò cernebatur occidentalis: Orientalis Ioui proxima diſtabat ab eo min. o. ſec. 50. ſequens ab hac aberat min. o. ſec. 20. ab hac verò orientalior min. 2. eratque reliquis maior: viciniores enim Ioui erant admodum exiguæ. Sed hora proximè quinta, ex Stel-

Ori. * ·O Occ.

lis Ioui proximis vna tantum cërnebatur à Ioue diſtans min. o. ſec. 30. Orientalioris verò elongatio à Ioue adaucta erat, fuit enim tunc min. 4. At hora ſexta præter duas, vt modo dictum eſt ab oriente

Ori. * ··O ·· Occ.

conſtitutas, vna verſus occaſum cernebàtur Stellula admodum exigua, à Ioue remota min. 2.

Die decimaſexta hora ſexta in tali conſtitutione ſteterunt. Stella nempe orientalis à Ioue min: 7. a-

G berat

Galileo Galilei, Sidereus Nuncius, Venice 1610, pp. 24v and 25r, New York copy.

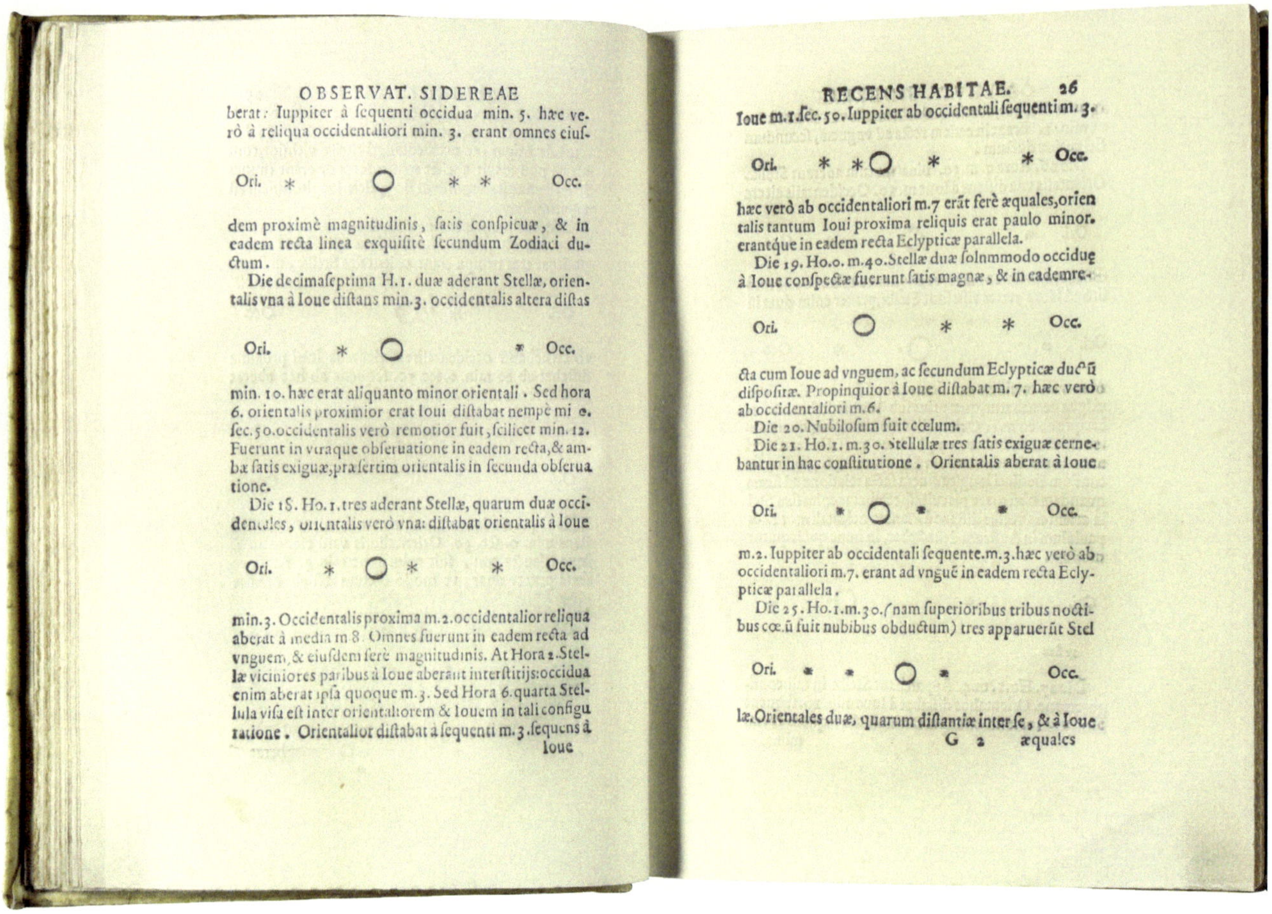

Galileo Galilei, Sidereus Nuncius, Venice 1610, pp. 25v and 26r, New York copy.

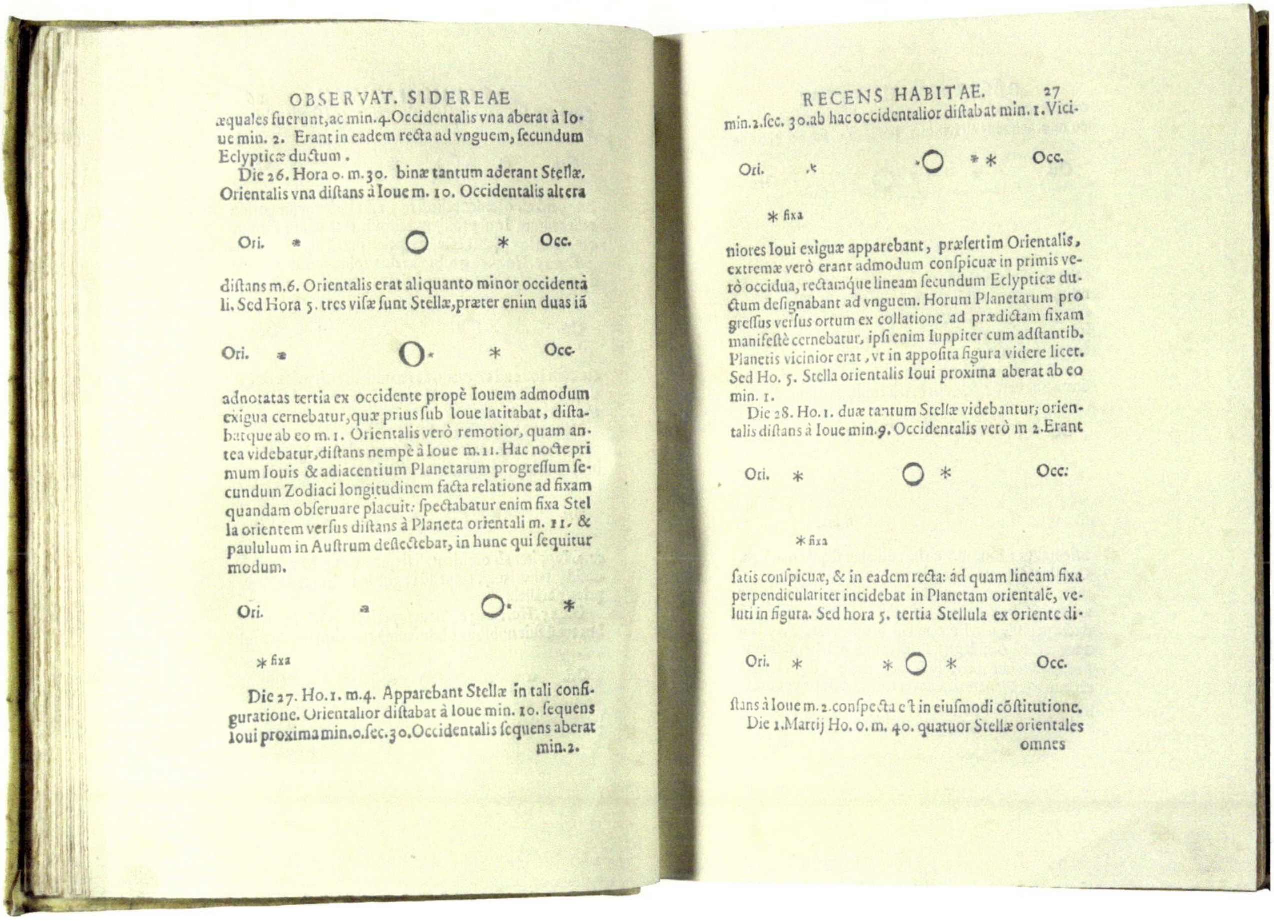
OBSERVAT. SIDEREAE

æquales fuerunt, ac min. 4. Occidentalis vna aberat à Ioue min. 2. Erant in eadem recta ad vnguem, secundum Eclypticæ ductum.

Die 26. Hora o. m. 30. binæ tantum aderant Stellæ. Orientalis vna distans à Ioue m. 10. Occidentalis altera

Ori. * O * Occ.

distans m. 6. Orientalis erat aliquanto minor occidentali. Sed Hora 5. tres visæ sunt Stellæ, præter enim duas iã

Ori. * O* * Occ.

adnotatas tertia ex occidente propè Iouem admodum exigua cernebatur, quæ prius sub Ioue latitabat, distabatque ab eo m. 1. Orientalis verò remotior, quam antea videbatur, distans nempè à Ioue m. 11. Hac nocte primum Iouis & adiacentium Planetarum progressum secundum Zodiaci longitudinem facta relatione ad fixam quandam obseruare placuit: spectabatur enim fixa Stella orientem versus distans à Planeta orientali m. 11. & paululum in Austrum deflectebat, in hunc qui sequitur modum.

Ori. * O* *

* fixa

Die 27. Ho. 1. m. 4. Apparebant Stellæ in tali configuratione. Orientalior distabat à Ioue min. 10. sequens Ioui proxima min. o. sec. 30. Occidentalis sequens aberat min. 2.

RECENS HABITAE. 27

min. 2. sec. 30. ab hac occidentalior distabat min. 1. Vici-

Ori. * *O * * Occ.

* fixa

niores Ioui exiguæ apparebant, præsertim Orientalis, extremæ verò erant admodum conspicuæ in primis verò occidua, rectamque lineam secundum Eclypticæ ductum designabant ad vnguem. Horum Planetarum progressus versus ortum ex collatione ad prædictam fixam manifestè cernebatur, ipsi enim Iuppiter cum adstantib. Planetis vicinior erat, vt in apposita figura videre licet. Sed Ho. 5. Stella orientalis Ioui proxima aberat ab eo min. 1.

Die 28. Ho. 1. duæ tantum Stellæ videbantur; orientalis distans à Ioue min. 9. Occidentalis verò m 2. Erant

Ori. * O * Occ:

* fixa

satis conspicuæ, & in eadem recta: ad quam lineam fixa perpendiculariter incidebat in Planetam orientalẽ, veluti in figura. Sed hora 5. tertia Stellula ex oriente di-

Ori. * * O * Occ.

stans à Ioue m. 2. conspecta est in eiusmodi cõstitutione.

Die 1. Martij Ho. o. m. 40. quatuor Stellæ orientales omnes

Galileo Galilei, Sidereus Nuncius, Venice 1610, pp. 26v and 27r, New York copy.

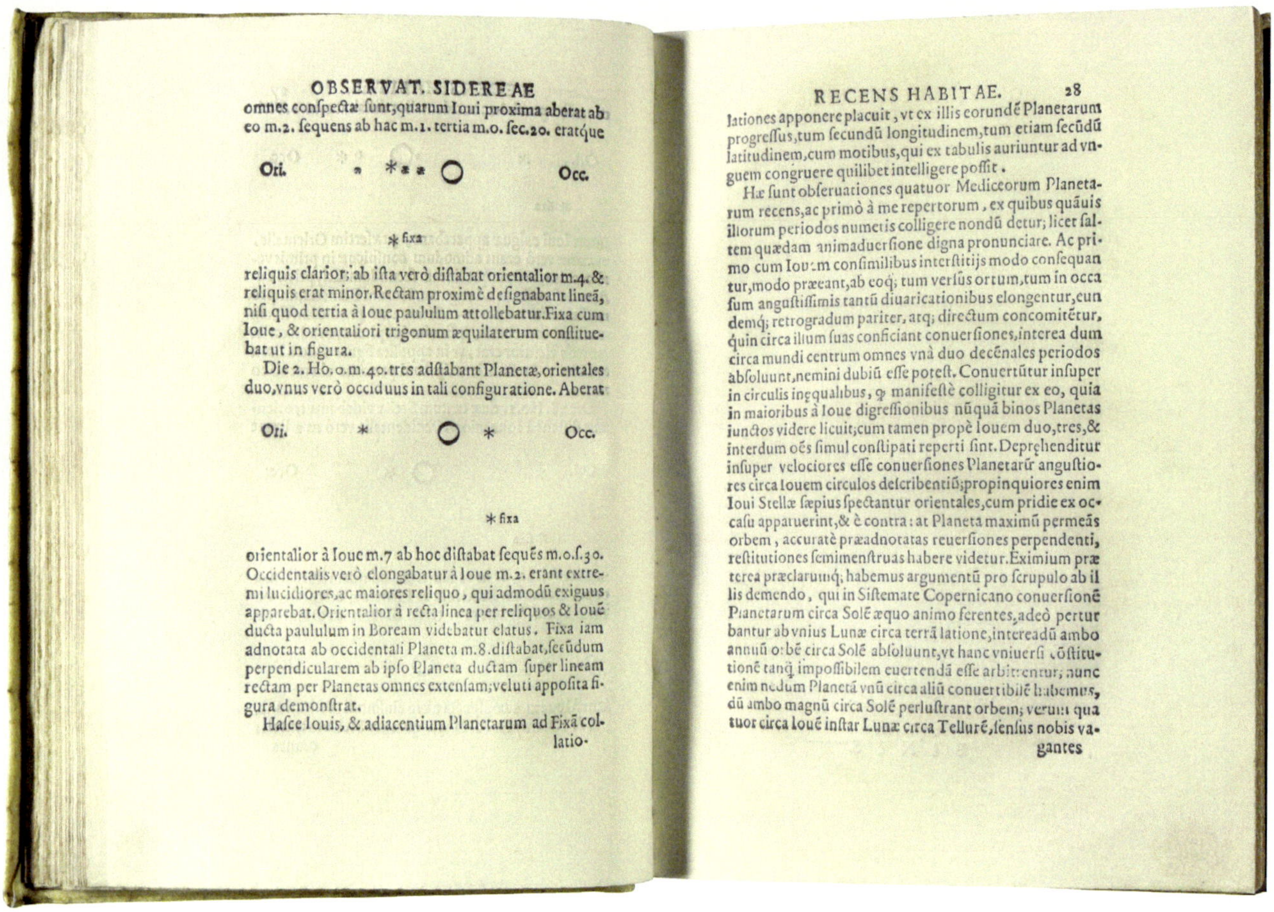

OBSERVAT. SIDEREAE

omnes conſpectæ ſunt,quarum Ioui proxima aberat ab
eo m.2. ſequens ab hac m.1. tertia m.o. ſec.20. eratq́ue

Ori. * * * * ○ Occ.

* fixa

reliquis clarior; ab iſta verò diſtabat orientalior m.4. &
reliquis erat minor. Rectam proximè deſignabant lineā,
niſi quod tertia à Ioue paululum attollebatur. Fixa cum
Ioue, & orientaliori trigonum æquilaterum conſtitue-
bat ut in figura.

Die 2. Ho. o. m. 40. tres adſtabant Planetæ, orientales
duo, vnus verò occiduus in tali configuratione. Aberat

Ori. * ○ * Occ.

* fixa

orientalior à Ioue m.7 ab hoc diſtabat ſequēs m.o.ſ.30.
Occidentalis verò elongabatur à Ioue m.2. erant extre-
mi lucidiores, ac maiores reliquo, qui admodū exiguus
apparebat. Orientalior à recta linea per reliquos & Iouē
ducta paululum in Boream videbatur elatus. Fixa iam
adnotata ab occidentali Planeta m.8. diſtabat, ſecūdum
perpendicularem ab ipſo Planeta ductam ſuper lineam
rectam per Planetas omnes extenſam; veluti appoſita fi-
gura demonſtrat.

Haſce Iouis, & adiacentium Planetarum ad Fixā col-
latio-

RECENS HABITAE. 28

lationes apponere placuit, vt ex illis eorundē Planetarum
progreſſus, tum ſecundū longitudinem, tum etiam ſecūdū
latitudinem, cum motibus, qui ex tabulis auriuntur ad vn-
guem congruere quilibet intelligere poſſit.

Hæ ſunt obſeruationes quatuor Mediceorum Planeta-
rum recens, ac primò à me repertorum, ex quibus quāuis
illorum periodos numeris colligere nondū detur; licet ſal-
tem quædam animaduerſione digna pronunciare. Ac pri-
mo cum Iouem conſimilibus interſtitijs modo conſequan
tur, modo præeant, ab eoq; tum verſus ortum, tum in occa
ſum anguſtiſſimis tantū diuaricationibus elongentur, cum
demq; retrogradum pariter, atq; directum concomitētur,
quin circa illum ſuas conficiant conuerſiones, interea dum
circa mundi centrum omnes vnà duo decēnales periodos
abſoluunt, nemini dubiū eſſe poteſt. Conuertūtur inſuper
in circulis inæqualibus, quod manifeſtè colligitur ex eo, quia
in maioribus à Ioue digreſſionibus nūquā binos Planetas
iunctos videre licuit; cum tamen propè Iouem duo, tres, &
interdum oēs ſimul conſtipati reperti ſint. Deprehenditur
inſuper velociores eſſe conuerſiones Planetarū anguſtio-
res circa Iouem circulos deſcribentiū; propinquiores enim
Ioui Stellæ ſæpius ſpectantur orientales, cum pridie ex oc-
caſu apparuerint, & è contra: at Planeta maximū permeās
orbem, accuratè præadnotatas reuerſiones perpendenti,
reſtitutiones ſemimenſtruas habere videtur. Eximium præ
terea præclarumq; habemus argumentū pro ſcrupulo ab il
lis demendo, qui in Siſtemate Copernicano conuerſionē
Planetarum circa Solē æquo animo ferentes, adeò pertur
bantur ab vnius Lunæ circa terrā latione, intereadū ambo
annuū orbē circa Solē abſoluunt, vt hanc vniuerſi cōſtitu-
tionē tanquā impoſſibilem euertendā eſſe arbitrentur; nunc
enim nedum Planetā vnū circa aliū conuertibilē habemus,
dū ambo magnū circa Solē perluſtrant orbem; verum qua
tuor circa Iouē inſtar Lunæ circa Tellurē, ſenſus nobis va-
gantes

Galileo Galilei, Sidereus Nuncius, Venice 1610, pp. 27v and 28r, New York copy.

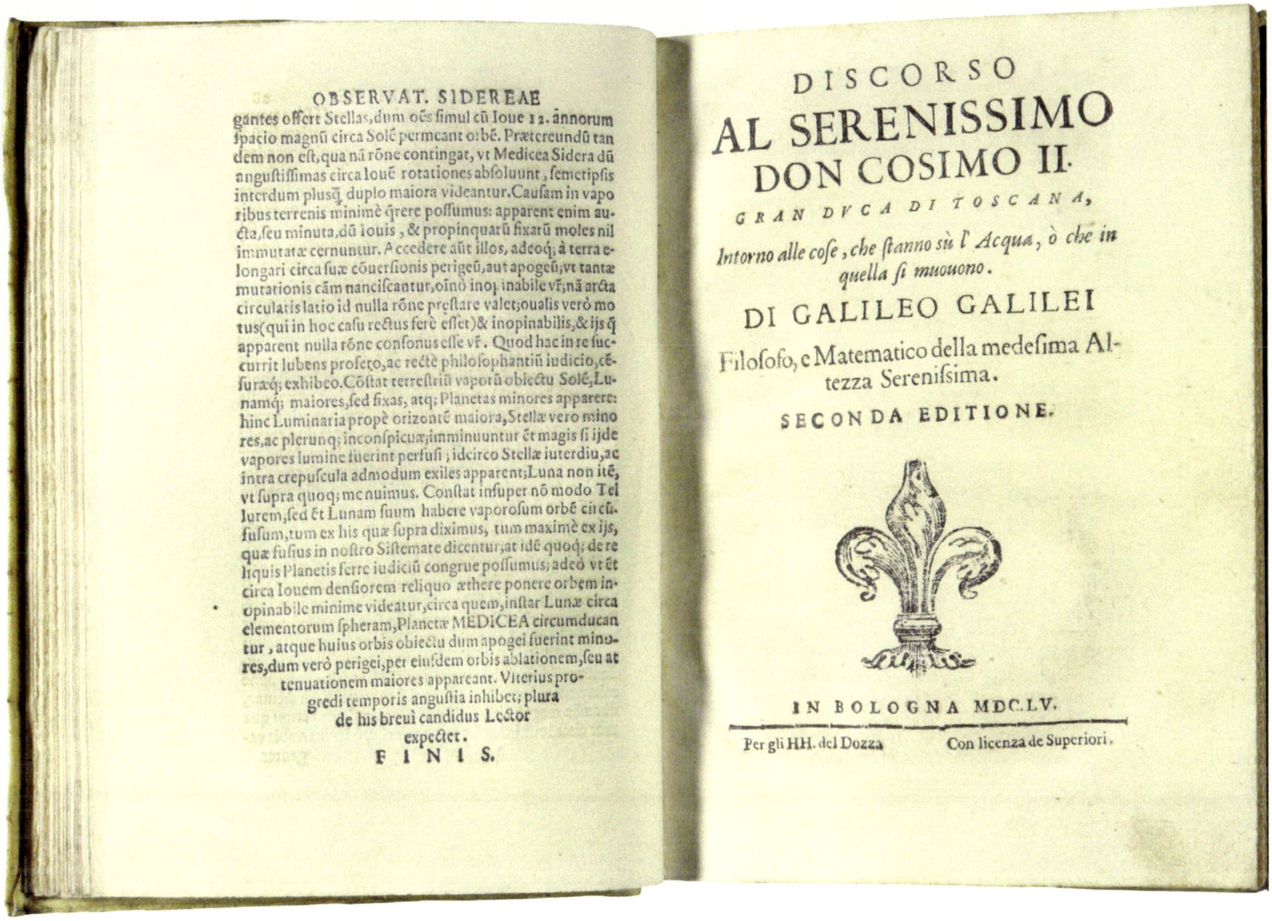

Galileo Galilei, Sidereus Nuncius, Venice 1610, p. 28v New York copy, bound with Galileo Galilei, Discorso, Bologna 1655, showing title page.

III

FROM THE DRAWINGS TO THE ETCHINGS

HORST BREDEKAMP

THE EXAMPLE OF THE LARGE CRATER

The rapid, energetic style of painting that was necessitated by the paper (see Brückle et al. p. 141) raises the further question of whether the drawings were simply illustrations for a book page empty of the etchings already printed in other copies, or if they provided the models for the etchings.

The basic design and the details already seem to indicate that the etchings responded to the drawings, rather than that the *disegni* responded to the printed pictures. Beginning on page 8r (Fig. 1), all of the drawings present forms that are more complex than were reproduced in the etchings, by using a mixture of precision and economy of effort (Fig. 2). The diagonal strokes in the lower nocturnal zone of the drawing have become the principle of the entire night side in the etching, in that parallel lines have been applied in the same direction. As a result of the haste of execution, the etched lines extending beyond the lower outline of the circle into outer space are not corrected. The second design element to note is the use of horizontal lines for the shading in the upper crescent. A third element is the characterization of the craters by curved lines, and finally the fourth concerns the zones of light on the night side of the drawing, of which at least the most marked ones have been adopted in the etchings.

A good example of how the etchings and drawings interlock is offered by the large crater that Galileo described several times and also drastically depicted in the etchings on pages 9v, 10r, and 10v, but that finds no correlation in reality. Reiterating his first excited

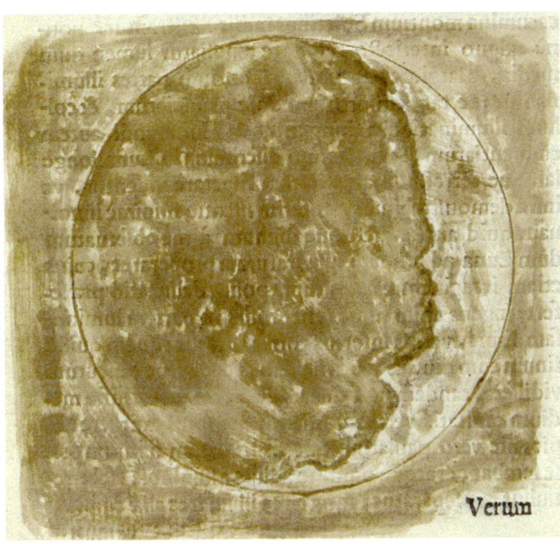

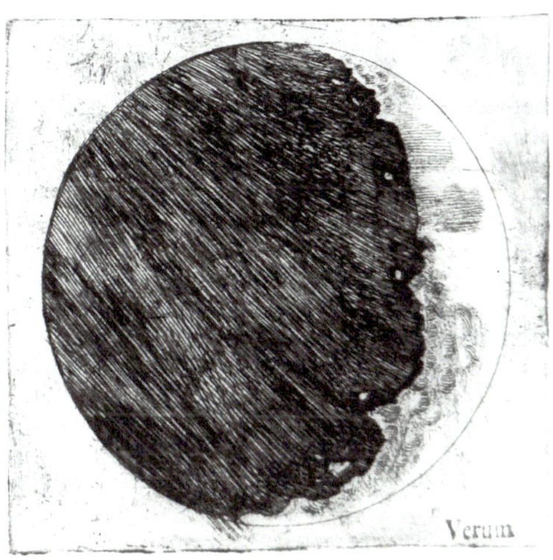

Fig. 1: Galileo Galilei, Sidereus Nuncius, Venice 1610, p. 8r, New York copy.

Fig 2: Galileo Galilei, Sidereus Nuncius, Venice 1610, p. 8r, Florence copy.

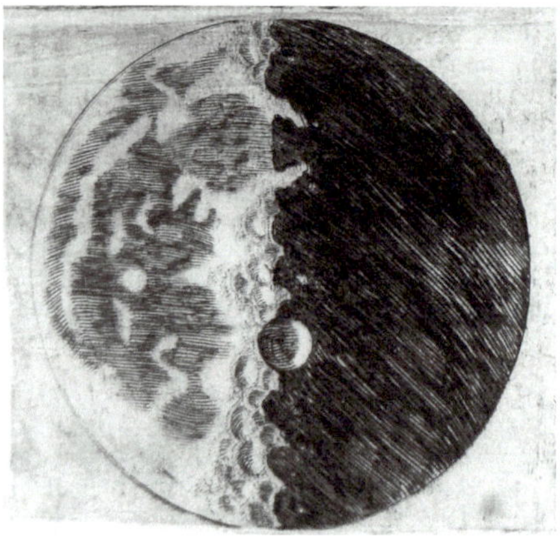

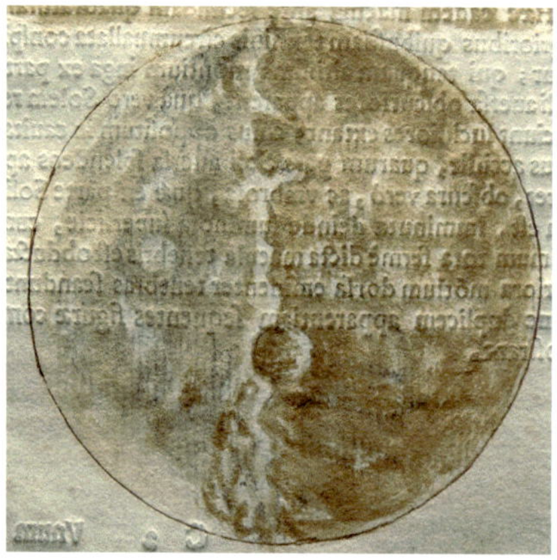

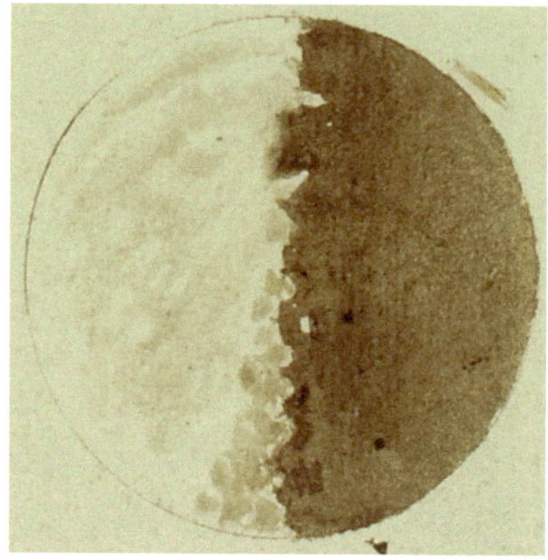

Fig. 3: Galileo Galilei, Sidereus Nuncius, Venice 1610, p. 10v, lower moon, Florence copy.

Fig. 4: Galileo Galilei, Sidereus Nuncius, Venice 1610, p. 10v, lower moon, New York copy.

Fig. 5: Galileo Galilei, moon phase "4", watercolor, 1610, BNCF, Gal. 48, fol. 28r.

depiction of this phenomenon,[1] in the *Sidereus Nuncius* Galileo underscored the astonishment and admiration that this crater inspired in him: "I definitely do not want to forget something else that I noticed, not without astonishment: Namely, approximately the middle of the moon is taken up by a hollow that is larger than all the rest and that shows a form of perfect roundness. I saw it around the time of both quarters and have imitated it, as well as I could, in the second position of the above depictions [*figurae*]."[2] Galileo was probably referring to the Albategnius, which would convey a breathtaking impression to any observer looking at it for the first time through a telescope comparable to Galileo's instrument.[3]

The impressive depiction offered at the bottom of page 10v (Fig. 3) characterizes this crater by leaving the glaringly-lit sunny side of the crater a bright crescent free of ink, while the shadow is formed by relatively thick horizontal lines whose gentle curves seek to portray the hollowing of the crater. In the corresponding drawing in the *New York copy*, a similar principle is applied (Fig. 4); upon closer examination, these shadow lines too show a slight curve (Fig. 6).

Initially seeming out of place, precisely such parallel lines depict the zone of the beginning of the nocturnal area in the Florentine drawings, and mainly in the moon numbered "5" (Fig. 4, p. 51). There, too, rather thick lines are drawn in parallel, forming a surface of deep darkness (Fig. 7). This detail indicates that the initially crude-seeming brush movements are also present in nuce in those of the Florentine drawings, on which Galileo was able to

1 *Opere*, vol. X, p. 275, lines 58–69.

2 *"Vnum quoque obliuioni minimè tradam, quod non nisi aliqua cum admiratione adnotaui: medium quasi Lunæ locum à cauitate quadam occupatum esse reliquis omnibus maiori, ac figura perfectæ rotunditatis; hanc prope quadraturas ambas conspexi, eandemque in secundis supra positis figuris quantum licuit imitatus sum"* (Galilei 1610, *Sidereus Nuncius*, p. 11r; *Opere*, vol. III/1, p. 67, line 3 to p. 68, line 4; translated after Galilei 1980, p. 93).

3 Gingerich 1975, p. 85; Casini 1984, p. 57.

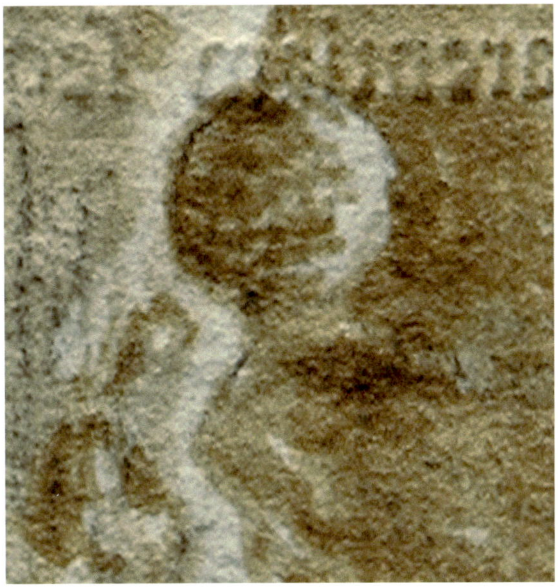

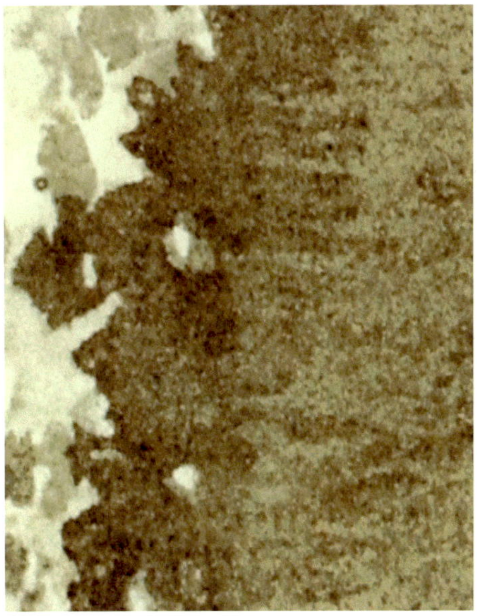

Fig. 6: Galileo Galilei, Sidereus Nuncius, Venice 1610, p. 10v, lower moon, New York copy, detail.

Fig. 7: Galileo Galilei, moon phase "5", watercolor, 1610, BNCF, Gal. 48, fol. 28r, detail.

expend more time and care. No imitator or forger could have repeated such a detail that is recognizable only on careful examination.

The large crater also permits a differentiated dating, though this can only be hinted at here. It is no longer present in the corresponding Florentine drawing showing equal day and night halves of the moon (Fig. 5). The difference from the depiction in the *Sidereus Nuncius* is so obvious that this drawing, as mentioned before, cannot have been created in the context of preparing and implementing the etchings. Not only because of its correction of the dimensions of the Albategnius, but also in the finesse of its technique, the Florentine watercolor differs from the bistre drawing of the *New York copy*. This is confirmed by the fact that Galileo never again depicted this huge crater in the sketches showing equal day and night halves of the moon.[4]

4 Bredekamp 2007, pp. 214–216.

SONJA KRUG / STEFAN SIMON

DIGITAL IMAGE ANALYSIS

Comparative digital analysis was performed on the five moon drawings from the *New York copy* of the *Sidereus Nuncius* and the copperplate etchings printed in the *Graz* and *Florence copies*. The sequences of figures for the drawings on pages 8r, 9v and 10v demonstrate the process of digitally enhancing the black material on the drawings, isolating that material and then overlaying it on the etchings. Using an image editing program (Image Pro® Plus by Media Cybernetics) the black material on the drawings and the letterpress-printed text were isolated and segmented from the brown drawing media (Figs. 1–3). As shown in Figs. 4, 6 and 8, the isolation was possible and the black material clearly overlaps the demarcation line in the drawings. To compare the drawn and printed images of each moon, the images were laid on top of each other with Adobe Photoshop® (Adobe Systems) (Figs. 5, 7 and 9). In these images, the printed text was used for alignment and scale. By decreasing the opacity of the image on top, the lines of each image were compared. This technique revealed an almost one-to-one correlation between the demarcation line in the drawing (as indicated by the black material) and that on the etching. In each case, the similarities and concurrences by far outnumbered the minor deviations in line. One precondition of this work was that all the photographs be at the same angle, plane-parallel, so that the drawings could be compared with the etchings. A slight aberration in the plane could distort the geometry of the illustrations. Rectification of the image plane was carried out by means of the program metigo® MAP 2.2 (Fokus GmbH Leipzig). To validate the method, photographs from the same detail of the *New York copy* taken with different photographic cameras and different parameters were rectified with metigo® MAP and compared in Adobe Photoshop®. The overlay of these photographs shows a good correlation in the layout of the lettering and moon figures with only small aberrations (Fig. 10). The distribution of the black material on the drawings can be followed as clearly in the rectified photographs as in the unrectified photographs (Fig. 11), the information is the same.

Significantly, the comparative digital analysis reveals that the isolated black material from the *New York copy* drawings follows the same pattern as the etchings, with or without rectification.

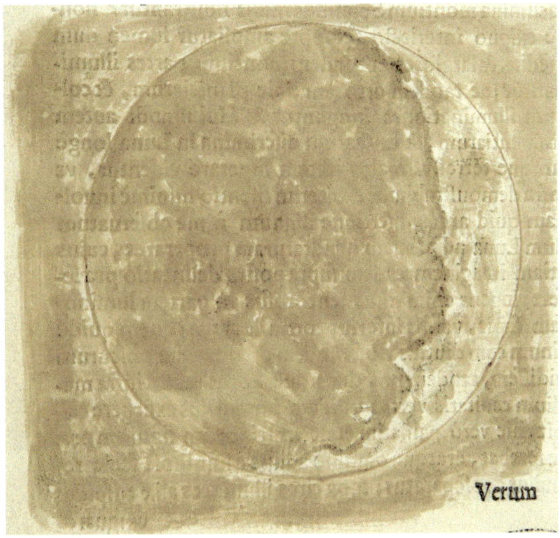

Fig. 1: Galileo Galilei, Sidereus Nuncius, Venice 1610, p. 8r, New York copy with black material enhanced.

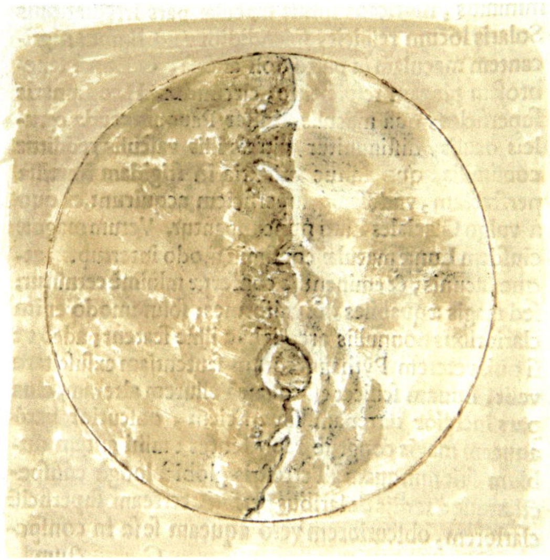

Fig. 2: Galileo Galilei, Sidereus Nuncius, Venice 1610, p. 9v, New York copy with black material enhanced.

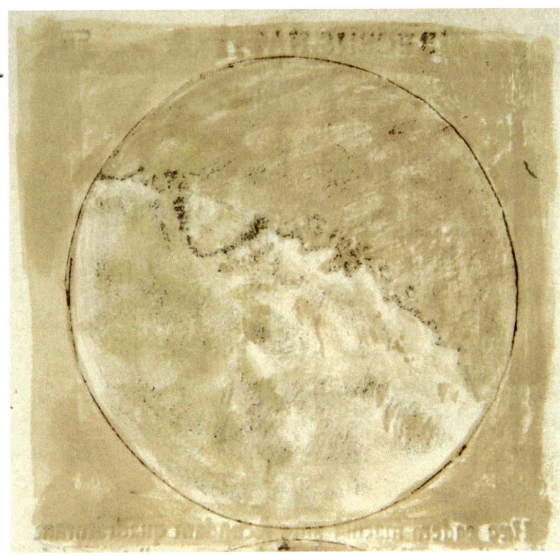

Fig. 3: Galileo Galilei, Sidereus Nuncius, Venice 1610, p. 10v, upper moon, New York copy with black material enhanced.

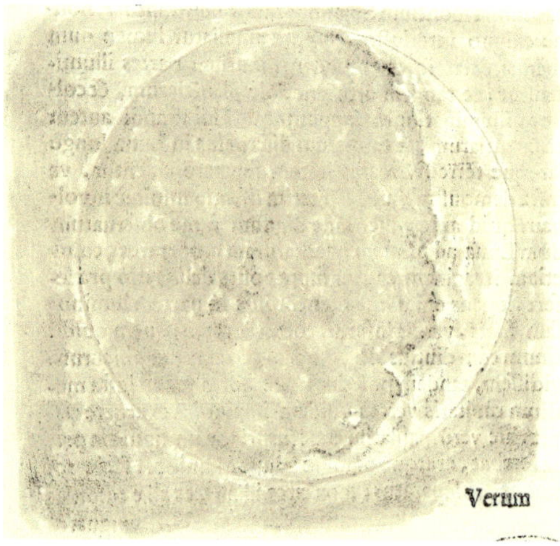

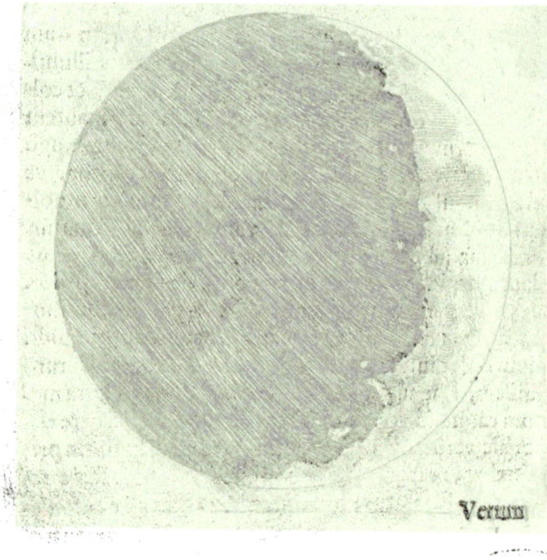

Fig. 4: Galileo Galilei, Sidereus Nuncius, Venice 1610, p. 8r, New York copy, black material isolated.

Fig. 5: Galileo Galilei, Sidereus Nuncius, p. 8r, the isolated black material form the New York copy overlaid on the etching from the Florence copy.

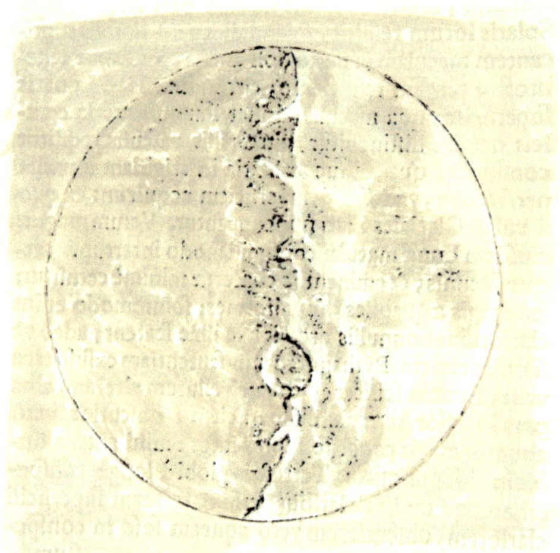

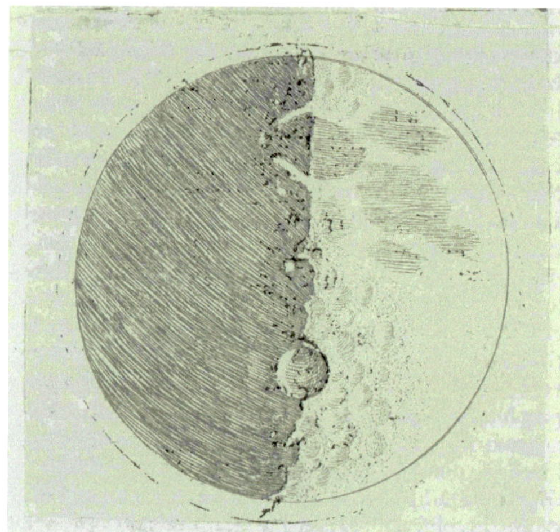

Fig. 6: Galileo Galilei, Sidereus Nuncius, Venice 1610, p. 9v, New York copy, black material isolated.

Fig. 7: Galileo Galilei, Sidereus Nuncius , p. 9v, the isolated black material form the New York copy overlaid on the etching from the Graz copy.

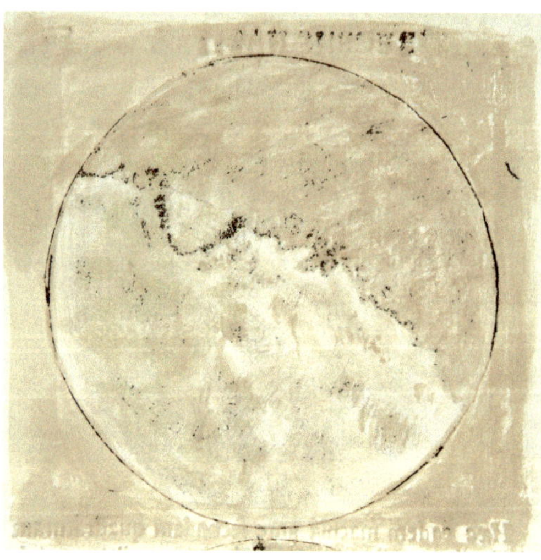

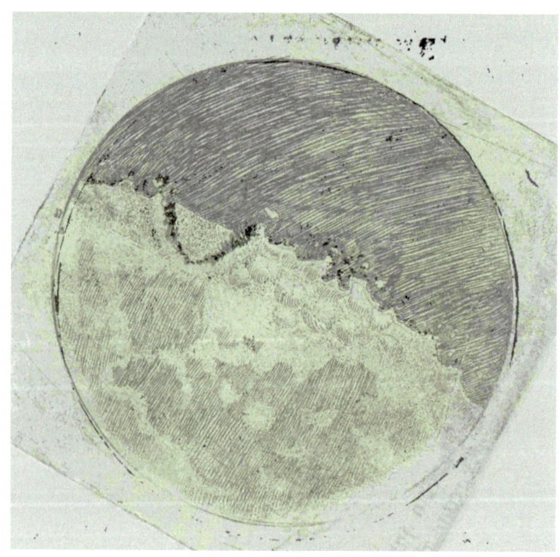

Fig. 8: Galileo Galilei, Sidereus Nuncius, Venice 1610, p. 10v, upper moon, New York copy, black material isolated.

Fig. 9: Galileo Galilei, Sidereus Nuncius, p. 10v, upper moon, the isolated black material form the New York copy overlaid on the etching from the Florence copy.

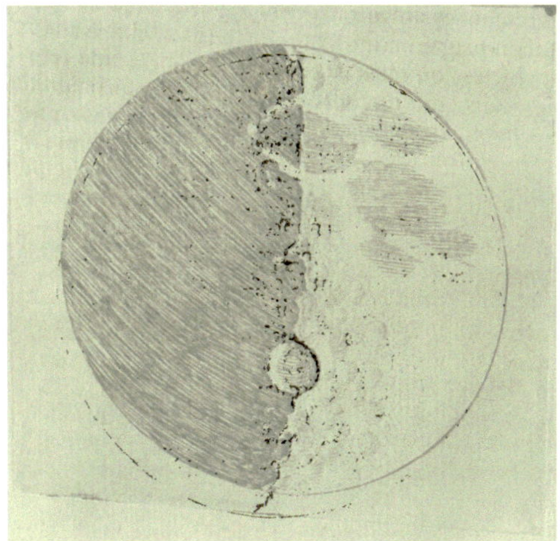

Fig. 10: Galileo Galilei, Sidereus Nuncius, p. 9v, the isolated black material from the New York copy, overlaid on the rectified photograph of the etching from the Graz copy.

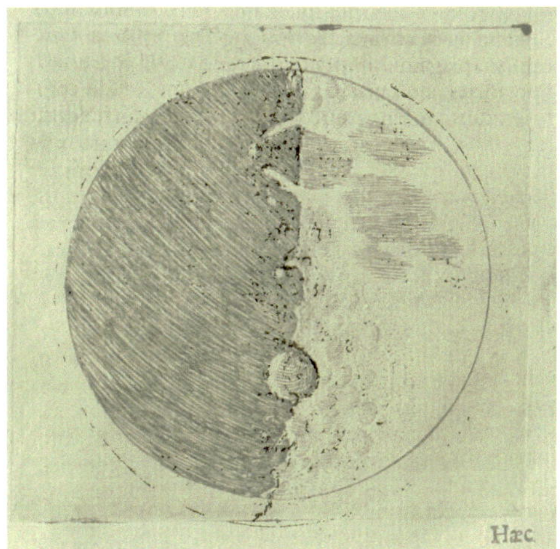

Fig. 11: Galileo Galilei, Sidereus Nuncius, p. 9v, the isolated black material from the New York copy, overlaid on the unrectified photograph of the etching from the Graz copy.

IRENE BRÜCKLE / RUTH TESMAR

MODELING THE TRANSFER

The mysterious black deposit on the drawings in the *New York copy* raised questions that were the subject of repeated discussion during the course of the study: what was its purpose, when and how was it applied? To review the evidence: the black color of the deposit results from the presence of carbon pigment, the most commonly used source of black then and today; the shiny appearance of the deposit suggests that it contains a significant amount of binding medium to keep the pigments adhered to each other and to the paper surface; the black deposit emphasizes the main features of the drawings; and, it only appears in the *New York copy* which has been identified as a working proof belonging to a book production stage that preceded the printing of the etchings.[1] Although we must consider it possible that the black material was placed on the drawings after March 1610, serving a later copying process of unknown destination, the historical context suggests that it may have had a function in the preparation of the etchings.

The way the black material was distributed over the drawings does not point to the slavish work of a copyist, although it is clearly connected with a purposeful action. To understand the relationship between the brown bistre washes and the black deposit, it is helpful to

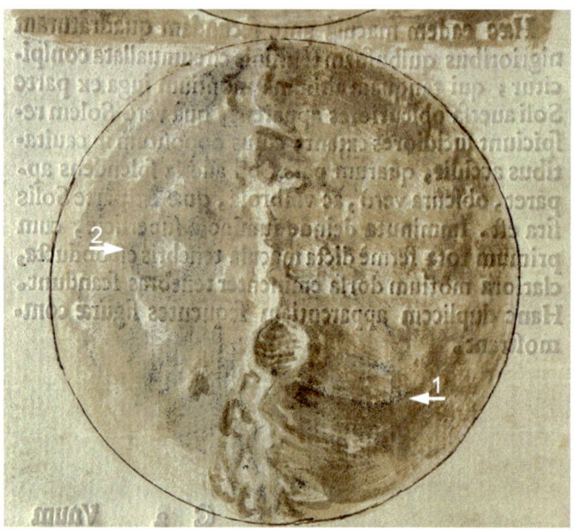

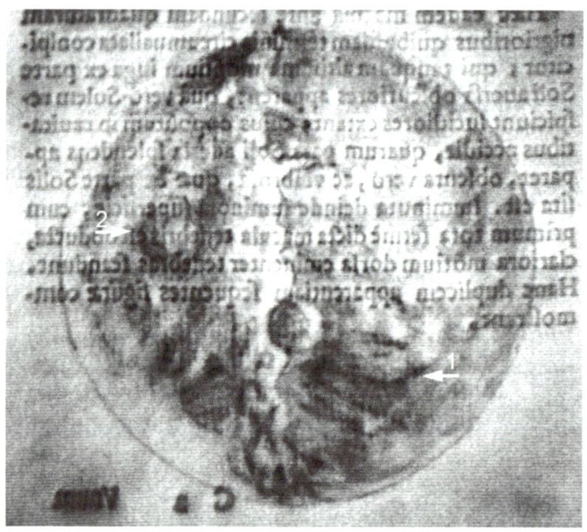

Fig. 1: Galileo Galilei, Sidereus Nuncius, Venice 1610, p. 10v, New York copy, lower moon; arrows point to selected areas of the black tracing.

Fig. 2: Galileo Galilei, Sidereus Nuncius, Venice 1610, p. 10v, New York copy, lower moon, infrared reflectography. The blackest features correspond to the black material on the drawing; arrows point to selected areas of the black tracing.

1 Bredekamp 2007, pp. 186–189; see Needham in vol. II of this publication, p. 173ff.

compare two photographs of the lower moon on page 10v, one taken under visible light (Fig. 1), the other taken by infrared reflectography, which makes the carbon-based deposit more distinctly visible (Fig. 2).[2] In setting down the black material, attention was selectively paid to key features of the drawing, while other features were virtually ignored. The moon outline, in particular, is barely indicated, perhaps because the moon radius in the etchings was to be made considerably smaller. Moreover, the entire night side of the moon received little attention with the exception of the area to the lower right of the large crater (arrow 1). Even here, however, the shading created by broad, angled bistre brush strokes was treated rather summarily; a few black marks were sufficient to identify this shadow zone. Looking at the corresponding etching of the lower moon on page 10v, a series of largely parallel lines indicates that on its night side, the moon surface is concealed (Fig. 3). This allows us to infer the significance of the corresponding areas in the drawing, assuming that continuous intention, if not authorship, was sustained through the three stages under discussion here. From drawing to black deposit to etching: the night sides of the moons are largely to be read as continuous shadow, interrupted only by a few topographical markings.

The fluid brush strokes on the day side of the moon (Fig. 1) received more attention than the night side during the black tracing action, as indicated by the more ample marks seen in the IR reflectography image (Fig. 2). Individual features of the wash drawing, such as the light roundel in the upper half (arrow 2) and the arcs at the bottom, are indicated in black and reappear in the etching. The day-and-night divide is traced by a relatively continuous black line that follows the bistre line quite closely. The significance of the moons' central divide – a major concern for Galileo (see Bredekamp p. 49) or any other person involved in their faithful rendering – is indicated by what seems to have been a correction made in the black material (Fig. 4). The drawing on page 9v shows the horn of a shadow zone traced in

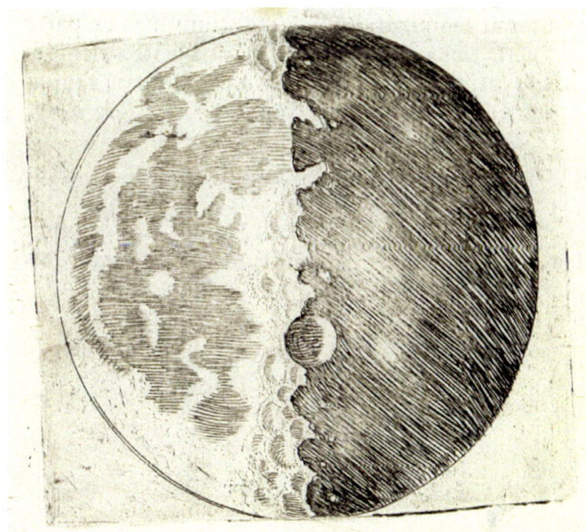

Fig. 3: Galileo Galilei, Sidereus Nuncius, Venice 1610, p 10v, Graz copy, lower moon.

2 The black deposit, presumably composed of fully carbonized organic material, can be differentiated from the soot of incompletely burned wood, of which bistre ink is made. Christopher Schmidt, Gemäldegalerie, Staatliche Museen zu Berlin, is thanked for taking the infrared reflectography image. The image was generated with a Hamamatsu camera and an IR Vidicon tube N2606-06 detector (spectral response 400–2200 nm, peak response ca. 1700–2000) and was photographed from the monitor.

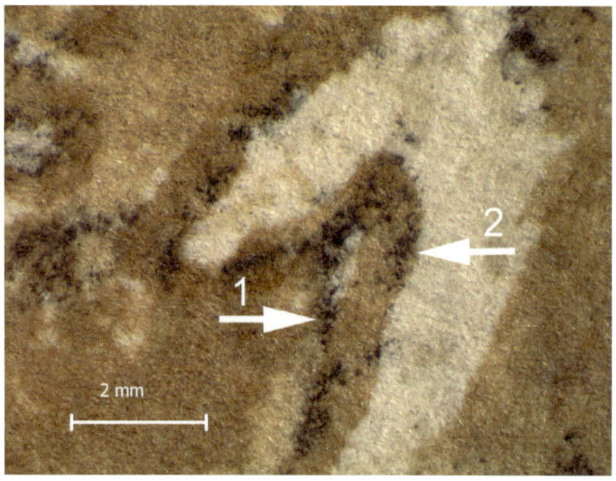

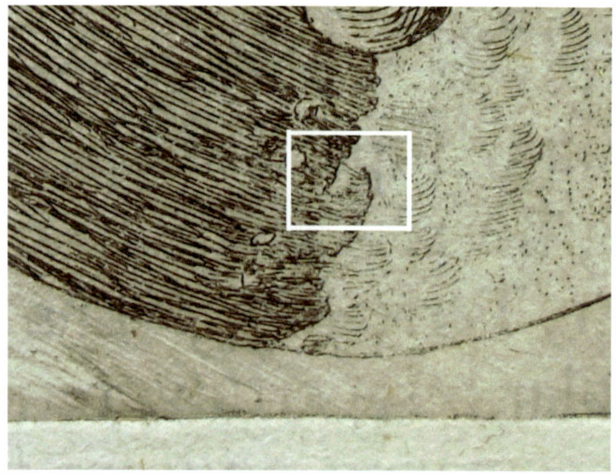

Fig. 4: Galileo Galilei, Sidereus Nuncius, Venice 1610, p. 9v, New York copy, detail of the corrected outline of the shadow zone.

Fig 5: Galileo Galilei, Sidereus Nuncius, Venice 1610, p. 9v, Venice copy, detail of etching; rectangle corresponds to area seen in Fig. 4.

black, however, there are two adjacent marks along the right edge of this shape. The line to the left (arrow 1) is more thickly drawn than the one the right (arrow 2), and it also matches the shape of the horn in the etching more closely (Fig. 5), as evidenced by the digital image analysis (Fig. 7, p. 96). If the drawing was used for the creation of the etchings, only the thick black line would have left a usable mark in a transfer process.

Overall, the black material on the moons was applied both in a judicious manner, suggesting a certain familiarity with their complex design, and with an ease of execution. Approximating the drawing directly *on* the drawing required dexterity and a confidence in picking out only important features if the original drawing was not to be spoiled – which was obviously a concern here. This attention was focused on the day-and-night divide and the dayside features, as supported by the digital image analysis (see Krug and Simon p. 94ff.).

The fact that the etchings do not match the design of the drawings in every aspect may require more than one explanation. First, some marks in the drawings are side effects caused by technical particularities of the drawing process that were never meant to be transferred into etching. Further, some final adjustments may have been made to the moons' salient features when they were transferred to the etching plates. Finally, translating the broad, fluid brush strokes into the linear system of etching necessarily resulted in a different form of representation, in which simplifications as well as elaborations were possible. If the black material is related to the preparation of the etching plates, we can assume that the divergences were tolerated or maybe even intended while the plates were prepared.

Turning to the method of its application, the deposit lacks any of the typical tool marks associated with common drawing, writing or printing processes, such as those seen elsewhere in the *Sidereus Nuncius*: letterpress printing struck deep impressions in the paper; the quill was used forcefully when Galileo signed the *New York copy*, creating a visible depression (Fig. 1, p. 31). Drawing materials that could have been used to create the black deposit, such as oiled charcoal or black chalk, would have impacted the paper physically, leaving distinct marks on its surface. Both would also be identifiable by the morphology of their deposited particles.[3]

3 Dietz and Brückle 2010, pp. 54–56.

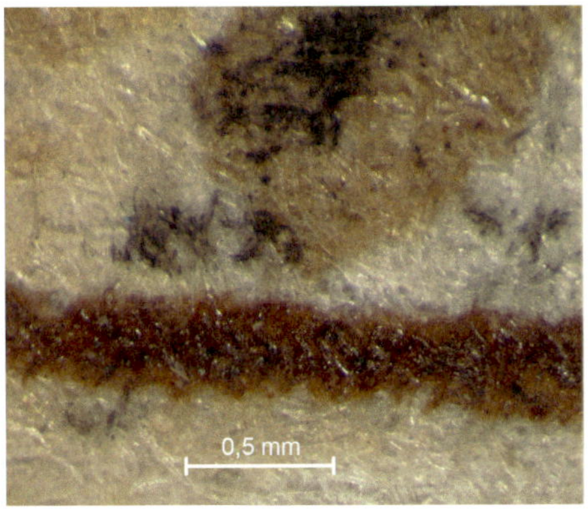

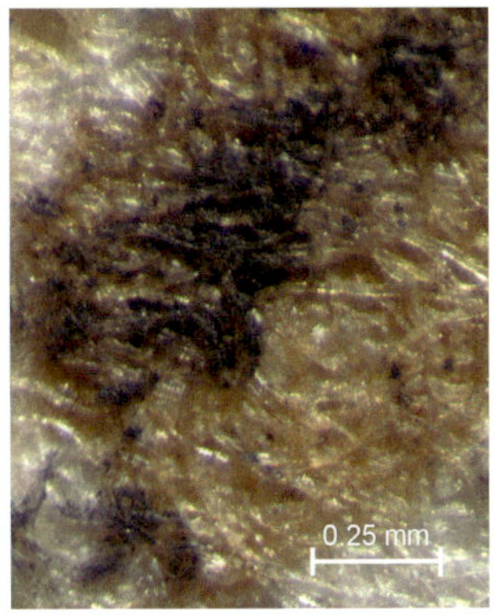

Fig. 6: Galileo Galilei, Sidereus Nuncius, Venice 1610, p. 9v, New York copy, details; small non-blackened area above the moon outline (left); adherence of the black material to the high points of the paper (right).

Instead, an unusual and intriguing pattern emerges. The black material only lies on the high points of the bistre drawing, as can be seen in the detail of page 9v (Fig. 6). Because the bistre ink was fully absorbed into the paper, it is the paper surface in these inked areas that determined the pattern of the black material. The paper received the black material preferentially on the protruding areas (see also Fig. 5, p. 47) and the recesses of the paper surface were not blackened. This indicates that the black material was deposited with only light pressure, and that, it had to be sticky enough to be transferred by such a comparatively delicate action. This detail of the moon on page 9v also bears evidence that the black material was applied after the drawing was completed. Where the pen had delivered the ink outline of the moon, the paper surface was depressed enough that the black material skipped over the line, leaving a small non-blackened area next to it. Furthermore, the impression created by the letterpress printing remained undiminished by the application of the black material (Figs. 7 and 8). The areas raised by the text printed on the other side were prefer-

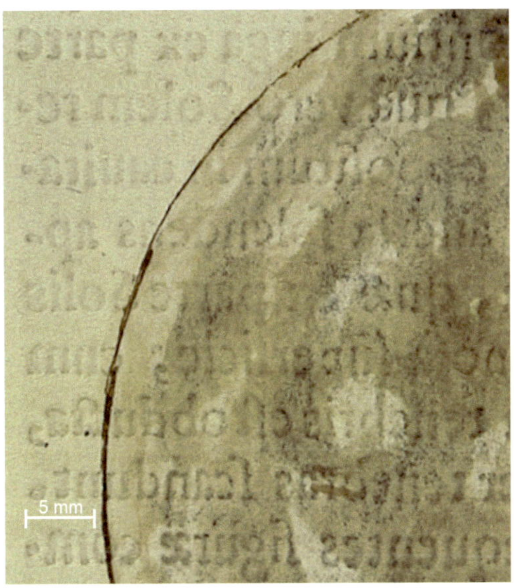

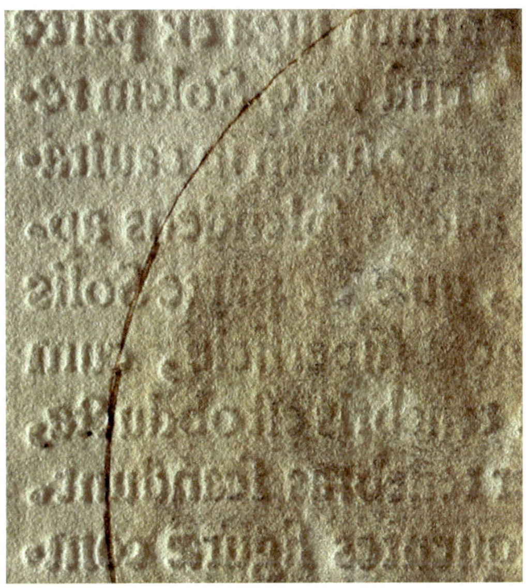

Fig. 7: Galileo Galilei, Sidereus Nuncius, Venice 1610, p. 10v, New York copy, detail.

Fig. 8: Galileo Galilei, Sidereus Nuncius, Venice 1610, p. 10v, New York copy, detail in ranking light showing black material adhered to the high points of the paper.

Fig. 9: Modeling the transfer, end of step 1. The drawing (a) lies face up on the table, and another sheet of paper (b) is lifted from it. This second paper has just received a tracing of the drawing (historically, a thin handmade paper would have been used instead of the tracing paper seen in this model).

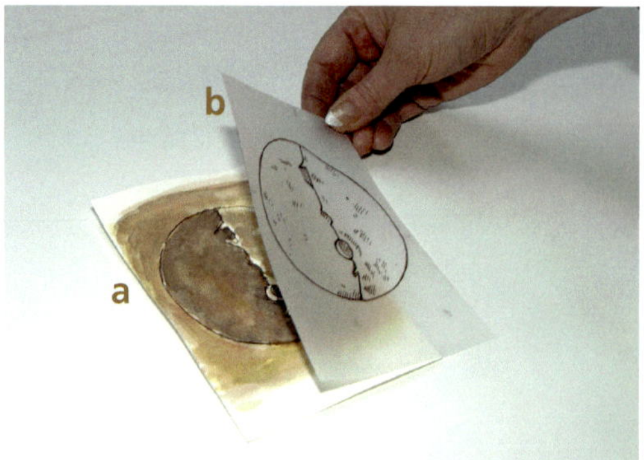

Fig. 10: Modeling the transfer, step 2 in progress. A carrier freshly coated with a sticky black ink film (a) receives the original drawing face down (b) and the traced drawing also face down and aligned (c). The drawing is placed so that it touches the printing ink only lightly and selective pressure can be applied.

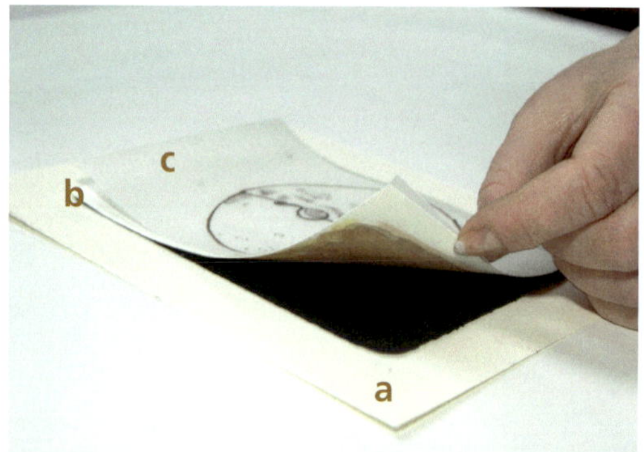

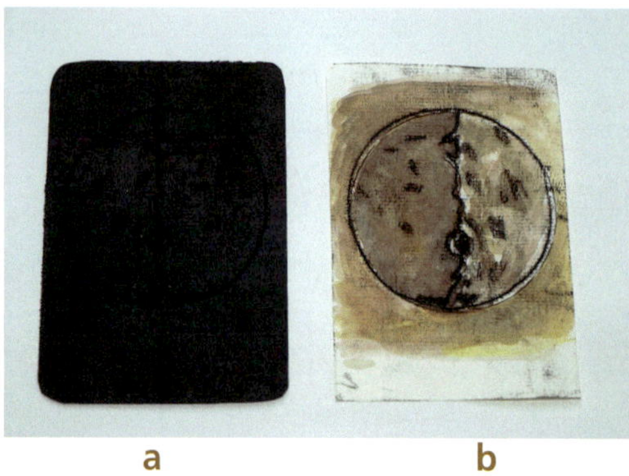

Fig. 11: Modeling the transfer, after completion of step 2. The drawing (b) shows a deposit of the black printing ink in key design areas that had been pressed against the sticky black film of printing ink (a).

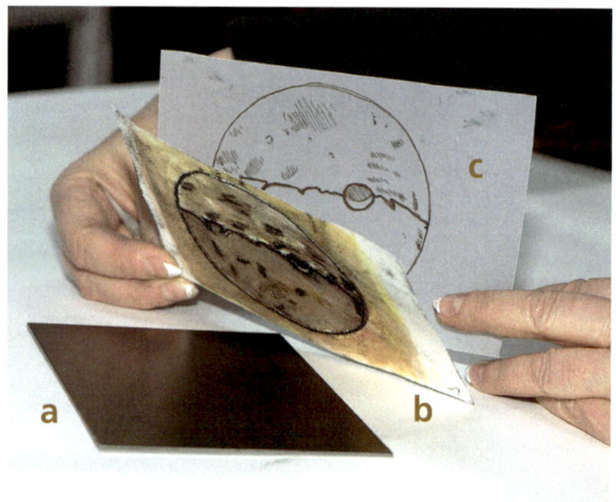

Fig. 12: Modeling the transfer, preparation for step 3. The copper plate (a) receives the drawing (b) placed face down so that the fresh black material can touch the etching ground on the copper plate. The orientation of the tracing (c) matches the drawing.

entially blackened, creating a secondary pattern in the black material on the drawing. The features described here are common to all the drawings.

Considering the evidence, it is likely that the black material was involved in the method of transferring the drawings to the etching plates during book production in the printing shop. At that point, the printed folios that now display the bistre drawings would have been separate sheets and easy to handle during the following hypothetical transfer process. In the first step, we assume that a blank piece of paper was laid on the drawing to trace it (Fig. 9). If the blank paper was not translucent enough to see through, both papers could have been held against a glass pane, allowing light to be transmitted. Galileo was well aware that paper could be conveniently used this way from his observations through the

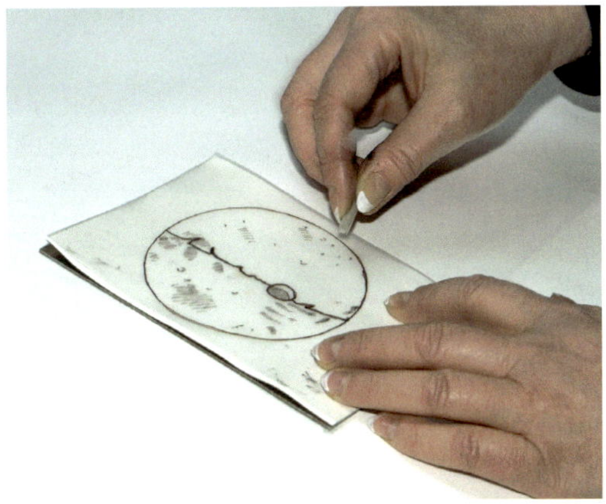

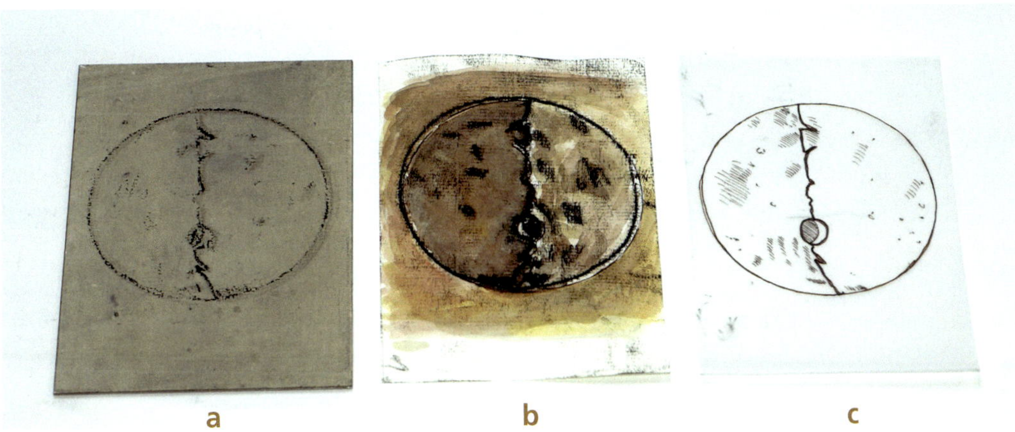

a b c

Fig. 13 (above right): Modeling the transfer, step 3 in progress. The drawing (not seen) and tracing lie face down on the copper plate. The tracing is used for orientation to apply the pressure that will transfer the design from the drawing to the copper plate.

Fig. 14: Modeling the transfer, after completion of step 3. The black deposit on the copper plate (a) is a mirror image of the drawing (b). The drawing shows a black deposit, similar to that on the original drawings in the New York copy. The tracing (c) aided the transfer process, ensuring that it could be done with minimal impact on the drawing.

telescope.[4] In the second step, the drawing would have been turned face down and the tracing also turned face down on the drawing's verso, with the images aligned. Then, both papers would have been put on a freshly prepared film of the black material, which would be sticky from a binding medium such as linseed oil (Fig. 10). The drawing thus lightly resting on the black material would now undergo the most delicate part of the procedure. Using the tracing for guidance, pressure would have been applied to bring the original drawing selec-

4 Because the view through the telescope reversed the image, he advised that the resulting drawing should be viewed from the verso. Held up to the light, the drawing could be viewed through the translucent paper so that the stars could be studied in their correct orientation. See Bredekamp 2007, p. 206, footnote 37.

tively in contact with the sticky black layer, causing bits of the material to offset locally on the face of the drawing. The modern reconstruction shows that this form of transfer is possible (Fig. 11). The replica drawing shows a black deposit only where pressure was applied. In the third step, while the black material on the drawing was still sticky, it would have been placed face down on a copper printing plate, with the tracing on the verso aligned with the drawing (Fig. 12). Again, the tracing would show where pressure was needed to transfer the black material from the face of the drawing to the copper plate. The tracing would have also protected the verso of the drawing from physical marks caused by this manipulation (Fig. 13), a familiar consideration to seventeenth-century artists.[5] A light-colored etching ground would have already coated the copper plate so the transferred lines would be easily visible, though a good deal of the sticky black substance would have remained on the drawing (Fig. 14).[6] These transferred marks would have provided sufficient guidance for working out the design before the etching process. This hypothetical process shows that, in principle, a transfer method using the means easily available at the time could have functioned well in preparation of the etchings and resulted in the deposit of the black material on the face of the drawings. Given the urgency of this printing job, it is quite imaginable that a somewhat unconventional method of transferring the drawings to the etching plates was used. Only slight variations in the established procedure would have been needed to accomplish the task as described here. The following chapter discusses how the etchings themselves bear evidence of hurried processing and printing, which supports the notion that time was at a premium, making creative operations a likelihood, if not a necessity.

5 The most famous seventeenth-century treatise on etching, published in 1645 by Abraham Bosse, describes how the transfer of a drawing to a copper plate can be accomplished by using two papers if the original drawing is to be preserved unaltered. See Bosse 1689, pp. 24–25. The same work also lists a collection of recipes for white grounds involving white wax, asphaltum, colophonium and mastic that required the additional step of blackening for regular plate processing.

6 Alexander Browne's 1669 *Ars pictoria* mentions preparing a copper etching plate with a white wax ground to receive black ink from a tracing, in order to preserve the original drawing.

THERESA SMITH / IRENE BRÜCKLE / OLIVER HAHN / TIMO WOLFF

PRINTING THE ETCHINGS

Examination of the printed moon images indicates that they were created in haste without the attention to fine execution befitting the otherwise expertly prepared lead type letterpress printing. Haste is evident in many details: the plates do not have the same dimensions, they are not square, the bevels are not even, and the blank plate areas are not cleanly polished. There are small random dots, deep scratches and even fingerprints (Figs. 1 and 2). Speed may have been the deciding factor in even the choice of technique. Instead of engraving, the more customary technique used to create book illustrations, the faster etching process was chosen. In this process, the copper plate was covered with a layer of material, a ground, to resist the corrosive action of an acid bath and thus protect the plate. Where the image was scraped through the ground with a needle or stylus, the acid contacted the bare metal and etched into the plate. In our case, the ground must have been quite soft because a fingerprint is recorded in the margin of the upper etching on p. 10v. The ground was lifted where it was accidentally touched, and the fingerprint was etched into the copper plate with the moon, and so repeats in every impression (Figs. 2–4). Abraham Bosse's comprehensive 1645 collection of etching recipes discusses both soft and hard grounds for preparing copper plates. He mentions that once a soft ground is applied, it requires particularly careful handling

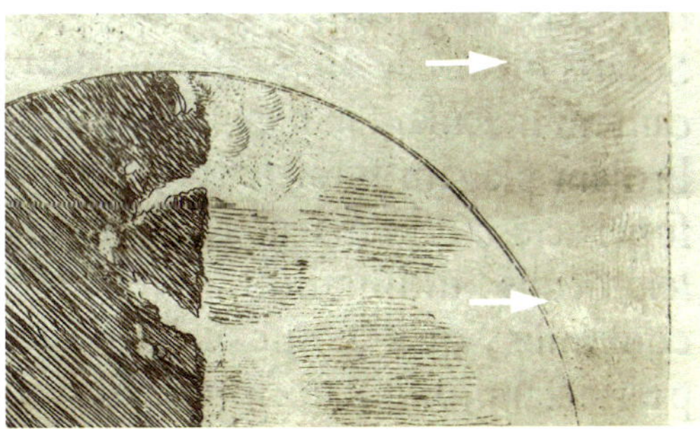

Fig. 1: Galileo Galilei, Sidereus Nuncius, Venice 1610, p. 9v, Venice copy, detail. Ink lifted from the plate by the printer's fingers shows as light areas in the etching.

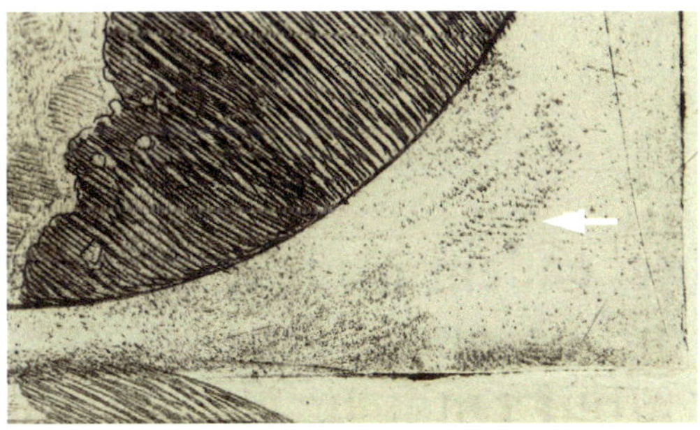

Fig. 2: Galileo Galilei, Sidereus Nuncius, Venice 1610, p. 10v, Washington copy, detail of upper moon. The arrow shows a fingerprint etched in the image which appears in every impression.

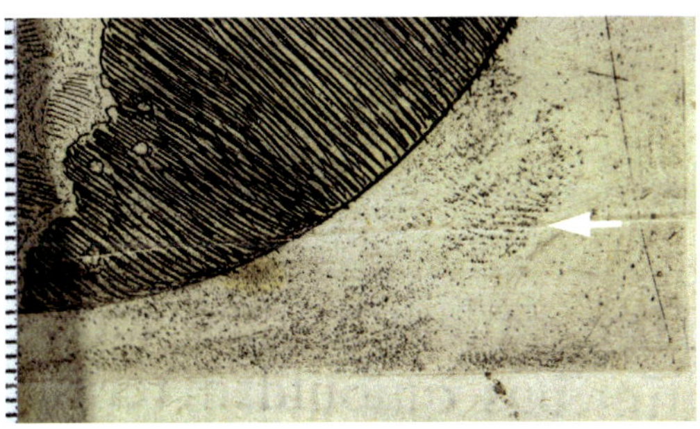

Fig. 3: Galileo Galilei, Sidereus Nuncius, Venice 1610, p. 10v, Graz copy, detail of upper moon (scale at left in mm).

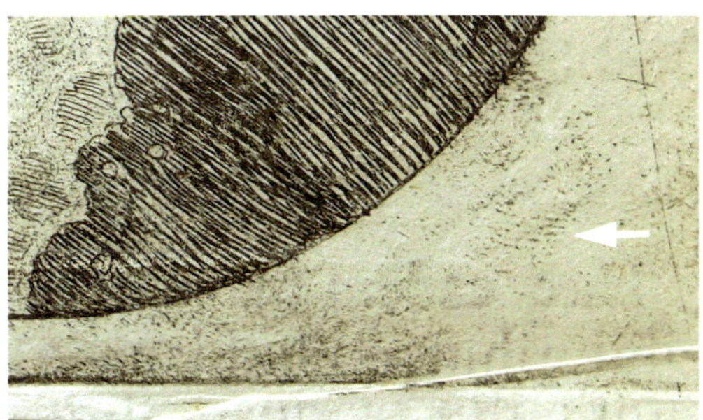

Fig. 4: Galileo Galilei, Sidereus Nuncius, Venice 1610, p. 10v, Venice copy, detail of upper moon.

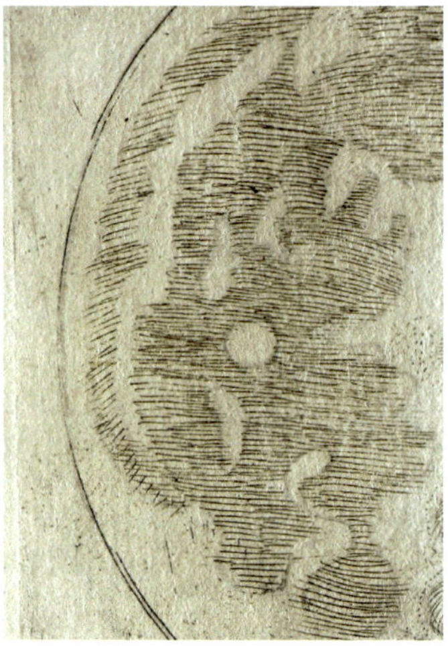

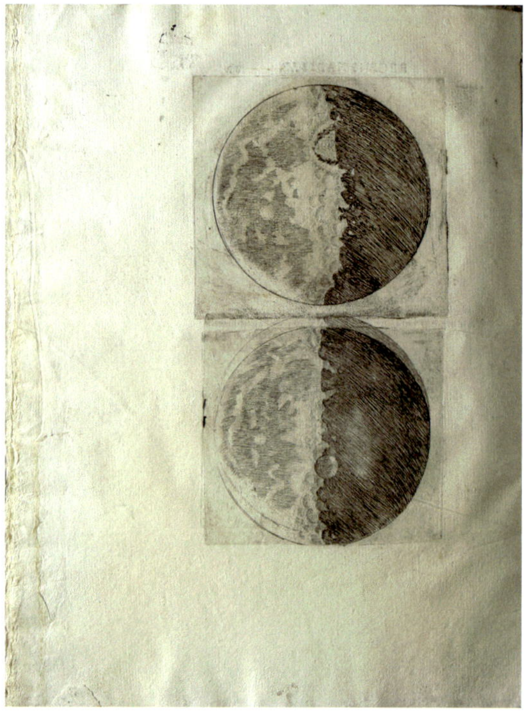

Fig 5: Galileo Galilei, Sidereus Nuncius, Venice 1610, p. 10v, Munster copy, detail of outline of lower moon.

Fig. 6: Galileo Galilei, Sidereus Nuncius, Venice 1610, p. 10v, Washington copy.

so the ground is not disturbed or marred.[1] It is therefore quite conceivable that a soft ground was used.

The plates were probably etched in only two stages to differentiate the lighter, day side from the darker, night side of the moons. Unlike the *New York copy* drawings, in the etchings the moon outlines may have been drawn with a compass, but the lines do not match up in several places (Fig. 5), with a similar end result (see Fig. 1, p. 42). The first etching bath would have established all of the lines that make up the moon images. A second layer of ground would then have been applied to protect the lighter areas: the moon outlines and the light mottling of the day sides. The second etch would have enhanced the dark areas on the night side. No attempt was made to eliminate the scratches or fingerprints created during the etching process. In addition, the plates bear marks of extensive, though incomplete polishing, indicating that they may have been used as engraving or etching plates before they were used for the *Sidereus Nuncius* (see Meier p. 113). Even after the completion of the etching process, the fingerprints and miscellaneous dots and scratches could have been eliminated from the plate by burnishing. That this was not done gives further indication of the speed with which these plates were produced.

It must remain speculation whether the plates were etched or even drawn by Galileo himself. Given the complexity of the etching process, it seems plausible that they are the work of a more experienced printmaker. If, however, Galileo was directly involved in any step of the etching preparation, he would have most likely taken charge of drawing the

1 Bosse 1689.

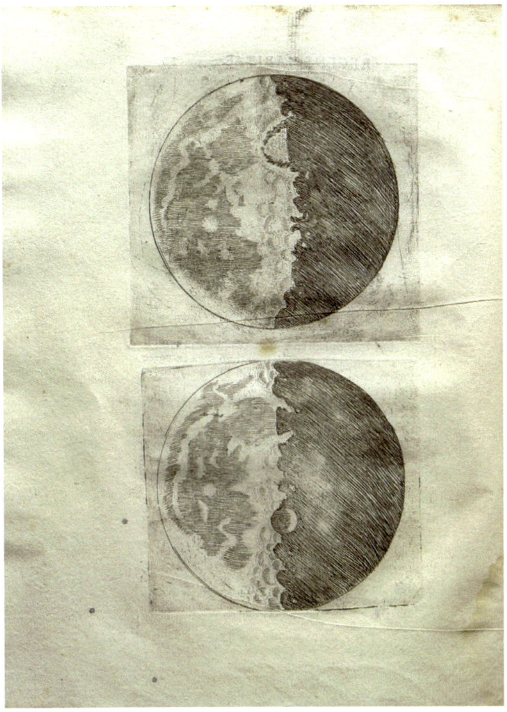

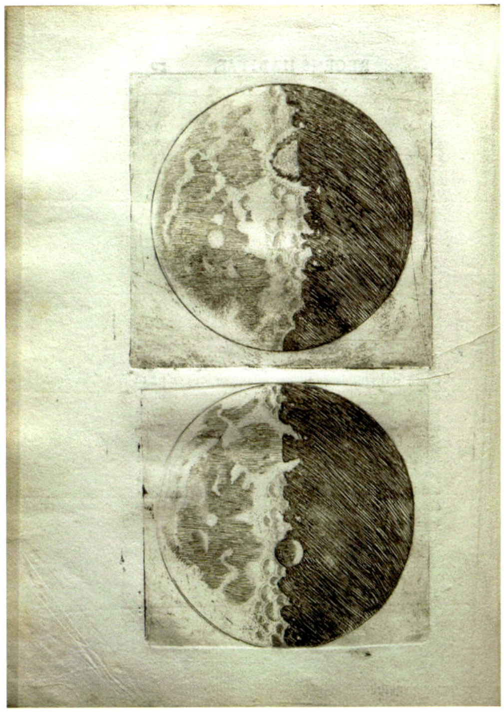

Fig. 7: Galileo Galilei, Sidereus Nuncius, Venice 1610, p. 10v, Graz copy.

Fig. 8: Galileo Galilei, Sidereus Nuncius, Venice 1610, p. 10v, Munster copy.

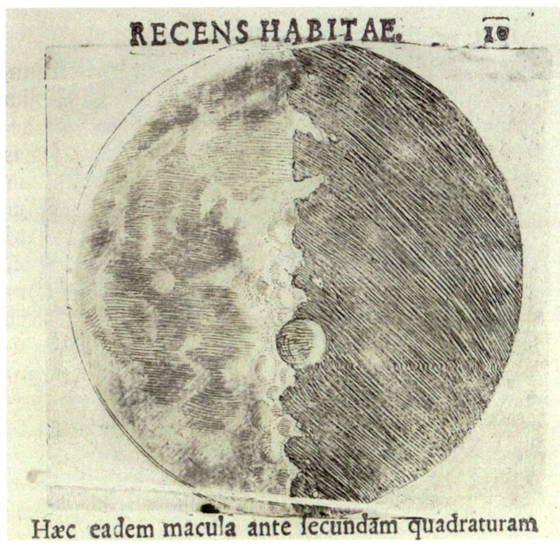

Fig. 9: Galileo Galilei, Sidereus Nuncius, Venice 1610, p. 10r, Graz copy. Creases correspond to those seen on p. 10v (see Fig. 7).

images on the plates, leaving the other, more technical details of the etching process to the printer.

The readied plates were printed with considerable haste. They were cleaned and reinked for each impression and variations in the ink deposit created shades of plate tone that differ widely from one impression to the next. Plates carefully wiped to remove excess printing ink from the non-image areas would retain a blank background around the image. In the *Venice copy*, fingerprints were also recorded where the printer's hand smudged the inked plate (Fig. 1). Futhermore, a comparison of p. 10v in the *Washington* and *Graz copies* (Figs. 6 and 7) shows the lack of uniformity in these two early printings. The basic vertical and horizontal placement of the plates was determined by the alignment of the images with the

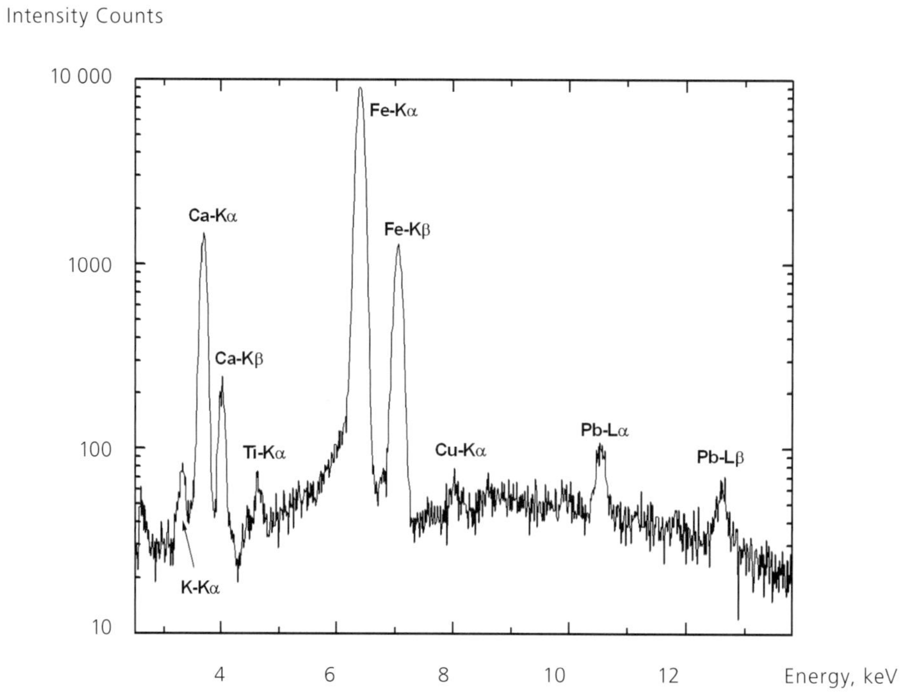

Fig. 10: Micro X-ray fluorescence spectrum of the fingerprints on p. 21r of the Washington copy. The elements potassium (K), calcium (Ca), and iron (Fe) result from the paper, whereas the lead (Pb) is likely due to the printer handling lead type before touching the paper.

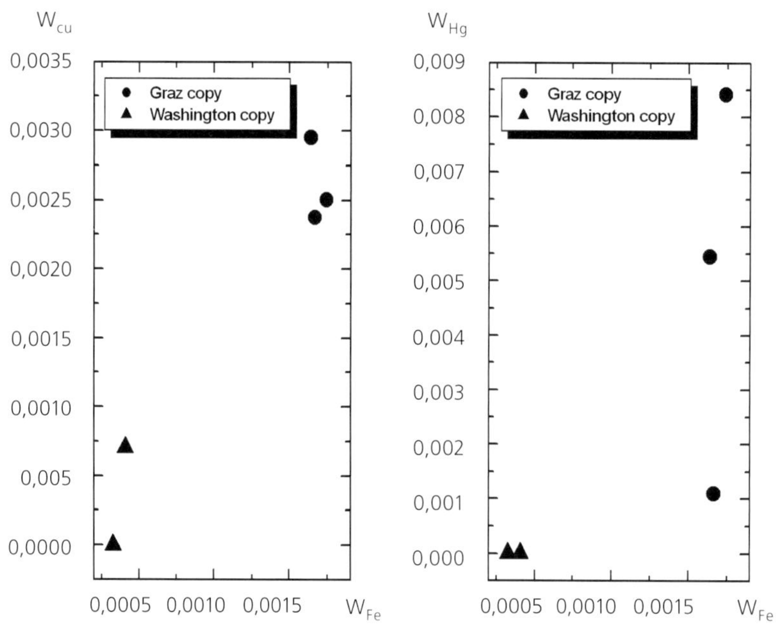

Fig. 11: Concentrations of copper (Cu) and mercury (Hg) in relation to iron (Fe), in the printing inks of the Graz and Washington copies.

text column. After it was first set down, but before printing, the lower plate in the *Washington copy* had to be repositioned to avoid overlapping the upper etching. This is why a faint first impression is visible slightly above the final impression. It indicates that the plates were printed in two passes through the press, as the overlap of the two images would not have occurred if the upper plate had been in place. In the *Graz copy*, the plates were printed slightly askew (Fig. 7), and the upper etching has creases where folds in the paper were partly eased open after the plate on p. 10r was printed (Fig. 9). These earlier printings also show that stray printing ink was deposited around the plates while they were positioned on the paper. In comparison, the later *Munster copy* is a special exemplar on fine paper, which shows better alignment between the two plates on p. 10v (Fig. 8), though there is also printing ink around the edges of the plates.

Several fingerprints were deposited in the lower margin of p. 21r of the *Washington copy* (see Fig. 2, p. 159). Micro X-ray fluorescence analysis showed that the fingerprints contain lead (Pb), probably from handling the metal type. It may be that these marks stem from the fingers of the letterpress printer (Fig. 10).

The printing inks of the etchings in the *Washington* and *Graz copies* were also examined using micro XRF, which showed that they differ in composition with regard to their trace elements. Considerable amounts of mercury (Hg) were found, in addition to iron (Fe) and copper (Cu), in the *Graz copy*. The ink of the etchings in the *Washington copy* contained only iron and copper (Fig. 11). Variations in the composition of printing inks can be plausibly expected with a change in batches of ink or the extension of an existing batch through the addition of replenishing components. Analysis of the inks of more copies might reveal patterns in ink composition that would enhance our understanding of the significance of such variations.

HANS JAKOB MEIER

GALILEO'S BREAK WITH THE VENETIAN WOODCUT

When Galileo Galilei had the *Sidereus Nuncius* printed in Venice, he chose the city that had developed the most significant tradition of classical lead letterpress printing within Italy, and whose unequaled achievements in the field of the illustrated book had set high standards since 1480. In Venice, through Aldus Manutius (1440–1515), the towering publisher of ancient and illustrated humanistic literature, the seamless integration of letterpress and woodcut for the combination of text and image had grown to great perfection and its typeface had become almost canonical.[1] When the tradition of the Venetian woodcut threatened to become conventional around 1510, Ugo da Carpi pioneered experiments with multicolored woodcuts and Titian (Tiziano Vecellio) discovered the woodcut for his conception of large-format pastoral landscape.[2] With a whole suite of elaborately worked woodcuts, Titian was the first Venetian artist to completely plumb this medium's artistic possibilities for constructing a wide variety of large-format line concepts.[3] For almost two decades, he had various masters work in woodcut. Several of these woodcut artists trained in Titian's new line conception also worked at the same time as printers and illustrators.[4] In this way, Titian restored such a fresh artistic potential to the Venetian woodcut in book printing that this medium predominated over engraving for illustrating texts in Venice until the early seventeenth century. Outside of the genre of book printing, however, the woodcut increasingly lost ground to engraving in every important European center of printing.

Galileo's decision to have the *Sidereus Nuncius* printed in Venice, but to use the new and, in Venice, still unfamiliar technique of etching instead of woodcut for the depictions of the moon drastically abrogated the traditional technical toolbox that Venetian book printing had developed for the interplay between text layout and illustration. Uncertainties in the registration between text layout and image in the printing of the *Sidereus Nuncius*, as well as differing printing plate formats on the pages with double illustrations, indicate the printer's unfamiliarity with the combination of lead typesetting and etching. One of Galileo's probable motives for choosing etching, despite foreseeable visual losses in the appearance of the printed page, was initially his hurry to get the book in print, as discussed in other sources.[5] A particular advantage of an etching is that it takes much less time to execute than a similar work in woodcut or engraving. A second consideration was doubtless Galileo's observation of the special graphic aptness of etching for depicting the moon. Re-

1 On Manutius: Zeidberg 1998.
2 Gramaccini and Meier 2009, p. 83, Cat. nos. 192, 204, 205.
3 Rosand and Muraro 1976, p. 101, Cat. no. 28.
4 For example, the woodcut artist and publisher Gregorio de' Gregoriis; Bury 2001, p. 170.
5 On the temporal sequence of the printing: Bredekamp 2007, pp. 184, 196.

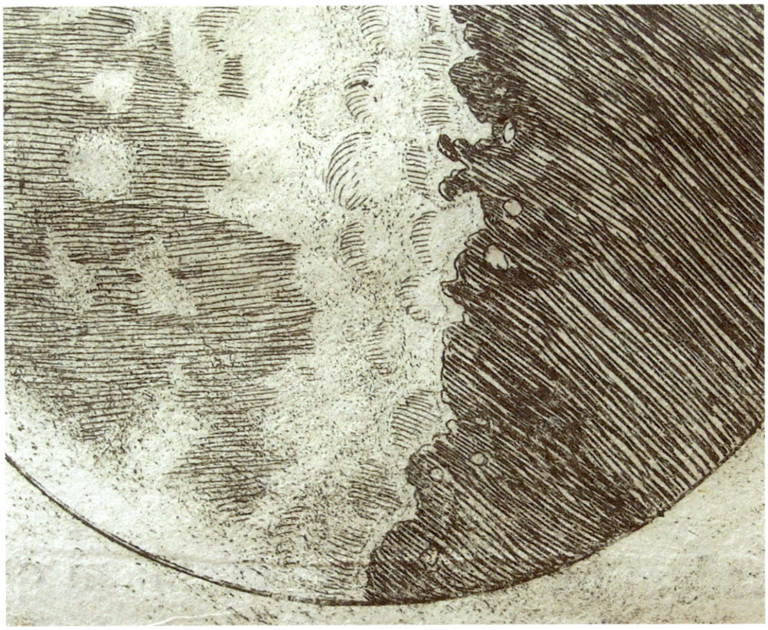

Fig. 1: Galileo Galilei, Sidereus Nuncius, Venice 1610, p. 10v, Venice copy, upper moon, detail. The modeling of the shadows falling in the hollows of the craters and the slight rise in the terrain on the sunlit lunar hemisphere show a combination of strewn points and unregulated line drawing that can be executed in this delicacy only by means of etching.

producing in print the delicate washes of the preliminary drawings, with their extremely finely executed demarcation lines, crater edges, valleys of light and shadowy hollows, demands a print medium in which a correspondingly nervous, meticulously descriptive, unregulated execution of lines can be built up (Fig. 1). Beyond that, etching offered to artists the advantage of being able to work in a print medium without having to abandon the manual principles of developing linear styles that were familiar to them from drawing. In Parma around 1530, Francesco Parmigianino was the first to comprehensively experiment with this technique, thereby removing, with virtuosity, the limits that had previously separated the aesthetics of drawing from that of classic printmaking.[6] In Venice, by contrast, etching was approached only as a *capriccio* throughout the entire sixteenth century.[7] This technique, which avoided direct work on the hard metal by taking recourse to a process involving an acid bath, was cultivated much more intensely in Parma, Urbino, and Bologna.

Because etching made it possible to work in a print medium without going through the notoriously laborious training in the technique of engraving, it drew into its orbit all the great draftsmen among the artists. An engraver simultaneously pushes and presses the graver while turning the metal plate to incise a line in the metal; this is practically the opposite of the gesture of drawing. The advantage of being able to use the etching needle in the same way as the accustomed drawing implements instead of learning engraving made etching the preferred print medium of dilettantes for centuries, up to the Comte de Caylus (Anne Claude de Caylus).[8] Though even the amateur etcher had to master one thing very precisely: the technique of biting the copper plate with acids.[9] Etching is an intaglio printing

6 Gnann 2007.
7 For example by Palma il Giovane; 27 of his engravings have come down to us.
8 On Caylus, cf. Rocheblave 1889; Gramaccini and Meier 2003, p. 118, Cat. no. 85; p. 119, Cat. no. 86.
9 Koschatzky 1972, p. 189.

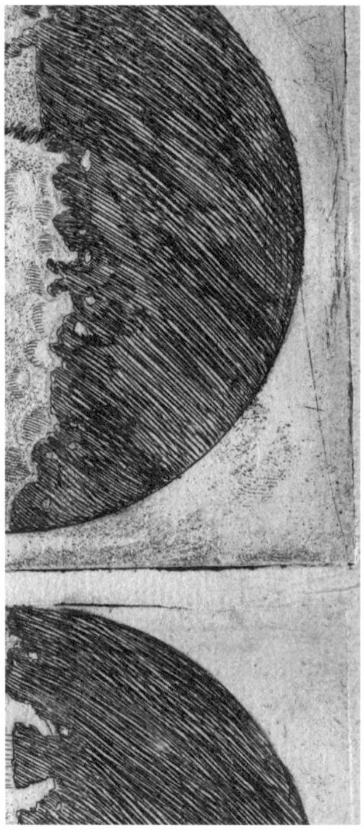

Fig. 2: Galileo Galilei, Sidereus Nuncius, Venice 1610, p. 10v, Munster copy, detail.

Clearly recognizable near the right edge of the upper moon is the trace of a vertical line from an earlier work.

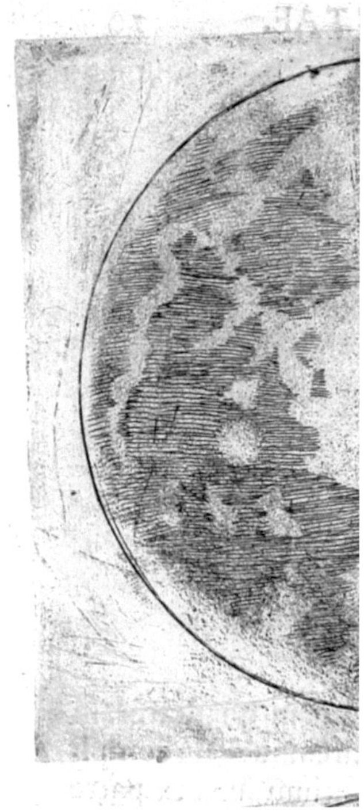

Fig. 3: Galileo Galilei, Sidereus Nuncius, Venice 1610, p. 10v, Graz copy, detail.

At the left edge of the outline of the upper moon, traces of lines from the earlier composition can be recognized. In the lower part of the print they coalesce into almost representational structures.

method, like engraving, but before the artistic work begins, the copper plate is coated with a film of acid-resistant ground. The desired lines are scratched into this ground with a needle; then the plate is placed in an acid bath, which eats into the exposed passages on the plate. The depth of the lines etched into the plate depends on the duration of the acid bath. After the ground is removed, printing ink is rubbed onto the cleaned plate, then wiped off; the passages recessed by etching are left filled with ink. To successfully etch the plate presupposes precise experience with the etching time needed to produce various tonal values in the print. Even artists like Federico Barocci experienced initial failures in this aspect of etching and destroyed areas of the graphic structure of their pictorial concept by overetching the plate.[10] The difference in quality between Galileo's preliminary drawings in the New York copy of the *Sidereus Nuncius* and the conspicuously simple etchings is enormous. In comparison with the preliminary drawings, the etchings made after them have little interpretive force. The execution of the outline of the circle and the interior drawing of the moons is often lacking in precision and displays no mentionable artistic inspiration, but the etchings show no overetched passages. Neither the transfer of the preliminary drawings onto the ground-covered plate (a process reconstructed by Brückle and Tesmar, p. 102), nor the execution of the linear concepts on the ground was necessarily carried out by a professional master, but the process of etching itself most definitely must have been

10 Bury 2001, p. 81, Cat. no. 43; Gramaccini and Meier 2009, n.p., plate XVa; p. 203, Cat. no. 169.

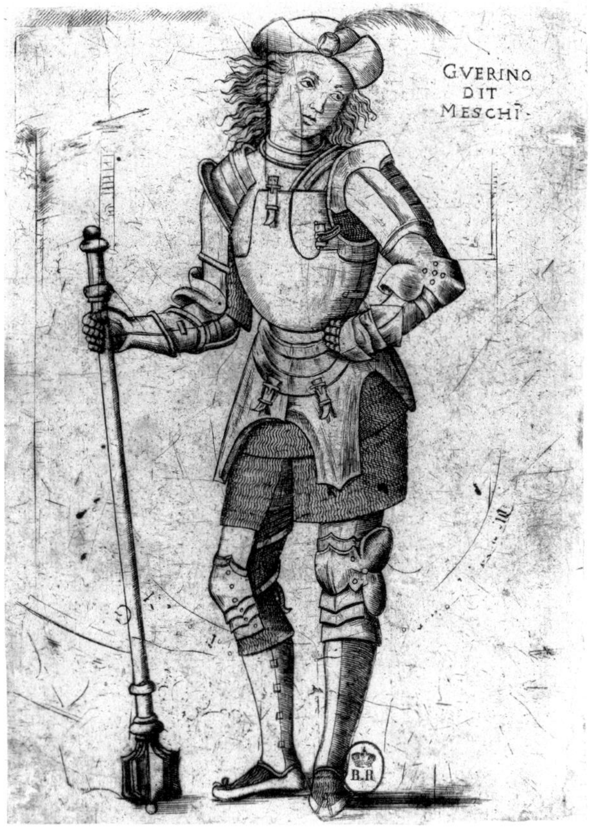

Fig. 4: Giulio Campagnola (?), Standing Knight (ca. 1490), engraving.

The composition was executed on a plate owned by the prominent German engraver Master E.S. (ca. 1420–1468), who worked in the Upper Rhine region. He executed his famous Large Virgin of Einsiedeln (1466) on it. Campagnola polished the older composition away, but inadequately. He then turned the plate around 180 degrees and used it anew. The line structures of the earlier composition, especially the arc of the canopy and even remnants of the inscription, are clearly visible in the print.

executed by an already practiced hand. This is all the more true when we consider that the etchings were produced in a hurry and no time would have remained for significant corrections of visually disturbing etching mistakes.[11]

The impression of great haste is intensified by the fact that the background areas around the etched moons display many traces of lines that clearly derive from a prior use of the copper plates for printing (Figs. 2 and 3). The passages in question are those in the *New York copy* drawings in which washes sometimes indicate the darkness of space surrounding the moon.

The character of these traces suggests that the original lines in these areas of the plates were carried out in engraving. Apparently, there was not sufficient time to remove these visible remnants of a previous design from the plates, even though they impaired the new

11 Bredekamp 2007, pp. 184, 196.

images. Copper plates were expensive, so it was customary practice in workshops to use them repeatedly. Copper plates also circulated between workshops and copper engravers, and it was not unusual for engravers to acquire plates that had already been used by other masters. To ensure a blemishless print when the plates were used again, they had to be polished thoroughly enough to eradicate all recognizable traces of the lines previously engraved in the plates. Since the work of polishing was time consuming, it was often inadequately carried out, even in the workshops of leading engravers. For this reason, even master prints by prominent engravers like Giulio Campagnola (Fig. 4), Antonio da Brescia, and Marcantonio Raimondi repeatedly show traces of earlier compositions.[12] In the case of the *Sidereus Nuncius*, the remnants of earlier lines on the plates were only fleetingly polished and their traces were reproduced in the new prints. In this way, Galileo's moon drawings in the *New York copy* had to stand up to two challenges, one interpretive and one print-technical: first, the translation of the original artistic statement into the less-nuanced intaglio printing method of etching, which was unfamiliar in Venetian book printing; and second, the visual contamination of the background areas surrounding the moons, and the subsequent diminishment of this powerful spatial effect of outer space surrounding the moons. For the practiced draftsman and art connoisseur Galileo, this was reason enough to comment at length in a letter to a confidant on these graphic inadequacies.[13]

12 Gramaccini and Meier 2009, n.p., plate Ib; p. 127, Cat. no. 1; p. 145, Cat. no. 43.
13 Bredekamp 2007, pp. 208, 341ff., Figs. 378 and 379.

ROBERT FELFE

PERFORMANCE OF THE ETCHINGS

As famous as they have become, it is not at all obvious that the images of the moon in the *Sidereus Nuncius* should have been executed as they were. To integrate etchings or engravings into the type area of a printed book page was expensive and complicated in terms of technology. After the text was printed, the copperplates had to be printed using a different type of press and fitting them exactly into the designated gaps on each sheet. Yet the pictures of the moon were etched despite these disadvantages with respect to woodcut, and despite extraordinary time pressure. One reason for this decision, as has frequently been noted, was the higher degree of precision this technique afforded over woodcut. But what does that mean in regard to *Sidereus Nuncius*? What are the concrete circumstances under which this precision could develop its own argumentative value as scientific evidence?

The images of the moon in the *Sidereus Nuncius* are framed by geometric diagrams. These abstract images explain the function of the telescope on p. 7r and they demonstrate how to ascertain the thickness of the foggy atmosphere of the planet (p. 12v), as well as the height of the mountains on its surface (p. 13v). These images are executed in woodcut. Their woodblocks were printed together with the text, integrated in the layout of movable type. This technological integration into the text accords with the aesthetic appearance of the woodcuts: the pure lines and the index letters have to be read in the same way as the explanatory text. They are primarily signs, printed on the neutral surface of the sheet of paper. At the same time, the first woodcut showing the schema of the telescope transcends these conditions of perception (Fig. 1). The eye of a hypothetical observer and the object of observation are explicitly placed outside the text block. In this way, by visualising the optical fundamentals of the instrument, the image evokes a disposition of seeing with its own rules, horizons and performances.

After this introduction to the technical and perceptive circumstances of the observations, on the next page the reader will look at the first picture of the moon as if it were the object itself (Fig. 2). The change from woodcut to etching is a highly effective way to realise this precarious switch of visual argumentation. The etchings render not only a description of the object, but they show it as part of a spatial situation that involves the reader as well. The combination and interplay of graphic techniques open the book onto a sphere of physical phenomena and suggest that this space and our visual experience can very suitably be analyzed by geometric operations. This mode of appearance is an effect of the etching. The woodcuts and etchings, therefore, not only illustrate the function of the telescope and what was seen through it, but the contrast of media and the suggestive qualities of the etchings restage the necessary interconnection between the instrument and the breathtaking discoveries made with it. For Galileo this was a crucial point in his claim to be the first human

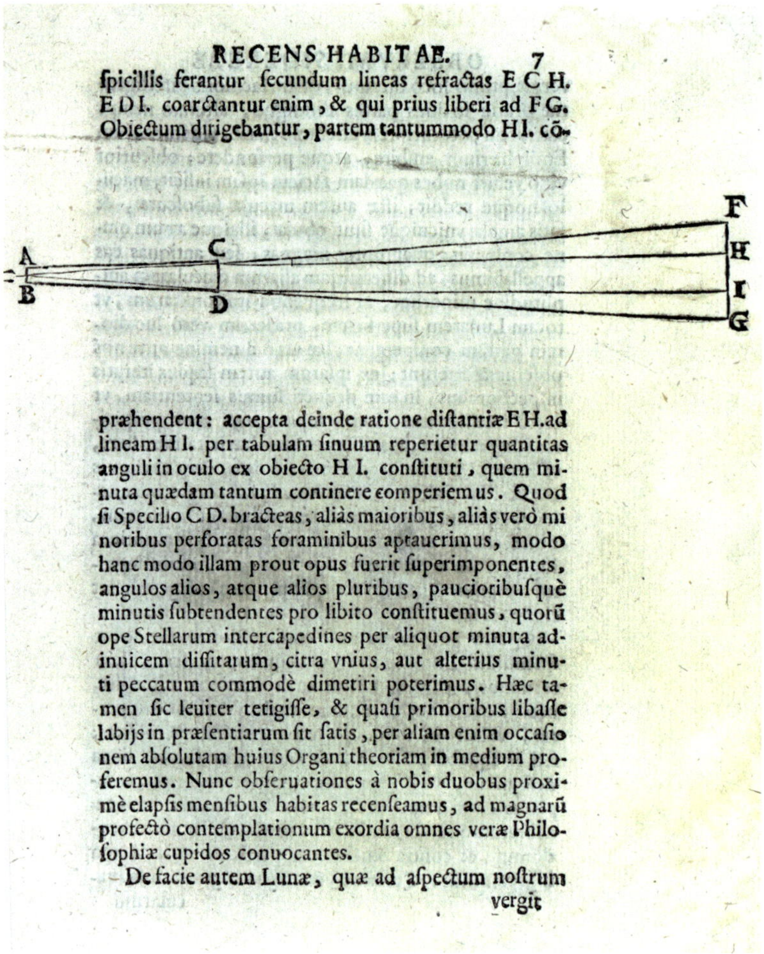

RECENS HABITAE. 7
ſpicillis ferantur ſecundum lineas refractas E C H. E D I. coarctantur enim, & qui prius liberi ad F G. Obiectum dirigebantur, partem tantummodo H I. cõ-

præhendent: accepta deinde ratione diſtantiæ E H. ad lineam H I. per tabulam ſinuum reperietur quantitas anguli in oculo ex obiecto H I. conſtituti, quem minuta quædam tantum continere comperiemus. Quod ſi Specilio C D. bracteas, aliàs maioribus, aliàs verò minoribus perforatas foraminibus aptauerimus, modo hanc modo illam prout opus fuerit ſuperimponentes, angulos alios, atque alios pluribus, paucioribuſquè minutis ſubtendentes pro libito conſtituemus, quorũ ope Stellarum intercapedines per aliquot minuta adinuicem diſſitarum, citra vnius, aut alterius minuti peccatum commodè dimetiri poterimus. Hæc tamen ſic leuiter tetigiſſe, & quaſi primoribus libaſſe labijs in præſentiarum ſit ſatis, per aliam enim occaſionem abſolutam huius Organi theoriam in medium proferemus. Nunc obſeruationes à nobis duobus proximè elapſis menſibus habitas recenſeamus, ad magnarũ profectò contemplationum exordia omnes veræ Philoſophiæ cupidos conuocantes.

De facie autem Lunæ, quæ ad aſpectum noſtrum vergit

Fig. 1: Galileo Galilei, Sidereus Nuncius, Venice 1610, p. 7r, Göttingen copy.

being who had seen the real quality of the surface of the moon and it was a motive for the haste driving the publication process (see Bredekamp p. 157).

Galileo may have been inspired to use different techniques of printmaking in this way by a prominent example. A book published some years earlier was probably well known to the physicist and astronomer, and may have been of methodological interest for his own research. In 1583 the first edition of *Le due regole della prospettiva pratica* appeared in Venice. As soon as 1602, a further edition followed and the book advanced to be one of the most important classics in the literature of one-point perspective – a topic, as we know, that was highly interesting to Galileo.[1] One of its authors was the architect Jacomo Barozzi da Vignola. When he died in 1573, he left behind a corpus of manuscripts and engraved copper plates comprising large parts of the book. After Vignola's death, the mathematician, astronomer and goldsmith Egnazio Danti added extensive commentaries and arranged the opus for

1 Presumably 10 editions had appeared by 1770, all except the last printed with the original engravings by Vignola; see Fiorani 2003, pp. 127–130, 151, 152 footnote 4. On Galileo, Ludovico Cigoli, Vignola and Danti, and one-point perspective, see Camerota 2009, pp. 143–151.

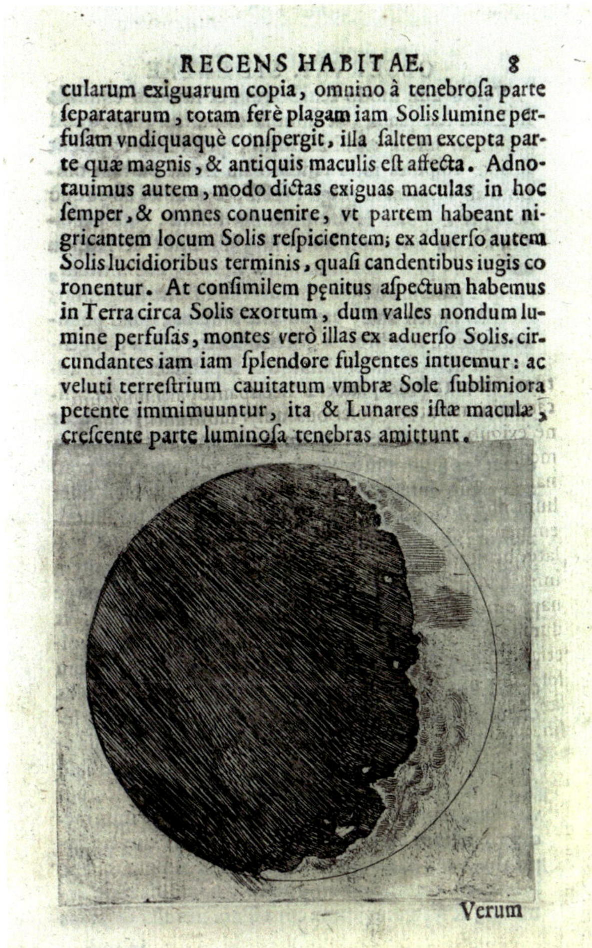
RECENS HABITAE. 8

cularum exiguarum copia, omnino à tenebroſa parte ſeparatarum, totam ferè plagam iam Solis lumine perfuſam vndiquaquè conſpergit, illa ſaltem excepta parte quæ magnis, & antiquis maculis eſt affecta. Adnotauimus autem, modo dictas exiguas maculas in hoc ſemper, & omnes conuenire, vt partem habeant nigricantem locum Solis reſpicientem; ex aduerſo autem Solis lucidioribus terminis, quaſi candentibus iugis coronentur. At conſimilem penitus aſpectum habemus in Terra circa Solis exortum, dum valles nondum lumine perfuſas, montes verò illas ex aduerſo Solis. circundantes iam iam ſplendore fulgentes intuemur: ac veluti terreſtrium cauitatum vmbræ Sole ſublimiora petente imminuuntur, ita & Lunares iſtæ maculæ, creſcente parte luminoſa tenebras amittunt.

Verum

Fig. 2: Galileo Galilei, Sidereus Nuncius, Venice 1610, p. 8r, Göttingen copy.

print. Danti was a central figure in Florentine mathematical culture in the time of Galileo's youth. In the 1560–70s he taught mathematics primarily in court circles, gave public lectures and was probably involved in the educational activities of the *Accademia del Disegno* as well as the *Accademia Fiorentina*. That Galileo met him in Florence seems likely.[2] Instruments designed by Danti have survived in the Medici collection; one large astrolabe attributed to Danti is called "di Galileo" because he used it by permission of the Grand Duke for his studies.[3]

Some of the most intriguing parts of *Le due regole* are the illustrations on pages 54 and 55 (Figs. 3 and 4). Geometrical definitions and explanations added by Danti extend over the first fifty plus pages. They are intended as a theoretical introduction to the original treatise on perspective written by Vignola. On p. 54 Danti explains the process of seeing based on a double visual pyramid – one for each eye. This model is a simple image embedded in the text block and executed in woodcut. On the opposite page, the first copper plate by Vignola was

2 Settle 1990, pp. 33–37.

3 Galluzi 2009, p. 314, no. VI.3.2; Turner 1987, pp. 134–139, 157–160.

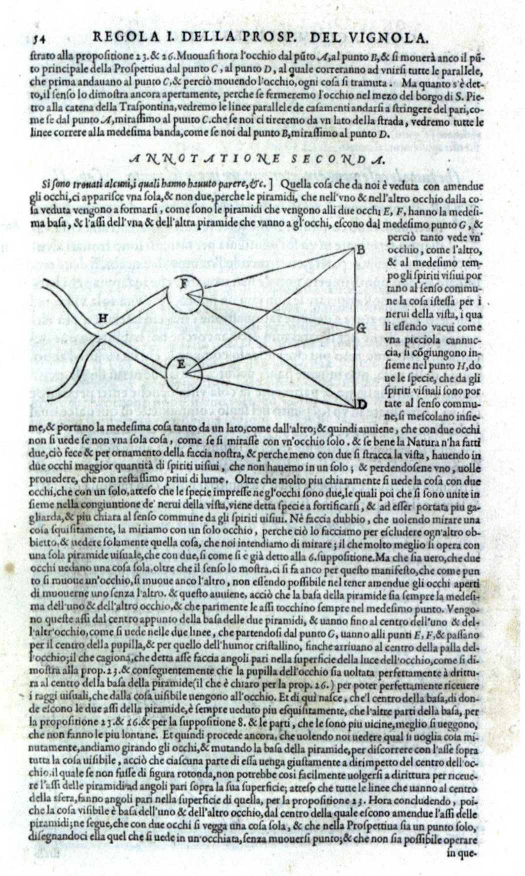

54 REGOLA I. DELLA PROSP. DEL VIGNOLA.

ſtrato alla propoſitione 23.& 26.Muouaſi hora l'occhio dal pũto *A*,al punto *B*,& ſi mouerà anco il pũto principale della Proſpettiua dal punto *C*, al punto *D*, al quale correranno ad vnirſi tutte le parallele, che prima andauano al punto *C*,& perciò mouendo l'occhio, ogni coſa ſi tramuta. Ma quanto s'è detto, il ſenſo lo dimoſtra ancora apertamente, perche ſe fermeremo l'occhio nel mezo del borgo di S. Pietro alla catena della Traſpontina, vedremo le linee parallele de caſamenti andarſi a ſtringere del pari, come ſe dal punto *A*, miraſſimo al punto *C*. che ſe noi ci tireremo da vn lato della ſtrada, vedremo tutte le linee correre alla medeſima banda, come ſe noi dal punto *B*, miraſſimo al punto *D*.

ANNOTATIONE SECONDA.

Si ſono trouati alcuni, i quali hanno hauuto parere, &c.] Quella coſa che da noi è veduta con amendue gli occhi, ci appariſce vna ſola, & non due, perche le piramidi, che nell'vno & nell'altro occhio dalla coſa veduta vengono a formarſi, come ſono le piramidi che vengono alli due occhi *E, F*, hanno la medeſima baſa, & l'aſſi dell'vna & dell'altra piramide che vanno a gl'occhi, eſcono dal medeſimo punto *G*, & perciò tanto vede vn' occhio, come l'altro, & al medeſimo tempo gli ſpiriti viſiui portano al ſenſo commune la coſa iſteſſa per i nerui della viſta, i quali eſſendo vacui come vna picciola cannuccia, ſi cõgiungono inſieme nel punto *H*, doue le ſpecie, che da gli ſpiriti viſuali ſono portate al ſenſo commune, ſi meſcolano inſieme, & portano la medeſima coſa tanto da un lato, come dall'altro; & quindi auuiene, che con due occhi non ſi uede ſe non vna ſola coſa, come ſe ſi miraſſe con vn'occhio ſolo. & ſe bene la Natura n'ha fatti due, ciò fece & per ornamento della faccia noſtra, & perche meno con due ſi ſtracca la viſta, hauendo in due occhi maggior quantità di ſpiriti uiſiui, che non hauemo in un ſolo; & perdendoſene vno, uolle prouedere, che non reſtaſſimo priui di lume. Oltre che molto piu chiaramente ſi uede la coſa con due occhi, che con un ſolo, atteſo che le ſpecie impreſſe negl'occhi ſono due, le quali poi che ſi ſono unite inſieme nella congiuntione de' nerui della viſta, viene detta ſpecie a fortificarſi, & ad eſſer portata piu gagliarda, & piu chiara al ſenſo commune da gli ſpiriti uiſiui. Nè faccia dubbio, che uolendo mirare una coſa ſquiſitamente, la miriamo con un ſolo occhio, perche ciò lo facciamo per eſcludere ogn'altro obbietto, & uedere ſolamente quella coſa, che noi intendiamo di mirare; il che molto meglio ſi opera con una ſola piramide uiſuale, che con due, ſi come ſi è già detto alla 6. ſuppoſitione. Ma che ſia uero, che due occhi uedano una coſa ſola, oltre che il ſenſo lo moſtra, ci ſi fa anco per queſto manifeſto, che come punto ſi muoue un'occhio, ſi muoue anco l'altro, non eſſendo poſſibile nel tener amendue gli occhi aperti di muouerne uno ſenza l'altro. & queſto auuiene, acciò che la baſa della piramide ſia ſempre la medeſima dell'uno & dell'altro occhio, & che parimente le aſſi tocchino ſempre nel medeſimo punto. Vengono queſte aſſi dal centro appunto della baſa delle due piramidi, & uanno fino al centro dell'uno & dell'altr'occhio, come ſi uede nelle due linee, che partendoſi dal punto *G*, uanno alli punti *E, F*, & paſſano per il centro della pupilla, & per quello dell'humor criſtallino, finche arriuano al centro della palla dell'occhio; il che cagiona, che detta aſſe faccia angoli pari nella ſuperficie della luce dell'occhio, come ſi dimoſtra alla prop. 23. & conſeguentemente che la pupilla dell'occhio ſia uoltata perfettamente à drittura al centro della baſa della piramide (il che è chiaro per la prop. 26.) per poter perfettamente riceuere i raggi uiſuali, che dalla coſa uiſibile uengono all'occhio. Et di qui naſce, che'l centro della baſa, di donde eſcono le due aſſi della piramide, è ſempre ueduto piu eſquiſitamente, che l'altre parti della baſa, per la propoſitione 23. & 26. & per la ſuppoſitione 8. & le parti, che le ſono piu uicine, meglio ſi ueggono, che non fanno le piu lontane. Et quindi procede ancora, che uolendo noi uedere qual ſi uoglia coſa minutamente, andiamo girando gli occhi, & mutando la baſa della piramide, per diſcorrere con l'aſſe ſopra tutta la coſa uiſibile, acciò che ciaſcuna parte di eſſa uenga giuſtamente a dirimpetto del centro dell'occhio, il quale ſe non fuſſe di figura rotonda, non potrebbe coſi facilmente uolgerſi a dirittura per riceuere l'aſſi delle piramidi ad angoli pari ſopra la ſua ſuperficie; atteſo che tutte le linee che uanno al centro della sfera, fanno angoli pari nella ſuperficie di quella, per la propoſitione 23. Hora concludendo, poiche la coſa viſibile è baſa dell'uno & dell'altro occhio, dal centro della quale eſcono amendue l'aſſi delle piramidi; ne ſegue, che con due occhi ſi vegga una coſa ſola, & che nella Proſpettiua ſia un punto ſolo, diſegnandoci ella quel che ſi uede in un'occhiata, ſenza muouerſi punto; & che non ſia poſſibile operare in que-

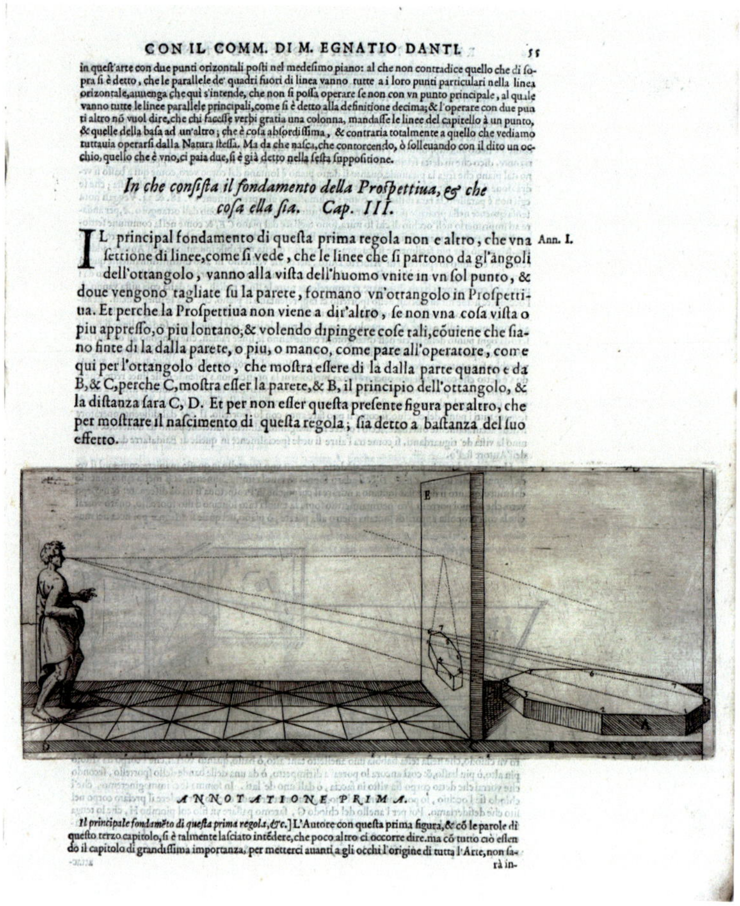

CON IL COMM. DI M. EGNATIO DANTI. 55

in queſt'arte con due punti orizontali poſti nel medeſimo piano: al che non contradice quello che di ſopra ſi è detto, che le parallele de' quadri fuori di linea vanno tutte a i loro punti particulari nella linea orizontale, auuenga che quì s'intende, che non ſi poſſa operare ſe non con vn punto principale, al quale vanno tutte le linee parallele principali, come ſi è detto alla definitione decima; & l'operare con due punti altro nõ vuol dire, che chi faceſſe verbi gratia una colonna, mandaſſe le linee del capitello à un punto, & quelle della baſa ad un'altro; che è coſa abſordiſſima, & contraria totalmente a quello che vediamo tuttauia operarſi dalla Natura ſteſſa. Ma da che naſca, che contorcendo, ò ſolleuando con il dito un occhio, quello che è vno, ci paia due, ſi è già detto nella ſeſta ſuppoſitione.

In che conſiſta il fondamento della Proſpettiua, & che coſa ella ſia. Cap. III.

IL principal fondamento di queſta prima regola non e altro, che vna ſettione di linee, come ſi vede, che le linee che ſi partono da gl'angoli dell'ottangolo, vanno alla viſta dell'huomo vnite in vn ſol punto, & doue vengono tagliate ſu la parete, formano vn'ottangolo in Proſpettiua. Et perche la Proſpettiua non viene a dir'altro, ſe non vna coſa viſta o piu appreſſo, o piu lontano; & volendo dipingere coſe tali, cõuiene che ſiano finte di la dalla parete, o piu, o manco, come pare all'operatore, come qui per l'ottangolo detto, che moſtra eſſere di la dalla parte quanto e da B, & C, perche C, moſtra eſſer la parete, & B, il principio dell'ottangolo, & la diſtanza ſara C, D. Et per non eſſer queſta preſente figura per altro, che per moſtrare il naſcimento di queſta regola; ſia detto a baſtanza del ſuo effetto. Ann. I.

ANNOTATIONE PRIMA.

Il principale fondamẽto di queſta prima regola, &c.] L'Autore con queſta prima figura, & cõ le parole di queſto terzo capitolo, ſi è talmente laſciato intẽdere, che poco altro ci occorre dire. ma cõ tutto ciò eſſendo il capitolo di grandiſſima importanza, per metterci auanti a gli occhi l'origine di tutta l'Arte, non ſarà in-

Fig. 3: *Jacomo Barozzi da Vignola and Egnazio Danti, Le due regole della prospettiva pratica, Rome 1583, p. 54, Niedersächsische Staats- and Universitätsbibliothek Göttingen, 2 BIBL UFF 99.*

Fig. 4: *Jacomo Barozzi da Vignola and Egnazio Danti, Le due regole della prospettiva pratica, Rome 1583, p. 55, Niedersächsische Staats- and Universitätsbibliothek Göttingen, 2 BIBL UFF 99.*

printed, visualizing the fundamentals of perspective in general. Especially with this first engraving, the pictures start to open up a new space for demonstration in the book. The subject is an observer looking through a picture plane at an object that will be represented on this plane. Thus the engraving by Vignola is the first application of the optical schema presented by Danti. At the same time, in contrast to the schematic woodcuts, the engraving creates a pictorial space that is vitalised by its own atmosphere of light and shadow and filled with a smooth continuum of tonal gradation.

The combination of two different printmaking techniques functions in a similar way in the treatise on perspective by Vignola and Danti and in the *Sidereus Nuncius*. In both books the interplay implies a transformation of preliminary conceptual reflections into an experimental situation. The higher degree of sensual participation triggered by etching and engraving encourages the reader to follow this cognitive movement before each picture reveals its information in cooperation with the text.

In addition to the structural similarity, it is more than likely that Galileo was inspired by publications on one-point perspective in a more concrete way, especially in regard to the moon as his subject. Since the middle of the sixteenth century, stereometric bodies had been a widespread subject for demonstrating different means of construction and producing the stupendous effects of an illusionary presence. Ludovico Cigoli for example, a painter and

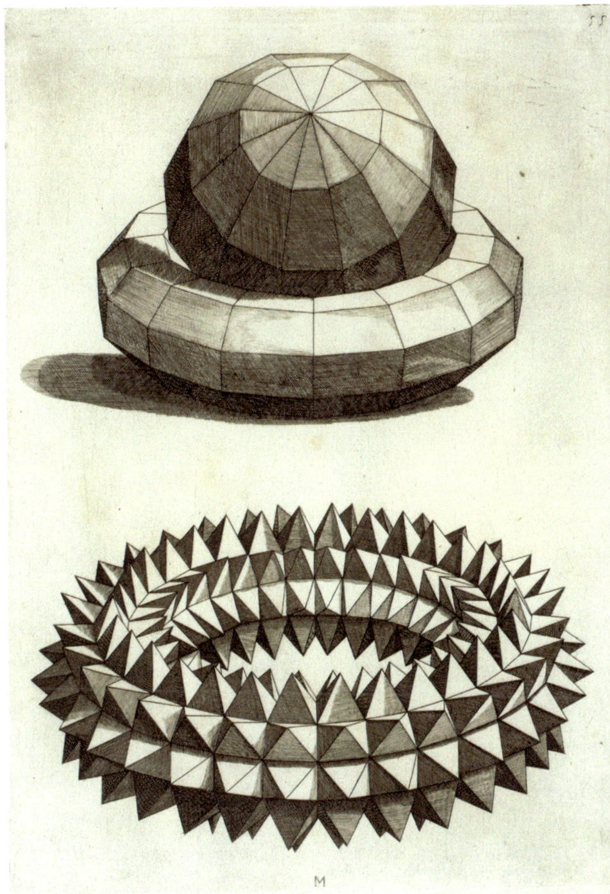

Fig. 5: Lorenzo Sirigatti, La Prattica di prospetiva, Venice and Florence 1596, pl. 55, copy of the Herzog-August-Bibliothek Wolfenbüttel, 5 Geom 2° [3].

friend of Galileo, followed this practice in some parts of his treatise on perspective.[4] But in this context another contemporary treatise might be more significant, because it was already published and included a large number of very ambitious copper engravings. Based on the methods of construction demonstrated by Vignola and Danti, Lorenzo Sirigatti published *La Prattica di prospetiva* in 1596. The book was dedicated to Grand Duke Ferdinando I de Medici and it has to be recognized as the most virtuosic printed publication of this genre in Italy. The second part contains a collection of engravings without text, most showing more or less complex stereometric bodies (Fig. 5). The figure rendered on the upper half of the illustration demonstrates how exercises in perspective like these by Sirigatti may have been of great interest to Galileo. Intending a high degree of three-dimensional illusionism, draftsman and engraver elaborated variations of hatching that made it possible to shade very subtle tonal differences. But it is not the suggestive presence of the geometric bodies alone that matters. Giving each plane of the surface its own shadow and tonal degree, this systematic method of hatching implies an analysis of the concrete shape of the body. This is precisely what Galileo would do in his etchings. Sirigatti was forced to construct a polyhedron that approximates a real globe, because it was impossible to construct an exactly per-

4 Tongiorgi Tomasi 2009, p. 142.

spectival view of a perfectly round body. Galileo however approached the issue from another angle. His intention was to show that the surface of a heavenly body, which was thought to be perfectly spherical, was in fact uneven – an irregularly shaped landscape with mountains, valleys, craters and plateaus.

The modelling of surface has one of its origins in the technique of printmaking. Looking at treatises on perspective, especially their illustrations, is instructive with respect to the methodological and aesthetic devices Galileo relied on. However, the innovation of his moon images must be emphasized. These prints were the beginning of a long tradition of visual representation of the moon. In order to sketch some aspects of this development it is helpful to compare the etchings in the *Sidereus Nuncius* to a series of engravings by Claude Mellan executed in 1636–37.[5] Some direct interconnections between these prints are well known. The series printed by Mellan was based on observations made by Pierre Gassendi and Nicolas-Claude Fabri de Peiresc. Both were correspondents of Galileo and they started their observations using a telescope sent by the Florentine astronomer.[6] Considering this cooperative relationship and the undeniable parallels between the prints, it seems surprising that Galileo was obviously annoyed by two engravings sent to him in the autumn of 1637. He described them as very clumsy, done by somebody who had never seen the moon with his own eyes.[7]

The etchings in the *Sidereus Nuncius* render concrete observations through the combination of the suggestive corporeality of the planet and the strong emphasis on the border between light and shadow on its surface (Fig. 6). In parallel to Galileo's verbal description, this demarcation line was stressed in every print. This is a significant difference between the prints and the Florentine drawings and it seems possible that the drawings in the *New York copy* were a first step toward this accentuation (see Smith and Brückle p. 41ff.; Bredekamp p. 91ff.). An important means to obtain this effect in the etchings was to let the shadowy parts of the moon sink into nearly complete darkness.

In contrast to the almost empty flatness of the dark side of the moon, the etchings show a large number of relief details on the bright side. Different modes of subtly elaborated hatching modulate a stupendous diversity of shapes. Often, each mark simultaneously describes a specific detail of the moon's surface and, with other marks, produces a tonal value that characterizes more extended formations and whole regions. The illuminated parts of the moon were etched in such a way that the practice of hatching and the delineation of shape are inseparably interwoven; in some parts they are one and the same. This unity is a cognitive nucleus *sui generis* and, in addition to some mistakes that would not have left the workshop of a professional printmaker, is a further indication that Galileo may have executed the etchings. Even the drawings in the *New York copy* suggest a very close interaction between the observation mediated by the telescope, and the drawing hand that is able to analyze the phenomena by interpreting effects of light in terms of the three-dimensional shape of a solid body.

5 Préaud and Lavergnée 1988, p. 115ff.; Fiacci, 1989, p. 44. In most publications the series contains three prints, Maxim Préaud mentions a fourth. Préaud 1988, pp. 172–174.

6 Galluzi 2009, p. 363 no. VII.4.7; Ashworth 1994; Jaffé 1990.

7 See the letter by Galileo to Benedetto Castelli (24. 10. 1637), and the letter by Benedetto Castelli to Galileo (26. 9. 1637), in *Opere*, vol. XVII, pp. 185, 186, 203, 204; Bredekamp 2007, p. 315.

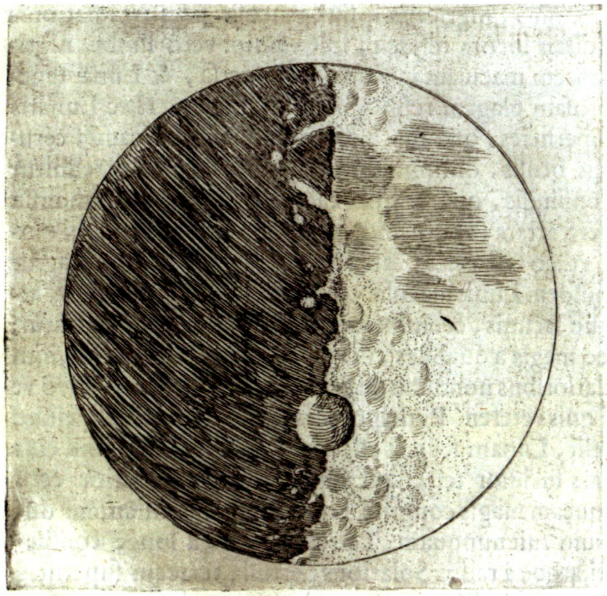

Fig. 6: Galileo Galilei, Sidereus Nuncius, Venice 1610, p. 9v, Göttingen copy.

There are fundamental differences between the printed moon images in the *Sidereus Nuncius* and the engravings by Mellan that might explain Galileo's harsh judgement and underline the singularity of his etchings. The first difference is in respect to the precise intention of the process of depiction. Soon after the publication of the *Sidereus Nuncius* the first images were printed that suggested a mapping of the moon's surface.[8] In this trend of pictorial representation, the plasticity of the moon's spherical body was subdued to a high degree in favour of rendering the details of the relief in their position to each other as precisely as possible. Having been described as early examples of a move toward a real cartographic investigation of the moon,[9] the engravings by Mellan could be seen as an attempt to harmonize two divergent intentions (Fig. 7). The engravings combine an impressive density of detail, drawn immediately under projection, with an astonishing illusionism of physical presence. The mapping of the moon is linked to the changing appearance of the orb due to its movement as a celestial body in space. This synthetic performance with an accent on the three-dimensional shape can be understood as reminiscent of Galileo. Yet Mellan's handling of the graphic technique was very unlike that of the etchings in Galileo's book, in that it was at one and the same time very rationally reduced and exhibited as a gesture of proud self-consciousness.

In the context of the history of printmaking, the method of rendering relief, light and shadow in the etchings of *Sidereus Nuncius* was just as effective as it was peculiar and wilful. Since the late fifteenth century, increasingly systematic practices of hatching had been developed. It was a widespread tendency, particularly in engraving and etching, to elaborate procedures based on abstract structures like single or multiple layers of parallel lines. These structures were to a high degree independent of the concrete subject. Variable in the course,

8 A famous example is the "Luna Facies" in Johannes Kepler's *Dioptrice* 1611, book 5, p. 176.

9 Withaker 2009, p. 256.

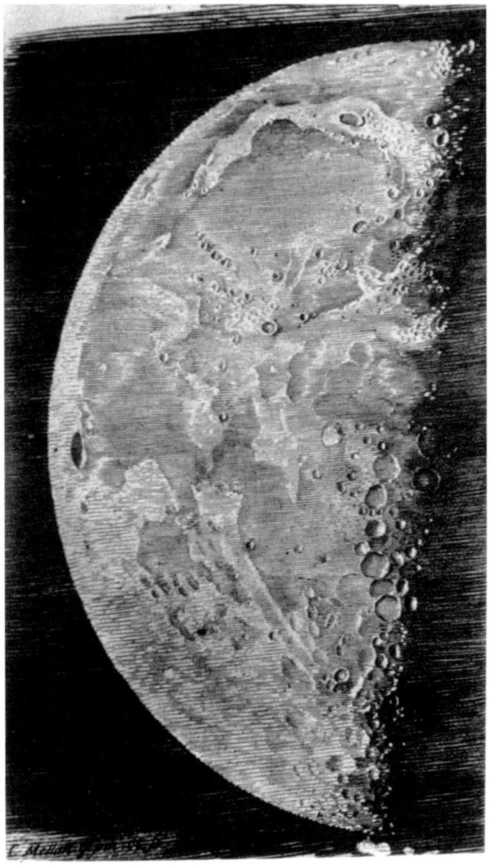

Fig. 7: Claude Mellan, moon engraving, 1636.

thickness and interval of the lines, these abstract grids could be transformed into the surface of every object represented. The virtuoso moiré effects in engravings from Netherlandish Mannerists around 1600, like Hendrik Goltzius, signify a remarkable zenith of this tendency. Claude Mellan was maybe the most famous proponent in the seventeenth century and his moon engravings are masterpieces in this context. There is no graphic element other than nearly straight horizontal lines. As a dense grid they open the darkness of cosmic space and they alone let the body of the planet become visible. Every line is set parallel to the next. The suggestive relief of the moon as well as every detail were generated by no other means than interruptions of the uniformly summarized lines and by variation of their width and spacing.

The basic principle in such methods of hatching is that the graphic mark is completely independent of the subject and its corporeal qualities. Some years later the French engraver and author of treatises on perspective, Abraham Bosse, tried to mechanise this principle (Fig. 8).[10] In 1653 he published an image of a grid of parallel strings positioned between a strong source of light and an object that is to be represented. The grid casts shadows on the surface of the object and the shadows delineate a complete system of hatching lines for an

10 Bosse 1653, pp. 73–75.

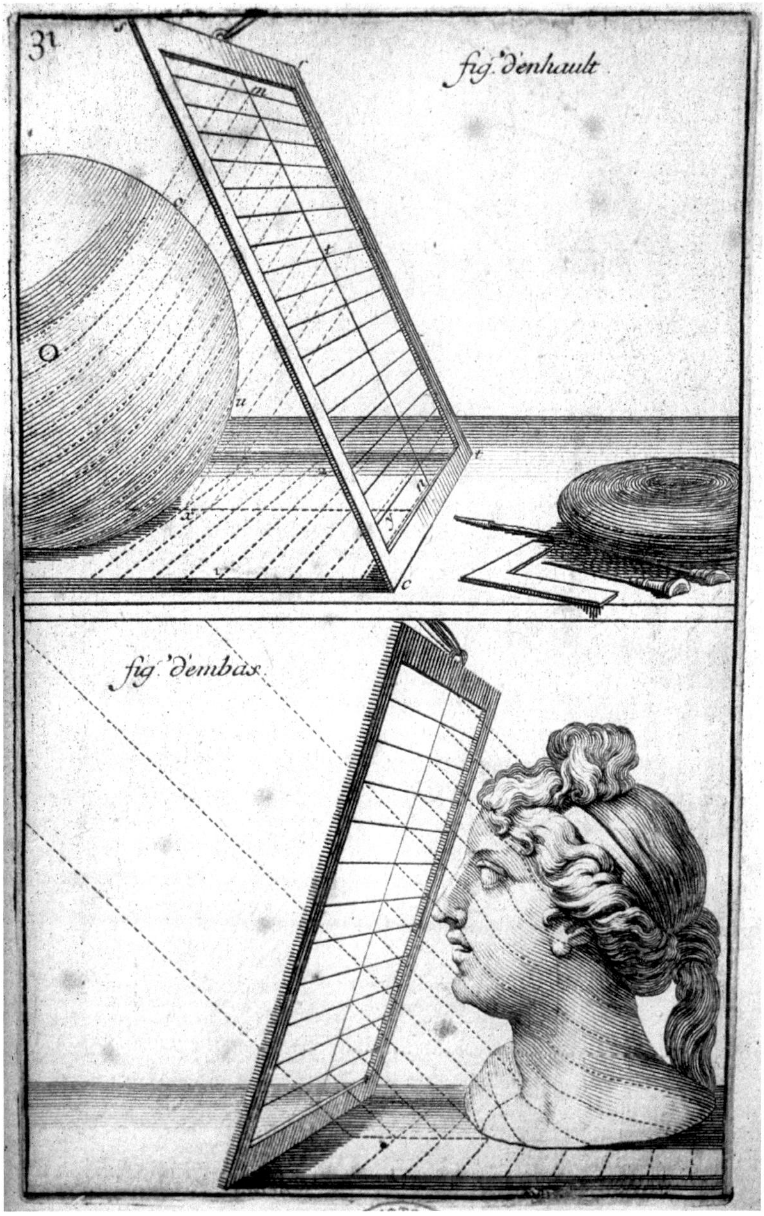

Fig. 8: Abraham Bosse, Moyen universel de pratiquer la perspective sur les tableavx, ou Surfaces Irregulieres, Paris 1653, pl. 31, Staatsbibliothek Preußischer Kulturbesitz Berlin, Nu 8131.

exact picture.[11] Bosse's experiments with the grid were undertaken in connection with his effort to systematically incorporate the problem of tonal degrees into the theory and practice of one-point perspective. In the study of perspective, regular bodies and their variants, which had preoccupied Sirigatti in the sixteenth century and so many other authors about fifty years earlier, were almost completely marginalized. In the engravings by Mellan, the lines of hatching scan a visual field again and again, without any involvement in the concrete shape. The burin of the engraver is free of every tie to the *gestalt* and the surfaces of the things to be rendered. As if he wanted to underline this freedom as an indication of his

11 Leonard and Felfe 2006, pp. 103–108.

artistic virtuosity, a single line swings in and leaves some sloppy traces in the upper left corner of the first engraving in this series, before it creates the darkness of the endless space.

This virtuosity has not much in common with Galileo's observations, his drawings or the impetus of the whole *Sidereus Nuncius*. Of course, he was neither a professional draftsman nor printmaker. Nevertheless, if the etchings of the moon had, and still have even today a quality that captures the imagination, it is because they are artefacts with a rarely achieved amount of density and concentration. With every tiny stroke, mistakes included, they are modelling the uneven surface of the moon as a physically existing body. And they declare it – sometimes even by "pasting" details observed at different times and different places onto one and the same copper plate – as scientific fact.[12] This intersection of a scientist's precise observation and his ambition to establish the discovery with a certain degree of artistic education and artisanal skill was a singularly productive and unique moment in the history of science and art.

12 See Needham in vol. II of this publication, p. 97.

IV

THE PRINTING MATERIALS

IRENE BRÜCKLE / MANFRED MAYER / THERESA SMITH

THE PAPER

The paper of the *Sidereus Nuncius* tells its own subtle story about the making of the book. It reveals material characteristics that are particular to this publication while casting light on the standard craft methods of paper- and bookmaking at the time. From papermaking to book production many intricate steps were taken which shaped the physical paper that we appreciate today when we examine individual *Sidereus Nuncius* copies. Many of these steps can be deduced from what is known in principle about historical production methods. Some of them are borne out by the material evidence of the paper, which reveals marks of its making that discretely bring to life decisions made by the artisans who, more or less unwittingly, supported the making of this seminal publication. During our technical examination, the paper of several exemplars was examined in addition to the *New York copy*, including copies in Graz, Washington, Venice, and Munster. The results of this examination support the historical evidence gained elsewhere in the study. On the basis of the paper, strong connections can be made between the *New York copy* and the fully printed copies, while supporting the unique status of the former.

Opening individual exemplars, one first appreciates the differences of their page dimensions (Fig. 1). The page margins of the *New York*, *Graz*, and *Munster copies* are trimmed closer to the text than the *Washington copy*, which still retains the deckle edges of the paper sheets. Most of the other extant copies lost their deckle edges when the book blocks were trimmed, probably the most invasive alteration that paper sustains when a book is first bound. How much paper was removed was determined by the desired size of the final binding, a decision made by the binder or the purchaser of a particular copy. If the copy was rebound later, the pages were usually trimmed again to even out the edges. These bookbinding decisions have determined the dimensions of the trimmed *Sidereus Nuncius* copies. In comparing these page spreads, one also notices that the paper of the *New York copy* is a bit darker than that of the *Graz* and *Washington copies*, though all three are of a similar stock, while the *Munster copy*, which is printed on a different paper, is brighter in tone.

The story of the book pages begins with paper production.[1] By 1600, the craft of papermaking had developed into an intricate, highly regulated profession that required well over thirty individual production steps to turn various kinds of worn rags into fine white paper.

1 Detailed accounts of the process, briefly reviewed here for the essential aspects that are significant for the description of the *Sidereus Nuncius* papers, can be found in the seminal and now classic book by Dard Hunter (1978). Timothy Barrett (1989) described the technical aspects of the process furthered by instrumental analysis of historic specimens.

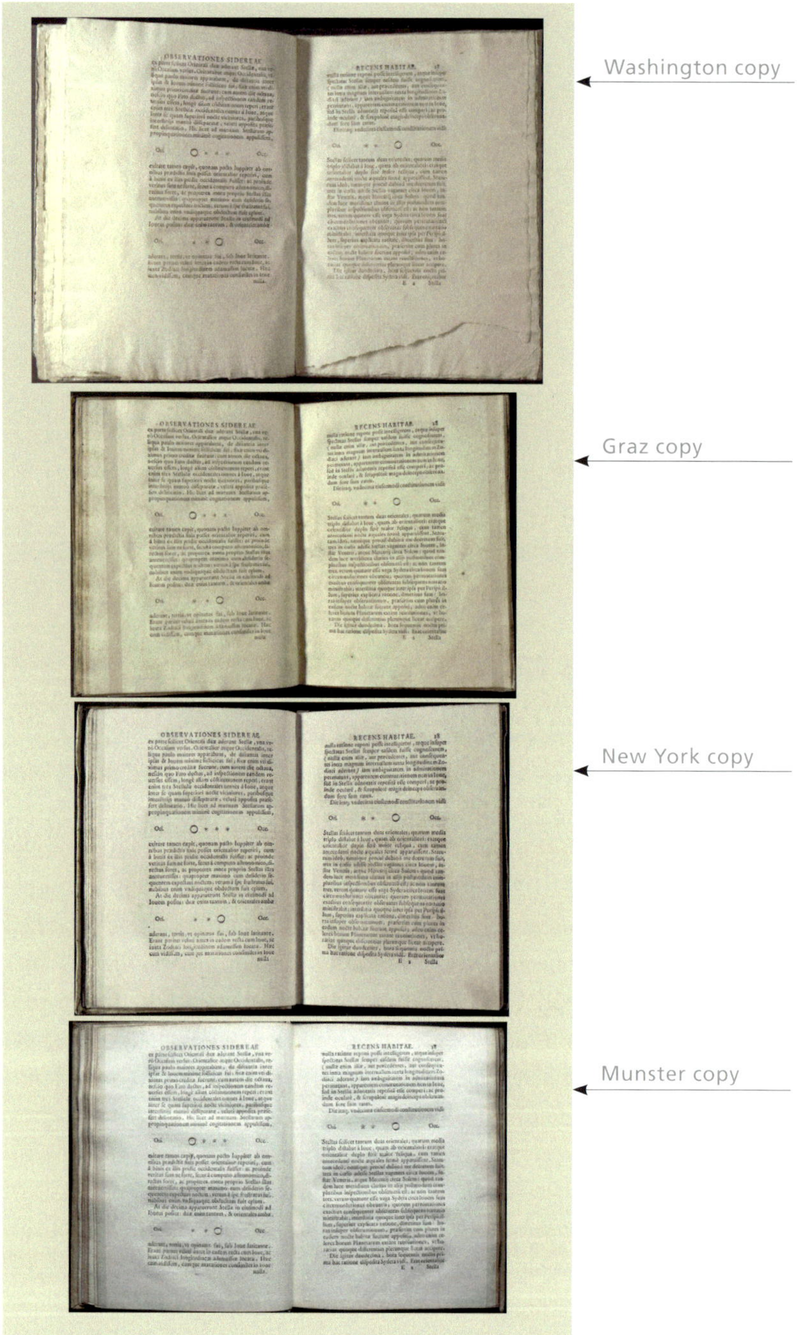

Fig. 1: Galileo Galilei, Sidereus Nuncius, Venice 1610, pp. 17v and 18r, Washington, Graz, New York, and Munster copies, slightly raking illumination. Note the differences in the page dimensions, drape and color of the papers.

Cast-off textiles, at the time made mainly from linen or flax and other bast fibers, were collected, cleaned in the paper mill by fermentation, or retting, and extensive rinsing, and then beaten to separate the cellulosic fibers. This pulp, or stuff, was diluted to create a watery fiber suspension in the vat, a tub from which thin mats of the fiber material were lifted on a papermaking mould. The mould deserves some attention because its structure caused distinctive marks in paper. It consisted of a rectangular wooden frame covered in the long direction with narrowly spaced copper wires that formed the so-called laid lines seen in the finished paper. Wooden ribs running at regular intervals across the short side of the mould supported the wires underneath. To fix the laid wires in position, they were fastened at each rib by stitches of fine wire. These fastening stations gave rise to chain lines perpendicular to the laid lines in the sheet. The wire-covered mould was fitted with a wooden frame, the deckle. When the mould was dipped into and withdrawn from the vat, the deckle provided a boundary to prevent the fiber suspension from spilling over the edges of the mould. The deckle was shared by two moulds. While one freshly made sheet was drained, then separated from the mould and transferred or couched onto a woollen felt, the shared deckle would already be in use, forming a new sheet on the second mould. The mould pairs carried the same watermark design in the form of shaped wires stitched to the wire mould coverings. Finished paper sheets interleaved by felts were assembled in a stack, or post, and pressed. Then the sheets were separated from the felts, piled up again, pressed once more, and transferred to lofts to dry. The airflow in the drying lofts was regulated by opening and closing louvers to ensure even water removal. Several sheets forming a set, or spur, were hung as one unit across a rope, which caused soft broad creases across the center of the sheet. These creases may remain noticeable in the dried sheets even after burnishing. After drying, surface sizing was a common, though optional step that imparted water-resistance and strength to paper, the former a prerequisite for writing paper, the latter beneficial to its handling properties. Sizing involved dipping the paper in a warm, dilute solution of gelatin, obtained by simmering waste animal parts such as cartilage or skins in water. Gelatin was the only sizing agent used in European papermaking at the time.[2] This treatment could be performed as part of paper production or at a later stage, for example after printing. Adding alum to the gelatin solution additionally enhanced the water resistance of the paper.[3] Variables in several of these paper production steps – the sheet formation, the watermarks, the sizing – play a role in differentiating the papers in *Sidereus Nuncius*.

Because paper was an expensive material and book producers were generally intent on making economical use of it, sheets were manufactured according to standard dimensions. These dimensions were determined by the technical necessities of paper production, which evolved in tandem with the dimensional requirements of paper use, including book production. The pages were laid out so that the finished book was in the quarto format. This required the full sheet of paper to be folded twice to make four folios, i.e. eight pages (Fig. 2). There are four folios in a typical quire or gathering in *Sidereus Nuncius*; seven full sheets were used for quires A–G, plus one half sheet inserted in quire D (see Mayer et al. p. 149ff.). To print four folios on one sheet of paper, the pages had to be laid out in a particular arrangement, as can be seen in the diagram of quire B, which consists of folios 5 through 8. The full dimensions of the most prevalent paper employed in the *Sidereus Nuncius* are known because at least one copy, the *Venice copy*, features paper that was never trimmed along any side of the book block (Fig. 3). Its folios were sliced open individually, page by page, along their top edges to open the text block for reading, leaving an undulating path

2 Kolbe 2002; Barrett 1992, pp. 228–233.
3 Brückle 1992.

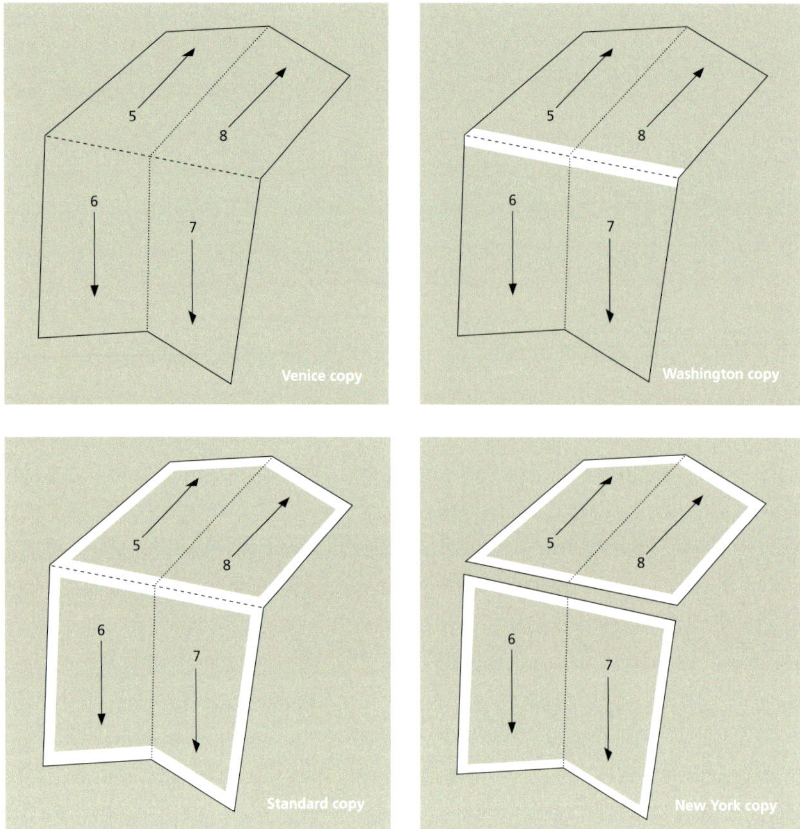

Fig. 2: Arrangement of folios in quarto format for quire B of the Venice, Washington and New York copies compared to a standard copy. Arrows point to the tail, dashed lines indicate the head of the book block and white areas were trimmed during binding.

from the knife (Fig. 4). The diagrams in Fig. 2 also show where the pages of other copies were trimmed during binding. The full dimensions of the standard paper are about 38 × 54 cm (Fig. 5).[4] The *New York copy* is a distinct exception from the other standard copies insofar as each of its quires was printed from two separate half sheets instead of one full sheet of paper, a point that will be considered later.

With the help of a digital reconstruction, the full paper sheet can be seen with the pages arranged as they were printed. The pages of quire B in the *Washington, Graz, New York* and *Munster copies* were photographed individually in transmitted light and these photographs were then digitally assembled to reconstruct the original sheets (Fig. 6). The resulting images recapture the moment when the printer had finished the recto and verso of the full sheet that would become quire B. In the transmitted light view, the varying

4 Variations in the dimensions of the standard full sheet of paper may be due to the use of several sets of twin moulds with slightly different dimensions, but the same watermarks. Although paper sizes were standardized, permutations included, to mention one example, reductions in size as (illegal) cost-cutting measures when rising production costs were not offset by sufficient revenue increases; see Weiss, p. 51. The approximate dimensions of the pages in selected exemplars (height × width, cm): *New York copy* 22 × 16; *Graz copy* 22.5 × 17; *Venice copy* 27 × 19; *Munster copy* 21.5 × 16.3. Paul Needham recorded 38 × 52 cm for the paper dimensions in exemplars he examined; see Needham, p. 23 in vol. II of this publication.

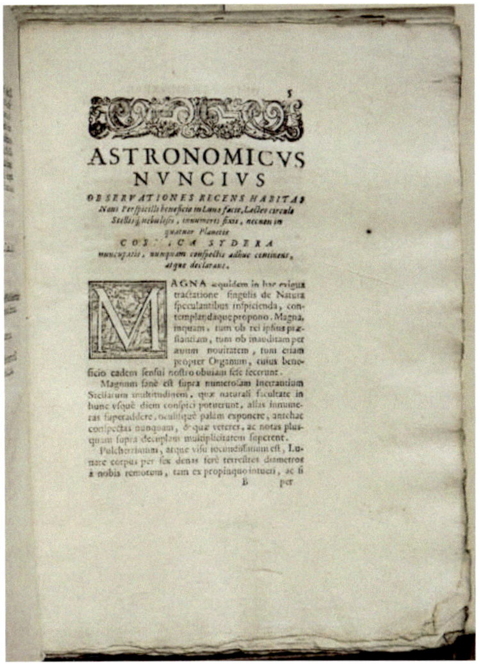

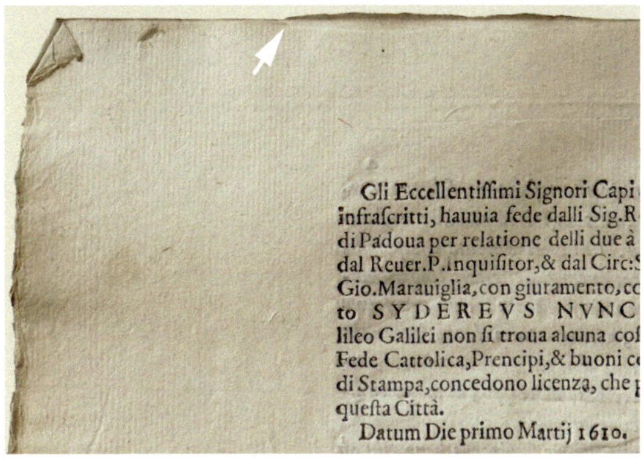

Fig. 3: Galileo Galilei, Sidereus Nuncius, Venice 1610, p. 5r, Venice copy. The untrimmed page features the original deckle edges.

Fig. 4: Galileo Galilei, Sidereus Nuncius, Venice 1610, p. 4v, Venice copy, detail. The untrimmed sheets were slit along the top edge (see also Fig. 2).

qualities of the sheets and the positions of the watermarks, digitally enhanced to make them more easily discernible, can be differentiated.

The images taken under transmitted illumination allow a more careful comparison of the papers used in the book production. The *Washington copy* serves as a good starting point because it allows us to consider the full sheet with its deckle edges still in place. The most striking features are the various flaws along the margins of the sheet. In several places, the edges are folded over, and some folds extend into the center of the sheet. These folds were created during paper production, not during printing. They occurred either while couching the sheet, i.e. transferring the wet fiber mat from paper mould to felt, or laying the sheet in the post after felt removal. This is not the only flaw: on p. 18r, shown in Fig. 1, a large section of the lower margin of the paper is torn off. Judging by the condition of the

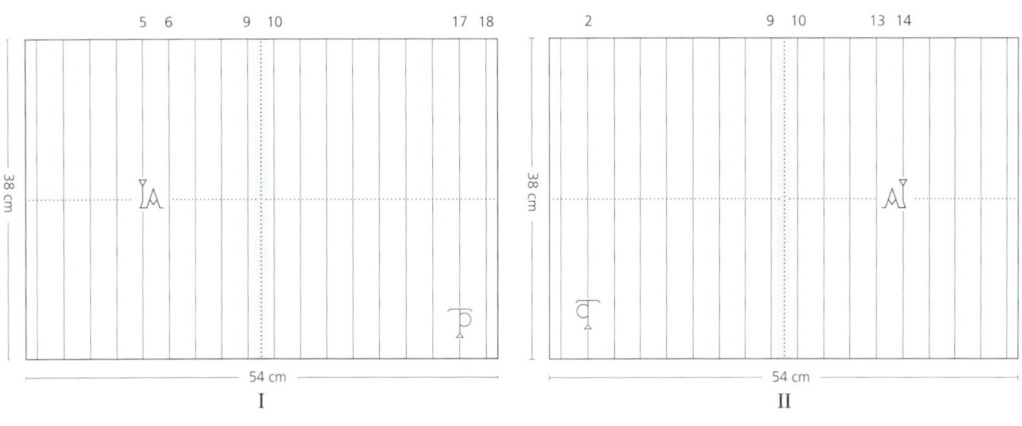

Fig. 5: Dimensions of standard paper with watermarks "A" and "Tp" indicated on the chain lines. Seen from the wire side, the main watermark is on the left of paper I, while it is on the right of paper II. The grey area indicates the mark that results from drying during production; the vertical dotted line corresponds to the fold at the head of the book block; the horizontal dotted line indicates the fold at the gutter (see also Fig. 2).

tear edge, this accident must have occurred during the same production steps, perhaps when the paper was lifted from the felt after emerging from the dewatering press, a step that also required gentle handling of the wet, and therefore weak, sheet. Considering the likely prevalence of errors such as these during rapid paper production, it is quite evident that removing part of the margin before binding was a desirable step that resulted in the more or less close trimming of the papers of the *Graz, New York* and *Munster copies*. It should be remembered that irregularities in the paper structure were not considered a mark of interest, as they might be in the appreciation of historical handmade paper today, when individual exemplars such as the *Washington* and *Venice copies* have acquired aesthetic and possibly monetary value for their untrimmed deckle edges.

The *Washington* and *Graz copies* also allow us to locate the two most important watermarks in the standard printed copies of *Sidereus Nuncius* (of the five watermarks identified in the book). To review, watermarks were created by bending wire into the shape of symbols or letters and sewing them onto the brass wire screen covering the mould surface. Standing proud above the mould surface, the wire shape diminished the amount of fiber material that settled above it, causing the paper to be thinner in this area, which is the reason that these marks appear lighter than the rest of the paper when viewed in transmitted light. Watermarks and laid lines, though visible from either side of the paper in transmitted light, are a feature of the side that was formed in contact with the mould. The so-called wire side of the paper is thus a sculptural cast of the mould surface, while the felt side is dominated by the impressed morphology of the woollen felts against which the freshly made sheet was couched.[5] Watermarks at the time were placed either in reverse on the mould surface, which means that they read correctly from the wire side of the finished sheet, or placed to read correctly on the mould and from the felt side of the sheet.[6]

The standard paper in the *Sidereus Nuncius* bears two watermarks that are a pair, a watermark and its countermark (Fig. 5, I). The watermark consists of the letter "A" with a V-shaped cross stroke, flanked on the left by an unidentified shape that could be a staff or possibly the letter "L" or "I". This watermark is accompanied by a countermark that can be read as the joined letters "Tp" sharing a vertical stroke and resting on a triangular base.[7] The position of the watermarks can be reconstructed with regard to their chain line stations. The "A" sits between chain lines 5 and 6, and the "Tp" is centered on 17, next to the final, 18th chain line.[8] Even when the deckle edges were preserved in the book, the vertical center of the sheet (here between chain lines 9 and 10) was occasionally removed. Once folded, this portion of the sheet became the top edge of the text block, which could be conven-

5 The difference between wire and felt sides may be more or less pronounced depending on how strongly the paper was pressed during dewatering and after sizing, and how well it was burnished as a finishing treatment. In the descriptive nomenclature for paper, we use the term wire side instead of mould side because it can be applied universally to both hand- and machine-made papers.

6 Examples of seventeenth-century papers with both variations of watermarks are featured in the catalogue of Rembrandt's watermarks; see Ash et al., 1998.

7 The occurrence of a watermark and countermark within the same sheet begins in Venice in the late fifteenth century. Subsequently, papermakers advertised their mills by initialling their names as countermarks; see Weiss 1962, p. 88.

8 The location of the "A" and "Tp" watermarks was reconstructed from the *Venice copy*.

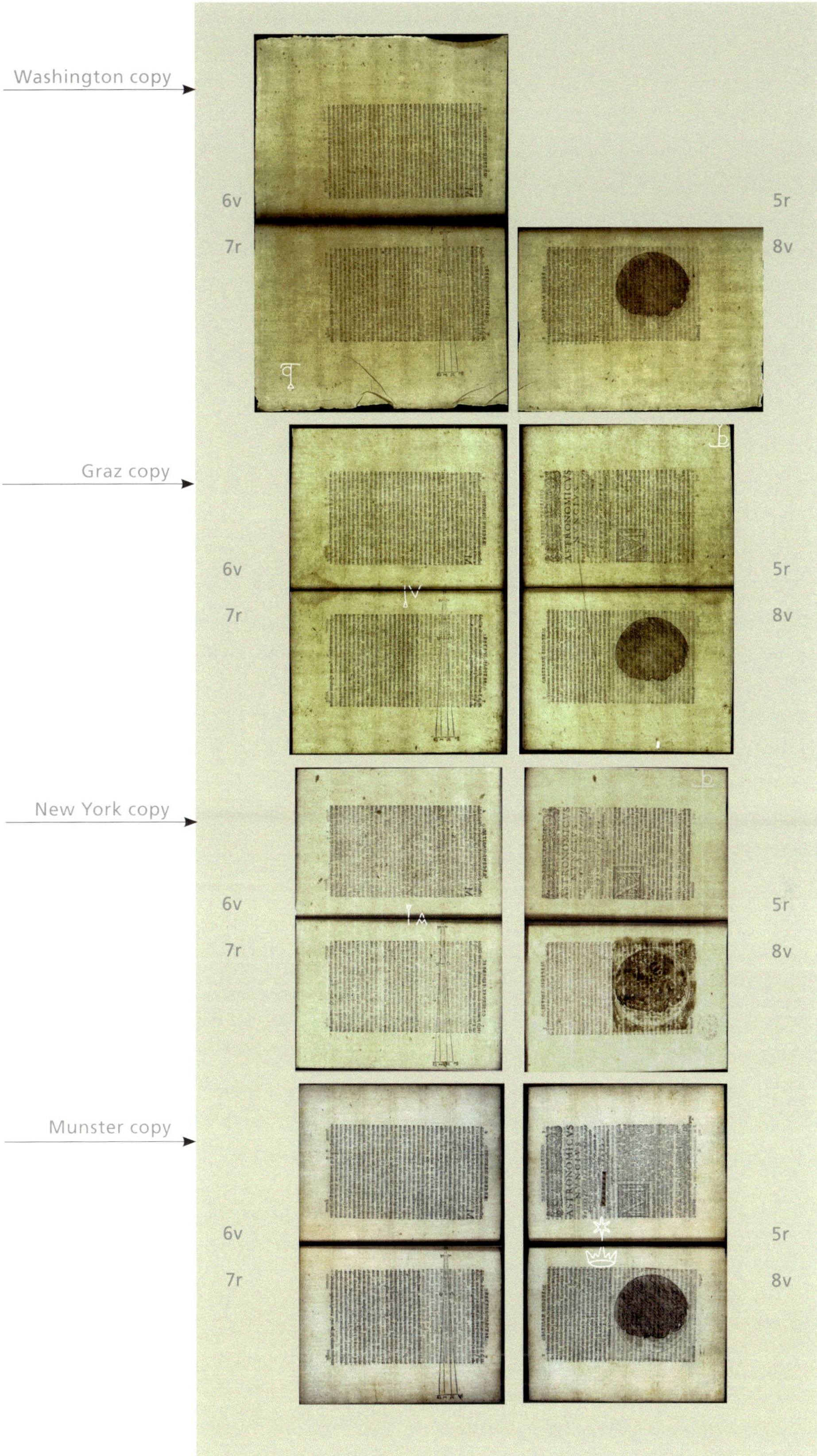

Fig. 6: Galileo Galilei, Sidereus Nuncius, Venice 1610, reconstruction of the paper sheets for quire B of the Washington, Graz, New York and Munster copies, viewed in transmitted light from the wire side. The New York copy is made of two separate half sheets. The folio numbers appear in the margin; folio 5r of the Washington copy was not documented. The watermarks are enhanced for better visibility.

iently trimmed to open the quires. This is why a narrow strip of paper is missing in the center of the sheet in the *Washington copy*, though the deckle edges were preserved (Fig. 6).

As pointed out by Paul Needham, the standard paper was made on twin moulds with the watermarks in reverse orientations, both in their design and placement on the mould (Fig. 5).[9] Paper I reads correctly from the wire side (corresponding with Needham's classification of the main watermark located on the right half of the mould, mR), while paper II reads correctly from the felt side of the sheet (main watermark placed on the left half of the mould, mL). The alternation of twin moulds naturally generated alternating sheets during the papermaking process (I-II-I-II etc.). This order may have been roughly preserved until each quire was printed, but became random when sheets were subsequently assembled, as shown in quires B through G in the *Graz copy*:

B: II (Fig. 6)
C: I (Fig. 7)
D: II, inserted D bifolium 15/16 I
E: bifolium 17/20 II, bifolium 18/19 I
F: II
G: bifolium 25/28 I, bifolium 26/27 II.

Not surprisingly, a thrifty employment of paper also occasioned the use of half sheets in printing the regular edition, as evidenced here in quires E and G. One should note that the orientation of the paper in the press was not important, which is why the watermarks do not always appear in the same orientation in relation to the text in the final book pages. The "A" and "Tp" watermarked paper is featured in quires B-G of most extant copies of the *Sidereus Nuncius*.

Among those copies of *Sidereus Nuncius* studied were two finely printed, luxury exemplars. The *Munster* (Fig. 1) and *Florence copies* were printed on an exceptionally fine, pristine white paper that bears a single watermark, a simple crown with a shank and a six-pointed star (Fig. 6). It is, however, the standard paper to which we must return to understand the use of paper in the *New York copy*.

Washington and *Graz*, our two references for the standard edition, show papers of similar quality. The transmitted light view allows us to study the general structural characteristics of the standard "A" and "Tp" watermarked papers. In this period, all paper had accumulations of fibers along the chain lines that created shadowy areas in transmitted light.[10] What is important to note, however, is the relatively consistent formation of both papers. The term formation here refers to the general distribution of fibers across the paper sheet. It

9 See Needham, in vol. II of this publication, p. 21ff. Twin moulds were differentiated as I and II by Karl Theodor Weiss in his seminal watermark study from 1962, pp. 115–118. Watermarks could be positioned in identical configuration on both moulds, but a number of variations were possible. While Weiss noted that watermark designs were moved around, the reversal of their design was not noted for the time period in question. Reversed twin marks dating from the fifteenth century are illustrated by Stieglecker 2009, p. 36.

10 In all paper moulds until the late eighteenth century, the wire was fastened directly to the wooden supporting ribs by the chain stitch. This enhanced water drainage and preferentially drew fibers to the ribs, resulting in a greater fiber mass accumulation and the characteristic shadow areas seen in transmitted light. Today this type of paper is known as "antique laid" to distinguish it from later "modern laid" paper, the result of changes in mould construction.

results from the interplay of several factors, chief among them the condition of the fiber, the concentration of the pulp slurry, and the vatman's dip and shake of the mould, which sloshed the fibers across its surface. These papers show occasional cloudy areas and slight variations in thickness, indicating that the fibers were distributed somewhat unevenly when they settled on the mould (Fig. 6). Such slight overall variations in the production of fine printing papers were neither unacceptable nor surprising, as papermaking was fast, repetitive, and backbreaking work. The papers of the *Washington* and *Graz copies* show these expected variances though they can be considered the same quality.

Evidence that the paper from the *New York copy* deviates from the standard edition comes, first of all, from the position of the watermarks. In Fig. 6, the letter "A" on bifolium 6/7 can be read upright, while the "Tp" on matching bifolium 5/8 is upside down. Obviously, two separate half sheets of paper were brought together for printing. Not only that, the two bifolia distinctly vary from each other in the quality of their overall formation. Bifolium 6/7 is a very thick, uncompressed and therefore voluminous paper, making the characteristic shadow areas along the chain lines virtually invisible. This paper also contains a great number of undispersed fiber knots that appear as roundish dark spots in transmitted light. Fiber knots result from clumps of pulp that dried on the stamper trough and were beaten into the next batch of pulp. By comparison, bifolium 5/8 is very thin and unevenly formed, as indicated by its greater light transmittance and mottled appearance. Additionally, water drops that fell on the fresh fiber mat caused local displacement of fibers, resulting in thin spots in the paper, sometimes called vatman's tears. The different tactile character of these papers can be appreciated when one turns the pages of the *New York copy*.

The thin-and-thick, half sheet combination repeats with consistency throughout the *New York copy*. All of its quires are constructed of half sheets originating from the two different papers, as illustrated for quire C (Fig. 7). In terms of formation quality, the thinner paper is the odd paper because it differs most distinctly from the paper of the other exemplars examined, while the thick paper, in this regard, compares more closely with the paper in the standard copies. The thin paper, with its many structural irregularities, may have been a less expensive proofing paper or maybe it was simply at hand during trial printing when additional paper was needed. Whatever the reasons for these choices, it is evident that care was taken to combine one thin with one thick paper in assembling each quire, possibly to average out the thickness and structural characteristics of the book overall. Bifolium 9/12 contains additional evidence of idiosyncratic paper employment in the *New York copy*, as the "A" watermark is positioned close to the edge of the bifolium instead of in its center (Fig. 7). Apparently, only the center of the sheet was used for printing (for orientation, see Fig. 5). Perhaps the rest of the sheet was flawed and therefore unfit for use. This odd feature additionally reinforces our understanding of the economical attitude toward paper as an expensive commodity.

Let us return once more to the question of watermarks. In ideal situations, watermarks can support the dating of documents as well as the identification of the paper mill, paper-quality, and paper stock. A few comments about the *Sidereus Nuncius* can be made regarding these points. If a dated document bearing an exactly identical watermark can be located, the production date of the undated paper can be narrowed down to a range of several years

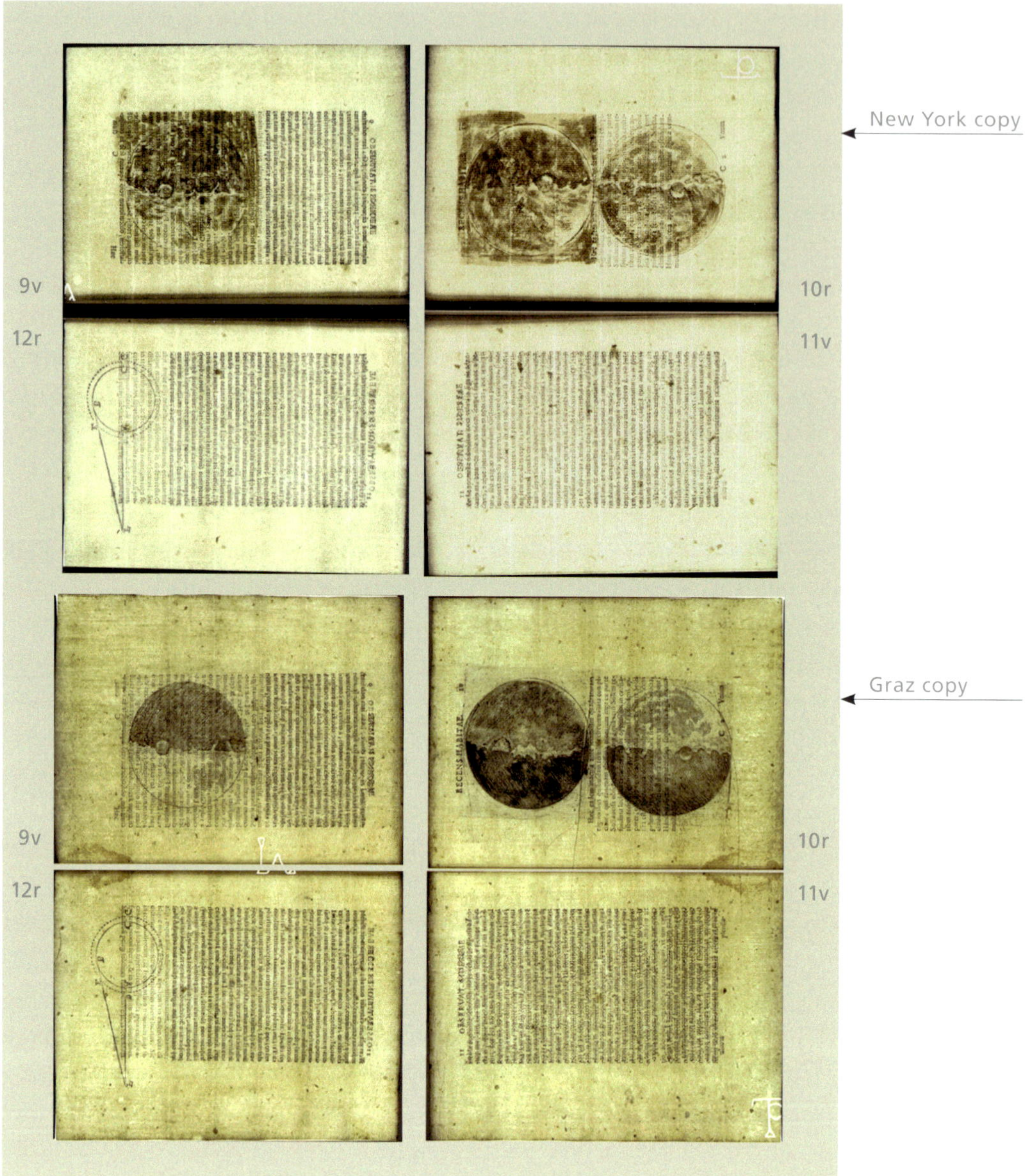

Fig. 7: Galileo Galilei, Sidereus Nuncius, Venice 1610, reconstruction of the paper sheets for quire C of the New York and Graz copies, viewed in transmitted light; the New York copy is seen from the feltside, the Graz copy is seen from the wire side. The watermarks are enhanced for visibility; the folio numbers appear in the margin.

before and after the dated document.[11] The book publication date of March 1610 fixes the *terminus ante quem* for the production of the paper. Also, differences between specimens of standard paper used in the *New York copy* and those found in regular copies allow the

11 The hypothesis established by the paper historian Gerhard Piccard in the 1950s for dating fourteenth- to seventeenth-century paper within four years of the occurrence of a dated paper with an identical watermark has been broadened. A case in point was made by Alois Haidinger in a survey of 300 watermarks in the medieval watermark database "Wasserzeichen des Mittelalters" (WZMA, http://www.ksbm.oeaw.ac.at/wz/wzma.php). He observed that 8% of the watermarks were found in manuscript papers used over a period of 15 to 37 years, either due to continued production or gradual depletion of previously manufactured paper (Haidinger 2004, pp. 18–19). The question of dating remains dependent on how closely one can gauge the paper production and consumption rates in a given historical context.

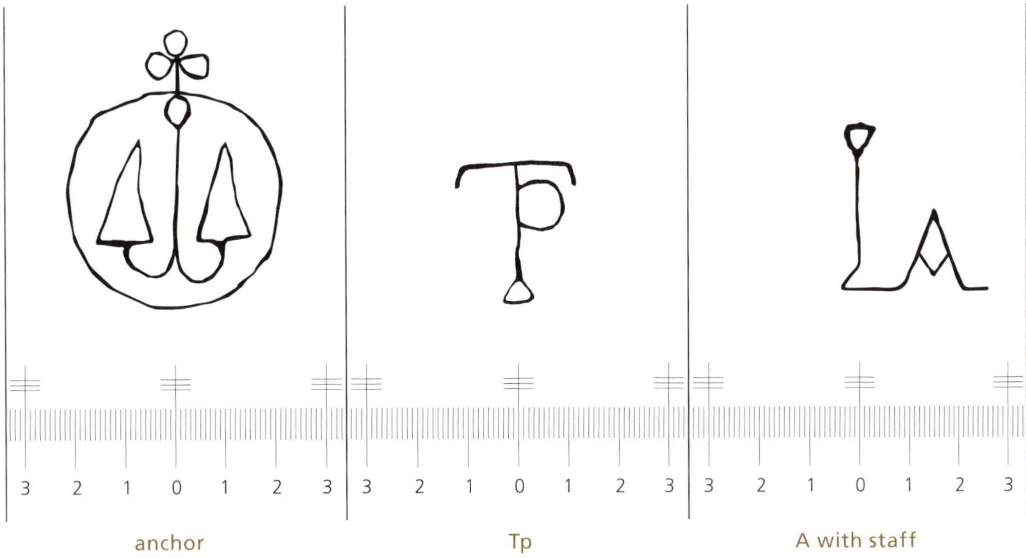

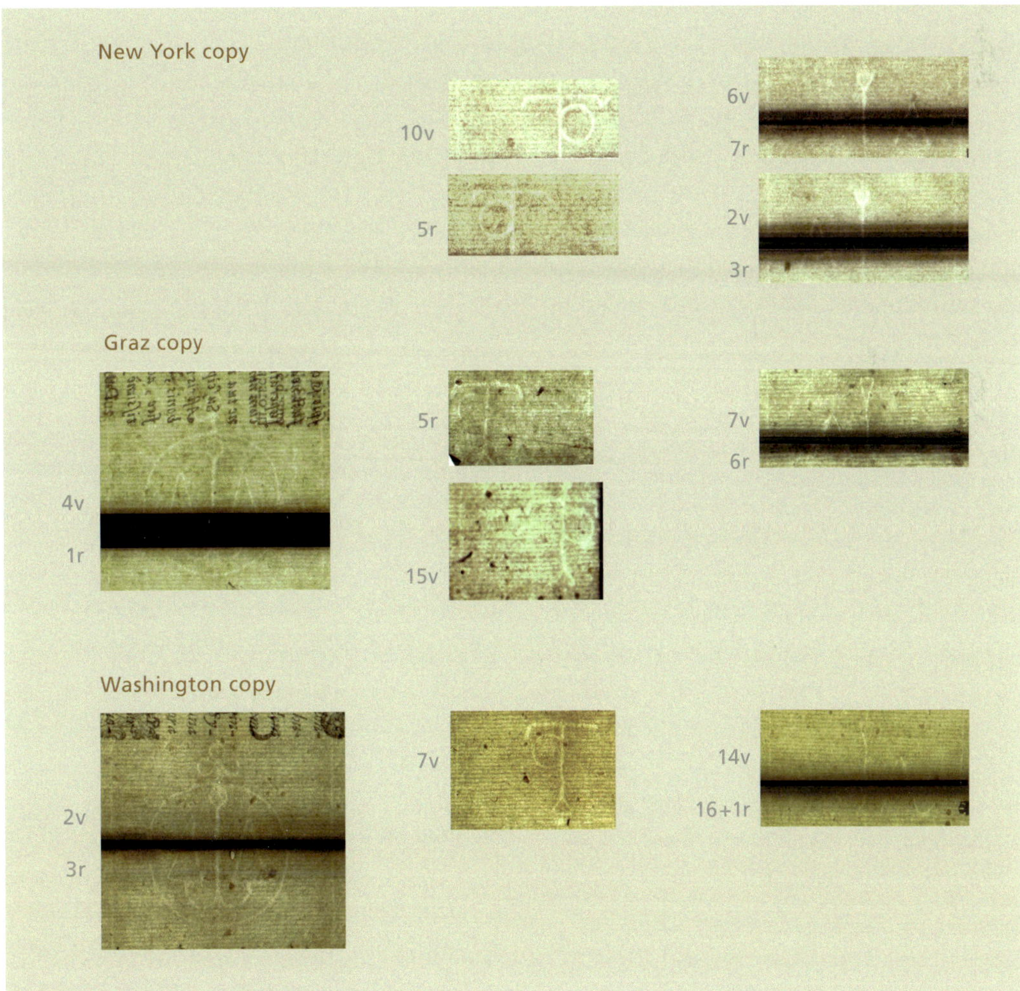

Fig. 8: Galileo Galilei, Sidereus Nuncius, Venice 1610, details of watermarks from the New York, Graz and Washington copies. The anchor watermark appears in quire A of the standard printed copies; the "A" and "Tp" watermarks appear throughout the New York copy and in quires B-G of the standard edition. Gutters that could not be captured photographically appear dark; folio numbers appear in the margin. The drawings indicate chain line spacing of the watermarks (above).

conclusion that the quantity of paper needed for printing the *Sidereus Nuncius* – even assuming a nearly simultaneous time frame for their printing – came from successive or separate production campaigns. Nonetheless, in terms of the paper's origin, the "Tp" countermark indicates that all of these sheets may originate from the same, as-yet-unidentified mill.[12] Furthermore, centermarks were often legal identifiers of paper dimension or quality. In our case, sheets bearing the same type of centermark vary sufficiently in their thickness and evenness to attest to the fact that watermarking is not necessarily a reliable indicator of paper quality. In other words, moulds bearing the same watermark cannot be necessarily associated with one narrowly defined paper type.[13] This serves as a reminder that historical craft production does not always conform to our desire for clear categorization. Finally, comparing similar and identical watermarks serves to differentiate paper stocks. Stock refers to paper that is made from a set of twin moulds with identified watermarks. Slight variances in shape usually exist between any two watermarks that make up twin moulds, even if they were designed as identical shapes. This is because, firstly, the wires were usually bent by hand; secondly, variances in sheet production could create a large number of distortions in the shape and appearance of watermarks; and, thirdly, each mould surface underwent individual alterations with continued use, causing wear, repair, and replacement of watermarks.[14] The "A" and "Tp" watermarks, though formally of the same type, show more than the two variations one would expect for only one set of twin moulds at a given moment in time (Fig. 8). In fact, the watermark shapes found in the sheets of the *New York copy* differ sufficiently enough from those in the regular copies to suggest that they stem from separate twin moulds. Considering this evidence, we account for three distinct qualities of the "A" and "Tp" paper in the *Sidereus Nuncius* production. Within the set of copies examined, two of the paper qualities, namely the thick, homogeneous paper and the thin, cloudy paper, may have been formed on the same twin moulds and are unique to the *New York copy*. The third quality, made on at least one other set of twin moulds, belongs to the standard copies.

The standard printed copies differ from the *New York copy* in another way. Quire A, which was printed last, consists of a different paper, but comparable in quality to the "A" and "Tp" paper. It bears a single watermark showing an anchor with single-lined arms and

12 That the paper was used in Venice does not necessarily indicate a local production. The letter "A" watermark with a V-shaped crossbar existed in Northern Italian papers as early as the fourteenth century (http://www.piccard-online.de, see letter "A" without additional motif). It has been pointed out that fifteenth- and sixteenth-century Venetian book production consumed much imported paper from Northern Italy; see Ornato 2009, pp. 40–41. In principle, identical watermarks on separate paper sheets are, however, no guarantee of a shared origin, as moulds could be transferred to other mills and used there without changing the watermarks. No additional examples (dated or not) of the "A" and "Tp" watermarked paper have yet been identified.

13 Paper qualities were differentiated into two broad categories, whites and wrapping. The former included subcategories that further distinguished paper according to qualities such as, in declining order, *"carta fine"*, *"fioretto"* and *"fiorettone"*, which today cannot be easily defined without identified specimens. Ream weight and, likely, the qualities of sample sheets to be matched in production would have formed the basis of contracts in Italian paper production according to Ornato 2009, pp. 33, 35.

14 The condition of the fiber, pulp consistency, mould shake, manipulation of the wet fiber mat during dewatering, and sheet shrinkage during drying all influenced the readability, shape and dimensions of the watermark; see Bower 2001, p. 8.

a shank within a circle, surmounted by a trefoil (Fig. 8). The watermark is related to an almost identical one featured in a map of Reggio di Lombardia that was published by Justus Sadeler, a late sixteenth- and early seventeenth-century Venetian print publisher.[15] This points to the plausible conclusion that different Venetian printers drew paper from common sources. Since the anchor-watermarked paper is of a quality quite compatible with the "A" and "Tp" paper, it is possible that it was used as a substitute when the latter ran out at the final stage of printing. Perhaps it was even foresight of the imminent shortage that stimulated the allocation of odd half sheets for printing the *New York copy*. The odd sheets could have been awaiting an appropriate application in storage at the printing shop employed by Galileo's publisher, Tommaso Baglioni. One additional paper in both the *New York copy* and the standard printed copies may be mentioned. As related elsewhere, quire D was expanded by one bifolium to accommodate additional text while the book was already being printed.[16] The additional paper stems from yet another supply, some sheets of which bear an as-yet-unidentified watermark.

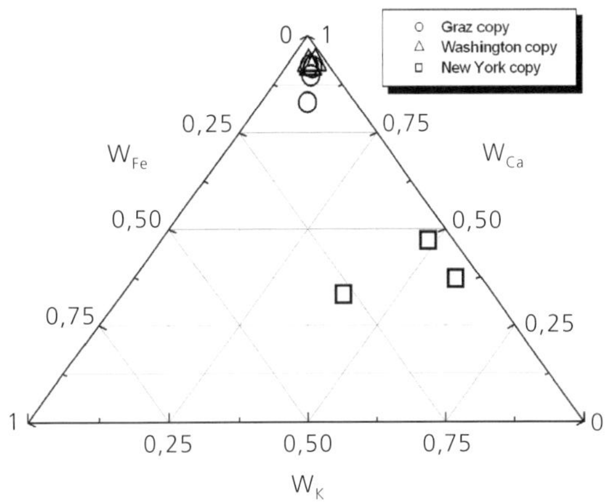

Fig. 9: Mass distribution by weight percentage of potassium (K), calcium (Ca) and iron (Fe) in selected papers, as determined by micro XRF analysis (data normalized/scaled to 1).

Some other factors must be mentioned that help to explain why the paper of the *New York copy* differs from that of the *Graz* and *Washington copies* both in color and in handling properties. It was shown through elemental analysis conducted with x-ray fluorescence that the *New York copy* contains a lower concentration of calcium, a similar amount of iron and a slightly higher amount of potassium than the paper of the *Graz* and *Washington copies* (Fig. 9). Calcium would have been imparted to the paper in the earliest production steps, when the rags were cleaned and softened by being steeped in slaked lime, or calcium hydroxide. Calcium ions promote the paper's long-term preservation.[17] Iron ions from the papermaking process water or metal equipment are ubiquitous in paper. They promote cellulose degradation and are associated with color formation. Potassium, perhaps most typically associated with the presence of aluminum potassium sulfate (alum) and commonly used as an additive to harden the gelatin surface sizing, may also be present due

15 Woodward 1996, p. 112, entry 177, Sadeler, n.pl., n.d., c. 1600: Newberry 2F128.
16 See Needham in vol. II of this publication.
17 Smith 2011.

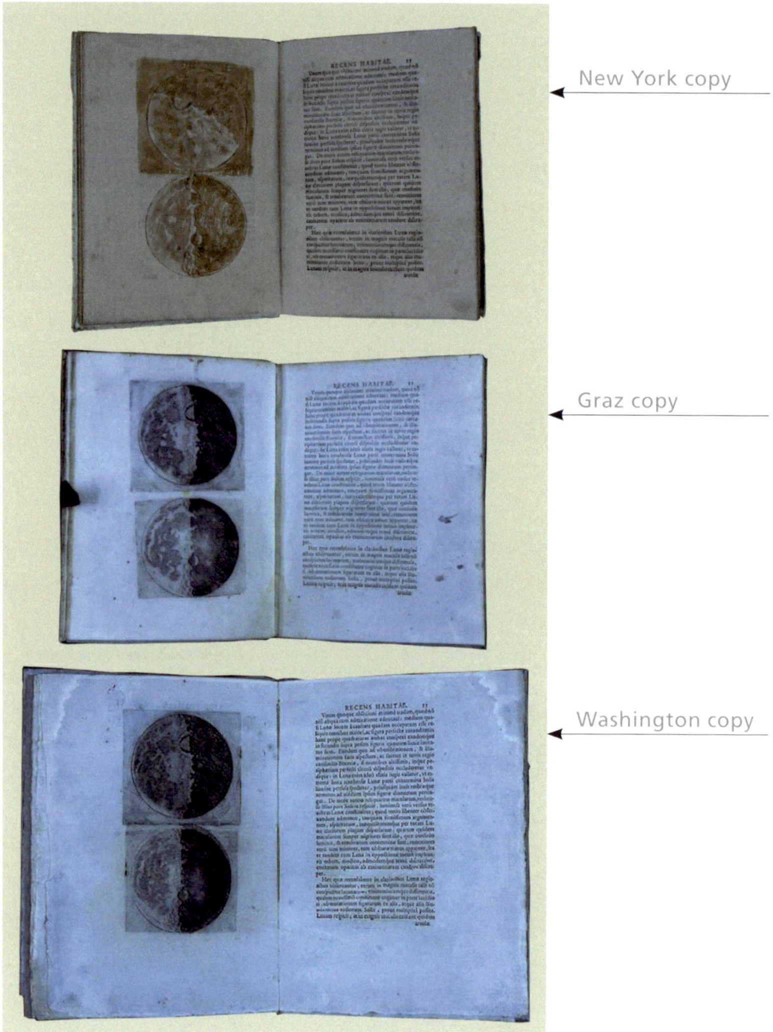

Fig. 10: Galileo Galilei, Sidereus Nuncius, Venice 1610, pp. 10v and 11r, New York, Graz and Washington copies. UVA-induced visible fluorescence indicates the presence of naturally aged gelatin sizing in the Graz and Washington copies.

to a contamination of process waters. The darker color of the *New York copy* in comparison to the *Washington* and *Graz copies* may be explained by its lower calcium content, which corroborates recent studies on the contribution of calcium to brightness retention during natural aging, even of papers containing alum and iron contaminants.[18]

Furthermore, the paper of a number of copies, including *Graz* and *Washington*, received surface sizing with a gelatin solution, which also accounts for their well-preserved state. Sizing was required for writing and drawing papers intended to receive aqueous inks without bleeding. Though optional for book papers, sizing enabled them to receive handwritten glosses, and made them sturdier due to the strengthening effect of the gelatin application. The paper of the *New York copy* apparently received no sizing, an ironic exception

18 Stephens et. al. 2008, pp. 201–215. More recent work was presented at the 38th Annual Meeting of the American Institute for Conservation in Milwaukee, WI, May 14, 2010 by T. Barrett, M. Ormsby, R. Shannon, M. Schelling, J. Mazurek, J. Wade, I. Brückle, J. Lang, J. White, Non-destructive analysis of fourteenth- to nineteenth-century papers.

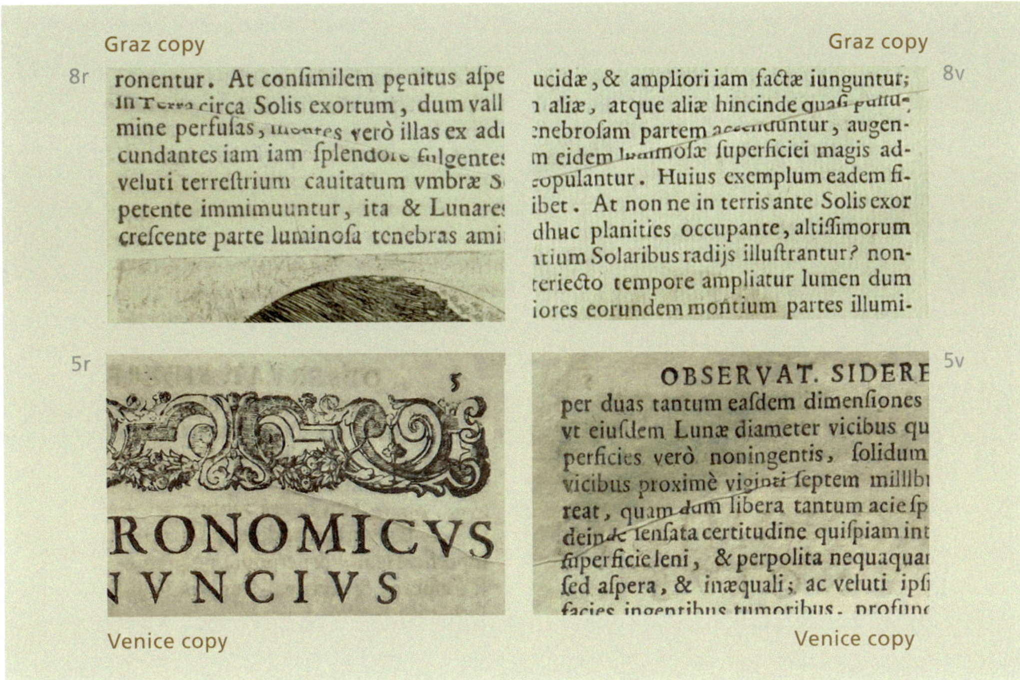

Fig. 11: Galileo Galilei, Sidereus Nuncius, Venice 1610; details; pages 8r and 8v of the Graz copy (top); pages 5r and 5v of the Venice copy (bottom); folds in the printed paper conceal part of the printing recto and verso of the same sheet.

because it is the only copy to receive aqueous ink drawings. The lack of sufficient sizing is evident in the seepage of the bistre ink into the paper, which exceeds the bleed-through effect potentially associated with this ink (see Smith and Brückle p. 41ff.). That the paper carries little if any gelatin sizing is shown by its lack of UV-induced visible fluorescence.[19] Under long wave ultraviolet radiation, the sized Graz and Washington papers (Fig. 10) show a distinctive fluorescence typical of naturally aged gelatin. Another piece of evidence suggests the sizing was administered after printing. Long creases extend across some pages in copies with sized paper. The printed text folds under the creases on both sides of the paper (Fig. 11), indicating that this was not a mistake made during the printing process (as seen, for example, in Fig. 9, p. 107). The creases more likely occurred when the printed sheets, wetted by the gelatin solution, were pressed to remove excess sizing. Thus, sizing was a post-printing operation possibly performed by the bookbinder, as requested by the purchaser of a particular copy.

Because the paper of the *New York copy* was not sized, it also did not have to undergo pressing to expel excess sizing which explains why the paper bears a particularly rich impression of the printed text (Fig. 12). The final "O" in "Galileo" on the title page shows this distinct difference in the imprinted paper topography when seen in raking light. In the *Graz* and the *Washington copies*, the paper surrounding the letter "O" is crushed, the effect of pressing after sizing.

Finally, it should also be mentioned that the pages of the *New York copy* bear more significant signs of use than the other copies examined. The lower page corners show the

19 UVA source: Sylvania Black Light Blue F18WBLB.

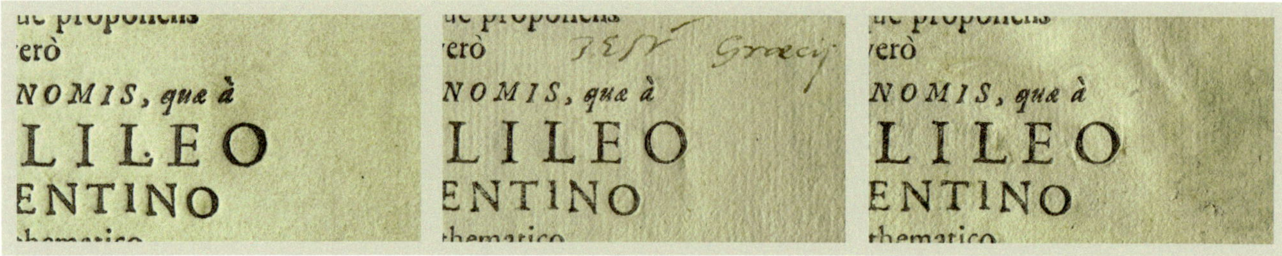

Fig. 12: Galileo Galilei, Sidereus Nuncius, Venice 1610, title page details, New York, Graz and Washington copies, raking light. The papers of the latter two were sized and pressed after printing.

typical ingrained, slightly glossy handling marks that can accumulate over time even with careful page turning. The pages, especially those featuring the drawings, are less compactly aligned than the other pages in the volume, which also contains several of Galileo's later publications. This indicates that the *Sidereus Nuncius* drew exclusive attention time and again.

Summarizing the evidence, the *New York copy* can be closely connected to the standard edition copies by their shared "A" and "Tp" watermarked paper. The *New York copy* is printed on two qualities of this paper while a third quality is found in other early printed copies of *Sidereus Nuncius,* such as the *Washington* and *Graz copies.* The *New York copy* is unconventional in that it systematically combines two half sheets in each quire, one of which is particularly thick and the other thin and inhomogeneous. It differs slightly in its elemental composition, and, unlike the paper in the examined standard copies, was not treated with gelatin sizing. Overall, the evidence of the paper fits well with other evidence presented in this volume indicating the unique status of the *New York copy,* which is best considered a proof copy.

OLIVER HAHN / TIMO WOLFF

THE LETTERPRESS PRINTING INKS

The fifty-seven printed pages of the *Sidereus Nuncius* reveal characters with a dark black appearance typical of printing ink. Printing inks were generally made of black pigment, such as carbon black, lampblack or soot, mixed with vegetable oil, such as boiled linseed oil. Only a few years after the first application of oil-based binders in paintings this invention was used for printing. These inks were able to stick both to the metal surface of the movable lead type and the damp surface of the paper used for printing.

Micro X-ray fluorescence (XRF) analysis showed that the printing ink of the *New York copy* consists mainly of organic materials, while infrared reflectography confirmed that the organic pigment in the ink is primarily made of carbon, an element too light to be detectable by XRF. Further comparative analysis of the black material deposited on the hand-drawn illustrations revealed that the materials differ in their chemical composition. In contrast to the printing ink, the black deposit contains small amounts of inorganic materials detectable by XRF, such as iron. The analyses confirmed that the black ink used to print the text was not the source of the black deposit on the drawings.

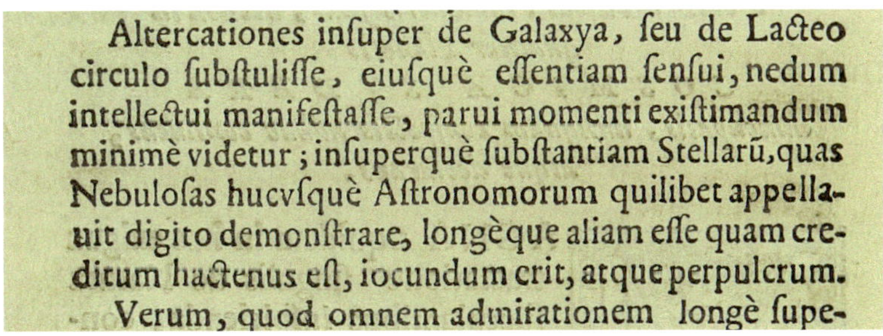
Altercationes inſuper de Galaxya, ſeu de Lacteo circulo ſubſtuliſſe, eiuſquè eſſentiam ſenſui, nedum intellectui manifeſtaſſe, parui momenti exiſtimandum minimè videtur; inſuperquè ſubſtantiam Stellarũ,quas Nebuloſas hucvſquè Aſtronomorum quilibet appellauit digito demonſtrare, longèque aliam eſſe quam creditum hactenus eſt, iocundum erit, atque perpulcrum.
Verum, quod omnem admirationem longè ſupe-

Fig. 1: Galileo Galilei, Sidereus Nuntius, Venice 1610, New York copy, p. 5v, letterpress printing.

MANFRED MAYER

THE PAGE LAYOUT

Long-established in codicological research is an awareness that in medieval books the page design is in many cases not incidental, and instead based on specific proportional relationships (harmonies, regularities, etc.). Referring to the proportions of the text space of the page, but in particular to the position of the text *within* the page, we use the general term layout. Observable at present is a pronounced tendency to incorporate investigations of page layout and related matters in codicological research. Opinion is divided, however, concerning the validity of such endeavors. To be sure, it would be possible to assert that research results to date are neither empirically verifiable nor rationally explicable. Concerning the density of some proportions (meaning those in close proximity to one another, that is to say closely related proportions), a degree of haphazardness on the part of layout designers can lead to erroneous interpretations. Unquestionably, however, medieval scribes and illuminators, and subsequently early printers, were aware of the ideas governing harmonious page design, and applied them purposefully.

The "ideal proportions," such as the square, the Golden Section, the Pythagorean Rectangle,[1] are reflected in modern standardized proportions, have been known since Antiquity, and have been applied in diverse forms – whether individually, or in conjunction with others. In many cases, these proportions pertain not only to the sheet format or text space, but to miniatures or entire picture cycles. Often, the positioning of individual images or text elements in relation to one another or in relation to the sheet as a whole are subject to certain regularities. Only in recent decades have a small number of researchers investigated issues related to the layout of manuscript and printed pages. Referred to here are published essays by Jan Tschichold and Leon Gilissen.[2] Tschichold identified two basic "ideal proportional frameworks" used in medieval manuscripts. These are distinguished, for instance, in the interrelationship of the margins. These proportional frameworks are referred to below as Type A and Type B. In both manuscripts and printed books, the term "margin" refers to the unprinted area found between the text or print space and the edge the page (Fig. 1).

The Type A proportions are inner margin (gutter): head margin: outer margin: tail margin = 2 : 3 : 4 : 6. In Fig. 2, the relationship is visualized by the small squares that are inscribed in the margins. One also speaks of a "canon of subdivision." It is essential that the outer margin be double the width of the inner margin (gutter), and that the tail margin be double the

1 The Pythagorean rectangle has the proportions 4:3, with a diagonal measurement of five units according to the Pythagorean Theorem.

2 Tschichold 1977; Gilissen 1977.

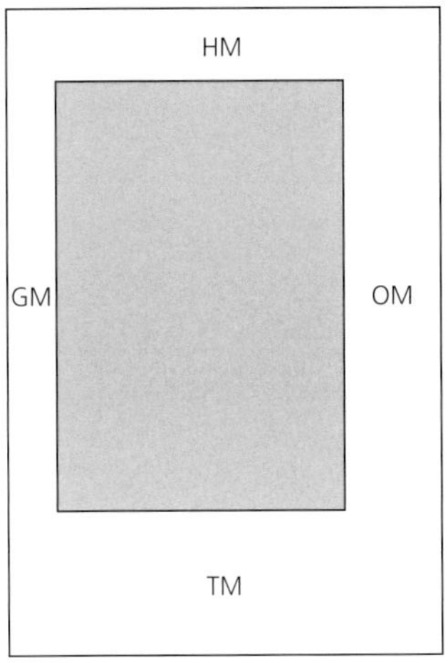

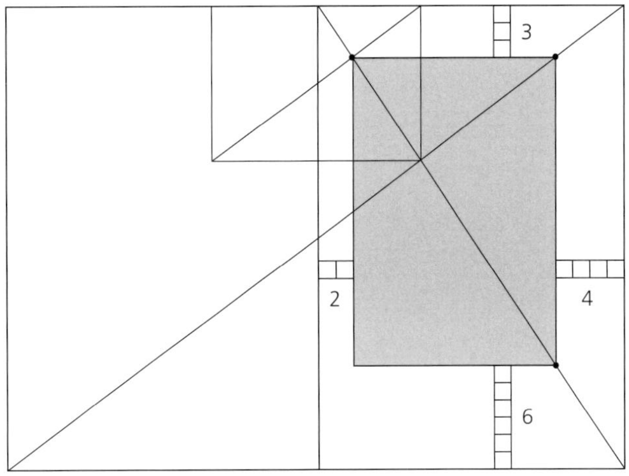

Fig. 1: The position of the margins: GM = gutter margin, HM = head margin, OM = outer margin, TM = tail margin.

Fig. 2: Type A proportional grid.

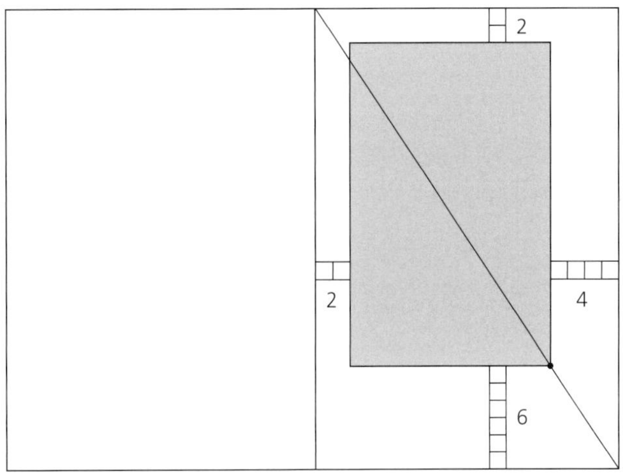

Fig. 3: Type B proportional grid.

width of the head margin. Two corners of the text space lie along the page diagonal, and one corner lies along the diagonal of the sheet.

Alongside the above-described relationship between the margins, Tschichold identified a second "ideal proportional grid" known as Type B, which corresponds closely to the page layout of the *Sidereus Nuncius*, making it highly interesting for this project (Fig. 3). With the Type B proportional grid, the inside and upper margins have equal widths. The proportions of the text or print space for the most part correspond to the Golden Section. Only one corner of the text space lies along the page diagonal.

Concerning the transfer of proportion studies onto the *Sidereus Nuncius*, only the *Graz* and *Washington copies* have been considered. The format and proportions of the text space and its position on the page were investigated. From p. 5r onward in the *Graz copy*, the print space is approximately equal in size, with an average format of 158 × 95 mm (height × width).[3] If

3 Measurements are taken from the base of the letters on the lowest integrated line up to the mid-height of the small lowercase letters of the uppermost integrated line. Ascenders and descenders are not taken into consideration.

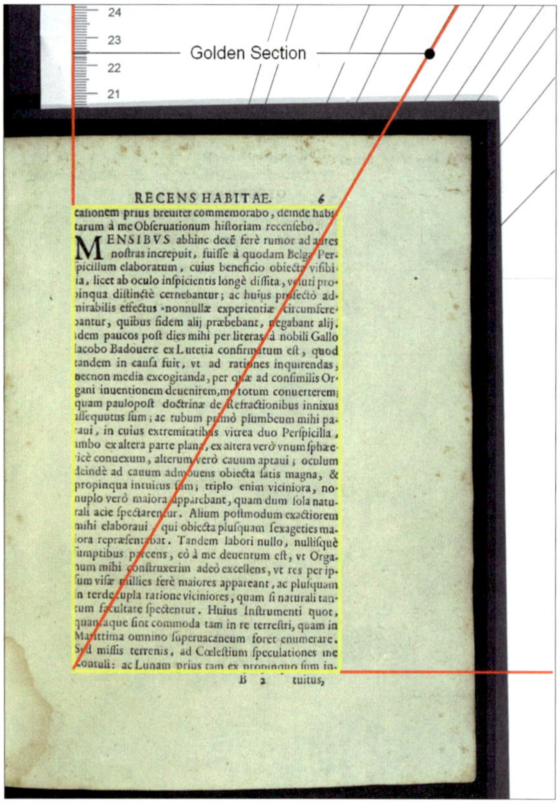

Fig. 4: Galileo Galilei, Sidereus Nuncius, Venice 1610, p. 6r with Type B proportional grid superimposed, Graz copy.

these measurements are standardized as a proportion, we arrive at the ratio of height: width = 1.65:1. This means that the proportions of the print space are very close to the Golden Section. Rounded off, the Golden Section has the proportion 1.62:1. Deviations in the proportions of the print space from the Golden Section are less than 2%.[4] Fig. 4 shows p. 6r of the *Graz copy*. The dimensions of the print space form the yellow rectangle. The sides were laid out electronically on a proportional schema without distortion. The red line connects the corners of the print area, and is nearly congruent with the markings for the Golden Section on the template.

The fact that the fore and bottom edges of the book block of the *Washington copy* have never been cut makes it an ideal object for studying the page size presumably intended by the printer. Used here are pages 11r and 12v, as well as the unnumbered pages, which open to show the Pleiades. Assuming that the sheets were folded by the bookbinder exactly at the center, the Type B linear framework can be laid out on the opened book. Type A can be distinguished from Type B by the minimal height of the letter's upper margin. In this case, the upper margin has the same width as the inside margin, and cannot have been conceived as being much taller. This conclusion results from a reconstruction of the print sheet (see Fig. 6, p. 133). In Figs. 5 and 6, the linear grid is superimposed on the *Washington copy*, pages 11v and 12r, as well as the pages showing the Pleiades. Notably, the outermost star on

4 (1.65 – 1.62) / 1.62 = 1.8% < 2%

the left lies approximately 1 cm inside of the edge of the sheet. Due to page trimming by the bookbinder, this star is absent from both the *Graz* and *New York copies.*

While the original proportional intentions cannot be determined with absolute certainty, it is the view of this author that the layout of the *Sidereus Nuncius* was based on a specific constructive principal. The close proximity of the proportions of the print space and the Golden Section is notable. Thanks to the *Venice copy,* furthermore, we have precise information concerning the original dimensions of the print sheets. The format of the uncut sheets is ca. 38 × 54 cm (h × w), a well-established ratio. If we assume that the book-

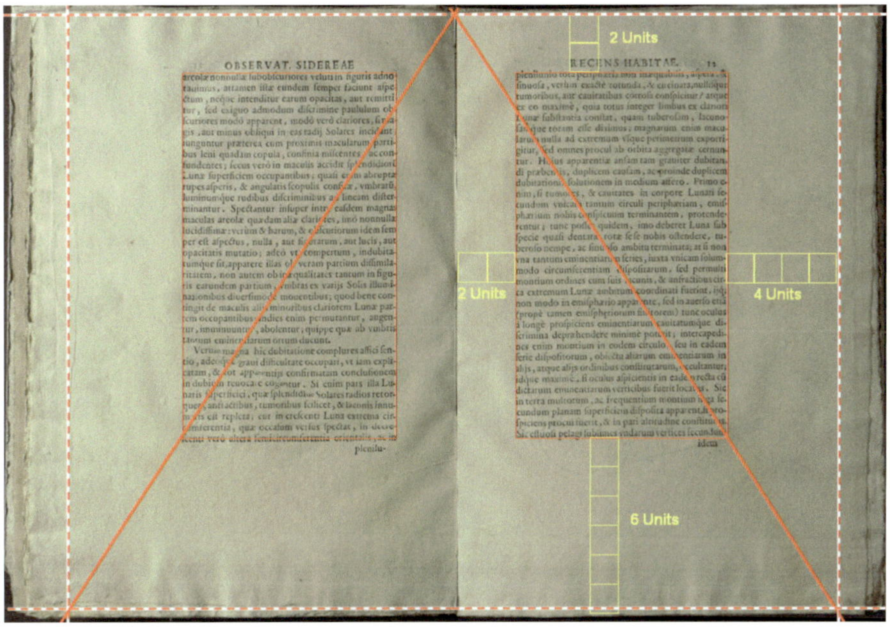

Fig. 5: Galileo Galilei, Sidereus Nuncius, Venice 1610, pp. 11v and 12r with Type B proportional grid superimposed, Washington copy.

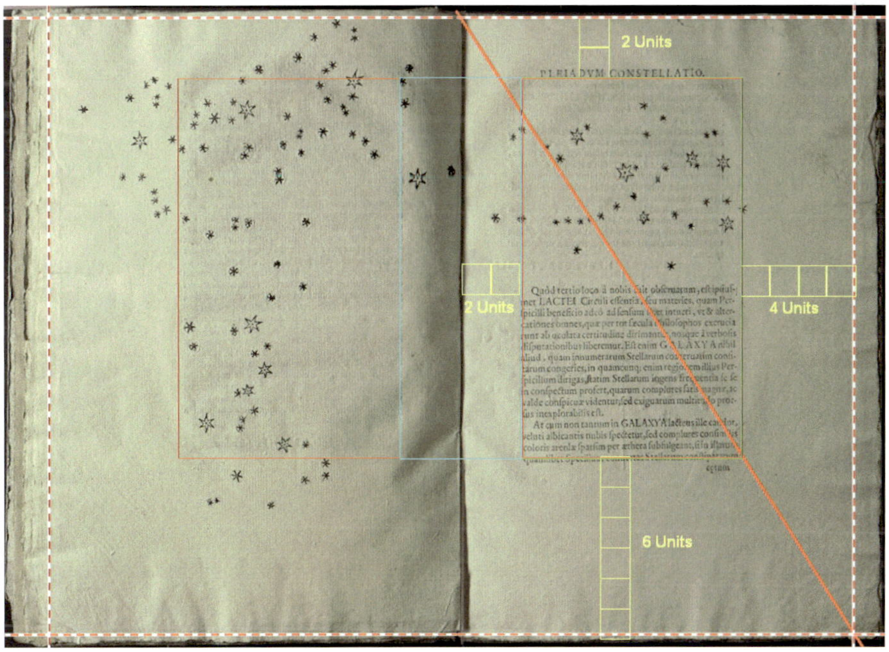

Fig. 6: Galileo Galilei, Sidereus Nuncius, Venice 1610, unnumbered pages showing the Pleiades with Type B proportional grid superimposed, Washington copy.

binder ideally trimmed only 1–2 cm from the edges of the full sheets when the book was bound for the first time, the outer and tail margins must actually have been much wider than they are in the *Graz* and *New York copies*. If one of the eligible constructive principles determining page design is superimposed, the result is a harmonious page layout, as can be seen in Fig. 7. It is debatable whether these deliberations can be regarded as consistent, and whether the original appearance of the pages was planned as presented here. It must be acknowledged that different copies were trimmed to different degrees, resulting in very different end proportions, which were the result of binding and, in some cases, rebinding the book, borne out of the necessity of creating a clean edge to which considerations of proportional design were sometimes secondary. In any event, we should certainly attempt to detect such concealed information when studying manuscripts and early printed books.

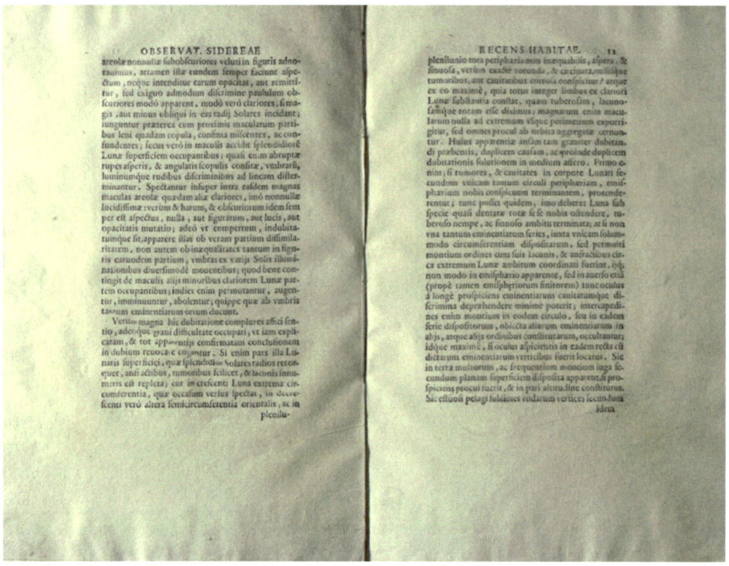
OBSERVAT. SIDEREAE

RECENS HABITAE. 12

Fig. 7: Galileo Galilei, Sidereus Nuncius, Venice 1610, pp. 11v and 12r of the Washington copy with cropping adjusted to Type B proportional grid.

MANFRED MAYER / IRENE BRÜCKLE / THERESA SMITH

WATERMARK DISTRIBUTION IN SELECTED COPIES

Five exemplars of *Sidereus Nuncius* were compared for the arrangement of the quires within the book and the watermarks on the folios: the *New York, Graz, Washington, Venice* and *Munster copies* (Fig. 1). The text block is composed of seven quires, lettered A through G. With the exception of quire D, each quire has four folios that are usually made from one folded sheet of paper (see Fig. 2, p. 130). The pagination as it appears in the books is given for each folio, for example, quire A is composed of folios 1 to 4 (according to bibliographic nomenclature, A1 to A4). Quire D contains an extra inserted paper sheet and two folios without pagination.[1]

Four watermarks were found in the five copies examined. The *Munster* copy is printed on a particularly fine paper bearing a watermark with a star over a crown. The paper that bears the "A" centermark and "Tp" cornermark appears in quires B through G of the standard edition copies, including the *Graz, Washington,* and *Venice* copies. In quire A, these copies all have paper watermarked with an anchor in a circle. While the *New York copy* is made up exclusively of the "A" and "Tp" watermarked paper, each quire is made up of two separate half sheets (see Brückle et al. pp. 135–138).

1 This anomaly is explained by Needham in vol. II of this publication.

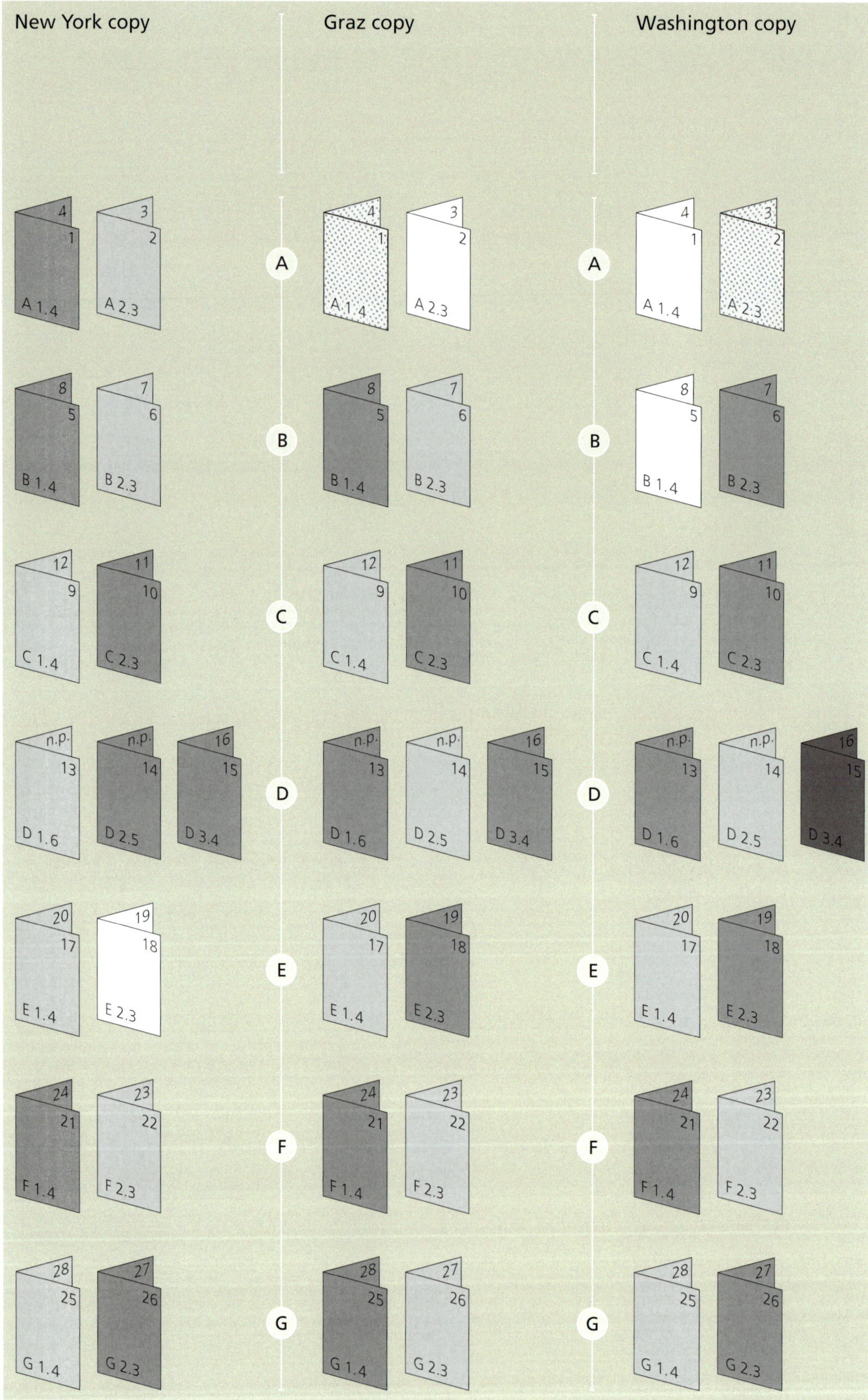

Fig. 1: Diagram of the quires and watermarks of the New York, Graz, Washington, Venice and Munster copies.

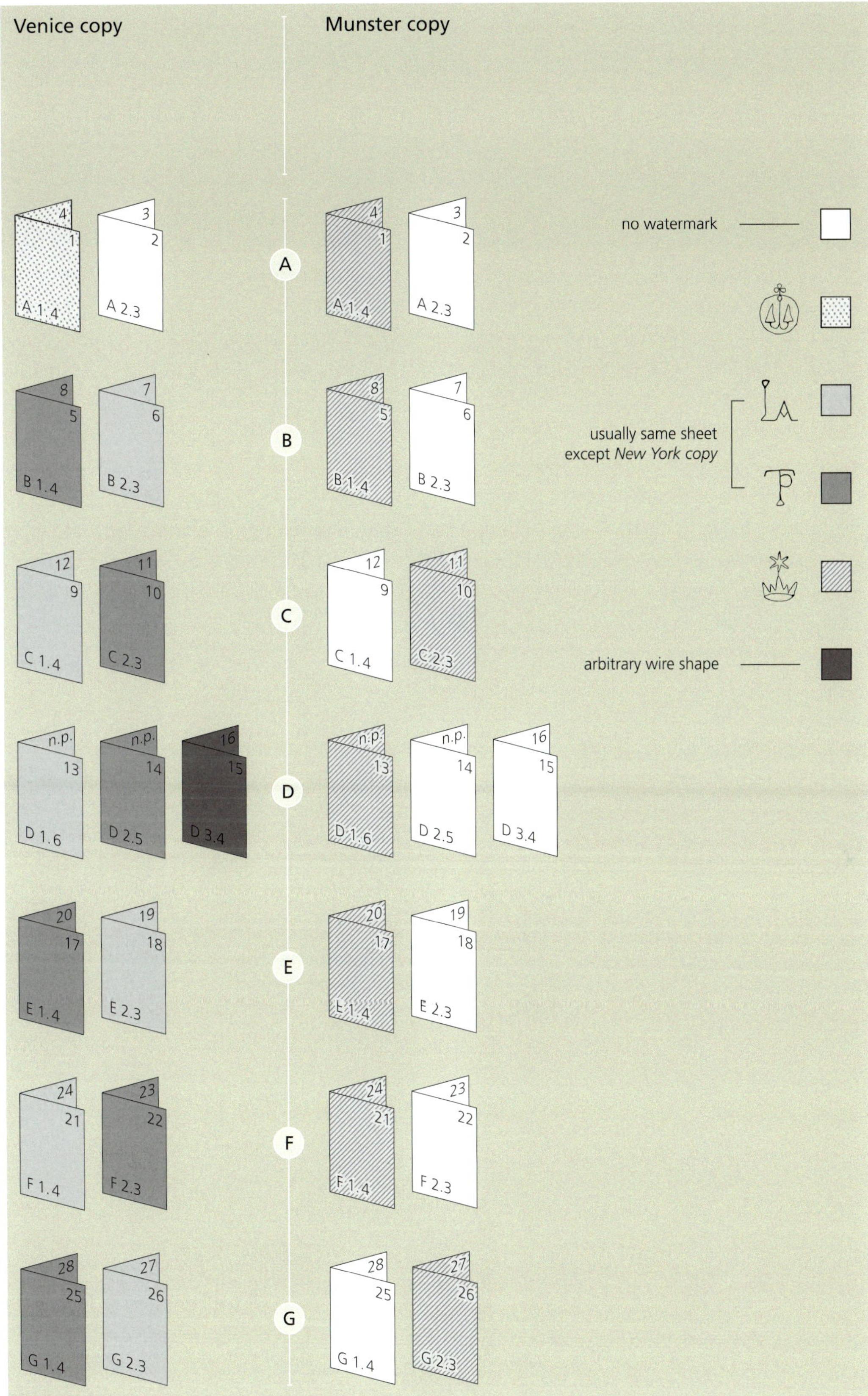
Venice copy
Munster copy
A
B
C
D
E
F
G
A 1.4
A 2.3
B 1.4
B 2.3
C 1.4
C 2.3
D 1.6
D 2.5
D 3.4
E 1.4
E 2.3
F 1.4
F 2.3
G 1.4
G 2.3
n.p.
no watermark
usually same sheet
except *New York copy*
arbitrary wire shape

MANFRED MAYER

THE EXAMPLE OF THE *GRAZ COPY*

What follows is a detailed description of the *Graz copy* as an example of the standard printed edition of *Sidereus Nuncius*. The copy is stored in the Department for Special Collections at the Graz University Library under the modern shelf number I 4099. On the spine of the inconspicuous and unadorned parchment binding is the remnant of a shelf number on a light blue paper label (Fig. 1). This and the numerous handwritten entries on the title page show that the book came from the collection of Paul Guldin.

Paul Guldin originally trained as a goldsmith, but in 1597 he entered the Jesuit order in Graz. There he was educated in mathematics and taught at schools in Rome, Graz and later the university in Vienna. He died in Graz in 1643. Guldin and Galilei are connected through Guldin's pupil Melchior Inchofer (1584–1648), who testified against Galileo Galilei at the latter's trial. After Austria's Emperor Joseph II carried out his monastery reform in 1782,

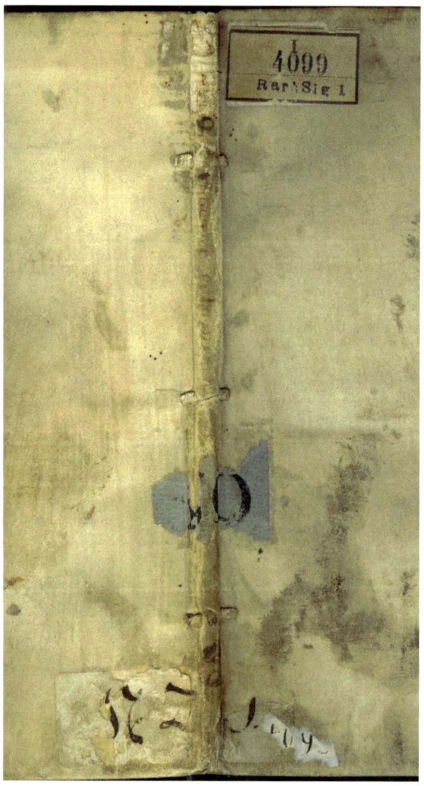

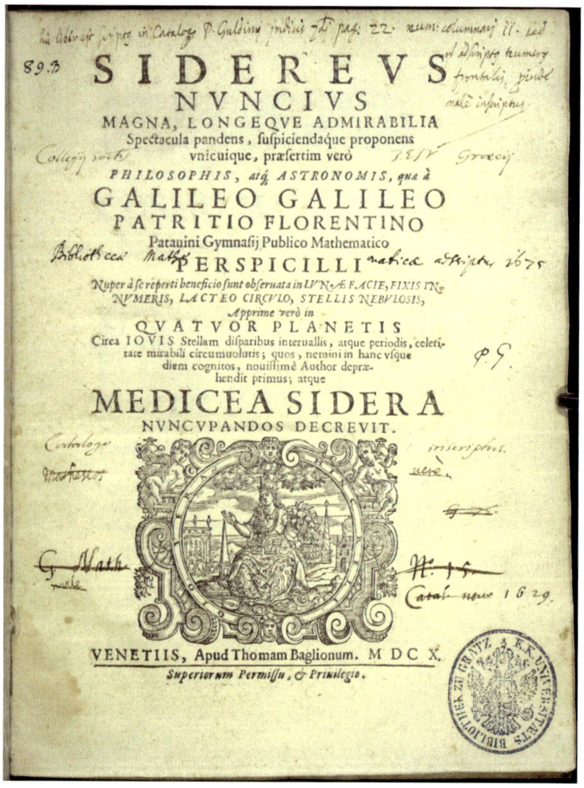

SIDEREVS
NVNCIVS
MAGNA, LONGEQVE ADMIRABILIA
Spectacula pandens, suspiciendaque proponens
vnicuique, præsertim verò
PHILOSOPHIS, atq; ASTRONOMIS, quæ à
GALILEO GALILEO
PATRITIO FLORENTINO
Patauini Gymnasij Publico Mathematico
PERSPICILLI
Nuper à se reperti beneficio sunt observata in LVNÆ FACIE, FIXIS INNVMERIS, LACTEO CIRCVLO, STELLIS NEBVLOSIS,
Apprime verò in
QVATVOR PLANETIS
Circa IOVIS Stellam disparibus interuallis, atque periodis, celeritate mirabili circumuolutis; quos, nemini in hanc vsque diem cognitos, nouissimè Author deprehendit primus; atque
MEDICEA SIDERA
NVNCVPANDOS DECREVIT.
VENETIIS, Apud Thomam Baglionum. M D C X.
Superiorum Permissu, & Priuilegio.

Fig. 1: Galileo Galilei, *Sidereus Nuncius*, Venice 1610, spine, Graz copy.

Fig. 2: Galileo Galilei, *Sidereus Nuncius*, Venice 1610, title page, Graz copy.

the book was transferred from the library of the Graz Jesuit College to the Graz University Library.

The previous owners' inscriptions on the title page are extensive (Fig. 2). The inscription at the top reads, *"Hic liber est scriptus in catalogo P. Guldini indicis 2di pag: 22. num: columnari ii sed / nunc adscriptus numeris / frontalis, pro inde / malè inscriptus"*. Other inscriptions include: *89. B; Collegij Societatis Jesu Graecij / Catalogo inscriptus; Bibliothecae Mathematicae adscriptus 1675; P. G.; Matheseos verè* [crossed out]; *G. 75* [crossed out]; *G. Math. No 15* [crossed out]; and *Catal. novo 1629*. On the verso there is a possible shelf number: *8*.

The letterpress text is printed on handmade, antique laid paper. The text consists of 60 pages with numbered folios, plus two endpapers. As in the other standard copies, two pages between fols. 16 and 17 are not numbered. The page format is 22.5 × 17.1 cm (h × w). The text is in its original binding, a flexible parchment binding without filling in the covers (a limp vellum binding). The cover format is 23.2 × 17.5 cm (h × w), and the spine is 0.8 cm thick. There are at least three paper labels present, two fragmentary. On the spine there is a handwritten inscription in brown ink, mostly faded. Under ultraviolet radiation it is clearly legible: *GALILEO nuncius sidereus* (sic). Two original green ribbon ties were attached to the covers, but are now missing.

The sewing is original and unaltered, with all-along stitching on three simple parchment cords that are sewn through the vellum cover at the hinge. Two kettle stitches are

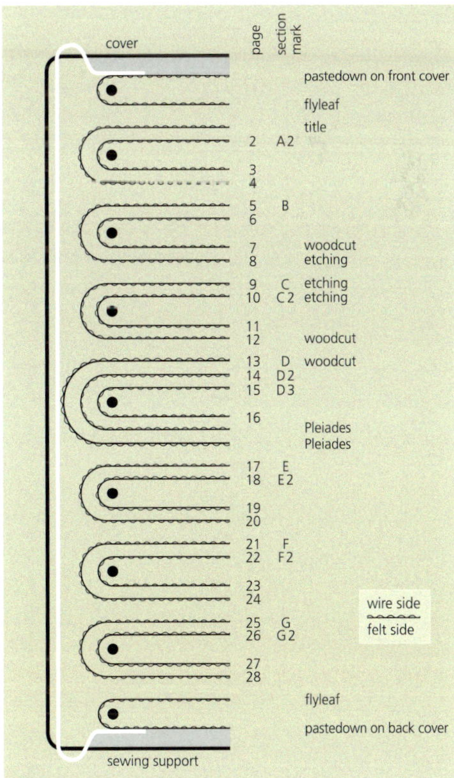

Fig. 3: Schematic U-diagramm of the signature structure of the Graz copy.

present, with neither headband nor tailband. The page edges are trimmed and speckled a delicate pink.

For greater clarity, see the diagram of the signatures (Fig. 3). The book block of the *Graz copy* consists of seven signatures, lettered A–G. Each signature consists of two bifolia, or eight pages. Signature D is an exception, consisting of three bifolia, or twelve pages. The two pages following folio 16 are not numbered. The numbering resumes with folio 17 in signature E.

Three watermarks are found in the text paper of the Graz copy and another on the endpapers. The paper of signature A is watermarked with an anchor (Fig. 4). Signatures B to G all display the "A" and "Tp" watermarks. The endpapers are watermarked with a "P".

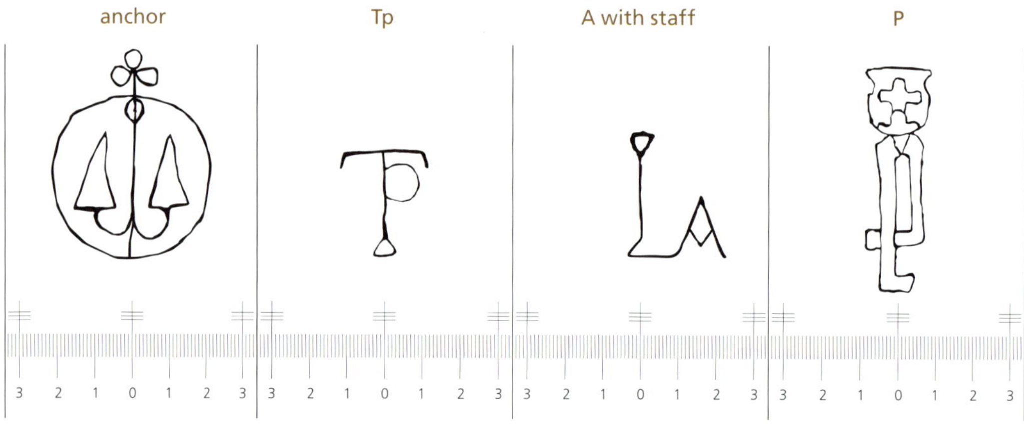

Fig. 4: Watermarks in the Graz copy.

V

AFTERWORD

HORST BREDEKAMP

THE FINGERPRINTS AND OTHER DISCOVERIES

GALILEO'S PRIDE AND FEAR

On March 19, a week after the book was completed, Galileo sent a remarkable letter to Belisario Vinta in Florence. It is known both as a draft and in the finished, sent form. With Galileo's fear that the weakness of the hastily printed book could rob him of the favor of Cosimo II de'Medici,[1] his explanation of that problem in the draft becomes a self-description of the fear of competition that was eating him up: "And the postponement and delay of the publication were connected with my too great danger and my risk that possibly another would preempt me: for this reason, I have decided to send this report ahead, together with the naming of the stars, in order to publish shortly hereafter many more special observations that I want to make of precisely these things."[2]

The fear of being too late in relation to competitors has probably seldom been expressed as openly as in this letter to the influential Vinta, who was indispensable for the desired position at the court of the Medici. The finally sent version of the letter contains once more the apology that the book did not get a perfect form, "because the lack of time did not allow it and because I absolutely did not want to postpone the publication, in order not to run the risk that someone else could have stumbled upon the same discoveries and preempted me. And for this reason, I have published it in the form of a report, most of which was written while the early parts were printed, with the intention that this [report] should be printed anew as soon as possible with many additions from other observations."[3]

The haste that took hold of Galileo between January and March 1610 could not have been more clearly described. The reason for it is shown in a letter of January 7, 1610, in which Galileo reports for the first time on his telescopic views of the sky. This letter is immensely important because in it Galileo laid the foundations for the first part of the *Sidereus*

1 *"Sarà ancora necessario che io sia scusato se l'opera non esce fuori stampata con quella magnificenza che alla grandezza del soggetto si saria richiesto"* (*Opere*, vol. X, p. 298, l. 13–15).

2 *"et l'indugiare et differire la publicazione era con mio troppo pericolo et risico che forse qualche altro non mi havesse preoccupato; onde mi sono resoluto mandare innanzi questo avviso, insieme con la denominazion delle stelle, per publicar poi in breve molte altre particolari osservazioni, le quali vo continuando di fare intorno a queste medesime cose"* (*Opere*, vol. X, p. 298, l. 15–21).

3 *"perchè l'angustia del tempo non l'ha permesso, nè io ho voluto punto prolungare la publicazione, per non correr risico che qualche altro non havesse incontrato l'istesso et preocupatomi; et per ciò l'ho mandato fuori in forma di avviso, scritto la maggior parte mentre si stampavano le cose precedenti, con proponimento di ristamparlo quanto prima con molte aggiunte di altre osservazioni"* (*Opere*, vol. X, p. 300, l. 86–92).

Nuncius, which was released just over two months later, on March 13. But above all, it clearly shows the dimensions of the pressure under which he placed himself, because in this letter Galileo laid claim to the first rights to the moon: "Not a single one of the aforementioned observations are seen or can be seen without an excellent instrument; for this reason, we may assume that we have been the first in the world to see the heavenly bodies at such close range and so clearly."[4]

This statement was based on the precision and the impatience of the instrumental eye. The gaze is rapid; while script must be taken in over time, the gaze, as a coup d'oeil, can grasp situations, objects, and problems in a fraction of a second. With the telescope, the speed of this organ had become the standard that inspired the competition. Telescopes circulated throughout Europe and soon also in Asia, and any moment someone else could beat Galileo to the punch by recognizing and publishing what he saw. The power of the eye was what granted Galileo the triumph of being the first person to gaze into the heavens as they were; the speed of the gaze, however, made him Argus-eyed in regard to potential competitors. Galileo's early use of the telescope is bound up with all of this; it produced not only a new image of the heavens, but also an equally magnificent and problematic culture of research. With the telescope, the need for increased speed and with it a previously unknown competitive pressure entered the world.

THE FINGERPRINTS

The sequence of the publication of the *Sidereus Nuncius* and the haste of its production reveal the psychology of the instrumental gaze. Through Paul Needham's chronology from the first to the last few days of the printing of the book (vol. II), the mind's eye can imagine the typesetters, constantly busy hanging up the just-printed quires to dry, at the same time looking for errors, correcting them for the following step of printing, and then looking for more errors until the final printing of the last copy. The time pressure to get this book printed is palpable.

These traces of haste also make the producers of this enigmatic little book downright physically palpable. On page 21r of the *Washington copy* are the black imprints of the four fingers of a right hand (Figs. 1 and 2). It has not been possible so far to determine whose they are, but as they contain traces of lead and presumably iron, the two metals that are used in printing shops, they must be those of one of the typesetters or even of the publisher.

The search for such chemical traces has strengthened the suspicion that Galileo produced the etchings himself in the last days before they were printed.[5] But this means the technique used is different than previously thought. These are obviously not drypoint prints, but etchings. The occasion for this suspicion was offered by the lower right corner of the upper moon square on page 10v (Fig. 3). Here, traces were found of lines that cannot have followed any conscious design, but an unconscious smudging with a finger. Next to

4 "*Di tutte le sopradette osservationi niuna se ne vede o può vedere senza strumento esquisito; onde possiamo credere di essere stati i primi al mondo a scuoprire tanto da vicino et così distintamente qualche cosa dei corpi celesti*" (*Opere*, vol. X, p. 277, l. 125–128).

5 Bredekamp 2007, pp. 189–207.

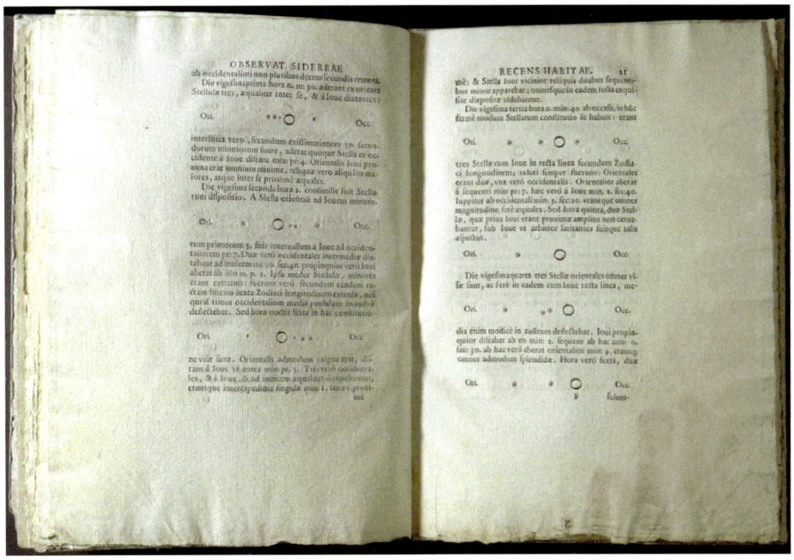

Fig. 1: Galileo Galilei, Sidereus Nuncius, Venice 1610, pp. 20v and 21r, Washington copy.

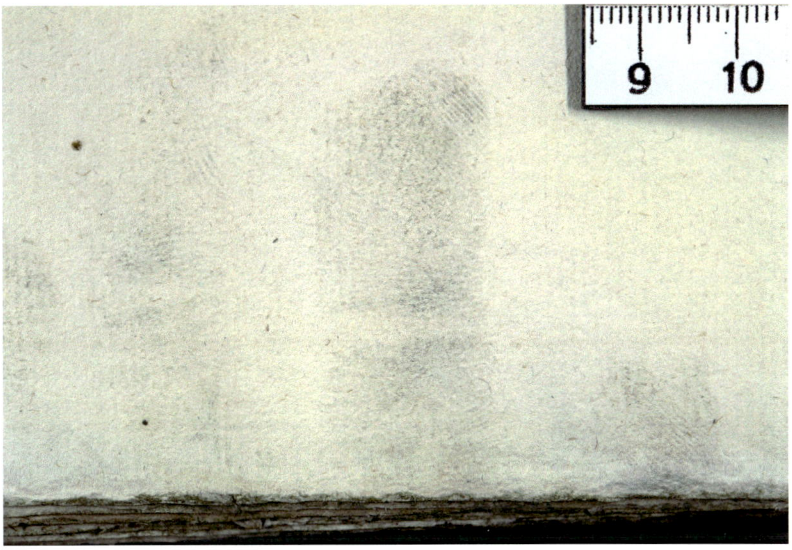

Fig. 2: Galileo Galilei, Sidereus Nuncius, Venice 1610, p. 21r, Washington copy, detail of the fingerprints.

the arc of the circle and also above the lower margin border are fields of two fingerprints that cannot have been incised, but can have come onto the plate only during the process of etching. Apparently, some person, probably Galileo himself, unintentionally grasped the coated plate, allowing the acid to penetrate here and etch these spots.

This makes the fidgety technique of depicting the moon explainable. The lines do not seem engraved and stripped out with a drypoint needle, but drawn with a stylus, as is customary with an etching. Many details, for example the lower left curve of the lower moon in the field where the curving circles run parallel (Fig. 4), make the scales fall from one's eyes. The neighboring field of shadow testifies equally clearly to the use of the etching technique. Here, too, the haste of the final days of the book's production is visible. All these details provide a glimpse of the motoric functions of Galileo's proceeding, and this indirectly provides a trail to the psychological disposition in which this book was made.

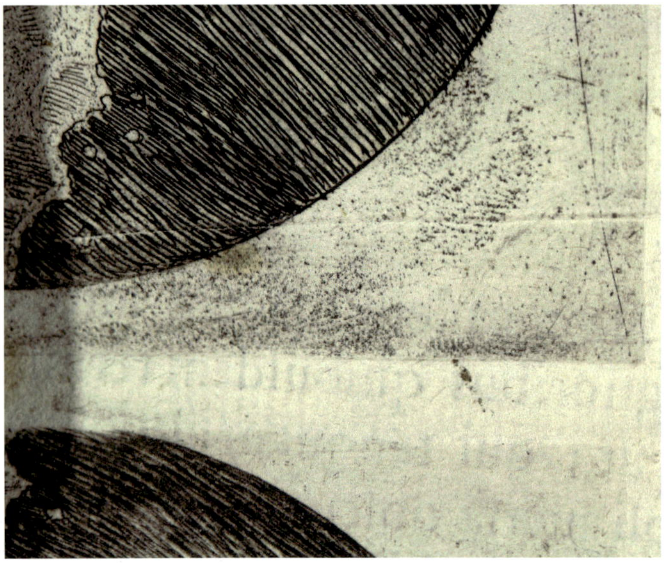

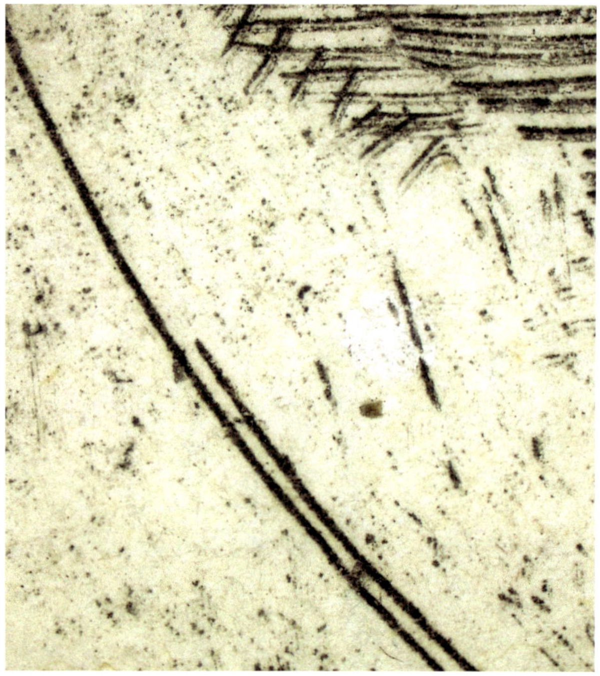

Fig. 3: Galileo Galilei, Sidereus Nuncius, Venice 1610, p. 10v, Graz copy, detail of the etched fingerprints in the upper moon.

Fig. 4: Galileo Galilei, Sidereus Nuncius, Venice 1610, p. 10v, Graz copy, detail of the outline of the lower moon.

JUPITER'S O

With its proof of the moons of Jupiter, the *Sidereus Nuncius* presented an additional small planetary system as evidence for the validity of the Copernican system. In the text, Galileo represented the various arrangements of this system by setting Jupiter as an oval in the middle, circled by the satellites, marked by asterisks (Fig. 5). Often, these depictions have been regarded as woodcuts, especially since Galileo noted in his January 15 entry to his Jupiter diary that he wanted to depict them with this technique: "The stars will be cut in wood, all in one block, the stars white and the rest black, and then the block will be cut into strips."[6] But they are, as Paul Needham has emphasized, not a woodcut, but a letterpress O turned ninety degrees (Needham, p. 129). A comparison through the microscope of these Jupiter symbols with the capital O used in the text has demonstrated that the capitals in Galileo's name on the title page, among others, were used as figures of the Jupiter. And this is true for both his first and last names (Fig. 6). Thus, the O of the last name (Fig. 7) was placed in the middle of page 20v, as is shown in particular by a detailed photo (Fig. 9).

Five years before the release of the *Sidereus Nuncius*, the Dutch art theorist Jan van den Velde published an engraving showing a circle flanked by two ovals (Fig. 8).[7] He referred to Giotto di Bondone's ability as a draftsman (as reported by Giorgio Vasari) to draw

6 *"per l'inanzi faransi intagliar in legno tutte in un pezzo, e le stelle bianche il resto nero e poi si segheranno i pezzi"* (*Opere*, vol. III/2, p. 428).

7 Lavin 2006, pp. 449–491.

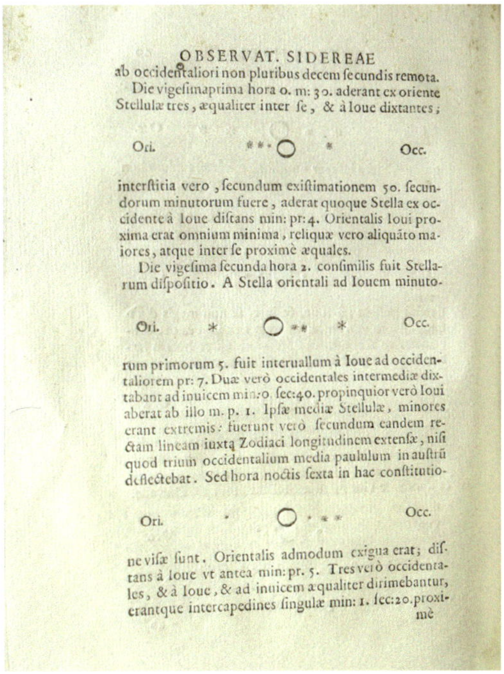

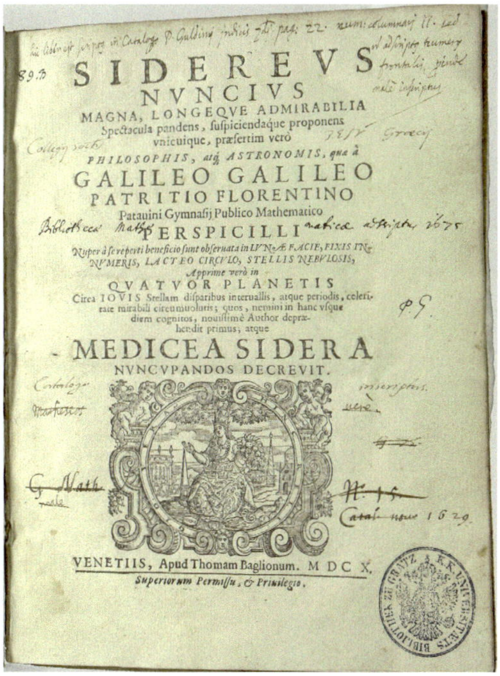

Fig. 5: Galileo Galilei, Sidereus Nuncius, Venice 1610, p. 20v, Graz copy.

Fig. 6: Galileo Galilei, Sidereus Nuncius, Venice 1610, title page, Graz copy.

a perfect circle on the wall without any mechanical aids.[8] Since this was associated with the two O's in Giotto's name,[9] a similar connection became possible in the *Sidereus Nuncius* with the two circles of the O in Galileo's name. In his 1661 self-portrait, now in Kenwood House, Rembrandt van Rijn depicted himself between two perfect circles (Fig. 10), which recall the two O's in Giotto's name and his art of the circle; and with an eye to this tradition, in 1921 the Viennese portraitist Ferdinand Schmutzer created a picture of Albert Einstein that characterized him as a new Giotto (Fig. 11).[10] The cosmological turn already implicit in Giotto[11] is equally underscored for Galileo and Einstein.

In some respects, Galileo's first biography, from the pen of his student Vincenzio Viviani, was made to resemble Vasari's vita of Giotto, thereby emphasizing the great nature researcher's disposition to artistic genius.[12] The biographer could have referred in particular to the Jupiter of the *Sidereus Nuncius*, whose form, taken from the "GALILEO" of the title page, connected the name of the celebrated person with a perfect circle, as van den Velde had already done with Giotto.

Galileo had dedicated the new spots of light in the sky to the Medici and named them for the Medici, in hopes of finally receiving the yearned-for position as court mathemati-

8 Vasari 1906, p. 381f.
9 Lavin 2006, p. 456.
10 Lavin 2003, pp. 37–43. On Rembrandt, cf.: Wellmann 2005, pp. 147–181.
11 Lavin 2006, p. 456f.
12 Segre 1989, pp. 207–231; Segre 1991, p. 112f.

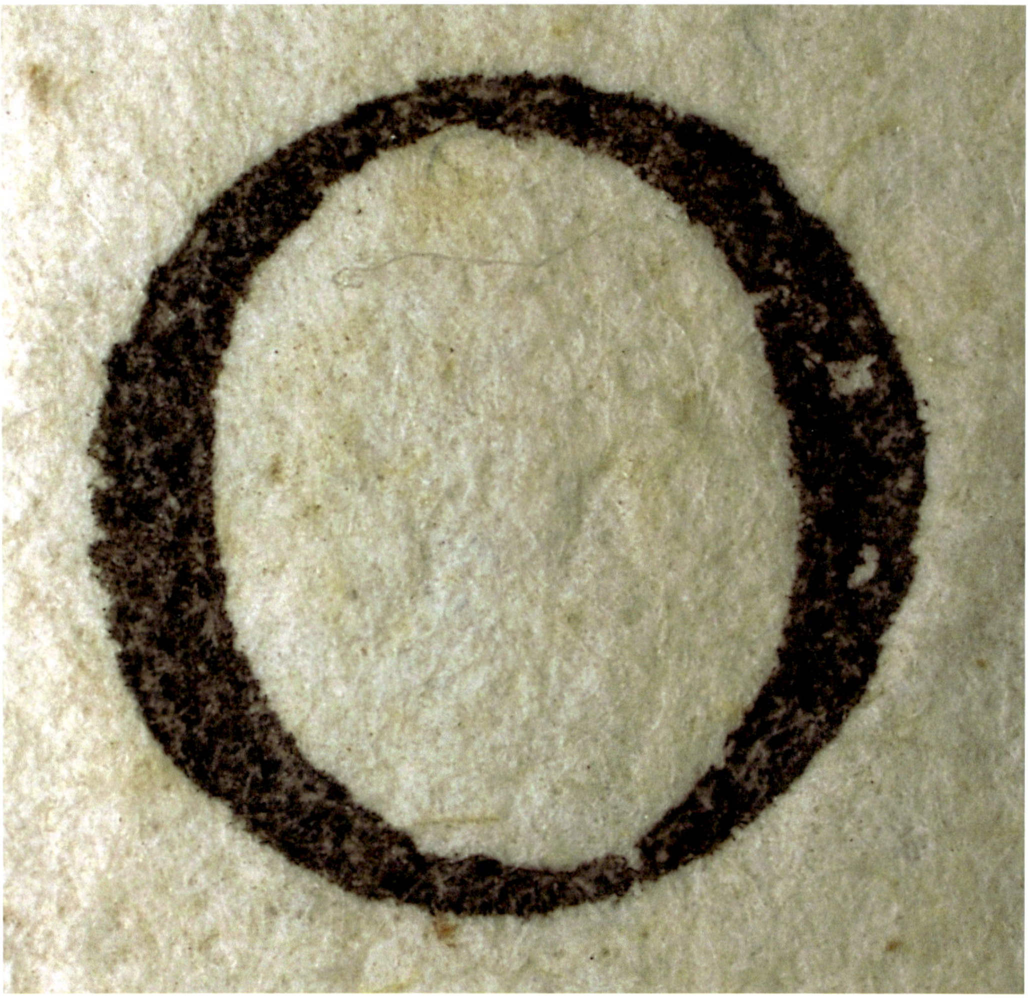

Fig. 7: Galileo Galilei, Sidereus Nuncius, Venice 1610, title page, Graz copy, Galileo's "O".

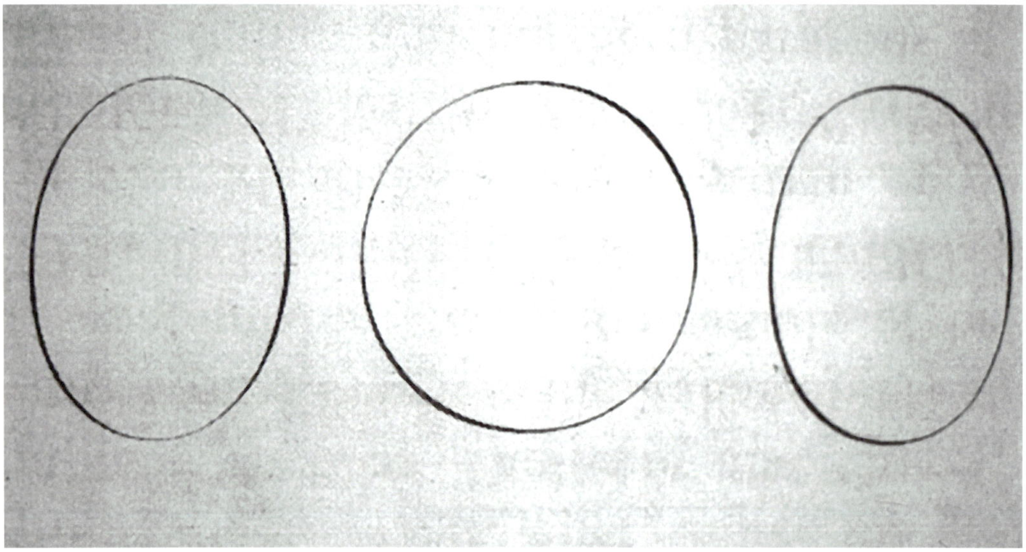

Fig. 8: Jan van den Velde, circle flanked by two ovals, engraving by Simon Frisius, 1605.

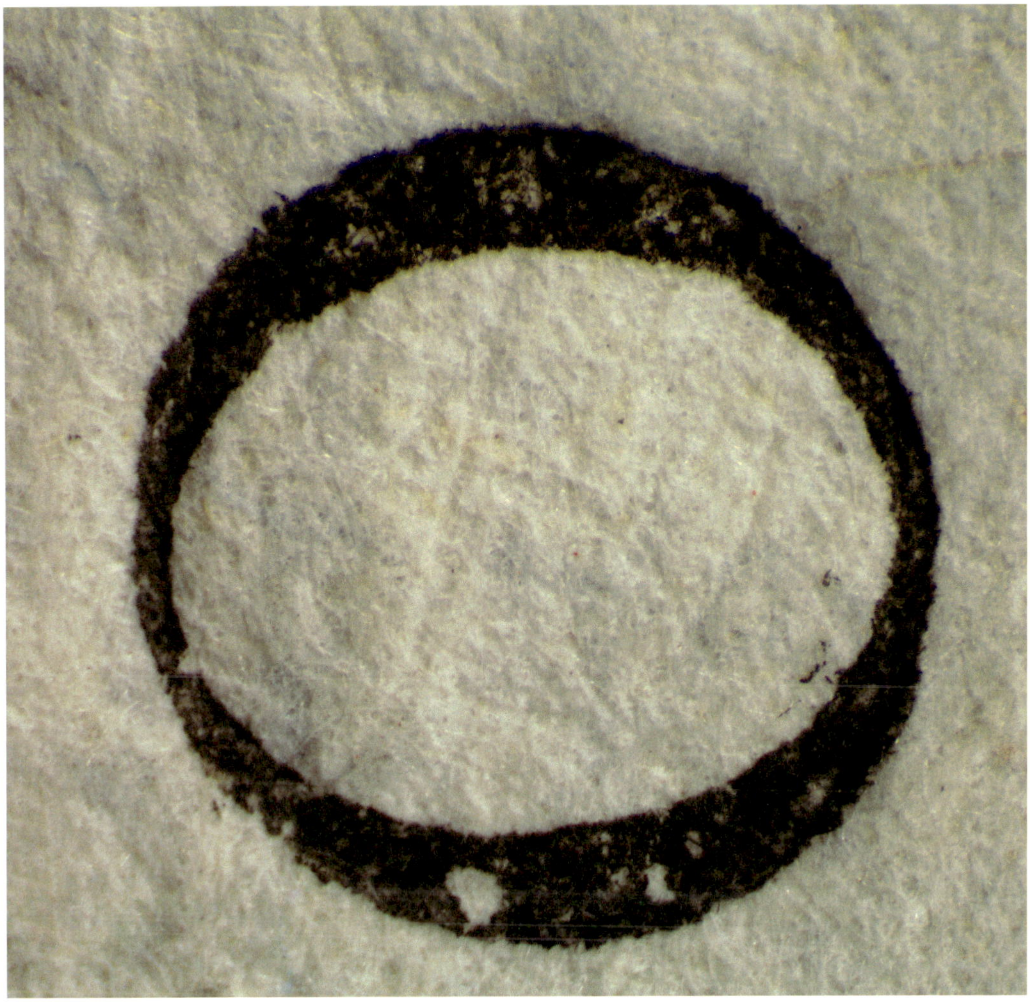

Fig. 9: Galileo Galilei, Sidereus Nuncius, Venice 1610, p. 20v, Graz copy, detail, Galileo's "O" used as Jupiter symbol.

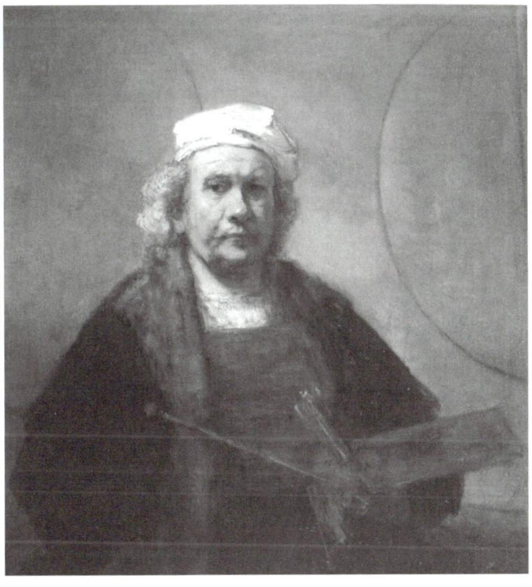

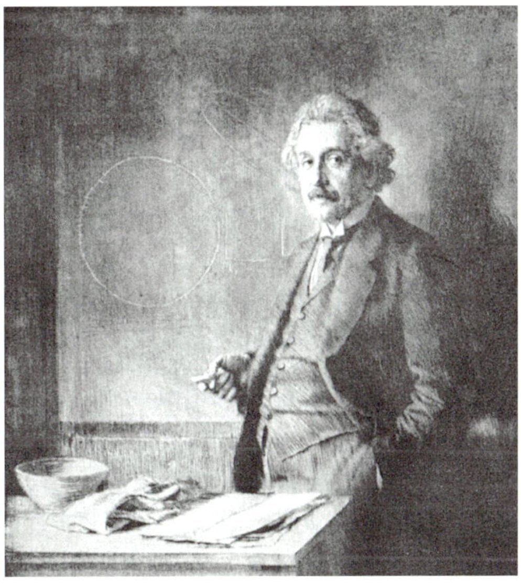

Fig. 10: Rembrandt, Self-Portrait, 1661, oil on canvas, detail, London, Kenwood House.

Fig. 11: Ferdinand Schmutzer, Portrait of Einstein, 1921, etching, sixth state, detail.

cian of Florence.[13] The calculation proved itself. With the *Sidereus Nuncius,* in which Jupiter is formed from the O of his name, he received this dream position. Already in his lifetime, Galileo was acclaimed as the reincarnation of Michelangelo Buonarroti.[14] But even earlier, by means of the Jupiter of the *Sidereus Nuncius,* Galileo had written his own name in the skies as the new Giotto.

13 Biagioli 1993, p. 47.
14 Bredekamp 2007, pp. 13–24.

VI

APPENDICES

OLIVER HAHN / IRENE BRÜCKLE

GLOSSARY OF SELECTED HISTORICAL INKS

BISTRE

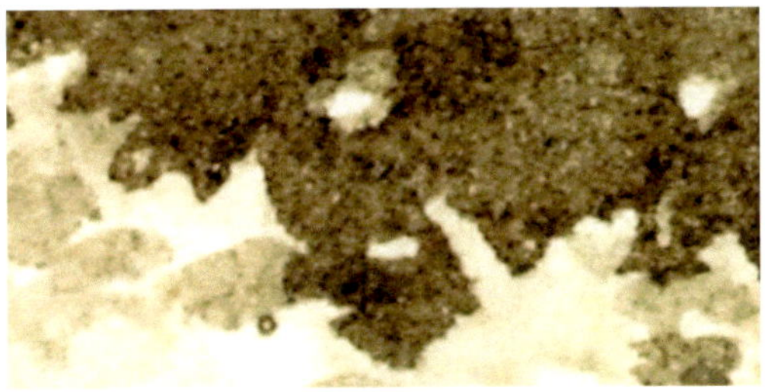

This brown ink prepared from wood soot has been used since early times.[1] As early as 1431, Jehan le Begue used the two terms "caligo" and "fuglio" to differentiate various shades of bistre.[2] To prepare bistre, soot was boiled in water and allowed to settle. While it was still hot, the clear part was separated from the sediment and the fluid was partly evaporated, leaving a fine sediment in dispersion. Bistre is available in many different shades of brown, depending on the predominant type of wood incinerated.[3] Bistre was used in watercolor painting and for wash drawings, but not for writing, as its fine particulates would clog a pen.

CARBON INK

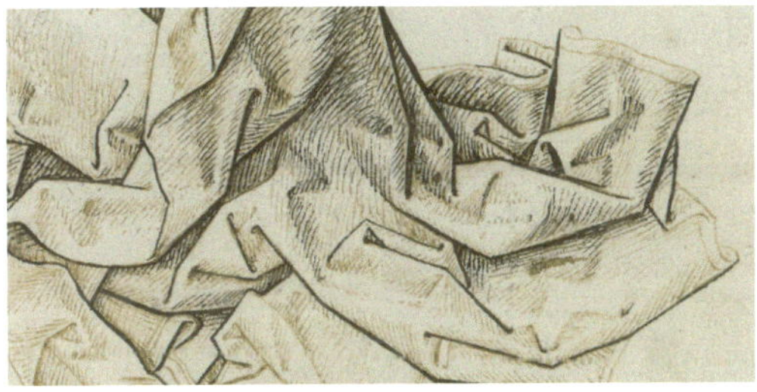

One of the oldest writing materials, black carbon ink has been used in Egypt, India and China for more than a millenium. It consists of carbon pigment mixed with a water-soluble binding medium. In general, the smoothest, finest carbon pigment came from burnt oil. Sesame oil was preferred in China. The ink could be dried and pressed into small bars, making it easily transportable and available for use. Considered valuable, it was imported from China to Europe in the fifteenth century, though confusion about its origin resulted in the term India Ink. Filippo Baldinucci mentioned the preparation of "inchiostro della China" in 1681, though many recipes for the preparation of carbon inks were known at that time.[4] Carbon inks were primarily used as writing inks, but were also used for pen and ink drawings, and washes.

1 Harley 2001.
2 Koschatzky 1996.
3 Meder 1923; Fuchs 1999.
4 Schopen 2006; Krekel 2005.

IRON GALL INK

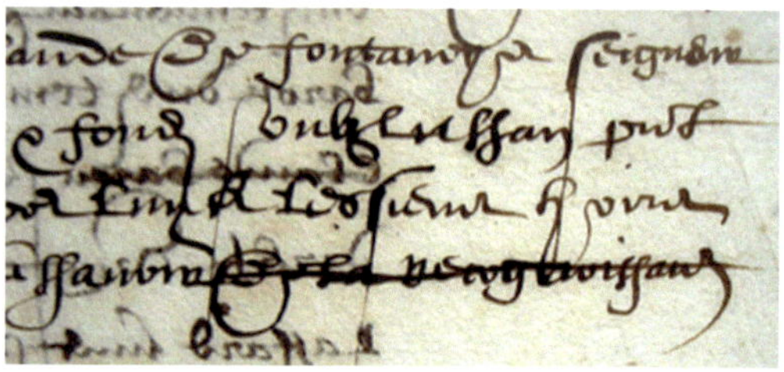

The mixture of iron salts and tannins to prepare a coloured compound was known by Pliny in Antiquity.[5] Iron gall ink is produced from four basic ingredients: oak galls, vitriol (a metal sulfate), gum arabic, and an aqueous medium such as wine, beer or vinegar. Gallic acid was usually extracted from oak galls. Vitriol (iron sulfate), the main inorganic component, was mined in different areas by various techniques and could be contaminated with other metals like copper, aluminium, zinc and manganese.[6] By mixing gallic acid with iron sulfate, a water-soluble ferrous gallate complex is formed. Due to its solubility, the ink penetrates the paper surface, making it difficult to erase. When exposed to oxygen, a ferric gallate pigment is formed. This complex is not water-soluble, which contributes to its indelibility as a writing ink. Gum arabic is a suspension agent that also modifies the viscosity of the ink. Due to the natural origin of the materials and the diverse ink preparation methods, there are a wide range of ingredients and impurities in historical iron gall inks. The permanence of iron gall ink contributed to its longevity as a writing ink; it was still in use in the twentieth century. From the fifteenth century onward, it was also a popular drawing ink, favoured by many artists.[7]

PLANT INK

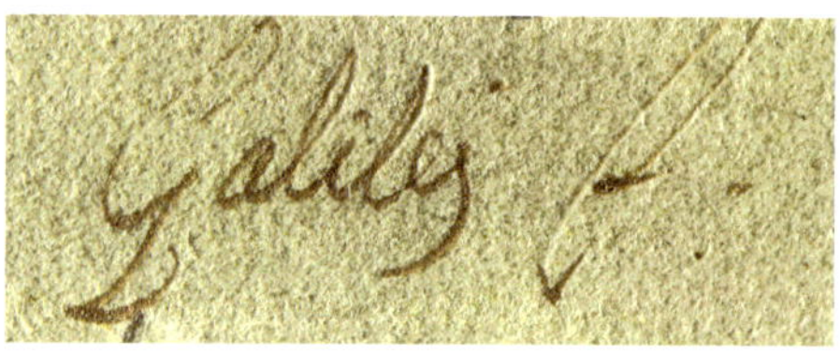

Coloured sap extracted from wood, bark and other plant materials has been used as brown ink since Antiquity. In Europe and Byzantium, brown inks were produced from the bark of oak, blackthorn and willow trees. The sap contains hydrolysable tannic acid and non-hydrolysable polymer tannins. The amount of gallic acid in the tannins can be increased through burning. Plant inks usually contain a binding medium to disperse the water-insoluble components. Plant inks can be used for writing but are primarily known in watercolour paintings and wash drawings.

5 The Iron Gall Ink Website, http://ink-corrosion.org.

6 Hickel 1963.

7 Muspratt 1905; Krekel 1999.

PRINTING INK

Printing with oil-based ink has been known in Europe since the eleventh century when it was used for decorating textiles. Printing on paper has been known since the end of the fourteenth century. Oil-based printing ink consists of a black pigment ground with varnish. The varnish is made by heating vegetable oil, usually linseed oil, until it thickens. Lamp black (carbon black) was the preferred pigment because of its high opacity and very small particle size, which allow very thin layers of ink to be used. A sixteenth-century source recommends that only lamp black be used for printing: *"L'inchiostro poi da stamparlettere si fa di solo fume di ragia."*[8] The essential properties of the printing ink are determined by the mixture of varnish and pigment, and are modified by resins, soaps, extenders and driers.

8 Piemontese 1555; Stijnman 2000, Vol. 1, Suppl., pp. 61–68.

OLIVER HAHN / IRENE BRÜCKLE / THERESA SMITH

ULTRAVIOLET AND INFRARED REFLECTOGRAPHY

UV and IR reflectography are non-destructive examination tools that can reveal information about objects of cultural heritage. Ultraviolet radiation is short wavelength energy outside the visible spectrum (400–10 nm). So-called "long wave UV" or UVA (400–320 nm), is most commonly used in the examination of artefacts. Some materials absorb UV radiation and re-emit it in the visible spectrum as UV-induced visible fluorescence. UV reflectography exploits the sensitivity of photographic emulsion to these wavelengths by capturing both visible fluorescence and any reflected (not absorbed) UV radiation using photographic filters that block specific wavelengths of radiation. Of most interest for our purposes, UV radiation causes visible fluorescence of proteinaceous materials such as parchment, animal glue, and paper sizing (Fig. 1), as well as organic materials used for adhesives, varnishes and binding media like gum arabic in watercolours and aged oils in printing ink.

Infrared radiation is long wavelength energy also outside the visible spectrum (700–20,000 nm). The wavelengths of use in examination of cultural artefacts are generally

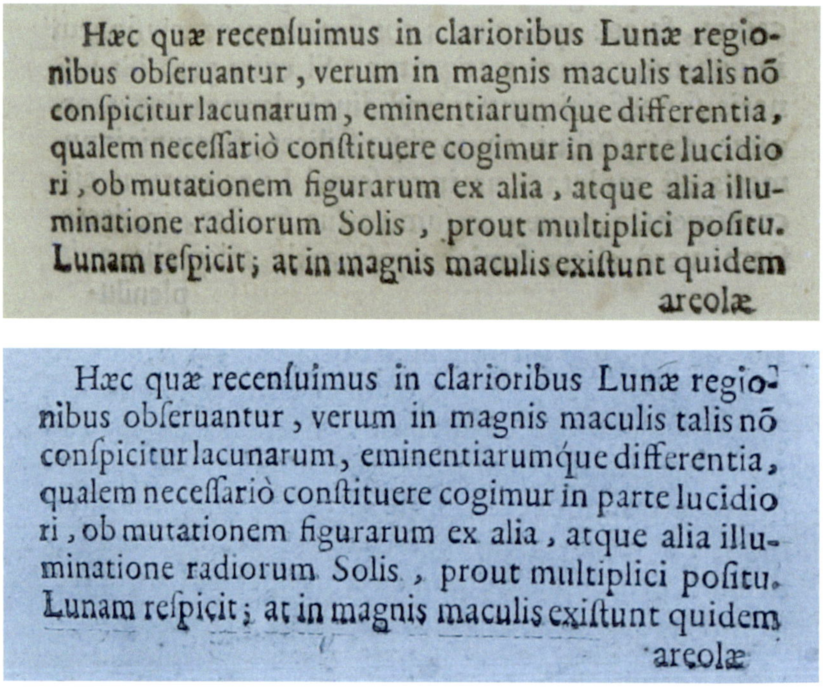
Hæc quæ recenſuimus in clarioribus Lunæ regionibus obſeruantur, verum in magnis maculis talis nō conſpicitur lacunarum, eminentiarumq́ue differentia, qualem neceſſariò conſtituere cogimur in parte lucidiori, ob mutationem figurarum ex alia, atque alia illuminatione radiorum Solis, prout multiplici poſitu Lunam reſpicit; at in magnis maculis exiſtunt quidem areolæ

Hæc quæ recenſuimus in clarioribus Lunæ regionibus obſeruantur, verum in magnis maculis talis nō conſpicitur lacunarum, eminentiarumq́ue differentia, qualem neceſſariò conſtituere cogimur in parte lucidiori, ob mutationem figurarum ex alia, atque alia illuminatione radiorum Solis, prout multiplici poſitu Lunam reſpicit; at in magnis maculis exiſtunt quidem areolæ

Fig. 1: Galileo Galilei, Sidereus Nuncius, Venice 1610, p. 11r, details of the New York copy (above) and Graz copy (below). UV-induced visible fluorescence indicates the presence of naturally aged gelatin sizing in the Graz copy.

divided into "near IR" or NIR (700–1000 nm), and "mid IR" or MIR (1000–3000 nm). Many materials exhibit different visual appearances under specific wavelengths of IR radiation, absorbing, transmitting or reflecting the radiation. Materials that absorb IR radiation appear dark, those that transmit are transparent, and those that reflect appear white, though the appearance of a material can change with different IR wave lengths. IR reflectography is the process of capturing images in the NIR range using photographic emulsions and digital cameras and in the MIR range using video cameras with specially calibrated monitors. The monitor images can be photographed using analog or digital photography. Carbon inks, graphite and charcoal all absorb strongly in IR, making them appear dark. Organic inks and iron gall ink only absorb shorter wavelengths. They transmit longer wavelengths, making the inks appear transparent (Fig. 2).

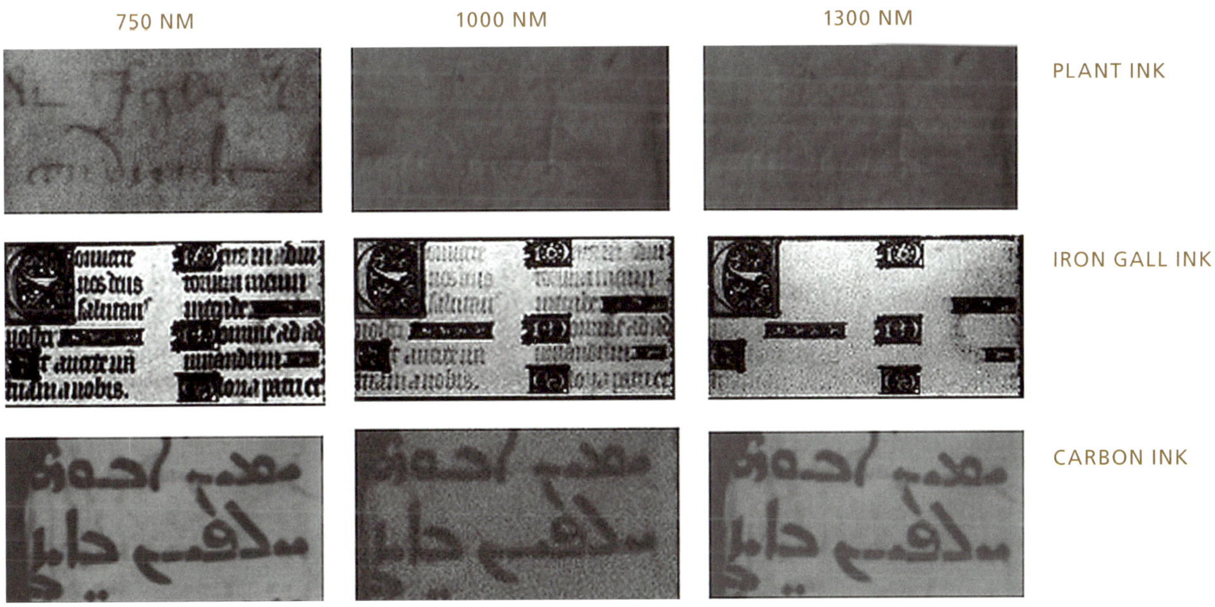

Fig. 2: Use of IR reflectography at different wavelengths aids in the classification of inks (Mrusek et al. 1995, p. 68).

ERICH JELEN / STEFAN SIMON

3D CONFOCAL MICROSCOPY

The 3D confocal microscope is an optical instrument for measuring and analyzing surfaces. It can be used to analyze three-dimensional structures, take measurements of geometry in the micrometer and nanometer range, such as layer thickness, and determine roughness compliance with international standards (DIN, ISO, EN) for paper and board. A wide range of materials from polymers to metals to composite materials can be analyzed. All types of surfaces can be recorded with precise measurement data. The advantage of this type of microscopy is that no sample preparation is required for measurement.

The method works on the principle of focusing a standard light microscope. As the microscope "scans" the surface, it moves the objective lens up and down to keep the surface in focus. This up-and-down movement is the change in height of the surface, the roughness, which is automatically recorded. All height data are combined to create an error-free three-dimensional image of the sample (Fig. 1). Within one measurement, all the data for creating different analyses are stored: height, reflection, roughness, profile, and three-dimensional picture. Reporting can be done at any time after the original measurement because all data are stored within the measurement file.

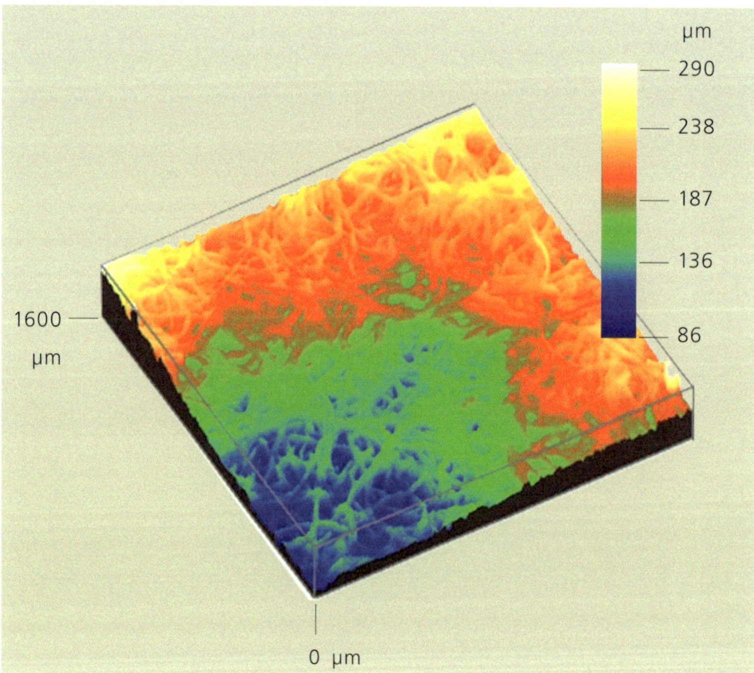

Fig. 1: Representative topographic profile obtained from 3D confocal microscopy.

The confocal technology oft the NanoFocus 3 D-Microscope msurf® is based on optical screening by means of a rotating multipinhole filter, which contains a multitude of microscopically small apertures. A white light source is projecting them on the object surface. Only the part of the reflected light, which lays exactly in the focus of lense, is depicted in the CCD camera. By a precisely computer-controlled lense drive, confocally filtered isolines of the height profile are generated in variable distances and continuously saved in a stacked pile. This is followed by the calculation of the 3D-topography of the object by a special software algorithm. So a complete and precise3D scan of an area of 0,064 mm^2 can be recorded in one measurement step.

This technique has been used to study many different surfaces (gilding, coatings, adhesive films) to characterize damaged and degraded materials and even to identify carving tools from the marks left behind.

OLIVER HAHN / TIMO WOLFF

MICRO X-RAY FLUORESCENCE ANALYSIS OF INK

A particular requirement for the investigation of cultural and historical objects is the use of techniques that require no sample or only minimal sampling. The methods should also be fast, universal, versatile, sensitive, and multi-elemental. The objects of interest are very fragile, usually have to be kept in a controlled environment and the investigations have to take place without touching the surface of the objects. X-ray fluorescence analysis (XRF) meets most of these requirements and can obtain qualitative and quantitative information from many different materials, such as iron gall ink, carbon ink, other drawing and writing media, as well as paper and parchment (Fig. 1).

The inorganic components of writing and drawing materials provide a basis for the differentiation of these materials. Because most writing and drawing materials used in the *Sidereus Nuncius* consist of light elements, such as carbon (C), hydrogen (H) and oxygen (O), which cannot be detected with ED-XRF, an absolute quantification of the chemical

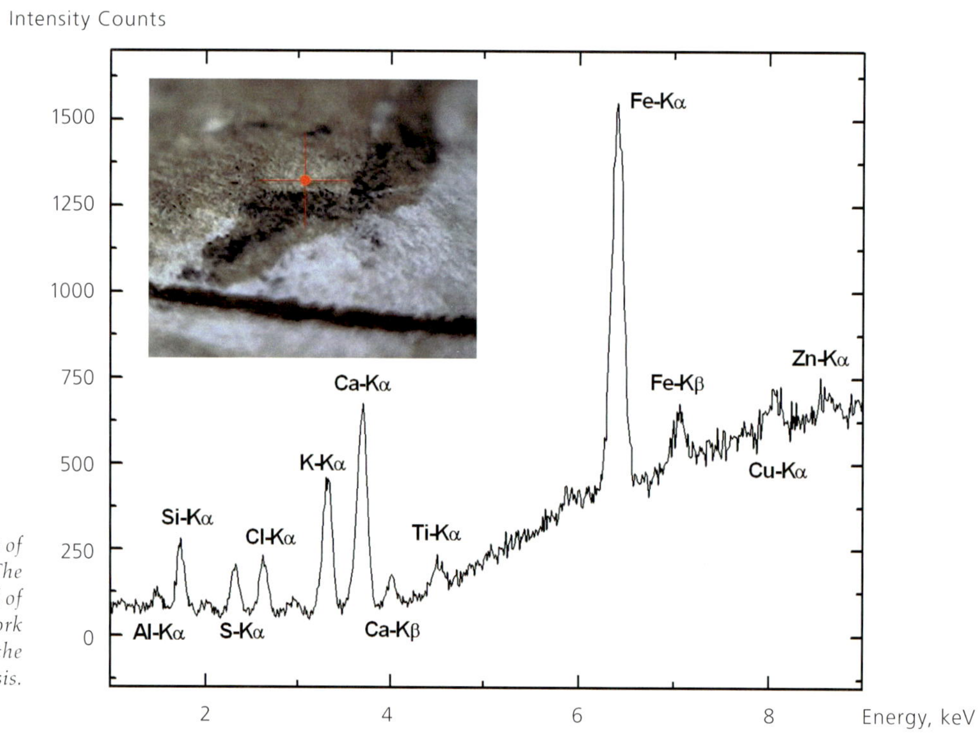

Fig. 1: XRF spectrum of bister ink and paper. The inset image is a detail of p. 9v of the New York copy captured during the analysis.

composition is not possible. However, the differentiation of various ink materials is possible by determination of the relative amount of minor constituents.

During this project, analyses were carried out with the mobile energy dispersive (ED) micro X-ray spectrometer ArtTAX Pro® (Bruker AXS, formerly Röntec-GmbH, Berlin, Germany), which consists of an air-cooled low-power X-ray tube, polycapillary X-ray optics (measuring spot size 70 µm diameter), an electrothermally cooled Xflash detector, and a CCD camera for sample positioning. Furthermore, additional open helium purging in the excitation and detection paths enables the determination of light elements from magnesium (Mg, $Z=12$) to calcium (Ca, $Z=20$) without vacuum. All measurements are made using the 30 W Mo tube, at 45 kV, 600 µA, and with an acquisition time of 15 s (live time) to minimize the risk of damage. For better statistics, at least fifteen single measurements were averaged for each data point.

Because standardized reference materials do not exist for these types of materials (low concentrations of lighter elements on a non-uniform substrate), the analyses were carried out using a fundamental parameter approach. The method is based on atomic databases and requires knowledge of the excitation conditions and other spectral parameters. The non-destructive X-ray spectrometry method used here is only sensitive for elements heavier than sodium (Na, $Z=11$). For this reason we assumed model compositions for the organic matrices. To calculate the primary fluorescence resulting from the printing and drawing materials, we postulate a two-layer-system. The thicknesses of the different layers were estimated from the Compton-scattering.

PROJECT PARTICIPANTS

HORST BREDEKAMP

was born in 1947 in Kiel, Germany, and studied art history, archaeology, philosophy and sociology in Kiel, Munich, Berlin and Marburg, where he received his Ph.D. in 1974. After an internship at the *Liebieghaus* in Frankfurt am Main, Germany (1974–1976) he became Assistent Professor at the *Universität Hamburg* (Hamburg University), where he tenured as Professor of Art History in 1982. Since 1993 he has been a tenured Professor of Art History at the *Humboldt-Universität* in Berlin and a permanent fellow of the *Wissenschaftskolleg zu Berlin* (Institute for Advanced Study in Berlin). Among other honors, he received the *Sigmund Freud Award* from the *Deutsche Akademie für Sprache und Dichtung* (German Academy for Language and Poetry) at Darmstadt, Germany (2001), the *Aby M. Warburg Award* from the City of Hamburg, Germany (2005), and the *Max Planck Research Award* from the *Max-Planck-Gesellschaft* and the *Alexander von Humboldt-Stiftung* (2006). He is a member of the *Berlin-Brandenburgische Akademie der Wisschenschaften* (Berlin-Brandenburg Academy of the Sciences and Humanities, since 1995) and member of the *Leopoldina* (German Academy of the Sciences Leopoldina, National Academy), Halle, Germany (since 2004).

IRENE BRÜCKLE

is Professor and Director of the Program for the Preservation and Conservation of Works of Art on Paper, Archives and Library Materials at the *Staatliche Akademie der Bildenden Künste Stuttgart* (ABK). After paper conservation training in Germany and internships at museums in San Francisco, New York, and Honolulu (1988–1990), she was a *J. Paul Getty Foundation* research fellow at the *Art Conservation Department, Buffalo State College*. She subsequently taught paper conservation there from 1994–2004 as full-time professor. In 2005, she was awarded the *Sheldon and Caroline Keck Award* for sustained excellence in conservation education from the *American Institute for Conservation*. She holds an M.A. in art history from the *State University of New York at Buffalo*, and a Ph.D. from the *Staatliche Akademie der Bildenden Künste Stuttgart*. From 2004 until 2008, she was head of paper conservation at the *Kupferstichkabinett Berlin*. She is co-editor of the journal *Restaurator: International Journal for the Preservation of Library and Archival Material*. Her research focuses on conservation treatment methodology and the technologies of paper and other artistic materials.

STAATLICHE AKADEMIE DER BILDENDEN KÜNSTE STUTTGART (ABK)

The *Studiengang Konservierung und Restaurierung von Graphik, Archiv- und Bibliotheksgut* at the *ABK Stuttgart* is a five-year graduate program offering an M.A. in the conservation of works on paper. The program focuses on developing, refining and standardizing methods for the preservation and conservation treatment of paper-based cultural materials founded on the study of their material composition and technology. In the courses, emphasis is placed on a sophisticated didactic approach that interrelates theoretical and practical learning and places attention on a thorough understanding of the significance of any conservation action on the immediate and long-term appreciation and use of the object. Through cooperation with public collections and museums in Germany and abroad, the program also provides students with many opportunities for contributing to projects that highlight a diversity of conservation pursuits and associated tasks. The program seeks to advance the understanding and care of artworks and other objects and to highlight the interdisciplinary quality of comprehensive preservation and conservation concepts.

ROBERT FELFE

received his Ph.D. in art history in 2000 at the *Humboldt-Universität* in Berlin. After this he worked in a research position at the *Research Center for Performing Cultures* founded at *Freie Universität Berlin*, sponsored by *Deutsche Forschungsgemeinschaft* and taught at Institutions in Germany, Great Britain, Italy and Switzerland. In 2010 he was guest fellow at the *Kunsthistorisches Institut in Florenz/Max-Planck-Institute*. A habilitation was submitted in 2011; the title is: *Naturform und bildnerische Prozesse. Elemente einer Wissensgeschichte der Künste im 16/17. Jahrhundert* (Natural Form and Image Processes. Elements of a History of Knowledge in the Art of 16/17th Century). Currently he is research fellow at the *Kolleg-Forschergruppe Bildakt und Verkörperung* (Collegium for the Advanced Study of Picture Act and Embodiment) at the *Humboldt-Universität*.

OLIVER HAHN

received his Ph.D. in chemistry in 1996 from *Universität zu Köln* (University of Cologne). Following employment as a Research Associate in the Department for Restoration and Conservation of Books, Graphic Arts and Archival Materials at the *Fachhochschule Köln* (Cologne University of Applied Sciences), he has worked at the *BAM Bundesanstalt für Materialforschung und -prüfung* (Federal Institute for Materials Research and Testing) in Berlin. Since May 2007, he has headed the research group "Analysis of Cultural Assets." His special fields of interest include the non-destructive investigation of art objects as well as the characterization of environmental conditions in museums and archives. He has taken part in many interdisciplinary research projects in conservation and archaeometry using a variety of scientific analyses ranging from synchrotron-based methods, such as X-ray absorption near edge spectroscopy (XANES) and Fourier transform infrared spectroscopy (FTIR) in reflection, to mobile investigative techniques like micro X-ray fluorescence spectroscopy (XRF) and visible spectrophotometry (VIS).

BAM BUNDESANSTALT FÜR MATERIALFORSCHUNG UND –PRÜFUNG

is a technical and scientific institution under the authority of the Federal Ministry of Economics and Technology. The responsibilities of the BAM encompass materials science, chemistry, the environment and safety. It is one of the leading German institutions for research in analytical chemistry and the safe handling, transportation and storage of dangerous materials and goods. BAM advises the Federal Government and industry on safety aspects of materials and chemical technology, and develops standards and technical regulations for the evaluation of substances, materials, and structures, including non-destructive damage assessment and environmental monitoring methods.

BARBARA HERRENKIND

was born in 1965 in Berlin. After her photographic education at the *Lette Verein Berlin* (1989), she worked as a freelance press and portrait photographer for agencies and publishers with a focus on art and culture. Since 1995 she has been a photographer at the Institut of Art and Visual Studies at the *Humboldt-Universität* in Berlin. She does photographic work for art historical publications, lessons and research projects as well as portrait photography and documentation of photographic and other exhibitions, e.g. in Berlin, Istanbul, Rome, and St. Petersburg. Her projects have included, among others, documentation of the long-term building site of the Sony Center at Potsdamer Platz (Berlin 1997–2000), the exhibitions "Theatrum Naturae et Artis" (Berlin 2000-2001), documenta XII (Kassel 2007), and the portrait exhibition, "Women on the Move" at Humboldt University (Berlin 2010–2011).

ERICH JELEN

studied chemistry and chemical engineering with a focus on environmental engineering at Bochum and Münster. He has additional qualifications as a cabinetmaker and a "building-related biologist" (Baubiologe IBN). Since 1999 he has worked at *Fraunhofer-Institut für Umwelt-, Sicherheits- und Energietechnik UMSICHT* on projects involving wood and wood-based materials, impregnation, supercritical carbon dioxide, environmental chemistry, instrumental chemical analysis and cultural heritage science, especially decontamination of wooden objects. Since 2006 he has been a member of the working group "Advanced Materials" for which he performs chemical and physical analysis for materials testing and characterization. He takes part in multidisciplinary research projects in restoration and conservation science, investigating the materials of cultural heritage objects and the indoor environment.

SONJA KRUG

received in 2002 her degree in conservation, art technology and conservation science at the *Technische Universität München* (Technical University of Munich). For her diploma thesis she was awarded the *Ibach-Denk-Mal-stiftung Award* (Ibach-memorial-trust Award). From

2002 on she works as a freelancer in Germany and France and within various projects at the *Rathgen-Forschungslabor* in Berlin. Her work focused on pesticide removal from wood and ethnographic objects. Since April 2007 she has been employed at the *BAM Bundesanstalt für Materialforschung und -prüfung* in Berlin, Division 4.2 Environmental Material and Product Properties. Her current projects involve historical tin-mercury mirrors and determining the pesticide content of cultural assets.

MANFRED MAYER

was born in 1954 and studied electrical engineering at the *Technische Hochschule Graz* (Graz University of Technology). His education includes several national and international courses, and study visits to the U.S. His particular area of responsibility is the treatment of medieval manuscripts and fifteenth to eighteenth century prints. Since 1986 he has been Head of the conservation workshop for special collections at the *Karl-Franzens-Universität* in Graz. He lectures on the codicology of the medieval codex, preservation of books and prints, and the digitization of books, e.g. in Switzerland, Russia and Estonia.

HANS JAKOB MEIER

earned his doctorate at the *Ludwig-Maximilians-Universität* (Ludwig Maximilians University) in Munich. This was followed by research residencies with fellowships at the *Herzog-August-Bibliothek Wolfenbüttel* (Herzog August Library Wolfenbüttel) and the *Zentralinstitut für Kunstgeschichte* in Munich. A three-year Assistant Professorship at *Universität Bern* (Bern University, Switzerland) led to research stays at the *Bibliothèque Nationale de France* (National Library of France) in Paris. In 2003, in collaboration with Norberto Gramaccini, he published *Die Kunst der Interpretation. Französische Reproduktionsgrafik 1648–1792* (The Art of Interpretation. French Reproductive Prints 1648–1792). A research stipend from the *Gerda Henkel Foundation* in 2005 took him to the *Kupferstichkabinett Berlin*. There, he curated the exhibition *The Art of Interpretation. Italian Reproductive Prints from Mantegna to Carracci* (2009). This and further research at the *Kupferstichkabinett Berlin*, the *British Museum* (London), and the *Bibliothèque Nationale* (Paris) was published in 2009, in collaboration with Gramaccini, as *Die Kunst der Interpretation. Italienische Reproduktionsgrafik 1485–1600* (The Art of Interpretation. Italian Reproductive Prints 1485–1600).

ENRICA SCHETTINI PIAZZA

was born in Rome where she currently lives. She has been a cultural consultant for the Italian Banking Association (A.B.I.), publishing *Banks and Books*, a catalogue of publications produced by the Italian Credit Companies and Institutes. As Curator of Manuscripts for the Corsini Library at the Lincean Academy in Rome, she carried out extensive studies on the history of the Academy. Her book, *Historical Bibliography of the National Academy of Linceans*, covers its origins to the present day. She has written extensively on the founder of the Academy, Federico Cesi, and various other Lincean subjects including the history of science in the seventeenth century. She has also taught the History of the Book at the Universities of Macerata and Cassino.

HEINRICH SCHULZE ALTCAPPENBERG

has been head of the *Kupferstichkabinett Berlin, Staatliche Museen zu Berlin – Preußischer Kulturbesitz* since 2002. He studied art history, literature, archaeology and journalism in Mainz, Hamburg and Bonn, where he received his Ph.D. in 1985. From 1986–1992 he was Assistant Professor at the Art Museum in Düsseldorf. Since 1992 he has worked at the *Kupferstichkabinett Berlin.* His numerous publications deal with the age of Goethe, Romanticism, the Italian drawing and print from the early Renaissance to the late Baroque and contemporary positions. Research projects, exhibitions and publications include *Le Voltaire de l'Art. Johann-Georg Wille und seine Schule in Paris* (Ph.D. 1987), *Facetten des Barock* (1990), *Die frühe Renaissance in Italien* (1995), *Tiepolo* (1996), *Sandro Botticelli* (2000), *Kunstsinn der Gründerzeit. Die Sammlung Adolf von Beckerath* (2002), *Caspar David Friedrich. Jahreszeiten* (2006), *Disegno. Der Zeichner im Bild* (2007/08), *Karl Friedrich Schinkel. Geschichte und Poesie* (research project and exhibition supported by BMBF 2009-2012), *Karl Friedrich Schinkel. Die Italienische Reise 1803–1805* (2010).

STIFTUNG PREUSSISCHER KULTURBESITZ (SPK)

The *Kupferstichkabinett Berlin,* which cared for the *Sidereus Nuncius New York copy* is part of the *Stiftung Preußischer Kulturbesitz* (SPK). The SPK, formed in 1957, was born out of the collections of the former state of Prussia. It is funded by the federal government and by all 16 states within Germany. With its many museums, libraries, archives and research institutions, the SPK is one of the largest cultural bodies in the world. The foundation forges a unique bond between art and culture and science and research. Its collections document the cultural development of humankind in all parts of the world, from its beginnings to the present day.

STEFAN SIMON

earned his Ph.D. in Chemistry at the *Ludwig-Maximilian Universität München* (Ludwig-Maximilian University Munich). He has worked as a conservation scientist with experience in scientific research and multilateral project coordination at numerous conservation sites in Europe, the Americas, the Middle East and China. He has specialized in material deterioration diagnostics, microanalytics, non-destructive testing, physico-chemical analysis and climatology. His other areas of specialization include the testing and evaluation of conservation products and the control of treatment efficacy and durability in the field of conservation. Founder and director of the private laboratory *KDC Konservierung & Denkmalpflege Consulting* in Olching, Germany, since 1993, Simon has also taught conservation science at *Technische Universität München* since 1998. From 2001 to 2005 he headed the "Building Materials" section at the *Getty Conservation Institute* (Los Angeles, USA), before being appointed Director of the *Rathgen-Forschungslabor* in February 2005. Since 2009 he has been Honorary Professor at *X'ian Jiaotong University*, China. He is currently Vice President of the *International Centre for the Study of the Preservation and Restoration of Cultural Property* (ICCROM) and President of the *International Council of Monuments and Sites* (ICOMOS), International Scientific Committee on Stone.

THERESA SMITH

received her M.A. in Art Conservation with a Certificate of Advanced Study from *Buffalo State College*. An interest in the history of conservation led to her thesis project, "Reconstructing Historical Conservation Treatments for Washing and Bleaching Paper". She has worked in conservation departments at the *Harvard Art Museum*, the *Kupferstichkabinett Berlin*, the *Andy Warhol Museum*, and the *Carnegie Museum of Art*. Since 2007 she has been a Paper Conservator for Special Collections in the *Harvard University Library, Weissman Preservation Center*. Her research focuses on conservation treatment outcomes (bleaching, alkaline degradation of aged papers) and the technical examination of artworks. She is on the editorial board of *Restaurator: International Journal for the Preservation of Library and Archival Material*.

RUTH TESMAR

was born in 1952 in Potsdam, Germany, and received her Ph.D. (1981) and diploma in painting and the graphic arts (1983) from the *Kunsthochschule Berlin-Weißensee* (Art Academy Berlin-Weißensee). Since 1993 she has been Professor for Artistic and Aesthetic Practice and Head of the Atelier *Menzel-Dach* at the *Humboldt-Universität* in Berlin. Trained in traditional printmaking, she teaches introductory relief, intaglio, lithographic and screen printing, as well as figure drawing and other techniques. Her work includes stage and costume design for theater, as well as book illustration. In 1986 and 1993 she received the award "Schönstes Buch" (Most Beautiful Book) and in 1988 she was also awarded first prize for her prints at the exhibition "100 ausgewählte Graphiken" (100 Selected Prints). Her work has been exhibited worldwide and can be found in museums and private collections.

TIMO WOLFF

worked in the research group "Analysis of Cultural Assets" at the *BAM Bundesanstalt für Materialforschung und -prüfung* in Berlin from 2004 to 2009. In 2009 he received his Ph.D. in physics from the *Technische Universität Berlin*. As part of his dissertation project, he developed a procedure for fundamental parameter based quantification for mobile micro X-ray fluorescence spectroscopy (XRF), which enables the reference-free quantitative analysis of different objects. His main focus is the non-destructive investigation of archeometric samples, for example the characterisation of iron gall ink corroded manuscripts by means of synchrotron-based X-ray absorption near edge spectroscopy (XANES). Since 2009 he has worked in the research group "Analytical X-ray Physics" at the *Technische Universität Berlin*.

BARBARA HERRENKIND

PROJECT IMPRESSIONS

FIRST CAMPAIGN (2006)

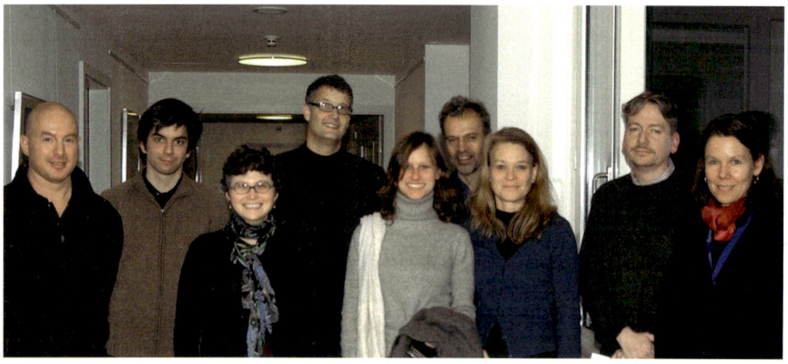

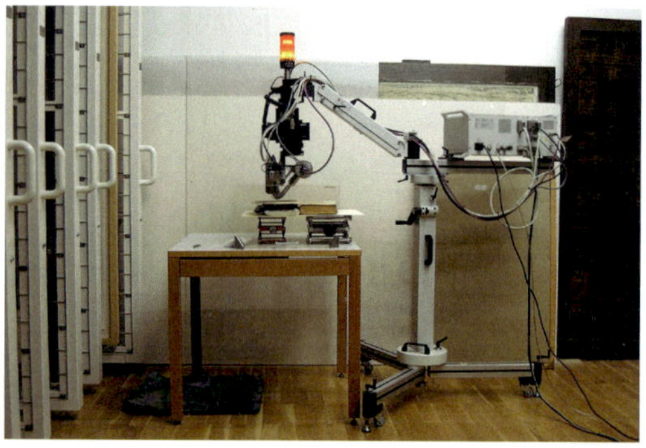

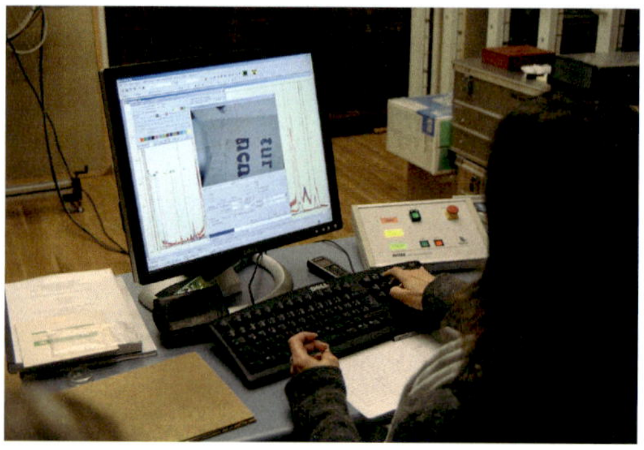

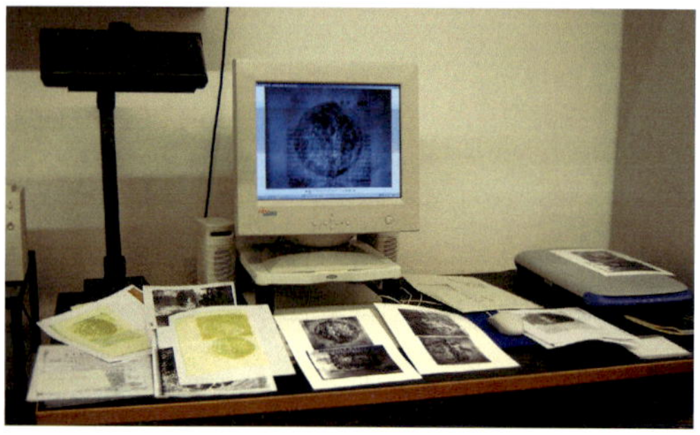

Fig. 1: The investigation team, February 2006 (from left): Boaz Paz, Timo Wolff, Theresa Smith, Stefan Simon, Sonja Krug, Horst Bredekamp, Birgit Kanngießer, Oliver Hahn, Irene Brückle.

Fig. 2: Set-up for micro XRF analysis (Bruker Nano GmbH).

Fig. 3: Analysis of printing ink with micro XRF by Sonja Krug.

Fig. 4: Digital image analysis in progress.

SECOND CAMPAIGN (2008)

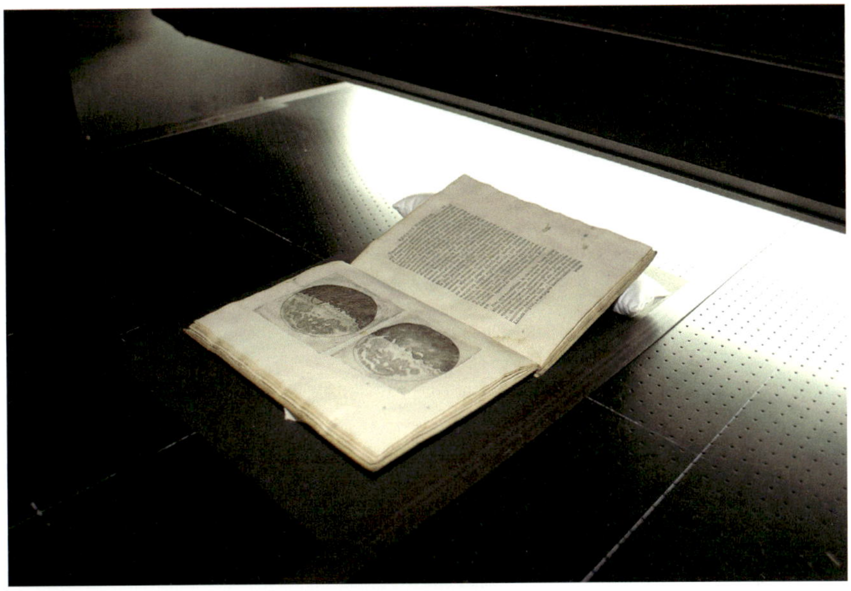

Fig. 5: The Washington copy.

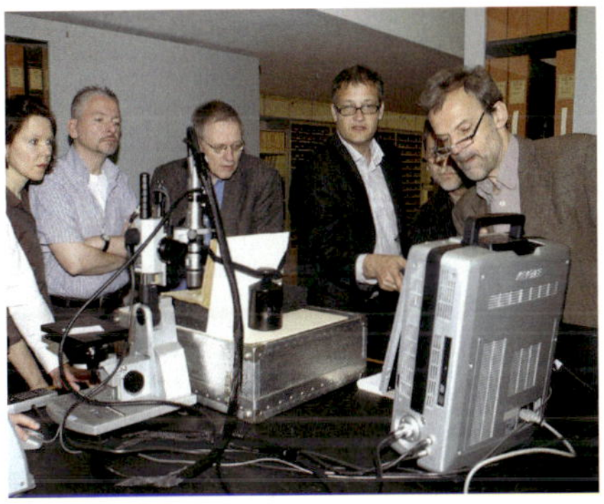

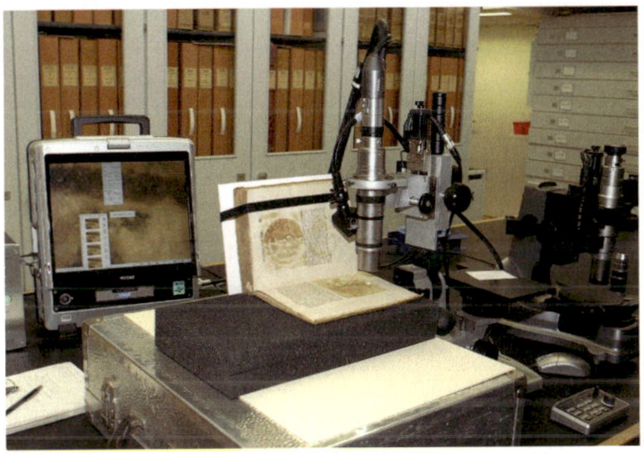

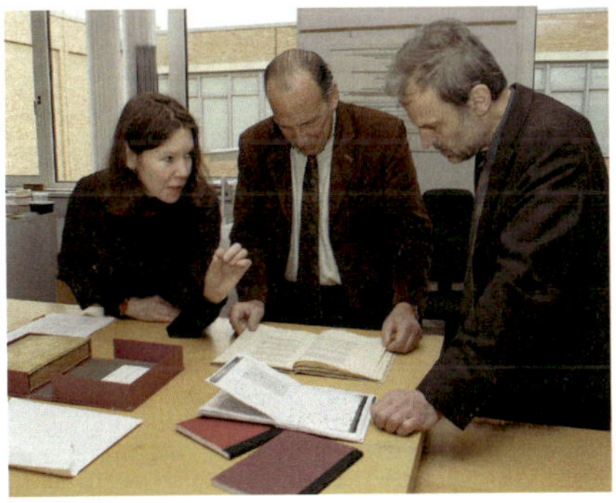

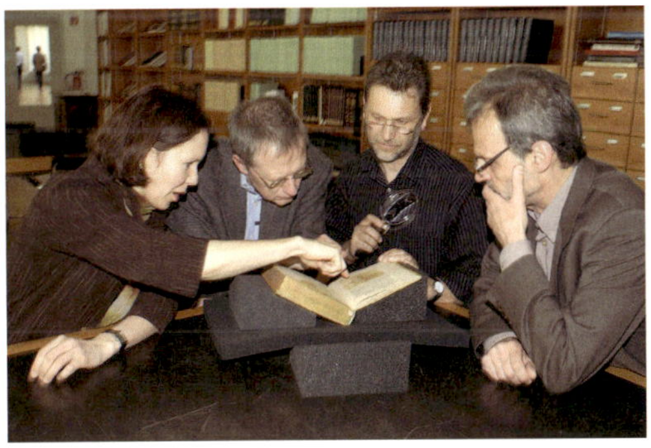

Fig. 6: Digital microscopy (Keyence; from left): Irene Brückle, Oliver Hahn, Paul Needham, Stefan Simon, Manfred Mayer, Horst Bredekamp.

Fig. 8: Discussing the Graz copy (from left): Irene Brückle, Richard Lan, Horst Bredekamp.

Fig. 7: Digital microscope set-up (Keyence) with the New York copy.

Fig. 9: Discussing the New York copy (from left): Irene Brückle, Paul Needham, Manfred Mayer, Horst Bredekamp.

BIBLIOGRAPHY

Accademia Nazionale dei Lincei. *Galileo Galilei: Celebrazioni del IV centenario della nascita.* Rome: Accademia Nazionale dei Lincei, 1965.

Ash, Nancy, Shelley Fletcher and Jan Piet Fieledt Kok. *Watermarks in Rembrandt's Prints.* Washington, D.C.: National Gallery of Art, 1998.

Ashworth Jr., William B. "The Map of the Moon of Gassendi, Peiresc and Mellan." In *Quadricentenaire de la Naissance de Pierre Gassendi 1592–1992.* Actes du Colloque international Pierre Gassendi, Digne-les-Bains 18–21 mai 1992. Société scientifique et littéraire des Alpes de Haute-Provence, pp. 341–352. Digne-les-Bains: Société scientifique et littéraire des Alpes de Haute-Provence, 1994.

Baker, Cathleen. "A Comparison of Drawings Inks Using Ultraviolet and Infra-Red Examination Techniques." In *Application of Science in Examination of Works of Art.* Eds. Pamela A. England and Lambertus van Zelst, pp. 159–163. Boston: [Museum of Fine Arts], 1985.

Barrett, Timothy. "Early European Papers/Contemporary Conservation Papers: A Report on Research Undertaken from Fall 1984 through Fall 1987." *The Paper Conservator* 13, pt. I (1989): pp. 7–27.

Barrett, Timothy. "Evaluating the Effect of Gelatin Sizing with Regard to the Permanence of Paper." In *Institute of Paper Conservation Conference Papers Manchester 1992.* Ed. Sheila Fairbrass, pp. 228–233. Manchester: IPC, 1992.

Benucci, Elisabetta, ed. *Galileo e l'universo dei suoi libri.* Florence: Valecchi, 2008. Exhibition catalog.

Berrie, Barbara H., ed. *Artists' Pigments: A Handbook of Their History and Characteristics.* Vol. 4. Washington, D.C.: National Gallery of Art, 2007.

Biagioli, Mario. *Galileo Courtier: The Practice of Science in the Culture of Absolutism.* Chicago: University of Chicago Press, 1993.

Bosse, Abraham. *Moyen universelle de pratiquer la perspective sur les tableaux ou surfaces irregulieres.* Paris: Abraham Bosse, 1653.

Bosse, Abraham. *Radier-Büchlein handelt von der Etzkunst.* Trans. Georg Andream Böcklern. Nuremberg: Fürst, 1689.

Bower, Peter. "The White Art: The Importance of Interpretation in the Analysis of Paper." In *Looking at Paper: Evidence and Interpretation.* Symposium Proceedings at the Royal Ontario Museum and Art Gallery of Ontario, May 13–16, 1999. Eds. John Slavin, Linda Sutherland, John O'Neill, Margaret Haupt and Janet Cowan, pp. 5–16. Toronto: Canadian Conservation Institute, 2001.

Bredekamp, Horst. "Gazing Hands and Blind Spots: Galileo as Draftsman." *Science in Context* 13, no. 3–4 (2000): pp. 423–462.

Bredekamp, Horst. *Galilei der Künstler: Der Mond. Die Sonne. Die Hand.* Berlin: Akademie Verlag, 2007.

Bredekamp, Horst. "I Molteplici Volti del Sidereus Nuncius." In *Il Cannocchiale e il Pennello: Nuova Scienza e Nuova Arte nell'Età di Galileo.* Eds. Lucia Tongiorgi Tomasi and Alessandro Tosi. Trans. Siliva Cattiti, pp. 164–171. Florence: Giunti, 2009. Exhibition catalog.

Browne, Alexander. *Ars Pictoria, or An Academy Treating of Drawing, Painting, Limning and Etching.* London: J. Redmayne, 1669.

Brückle, Irene. "Aspects of the Use of Alum in Historical Papermaking." In *Institute of Paper Conservation Conference Papers Manchester 1992.* Ed. Sheila Fairbrass, pp. 201–206. Manchester: IPC, 1992.

Bury, Michael. *The Print in Italy 1560–1620.* London: British Museum Press, 2001. Exhibition catalog.

Camerota, Felippo. "Cigoli e la Prospettiva." In *Il Cannocchiale e il Pennello: Nuova Scienza e Nuova Arte nell'Età di Galileo.* Eds. Lucia Tongiorgi Tomasi and Alessandro Tosi. Trans. Siliva Cattiti, pp. 143–151. Florence: Giunti, 2009. Exhibition catalog.

Casini, Paolo. "Il 'Dialogo' di Galileo e la Luna di Plutarcho." In *Novità celesti e crisi del sapere.* Convegno internazionale di studi galileiani in Pisa, Venice, Padua, and Florence. Ed. Paolo Galluzzi, pp. 57–62. Florence: Giunti, 1984.

Colbourne, Jane. "A Survey of Methods Used in the Technical Examination and Analysis of Brown Inks." In *The Iron Gall Ink Meeting, Postprints.* Ed. A. Jean E. Brown, pp. 37–46. Newcastle upon Tyne: Conservation of Fine Art, University of Northumbria, 2001.

Dietz, Georg J. and Irene Brückle. "Distinguishing between Charcoal and Natural Black Chalk in Old Master Drawings." In *Wege zur Konservierungswissenschaft: Projekte am Studiengang Restaurierung und Konservierung von Graphik, Archiv- und Bibliotheksgut, Staatliche Akademie der Bildenden Künste Stuttgart, 2000–2008.* Ed. Gerhard Banik, pp. 54–56. Munich: Siegl, 2010.

Edgerton, Samuel Y. "Galileo, Florentine 'Disegno', and the 'Strange Spottedness' of the Moon." *Art Journal* 44, no. 1 (1984): pp. 225–232.

Edgerton, Samuel Y. *The Heritage of Giotto's Geometry: Art and Science on the Eve of the Scientific Revolution.* Ithaca: Cornell University Press, 1991.

Eszer, Ambrosius. "Niccoló Riccardi, O.P. – 'Padre Mostro' (1585–1639)." *Angelicum* 60, fasc. 3 (1983): pp. 428–461.

Ficacci, Luigi, ed. *Claude Mellan: gli anni romani un incisore tra Vouet e Bernini.* Rome: Multigrafica, 1989. Exhibition catalog.

Fiorani, Francesca. "Danti Edits Vignola: The Formation of a Modern Classic on Perspective." *Studies in the History of Art* 59 (2003): pp. 127–158.

Fuchs, Robert. "Der Tintenfraß historischer Tinten und Tuschen – ein komplexes, nie enden wollendes Problem." *Werkhefte der Staatlichen Archivverwaltung Baden-Württemberg* A-10 (1999): pp. 25–36.

Gabrieli, Giuseppe. "Verbali delle adunanze e cronaca della prima Accademia lincea (1603–1630)." In *Contributi alla storia della Accademia dei Lincei.* 2 Vols. Ed. Giuseppe Gabrieli, Vol. 1, pp. 497–550. Rome: Accademia Nazionale dei Lincei, 1989.

Gabrieli, Giuseppe. *Il carteggio linceo.* Rome: Accademia Nazionale dei Lincei, 1996.

Galilei, Galileo. *Difesa Contro alle Calunnie & imposture di Baldessar Capra Milanese.* Venice: Tommaso Baglioni, 1607.

Galilei, Galileo. *Sidereus Nuncius.* Venice: Tommaso Baglioni, 1610.

Galilei, Galileo. *Discorso Intorno alle cose, che Stanno su l'aqua, ò che in quella si muouono.* Florence: C. Giunti, 1612.

Galilei, Galileo. *Dialogo sopra i due masssimo sistema del mondo tolemaico e copernicano.* Florence: Giovanni Batista Landini, 1632.

Galilei, Galileo. *Le opere di Galileo Galilei.* Reprint edition. 20 Vols. Ed. Antonio Favaro. Florence: Edizione Nazionale, 1968.

Galilei, Galileo. *Sidereus Nuncius: Nachricht von neuen Sternen.* Hans Blumberg, ed. Frankfurt am Main: Suhrkamp, 1980.

Galluzi, Paolo, ed. *Galileo: Immagini dell'universo dall'anticita al Telescopio.* Florence: Giunti, 2009. Exhibition catalog.

Gilissen, Leon. *Prolégomènes à la codicologie: recherches sur la construction des cahiers et la mise en page des manuscrits médiévaux.* Gent: Story-Scientia, 1977.

Gingerich, Owen. "Dissertation cum Professore Righini et Sidereo Nuncio." In *Reason, Experiment, and Mysticism in the Scientific Revolution.* Eds. Maria Luisa Righini Bonelli and William R. Shea, pp. 77–88. New York: Science History Publications, 1975.

Gingerich, Owen. "The Curious Case of the M-L *Sidereus Nuncius.*" *Galilaeana* VI (2009): 141–165.

Gnann, Achim. *Parmigianino und sein Kreis: Druckgraphik aus der Sammlung Baselitz.* Ostfildern: Hatje Cantz, 2007. Exhibition catalog.

Gramaccini, Norberto and Hans Jakob Meier. *Die Kunst der Interpretation: Französische Reproduktionsgrafik 1648–1792.* Munich: Deutscher Kunstverlag, 2003.

Gramaccini, Norberto and Hans Jakob Meier. *Die Kunst der Interpretation: Italienische Reproduktionsgrafik 1485–1600.* Berlin: Deutscher Kunstverlag, 2009.

Haidinger, Alois. "Datieren mittelalterlicher Handschriften mittels ihrer Wasserzeichen." *Anzeiger der philosophisch-historischen Klasse der Österreichischen Akademie der Wissenschaften* 139 (2004): pp. 5–32.

Haidinger, Alois. "Zum Verwendungszeitraum identischer Papiere." In *Ochsenkopf und Meerjungfrau: Papiergeschichte und Wasserzeichen vom Mittelalter bis zur Neuzeit.* Eds. Peter Rückert, Sandra Hodecek, Georg Dietz, and Emanuel Wenger, pp. 35–36. Stuttgart: Landesarchiv Baden-Württemberg, 2009.

Harley, Rosamond D. *Artists' Pigments c. 1600–1835: A Study in English Documentary Sources.* 2nd revised edition. London: Archetype Publications, 2001.

Hickel, Erika. "Chemikalien im Arzneischatz deutscher Apotheken des 16. Jahrhunderts unter besonderer Berücksichtigung der Metalle." PhD diss. Technische Hochschule Carolo-Wilhelmina Braunschweig, 1963.

Horatius Flaccus, Quintus [Horace]. *Poemata.* Venice: Aldum Romanum, 1509.

Hunter, Dard. *Papermaking: The History and Technique of an Ancient Craft.* Reprint edition. New York: Dover Publications, 1978.

Iron Gall Ink Website. "Iron Gall Ink – History." Accessed May 2011. http://ink-corrosion.org/igi/igi-history.

Jacoby, Joachim. *Die Zeichnungen von Adam Elsheimer: kritischer Katalog.* Frankfurt am Main: Städel Museum, 2008.

Jaffé, David. "Mellan and Peiresc." *Print Quarterly* 12, no. 2 (1990): 168–177.

Kepler, Johannes. *Ioannis Kepleri Sae. Cae. Mtis. mathematici Doptrice.* Augustae Vindelicorum: Typis Davidis Franci, 1611.

Kolbe, Gesa. "Gelatine in Historical Paper Production and Present-day Restoration of Paper Manuscripts." In *Care and Conservation of Manuscripts.* Proceedings of the Sixth International Seminar Held at the Royal Library, Copenhagen 19–20 October 2000. Eds. Gillian Fellows-Jensen and Peter Springborg, pp. 33–46. Copenhagen: Museum Tusculanum Press, 2002.

Koschatzky, Walter. *Die Kunst der Graphik.* Salzburg: Residenz Verlag, 1972.

Koschatzky, Walter. *Die Kunst der Zeichnung – Technik, Geschichte, Meisterwerke.* 8th edition. Munich: Deutsche Taschenbuch-Verlag, 1996.

Krekel, Christoph. "Chemistry of Historical Iron Gall Inks: Understanding the Chemistry of Writing Inks Used to Prepare Historical Documents." *International Journal of Forensic Document Examiners* 5 (1999): pp. 54–58.

Krekel, Christoph. "Schwarze Tinten und Tuschen." In *Der "Liber illuministarum" aus Kloster Tegernsee: Edition, Ubersetzung und Kommentar der kunsttechnologischen Rezepte.* Eds. and trans. Anna Bartl, Christoph Krekel, Manfred Lautenschlager and Doris Oltrogge, pp. 631–637. Stuttgart: Franz Steiner, 2005.

Lavin, Irving. "The Story of O from Giotto to Einstein." *Bildwelten des Wissens: Kunsthistorisches Jahrbuch für Bildkritik* 1 (2003): pp. 37–43.

Lavin, Irving. "Il Volto Santo di Claude Mellan: Ostendatque etiam quae occultet." In *L'Immagine di Cristo dall'Acheropita alla mano d'artista: Dal tardo medioevo all'età barocca.* Eds. Christoph L. Frommel and Gerhard Wolf, pp. 449–491. Vatican City: Biblioteca Apostolica Vaticana, 2006.

Leonhard, Karin and Robert Felfe. *Lochmuster und Linienspiel: Überlegungen zur Druckgrafik des 17. Jahrhunderts.* Freiburg im Breisgau: Rombach Druck- und Verlagshaus, 2006.

Meder, Joseph. *Die Handzeichnung: Ihre Technik und Entwicklung.* 2nd edition. Vienna: Schroll & Co., 1923.

Meder, Joseph. *The Mastery of Drawing.* 2 Vols. Ed. and trans. Winslow Ames. New York: Abaris Books, 1978.

Mommsen, Theodor. Letter to the Prussian Ministry of Culture and Education, 19 March, 1862. Staatsbibliothek – Stiftung Preußischer Kulturbesitz, Berlin. Unpublished.

Mrusek, Ralf, Robert Fuchs and Doris Oltrogge. "Spektrale Fenster zur Vergangenheit: ein neues Reflektographieverfahren zur Untersuchung von Buchmalerei und historischem Schriftgut." *Naturwissenschaften Berlin* 82 (1995): pp. 68–79.

Muspratt, James Sheridan, ed. *Encyklopädisches Handbuch der technischen Chemie.* 8 Vols. Braunschweig: Vieweg, 1905.

Nicoló, Anna. *Lynceographum Quo Norma Studiosae Vitae Lynceorum Philosophorum Exponitur.* Rome: Accademia Nazionale dei Lincei, 2001.

Ornato, Ezio. "Papier und Wasserzeichen in gedruckten Büchern: Das Beispiel Venedig." In *Ochsenkopf und Meerjungfrau: Papiergeschichte und Wasserzeichen vom Mittelalter bis zur Neuzeit.* Eds. Peter Rückert, Sandra Hodecek, Georg Dietz, and Emanuel Wenger, pp. 40–41. Stuttgart: Landesarchiv Baden-Württemberg, 2009.

Piccard Online. Accessed January 2010. http://www.piccard-online.de/sitemap.php?sprache=en.

Piemontese, Alessio [Girolamo Ruscelli]. *I Secreti del reverendo donno Alessio Piemontese*. Venice: Sigismondo Bordogna, 1555.

Préaud, Maxime. *Claude Mellan*. Vol. 17 of *Inventaire du fonds français: Graveurs du XVIIe siècle*. 17 Vols. Eds. Roger-Armand Weigert and Maxime Préaud. Paris: Bibliothèque Nationale, 1988.

Préaud, Maxime and Barbara Brejon de Lavergnée, eds. *"L'oeil d'or": Claude Mellan 1598–1688*. Paris: Bibliothèque Nationale, 1988. Exhibition catalog.

Rocheblave, Samuel. *Essai sur le Comte de Caylus: l'homme – l'artiste – l'antiquaire*. Paris: Hachette, 1889.

Rosand, David and Michelangelo Muraro. *Tiziano e la silografica Veneziana del Cinquecento*. Venice: Neri Pozza, 1976. Exhibition catalog.

Schettini Piazza, Enrica. " 'Più studio che passatempo': La libraria di Federico Cesi e le sue peregrinazioni." In *Federico Cesi e i primi lincei in Umbria*. Atti del convegno di studi nel IV centenario della fondazione dell'Accademia dei Lincei, Terni, pp. 24–25 ottobre 2003. Ed. Vincenzo Pirro, pp. 129–154. Arrone: Thyrus, 2005.

Schettini Piazza, Enrica. "I Barberini e i Lincei: Dalla mirabil congiuntura alla fine della prima Accademia (1623–1630)." In *I Barberini e la cultura europea del Seicento*. Atti del convego internazionale Palazzo Barberini alle Quattro Fontane, 7–11 dicembre 2004. Eds. Lorenzi Mochi Onori, Sebastian Schutze and Francesco Solinas, pp. 117–126. Rome: De Luca Editori D'Arte, 2007.

Schopen, Armin. *Tinten und Tuschen des arabisch-islamischen Mittelalters. Dokumentation – Analyse – Rekonstruktion: ein Beitrag zur materiellen Kultur des Vorderen Orients*. Göttingen: Vandenhoeck & Ruprecht, 2006.

Schwahn, Wolfgang. "Braune Tinten – Lavierungstechniken. Vergleichende Untersuchungen an Bister und Sepia." Diploma thesis. Akademie der Bildenden Künste, Wien, 1985.

Segre, Michael. *In the Wake of Galileo*. New Brunswick: Rutgers University Press, 1989.

Segre, Michael. "Viviani's Life of Galileo." *Isis* 80, no. 302 (1991): pp. 207–231.

Settle, Thomas B. "Egnazio Danti and Mathematical Education in Late Sixteenth-Century Florence." In *New Perspectives on Renaissance Thought: Essays in the History of Science, Education and Philosophy*. Eds. John Henry and Sarah Hutton, pp. 24–37. London: Duckworth, 1990.

Shea, Wiliam R. *Galileo's Sidereus Nuncius or A Sidereal Message*. Trans. William R. Shea. Sagamore Beach: Science History Publications, 2009.

Shea, Wiliam R. "Owen Gingerich's Curious Case." *Galilaeana* VII (2010): pp. 97–110.

Sirigatti, Lorenzo. *La Prattica di Prospettiva*. Venice: Girolamo Franceschi, 1596.

Smith, Anthony W. "Aqueous Deacidification of Paper." In *Paper and Water: A Guide for Conservators*. Gerhard Banik and Irene Brückle, pp. 341–388. Oxford: Elsevier Butterworth-Heinemann, 2011.

Stephens, Catherine H., Timothy Barrett, Paul M. Whitmore, Jennifer A. Wade, Joy Mazurek and Michael Schilling. "Composition and Condition of Naturally Aged Papers." *Journal of the American Institute for Conservation* 47 no. 4 (2008): pp. 201–215.

Stieglecker, Maria. "Verschiedene Klöster – Identische Papiere." In *Ochsenkopf und Meerjungfrau: Papiergeschichte und Wasserzeichen vom Mittelalter bis zur Neuzeit*. Eds. Peter Rückert, Sandra Hodecek, Georg Dietz and Emanuel Wenger, pp. 40–41. Stuttgart: Landesarchiv Baden-Württemberg, 2009.

Stijnman, Ad. "Oil-based Printing Ink on Paper." *Papier Restaurierung* 1, suppl. (2000): pp. 61 68.

Tabulae astronomicae [Adolphine Tables]. Venice: Johannes Hamman, 1492.

Tongiorgi Tomasi, Lucia and Alessandro Tossi, eds. *Il Cannocchiale e il Pennello: Nuova Scienza e Nuova Arte nell'Età di Galileo*. Trans. Siliva Cattiti. Florence: Giunti, 2009. Exhibition catalog.

Tschichold, Jan. *Leben und Werk des Typographen Jan Tschichold*. Dresden: Verlag der Kunst, 1977.

Turner, Antony. Early scientific Instruments: Europe 1400–1800. London: Sotheby's Publications, 1987.

Vasari, Giorgio. *Le Vite de' più eccelenti pittori scultori ed architettori*. 9 Vols. Ed. Gaetano Milanesi. Florence: G.C. Sansoni, 1906.

Vignola, Jacomo Barozzi da and Egnazio Danti. *Le Due Regole della Prospettiva Pratica*. Rome: Francesco Zannetti, 1583.

Watrous, James. *The Craft of Old Master Drawings*. Madison: University of Wisconsin Press, 1957.

Weiss, Karl Theodor. *Handbuch der Wasserzeichenkunde*. Ed. Wisso Weiss. Leipzig: Fachbuchverlag, 1962.

Wellmann, Marc. *Die Entdeckung der Unschärfe in Optik und Malerei: Zum Verhältnis von Kunst und Wissenschaft zwischen dem 15. und dem 19. Jahrhundert*. Frankfurt am Main: Lang, 2005.

Withaker, Ewen A. "Rappresentazioni e mappe della luna: I primi due secoli." In *Galileo: Immagini dell'universo dall'antichita al telescopio*. Ed. Paolo Galluzi, pp. 255–261. Florence: Gunti, 2009. Exhibition catalog.

Woodward, David. *Catalogue of Watermarks in Italian Printed Maps ca. 1540–1600.* Chicago: University of Chicago Press, 1996.

Zeidberg, David S., Fiorella Gioffredi Superbi and Franklin D. Murphy. *Aldus Manutius and Renaissance Culture: Essays in Memory of Franklin D. Murphy.* Acts of an International Conference, Venice and Florence, 14–17 June 1994. Florence: L.S. Olschki, 1998.

LIST OF COPIES MENTIONED

Berlin copy	Staatsbibliothek zu Berlin, Stiftung Preußischer Kulturbesitz Berlin (Mat F30/Mat F591)
Cologne copy	Staats- und Universitätsbibliothek Köln (N4/612)
Durham copy	Palace Green Library, Durham University (SB 0255)
Florence copy	Biblioteca Nazionale Centrale di Firenze (Postillata 110)
Graz copy	Universitätsbibliothek, Karl-Franzens-Universität Graz (I 4099)
Göttingen copy	Niedersächsische Staats- und Universitätsbibliothek Göttingen (4° Astron. II, 331 Rara)
Liège copy	Bibliothèque de l'Université, Liège (R.190C; Brussels, facsimile printed 1967). See Needham, pp. 152–153, 219 in vol. II of this publication
Munster copy	Universitäts- und Landesbibliothek Münster (4° M4 278 So. Mag)
New York copy	possession of Richard Lan and Seyla Martayan, firm of Martayan Lan New York
New York copy (II)	United States, private ownership (former Gudesche Bibliothek Rendsburg, Q 282)
Oklahoma copy	History of Science Collections, University of Oklahoma (Class 520, Book G 133s)
Paris copy	Bibliothèque Nationale de France (Rés.p.V.743)
Paris copy (II)	Observatoire de Paris (20174 (1))
St Andrews copy	Department of Special Collections, St. Andrews University (QB41.G2SA Rare Books)
Venice copy	Biblioteca Nazionale Marciana, Venice (Misc. 1119.10)
Washington copy	Library of Congress, Washington, D.C. (QB41.G17 1610)
Zurich copy	Bibliothek, Eidgenössische Technische Hochschule Zürich (Rar4342)

LIST OF INSTITUTIONS

Accademia Nazionale dei Lincei (National Academy of Linceans)
BAM Bundesanstalt für Materialforschung und -prüfung (Federal Institute for Materials Research and Testing)
Bibliotheca Nazionale di Firenze (BNCF, Central National Library of Florence)
Fraunhofer Institut für Umwelt-, Sicherheits- und Energietechnik UMSICHT (Fraunhofer-Institute for Environmental, Safety and Energy Technology UMSICHT)
Harvard University
Humboldt-Universität (Humboldt University)
Karl-Franzens Universität of Graz (Karl-Franzens University)
Kupferstichkabinett Berlin (Museum of Prints and Drawings)
Rathgen-Forschungslabor (Rathgen Research Laboratory)
Scheide Library (Princeton)
Staatliche Akademie der Bildenden Künste Stuttgart (State Academy of Art and Design Stuttgart)
Stiftung Preußischer Kulturbesitz (Prussian Cultural Heritage Foundation)
Technische Universität Berlin (Technical University Berlin)
Universitäts- und Landesbibliothek Münster (University and Regional Library Munster)

INDEX OF PROPER NAMES

PHOTO CREDITS

Brückle, Irene:

Fig. 8, p. 13; Fig. 9, p. 13; Fig. 1, p. 31; Fig. 15, p. 36; Fig. 19, p. 36; Figs. 1 and 2, p. 39; Figs. 1 and 2, p. 42; Figs. 3–5, p. 43; Figs. 2 and 3, S. 50; Fig. 7, p. 53; Fig. 6, p. 91; Fig. 1, p. 95; Figs. 4 and 5, p. 100; Figs. 6–8, p. 101; Figs. 1–4, p. 105; Fig. 5, p. 106; Fig. 1, p. 111; Fig. 1, p. 128; Fig. 2, p. 130; Figs. 3–5, p. 131; Fig. 6, p. 133; Fig. 7, p. 136; Fig. 8, p. 137; Fig. 10, p. 140; Fig. 12, p. 142; Fig. 11, p. 141; Figs. 1 and 2, p. 159; Figs. 3 and 4, p. 160; Fig. 7, p. 162; Fig. 9, p. 163; Fig. below, p. 168; Fig. 1, p. 170.

Brückle, Irene / Mayer, Manfred:

Fig. 1, pp. 150–151.

Hahn, Oliver:

Fig. 3, p. 39; Figs. 10 and 11, 108; Fig. 9, p. 139; Fig. below, p. 167; Fig. above, p. 168; Fig. 1, p. 174.

Herrenkind, Barbara:

Figs. 1–3, p. 8; Fig. 4, p. 9; Fig. 11, p. 14; Fig. 1, p. 25; Fig. 8, p. 34; Fig. 1, p. 50; Figs. in part II; Figs. 1 and 2, p. 91; Figs. 3 and 4, p. 92; Fig. 6, p. 93; Fig. 1, p. 98; Figs. 9–12, p. 102; Figs. 13 and 14, p. 103; Fig. 8, p. 107; Fig. 2, p. 112; Fig. 1, p. 128; Fig., p. 169; Figs. 1–4, p. 182; Figs. 5–8, p. 183.

Jelen, Erich:

Figs. 21–23, p. 37; Figs. 1–4, p. 46; Fig. 5, p. 47; Fig. 6, p. 48.; Fig. 1, p. 172.

Mayer, Manfred:

Figs. 6 and 7, p. 12; Fig. 3, p. 99; Fig. 6, p. 106; Fig. 7, p. 107; Fig. 9, p. 107; Fig. 3, p. 112; Figs. 1–3, p. 145; Fig. 4, p. 146; Figs. 5 and 6, p. 147; Fig. 7, p. 148; Figs. 1 and 2, p. 152; Fig. 3, p. 153; Fig. 4, p. 154; Figs. 5 and 6, p. 161.

Photo RMN (Réunion des musées nationaux):

Fig. 4, p. 113.

Publications. Benucci 2008: Fig. 5, p. 27; Fig. 2, p. 32; Fig. 6, p. 33; Figs. 12 and 13, p. 35. Jacoby 2008: Fig. 4, p. 27. Mrusek et al. 1995: Fig. 2, p. 171. Nicolo 2001: Fig. 3, p. 26.

Schmidt, Christopher:

Fig. 2, p. 98.

Archive of Authors:

Fig. 5, p. 10; Fig. 10, p. 14; Fig. 2, p. 26; Figs. 6 and 7, p. 28; Fig. 3, p. 32; Figs. 4, 5 and 7, p. 33; Figs. 9–11, p. 34; Fig. 16, p. 36; Figs. 18–20, p. 36; Fig. 4, p. 51; Figs. 5 and 6, p. 52; Figs. 8 and 9, p. 53; Fig. 2, p. 89; Fig. 3, p. 90; Fig. 5, p. 92; Fig. 7, p. 93; Figs. 2–3, p. 95; Figs. 4–9, p. 96; Figs. 10–11, p. 97; Fig. 1, p. 116; Fig. 2, p. 117; Figs. 3 and 4, p. 118; Fig. 5, p. 119; Fig. 6, p. 121; Fig. 7, p. 122; Fig. 8, p. 123; Fig. 1, p. 128; Fig. 6, p. 133; Figs. 1 and 2, p. 143; Fig. 8, p. 162; Figs. 10 and 11, p. 163; Fig. above, p. 167.

Galileo

Makes a Book

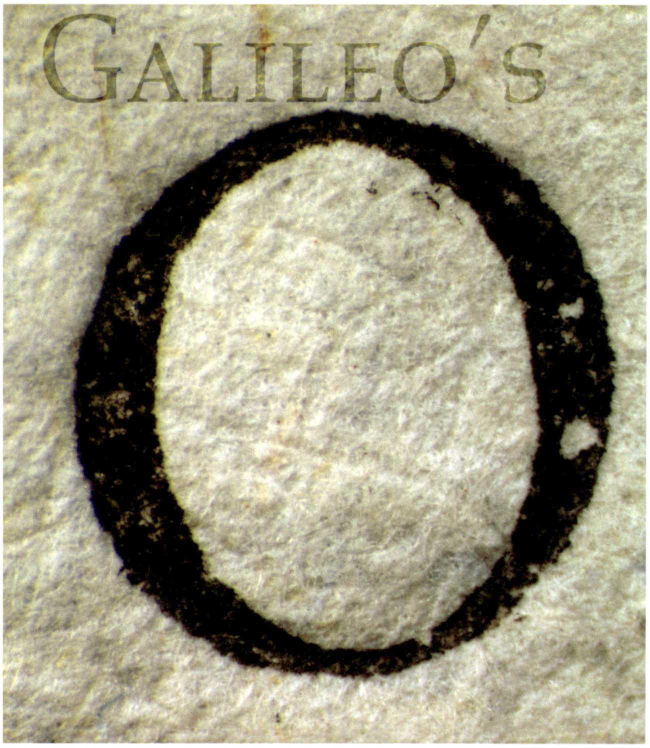

Edited by Horst Bredekamp Volume II

Paul Needham

GALILEO
MAKES A BOOK

The first edition of Sidereus Nuncius
Venice 1610

Akademie Verlag

Published with support by the Max-Planck-Gesellschaft,
the Alexander von Humboldt-Stiftung and the
BAM Bundesanstalt für Materialforschung und -prüfung

Bibliographical information of the German National Library

The German National Library lists this publication in the German national bibliography; bibliographic details are available on http://dnb.d-nb.de.

A scientific publisher of the Oldenbourg publisher group.

www.akademie-verlag.de

Organisation of both volumes by Alexis Ruccius

Graphic design and layout by Petra Florath, Berlin
Printing and binding: DZA Druckerei zu Altenburg GmbH, Altenburg

This paper is resistant to ageing (DIN/ISO 9706).

ISBN 978-3-05-005095-9

Ruth

Contents

Introduction

The publication in Venice in mid-March 1610 of Galileo's *Sidereus Nuncius*, reporting his recent observations of the night skies using a telescope of his own, has long been recognized as an epochal moment in the history of astronomy, and of the natural sciences in general. The element of surprise and grandeur in its statements, beginning with its very title-page, is so striking that it is hard to characterize it except by resort to superlatives which run the risk of sounding exaggerated. But even a superlative may, according to circumstances, be a sober statement. A distinguished historian of astronomy has written that the *Sidereus Nuncius* embodies "more discoveries that changed the world than anyone has ever made before or since."[1]

In recent decades, this perception has been reflected, in the world of antiquarian book collecting, by high prices for good copies of the *Sidereus Nuncius*, which is not, in the strictest sense, a book of great rarity. For any ambitious and well-off collector of the history of science, the *Sidereus Nuncius* is a necessary book, in the same way as Copernicus's 1543 *De revolutionibus orbium coelestium* and Newton's 1687 *Philosophiae naturalis principia mathematica* are: no rationale of collecting the history of science allows it to be ignored as lying outside the central theme. The *Sidereus Nuncius* is included therefore more or less automatically in catalogues of the highlight monuments of human thought: the influential *Printing and the Mind of Man* (1967: no. 113); Harrison D. Horblit's *One Hundred Books Famous in Science* (1964: no. 35); Bern Dibner's *Heralds of Science* (1955, 1980: no. 7); and many others. Perhaps the first significant exhibition in America on the history of science was that held at the University of California at Berkeley in June 1934, assembled by the medical researcher Herbert M. Evans and a group of his friends, the "History of Science Club". The *Sidereus Nuncius* featured in the exhibition, represented by Evans's own copy, now at the Institute for Advanced Study in Princeton.[2]

1 Swerdlow 1998, p. 245. Naturally, opinions vary; one historian has qualified *Sidereus Nuncius*'s lengthy title, which summarizes these discoveries, as "bombastic", a term that implies empty exaggeration (Biagioli 1993, p. 103).

2 Evans 1934, no. 18. The exhibition was held at the university library on the occasion of the meeting of the American Association for the Advancement of Science. Evans's preface is a manifesto on the importance of assembling and studying the original records of scientific research: "… it may be maintained that one cannot adequately understand any scientific subject without knowledge of the manner in which our present conceptions were established."

The fame of the *Sidereus Nuncius* goes back indeed to its origins. In the weeks before it appeared rumors had already leaked to the world of learning that Galileo had made sensational discoveries about the cosmos. On Friday 12 March 1612, Marcus Welser wrote from Augsburg to the Jesuit mathematician Christopher Clavius in Rome that reliable news from Padua had come to his ears: the mathematician Galileo, with his new instrument the *visorio*, had seen extraordinary things in the heavens including four new planets.[3] The next day, Saturday the thirteenth – the very day that a completed copy of the *Sidereus Nuncius* first reached Galileo's own hands – the English ambassador to Venice, Sir Henry Wotton, procured and posted a copy to King James I in England, with an accompanying letter to Robert Cecil, the leading figure at James's court, highlighting the revolutionary discoveries, which were "the strangest piece of news ever yet received from any part of the world."[4] The news was so unexpected, so revolutionary, Wotton feared that in retelling it he would be showed up for a credulous fool. Welser expressed a similar qualm.

We possess an unusually extensive contemporary body of information and documentation on the *Sidereus Nuncius*, from Galileo's first decision to publish such a work, to his drafting of the text, to its making in Venice under the auspices of the publisher Tommaso Baglioni – who three years earlier had published Galileo's *Difesa* against the plagiarisms and "calumnies" of Baldassare Capra regarding Galileo's calculating compass – to its publication and immediate reception. Much of our information, both direct and indirect, comes from Galileo's own pen. In one letter, he states the edition size, 550 copies. Other letters give information about Galileo's hopes, eventually successful, that the dedication of the work to the young Cosimo II de Medici, who had recently succeeded his father as Grand Duke of Tuscany, would procure a lucrative and prestigious appointment as philosopher and mathematician to the Tuscan court.[5] We have Galileo's heavily reworked autograph Draft for the *Sidereus Nuncius*, allowing in principle a detailed comparison, word by word, of the formation of his text with its result in print. There is also an incomplete autograph Fair Copy (as it has been called) whose position in the transmission of the *Sidereus Nuncius* text has not hitherto been clearly analyzed. The autograph notes of Galileo's daily telescopic observations of Jupiter survive, so that those observations can be followed from original observation record to rough redrafted text to revised printed text.[6] Contrast this wealth of compositional data with one of England's literary monuments, printed, like the *Sidereus Nuncius*, in quarto format just two years earlier: Shakespeare's *King Lear* (1608). We have about a dozen copies of the printed *Lear* itself; we have even a brief external record, the registration of its copy at the London Stationer's Company on 26

3 *Opere* X, p. 288 (Ep. 270**).

4 Sharratt 1994, pp. 1–2; Smith 1907, I, pp. 486–487.

5 Biagioli 1993 and 2006.

6 Most of Galileo's autograph notes, descended through his family, are preserved in the Biblioteca Nazionale Centrale, Florence (BNCF). The autograph drafts of *Sidereus Nuncius* were mostly bound in the early nineteenth century, with other extraneous materials, in BNCF Ms. Gal. 48. Other autographs of the same project strayed, and are now bound in Mss. Gal. 50 and 70. For the structure and a reconstruction of the original dossier, see chapter V.

November 1607. But we have no draft in Shakespeare's hand, nor letters he wrote to friends while he was creating it.

Despite a multitude of studies by learned scholars touching seemingly all aspects of Galileo's life and creative work, he remains neglected to a degree as an author and maker of books. Is it conceivable that a substantial draft for *Lear* should exist in Shakespeare's hand, should even have been reproduced in usable if far from ideal facsimile (as is the case for most of the *Sidereus Nuncius* draft), yet should never have attracted the closest attention of Shakespeare scholars? Yet it is widely acknowledged that not least among his cornucopia of talents, Galileo was an artist with words. The mathematician Bonaventura Cavalieri wrote Galileo, after receiving a copy of the 1632 *Dialogo*, that it gave him the same sensation as *Orlando Furioso*: wherever his eye happened to fall, he was compelled to keep on reading.[7] This is equally the case with the *Sidereus Nuncius*, whose prose embodies not just close reasoning, but also life and emotion.[8]

The following pages are devoted to Galileo not simply as a writer, but also, in intimate connection with his writing, as the maker of a book, conceived and brought to fruition in a Venice printing shop in the first two and a half months of 1610. Considerable emphasis is laid on the physical production of the *Sidereus Nuncius*, a brief but complex work. The physical production of *Sidereus Nuncius* has been little explored, beyond a useful preliminary outline published by Owen Gingerich and Albert Van Helden in 2003.[9] I try to go a little more deeply than those authors do into the physical evidence provided by various copies of the *Sidereus Nuncius* itself, and by a closer yet far from comprehensive scrutiny of the relation of the printed text to Galileo's preceding Draft, Fair Copy, and Maquettes. These are matters with which, of course, Galileo himself would have been intimately concerned. If the *Sidereus Nuncius* can, through study of the interrelations and interaction of written texts, paper, and printing become more richly understood, we will to some degree enter more actively into Galileo's mind in the winter weeks of 1610, a time when it was of the highest importance for him not only to study the heavens with his recently perfected telescope, but to bring his studies speedily into print in multiple copies, and thus into the public eye.

7 *Opere* XIV pp. 336–337 (Ep. 2251): "... sento in me ... l'effetto che mi ricordo havere esperimentato nel leggere il Furioso, che dovunque io dia principio a leggere, non posso ritrovarne il fine: così appunto me à accaduto ne'suoi Dialogi ..."

8 Cf. Pighi 1972, p. 548, stating that Galileo's Latin is inferior to his Italian, where he is "uno dei più grandi prosatori nostri," and inferior to Kepler's Latin as found in the latter's *Dissertatio cum Nuncio sidereo*. But there is, Pighi writes, literary significance to the *Sidereus*, in the "limpidissima semplicità e quasi impassabilità di quel diario d'entusiasmanti osservazioni, dove la gioia della scoperta è resa soltanto da qualche usuale e stanco vocabulario." In this latter reservation, Pighi is perhaps too much the classicist: the "tired vocabulary" is that of the scholastic tradition in which Galileo's readers would have been imbued; for them it might have had the simplicity of the thoroughly familiar.

9 Gingerich and Van Helden 2003. A more recent study, Wootton 2009, conflicts at several points with the physical evidence of Galileo's draft.

CALENDAR COVERING THE FIRST FOURTEEN WEEKS OF THE YEAR 1610

SUN	MON	TUE	WED	THU	FRI	SAT
					JAN 1	2
3	4	5	6	✎ 7 ♃	8 ♃	☁ 9
10 ♃	11 ♃	12 ♃	13 ♃	14 ♃	15 ♃	16 ♃
17 ♃	18 ♃	19 ♃	20 ♃	21 ♃	22 ♃	23 ♃
24 ♃	25 ♃	26 ♃	27 ♃	☁ 28	☁ 29	✎ 30 ♃
31 ♃	FEB 1 ♃	2 ♃	3 ♃	4 ♃	☁ 5	✎ 6 ♃
7 ♃	8 ♃	9 ♃	10 ♃	11 ♃	12 ♃	✎ 13 ♃
☁ 14	☆ 15 ♃	16 ♃	17 ♃	18 ♃	19 ♃	✎ ☁ 20
21 ♃	☁ 22	☁ 23	ASH WED ☁ 24	25 ♃	✎ 26 ♃	27 ♃
28 ♃	✎ MAR 1 ♃	2 ♃	3	4	5	✎ 6
7	✎ 8	9 ♃	10 ♃	11	✎ 12 ♃	PRINTING FINISHED ✎ 13
14	15	✎ 16	17	✎ 18 ♃	✎ 19	✎ 20 ♃
21	22	23	24	25 ♃	26	✎ 27 ♃
28 ♃	29 ♃	✎ 30	31	APR 1	2 ♃	✎ 3 ♃
PALM 4	5 ♃	6 ♃	7	8	9 ♃	10
EASTER ✎ 11 ♃						

♃ Jupiter observations ☁ cloudy night ✎ dated documents and letters ☆ Galileo's 46th birthday

I.
January 1610

Galileo first gave attention to the recent Netherlandish invention of the telescope in May 1609, when a former pupil, Jacques Badover, son of a Venetian Protestant emigré, sent a report from Paris that a spectacles merchant on the Pont Neuf was offering telescopes for sale. At some time thereafter – by his account, shortly thereafter – Galileo began his own experiments in telescope-making, his first result having a three-power resolution. By late August he had made a nine-power telescope which he presented to the Venetian Senate, demonstrating its virtues from the campanile of San Marco. He received in reward a large increase in his salary at the University of Padua. In the autumn of 1609 Galileo continued to improve his telescopes, making one of twenty-power with which, in the late autumn, he began to observe the skies, and particularly the Moon in its changing phases.[1]

At the beginning of January there is no sign that Galileo had an immediate plan to publish his astronomical observations, but by the middle of the month, he had formed such a plan, and by the month's end he had journeyed from Padua to Venice to make arrangements for printing. As is obvious from a letter he wrote on the thirtieth of January, he hoped to gain permission to dedicate his work to Cosimo II, the youthful Grand Duke of Tuscany. The precipitating event for all this, as is widely agreed, was not his observations of the Moon, but the surprising results of his subsequent observations of Jupiter, which began on Thursday January 7. On this day Galileo wrote, or finished, a long letter to an unnamed correspondent, describing his telescopic observations over the course of a lunar cycle. The letter survives in two later copies, one, since generations unfindable, at the Vatican Library, the other among Galileo's papers in the Biblioteca Nazionale Centrale, Florence. The latter retains the date of the letter. The Vatican copy, unfortunately of unrecorded transmission context, lacks a date. If the Florence copy descended by way of Galileo's papers, the lost Vatican copy presumably descended by way of its recipient's papers. It includes pen and wash sketches of the Moon based on Galileo's sketches in the original of the letter, and contains a coda not present in the Florence copy: "this very evening I saw Jupiter accompanied by three fixed stars, completely invisible [without the telescope] because of their minuteness … their configuration … did not occupy more than a degree of longi-

1 Van Helden 1977, pp. 26–27 and passim.

tude ..." (*... pur questa sera ho veduto Giove accompagnato da 3 stelle fisse, totalmente invisibili per la lor picciolezza ...*).[2]

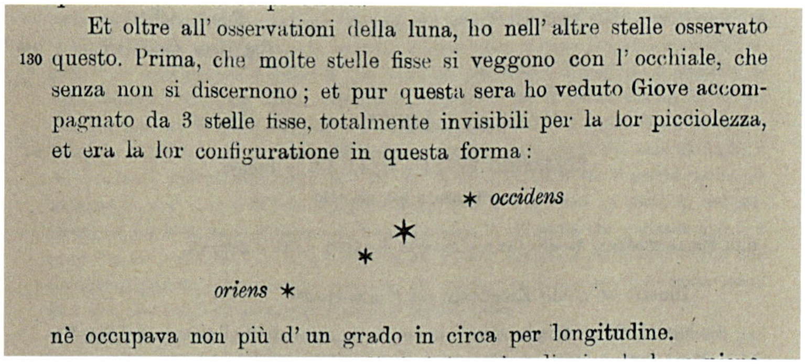
Et oltre all' osservationi della luna, ho nell' altre stelle osservato questo. Prima, che molte stelle fisse si veggono con l' occhiale, che senza non si discernono; et pur questa sera ho veduto Giove accompagnato da 3 stelle fisse, totalmente invisibili per la lor picciolezza, et era la lor configuratione in questa forma:

nè occupava non più d' un grado in circa per longitudine.

I.1: Opere X, p. 277.

On the next nights Galileo continued to observe Jupiter, and immediately discovered either that the small fixed stars were not fixed, or, as was his first thought, that Jupiter itself was in motion opposite to that calculated in the published *Ephemerides* he consulted. He began to enter his records of the changing positions of the small stars with respect to Jupiter sequentially on a fresh sheet of paper. On this sheet he noted on Monday January 11: "it appears that around [or, in the vicinity of] Jupiter are three other small wandering stars, invisible before now to everyone" (*... dal che appare intorno à Giove esser. 3. altre stelle erranti invisibili ad ogn'uno sin à questo tempo*) (figure I.2).

Two days later, Wednesday January 13, having improved the stability of his instrument, Galileo could see that there were in fact four small stars moving in close proximity to Jupiter. Another two days on, after an intervening cloudy night, he saw the four stars again. Having now recorded seven nights of observations, Galileo entered on his observation sheet a marginal remark that reveals his decision to publish his findings: "[The observation diagrams] are to be cut in wood together on a single block, the stars in white and the remainder in black, and then the block may be sawed into strips" (*... per l'inanzi faransi intagliar in legno tutte in un pezzo, et le stelle bianche il resto nero e poi si segheranno i pezzi.*) (figure I.3). In other words, on this evening Galileo was already looking forward to how his observations would be put into print. January 15 is the day of conception of *Sidereus Nuncius*, and its gestation was of eight weeks.

Stillman Drake pointed out a second strong hint on the same page of the observation sheet that indicates how rapidly Galileo was advancing his plans for a published book. Galileo's notes on his first observation of January 15, three hours after sunset, are written in Italian. His notes on further observations four and five hours later, that is between

2 *Opere* X, p. 277 (Ep. 259). Drake 1978, p. 143, hypothesized that the 7 January letter was addressed to Enea Piccolomini but had not been sent. However, the fact that a draft of the letter remained in Galileo's archive, as strongly indicated by the existence of the Florence copy, can hardly be taken as evidence one way or the other that an original was not sent. Antonio Favaro, editor of the *Opere*, believed the letter had been addressed to Antonio de' Medici, an older cousin of Cosimo II by illegitimate birth, who a few months earlier had congratulated Galileo on his invention of a telescope. But Drake p. 144 observes that the style of address differs from that of other letters from Galileo to Antonio.

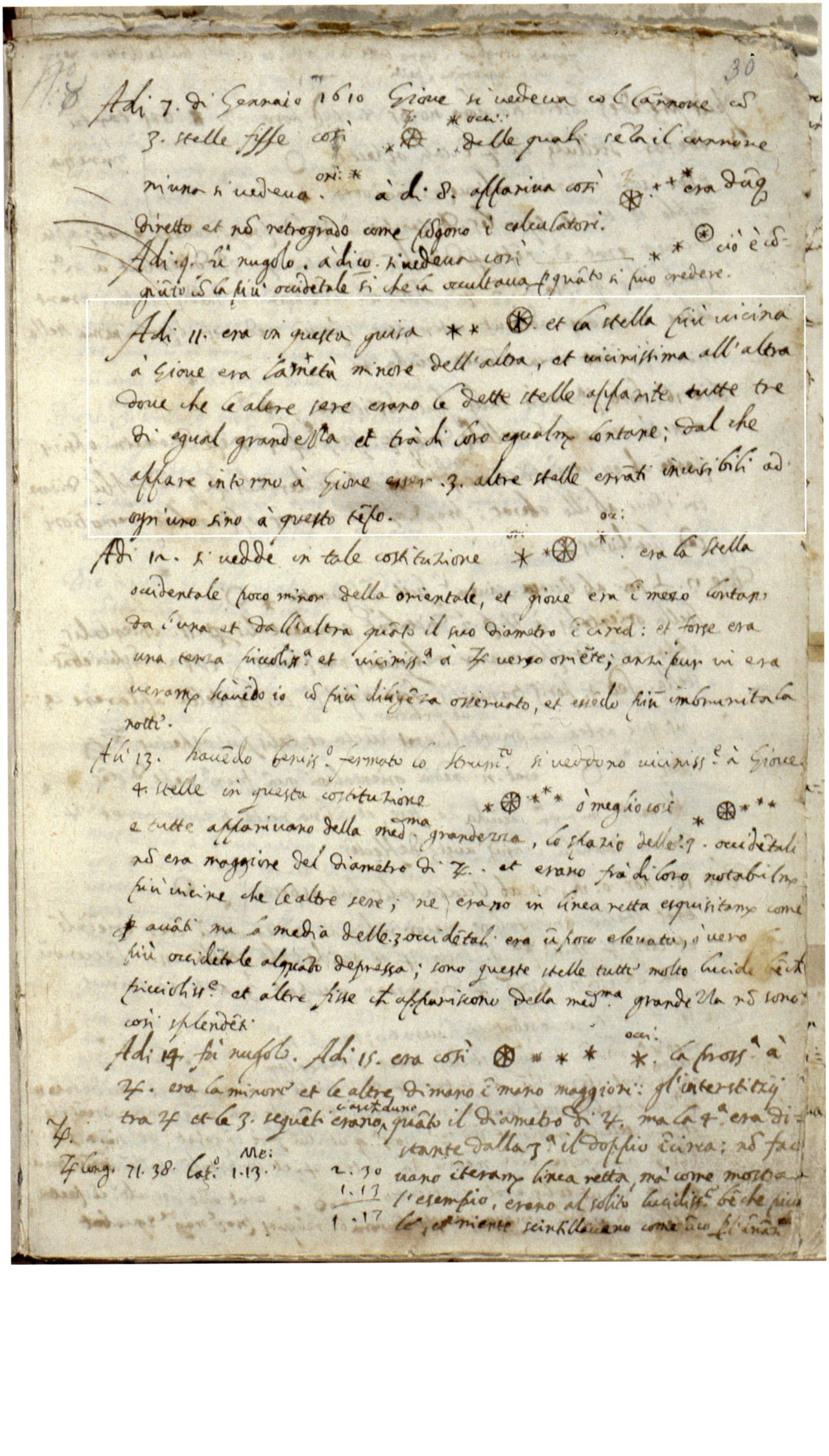
30
Adì 7. di Gennaio 1610 Giove si vedeva col cannone con
3. stelle fisse così * * ✲ * delle quali senza il cannone
niuna si vedeva. a dì 8. appariva così ✲ * * * era dunque
diretto et non retrogrado come vogliono i calculatori.
Adì 9. si nugolò. a dì 10. si vedevan così * * ✲ cioè congiunto con la più occidentale si che la occultava per quanto si può credere.
Adì 11. era in questa guisa * * ✲. et la stella più vicina
à Giove era la metà minore dell'altra, et vicinissima all'altra
dove che le altre sere erano le dette stelle apparite tutte tre
di equal grandezza et trà di loro egualmente lontane; dal che
appare intorno à Giove esser 3. altre stelle erranti invisibili ad
ogn'uno sino à questo tempo.
Adì 12. si vedde in tale costituzione * ✲ *. era la stella
occidentale poco minor della orientale, et Giove era in mezo lontano
da l'una et dall'altra quanto il suo diametro in circa: et forse era
una terza piccolissima et vicinissima à ♃ verso oriente; anzi pur vi era
veramente havendo io con più diligenza osservato, et essendo più imbrunita la
notte.
Adì 13. havendo benissimo fermato lo strumento si veddono vicinissime à Giove
4. stelle in questa costituzione * ✲ * * * ò meglio così ✲ * * *
e tutte apparivano della medesima grandezza, lo spazio delle 3. occidentali
non era maggiore del diametro di ♃. et erano frà di loro notabilmente
più vicine che le altre sere; ne erano in linea retta esquisitamente come
per avanti ma la media delle 3. occidentali era un poco elevata, ò vero
più occidentale alquanto depressa; sono queste stelle tutte molto lucide benche
piccolissime et altre fisse si appariscono della medesima grandezza non sono
così splendenti.
Adì 14. fu nugolo. Adì 15. era così ✲ * * * *. la prossima à
♃. era la minore et le altre di mano in mano maggiori: gl'interstitii
tra ♃ et le 3. seguenti erano tutti eguali, quanto il diametro di ♃. ma la 4a era di-
stante dalla 3a il doppio in circa; ne face-
vano interamente linea retta, ma come mostra
l'esempio, erano al solito lucidissime benche picco-
le, et niente scintillavano come fin' inanzi
♃.
♃ long. 71.38. Lat. 1.13. Me: 2.30 1.13 1.17

I.2: BNCF, Ms. Gal. 48, 30r.

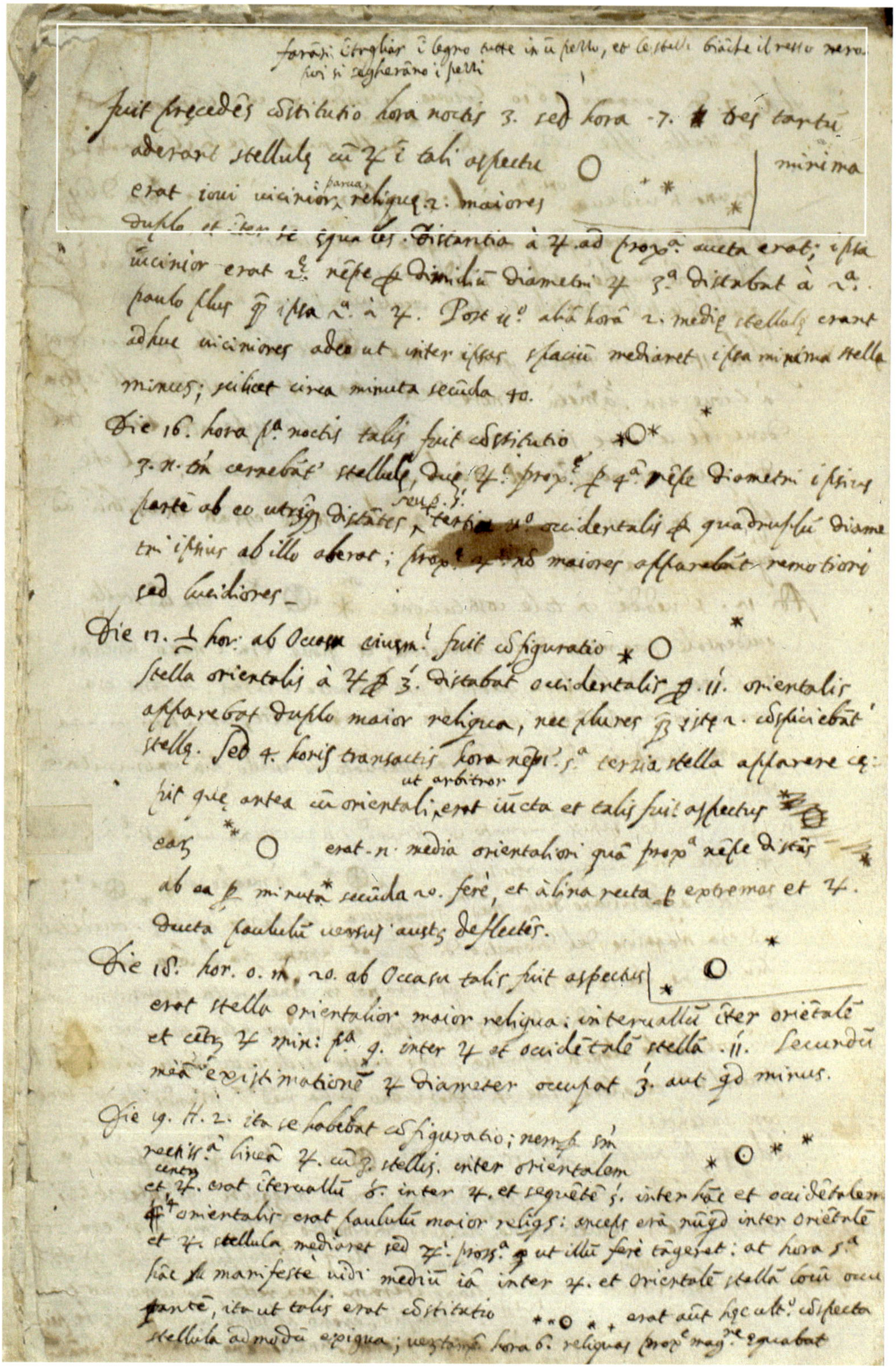

I.3: BNCF, Ms. Gal. 48, 30v.

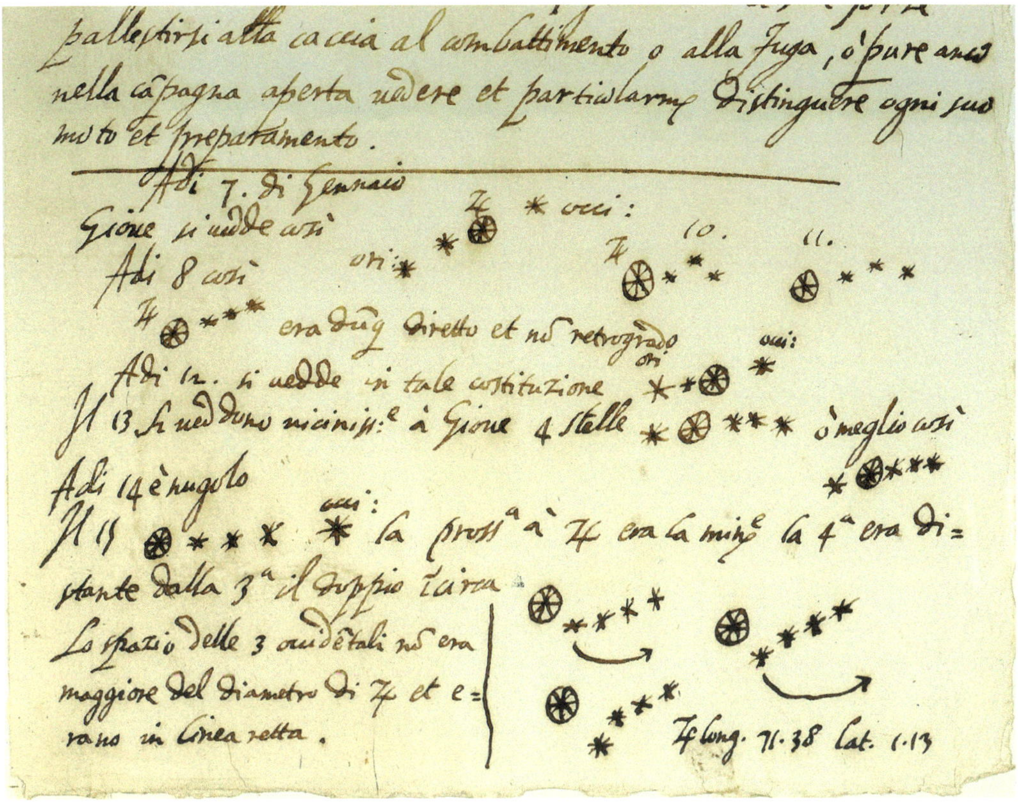
pallestiri, atta alla caccia al combattimento o alla fuga, o pure anco nella capagna aperta vedere et particularmente distinguere ogni suo moto et preparamento.

Adi 7. di Gennaio Giove si vedde così

Adi 8 così

era dunque diretto et non retrogrado

Adi 12. si vedde in tale costituzione

Il 13 si veddono vicinis.e à Giove 4 stelle o meglio così

Adi 14 è nugolo

Il 15 la pross.a à ♃ era la min.e la 4.a era distante dalla 3.a il doppio in circa

Lo spatio delle 3 occidentali non era maggiore del diametro di ♃ et erano in linea retta.

♃ long. 71.38 Lat. 1.13

I.4: *Galileo draft notes, Special Collections Library, University of Michigan.*

about 11 o'clock and midnight, are in Latin, as are all following notes. The book he then had in mind, directed to Europe's world of learning, would naturally and necessarily be in Latin.[3] Closely connected with this stage of Galileo's observations is a loose autograph leaf of sketches and notes on the "stars" near Jupiter preserved at the University of Michigan, which cover the same days 7 to 15 January. By Drake's interpretation, these sketches represent a stage in Galileo's thinking in which, although he observed the movements of these small stars, he did not at first assume that the movements represented rotation around Jupiter; they might be explained as movement back and forth along a straight line. In any case, it is telling that he entered Jupiter's longitude and latitude for 15 January on this leaf, exactly as he did on the first page of his observation sheet, on the basis of the Ephemeris that he had at hand.[4] There is a good presumption that the notes of the Michigan leaf were written on that same fateful day of 15 January 1610.

For the next two weeks Galileo continued his nightly observations of the moons (as we now call them) of Jupiter, but at the end of the month he traveled to Venice to begin overseeing the printing of his "observations." He wrote from there on Saturday January 30 to Belisario Vinta, private secretary to Cosimo, in Florence, with whom he was well acquainted: outlining briefly his discoveries, expressing his sense of the divine grace that had enabled him to be the first mortal to view what had been hidden from the preceding ages, and

3 Drake 1978, p. 152.
4 Drake 1976.

stating his wish to send copies not just to the natural philosophers and mathematicians of Europe, but also to Cosimo, who would moreover be presented with an excellent telescope by which the discoveries could be verified (*Opere* III.i, no. 262). It appears that Galileo wrote similarly to the Florentine courtier Enea Piccolomini for posting together, on the same day. A week later, Saturday 6 February, both men wrote back to Galileo, saying that they had read his letters to the Grand Duke, who expressed a great desire to see this work printed as quickly as possible (*Opere* III.i, nos. 263-4). In Vinta's words, the Grand Duke and his wife were quite bowled over by this latest example of Galileo's almost supernatural genius (*stupefatti di questa nuova prova del suo quasi sopranaturale ingegno* ...). As to the moons of Jupiter, Galileo was carefully evasive in his letter to Vinta: they moved about an unspecified "very large star," and thus could hypothetically be almost anywhere in the sky.

During the last two weeks of January, Galileo composed a major portion of the first part of his text (that is, exclusive of his Jupiter reports, which he continued to compile) – the part that was eventually printed as quires B–D of *Sidereus Nuncius*. Much of the description of the Moon's surface in this Draft is essentially a translation, more formal and with occasional explanatory expansions, of what Galileo wrote to his unnamed correspondent on 7 January.[5] Compare:

7 January 1610 letter (*Opere* X.i no. 259, p. 274): [Four or five days after the New Moon] ... *il confine che è tra la parte illuminata et il resto del corpo tenebroso, esser non una parta di linea ovale pulitamente segnata, ma un termine molto confuso, anfrattuoso et aspro, nel quale molte punte luminose sporgono in fuori et entrano nella parte oscura* ...

Sidereus Nuncius, 7v (small caps added): ... *iam terminus, partem obscuram à luminosa diuidens, non aequabiliter secundam oualem lineam extenditur,* VELUTI IN SOLIDO PERFECTÈ SPHAERICO ACCIDERET; *sed inaequabili, aspera, & admodum sinuosa linea designatur ... complures enim veluti excrescentiae lucidae vltra lucis tenebrarumquè confinia in partem obscuram extenduntur ...*

The remark "as would be the case for a perfectly spherical solid" was added in the Draft, to highlight the direct conflict between Galileo's observations and the Aristotelian concept of the perfection of the spheres. As will be examined in detail in chapter VIII, the Moon sketches in the 7 January letter are critical to understanding the genesis of the four etchings of the Moon that were created for *Sidereus Nuncius*.

5 Bredekamp 2007, p. 178 note 6, provides an extensive concordance of related passages in the 7 January letter and *Sidereus Nuncius*.

II.
Structure, Format and Paper of Sidereus Nuncius

A printed book is, scale apart, a structured material object in much the way that a building is. Without some knowledge of the materials and human actions that produced (in multiple copies) the structure of a printed book, theories about the textual interrelations and peculiarities of that book have a high probability of going wrong, even of being physically impossible. The first edition of *Sidereus Nuncius* is a quarto of thirty leaves, that is sixty pages, comprised of seven gatherings or quires, signed A through G. Specifically, it is, with

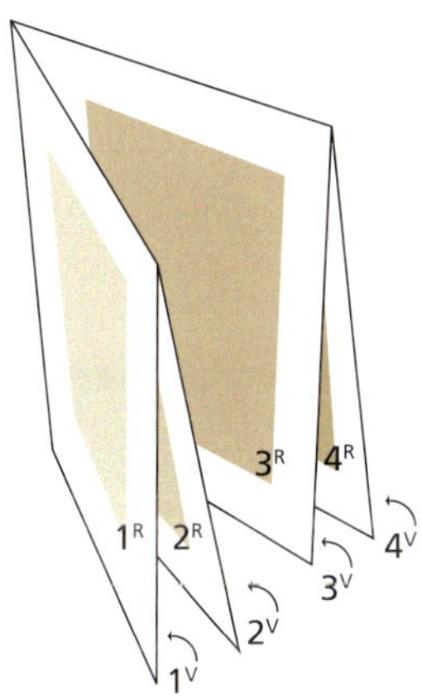

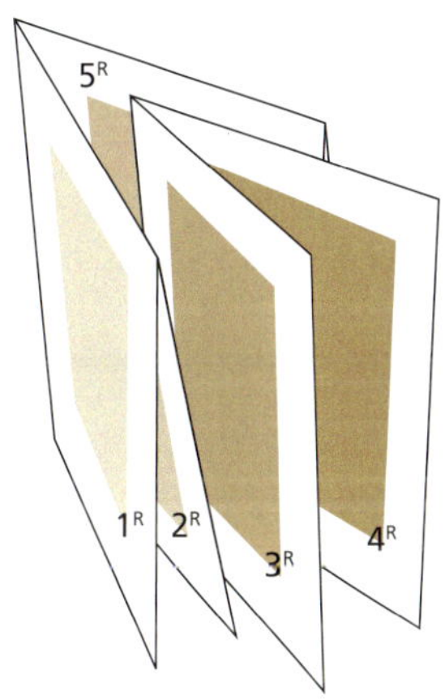

II.1: Full sheet folded in quarto.

II.2: Quarto full sheet plus half-sheet.

one exception, a full-sheet quarto: each quire of four leaves is a single sheet of paper, folded twice at right angles so that a given leaf is one quarter of the sheet it was printed on.

As is clear from the diagram, four printed pages of the gathering sit on one side of the sheet (1 recto, 2 verso, 3 recto and 4 verso), and four on the other (1 verso, 2 recto, 3 verso and 4 recto). Leaves 1 and 4 are attached across the inner margin, as are 3 and 4; leaves 1 and 2 were originally, before binding and trimming, attached across the top margin, as were 3 and 4. The exception is quire D in six leaves, where the third and fourth leaves are an additional half-sheet of paper inserted in the middle, leaves 1, 2, 5, 6 making up the full sheet.

This summary description – a quarto of seven and a half sheets of paper – is how printers, publishers and booksellers would have looked at the *Sidereus Nuncius* when it came on the market. We see this in the respective Italian and Latin wording of the colophons of the 1613 Rome quarto editions of Galileo's *Istoria e dimostrazioni intorno alle macchie solari*, and of Christoph Scheiner's *De maculis solaribus*. In the quire register of Galileo's *Macchie solari* the publisher Giacomo Mascardi listed the quires of a complete copy as "A B C D [...] Q R S T V. Tutti sono *fogli intieri*, eccetto A, & T, che ciascun è *un foglio, e mezo*" (italics supplied). In Scheiner's work, Mascardi listed the quires as "a C D E F G. Omnia Sunt *integra folia*, præter a, quod est *folium cum dimidio*". That is, the quires in both editions are full sheets, except for the designated ones constructed of a sheet plus a half-sheet. By the now-conventional bibliographical notation devised in the nineteenth century by Henry Bradshaw, the structure of the *Sidereus Nuncius* can therefore be expressed concisely as A^4; B–C^4 D^6; E–G^4. The semi-colons of this formulation mark off the work's three textual units.

Quire A (4 leaves, 8 pages) contains the title-page including Baglioni's imprint and device, with blank verso; Galileo's dedication to Cosimo on leaves A2r–4r (fos. 2–4); and on the verso of A4, a record of the license to print granted by the Venetian Council of Ten.

Quires B, C, and D (14 leaves, 28 pages) contain Galileo's account of his development of an improved telescope in the months after he first gained knowledge of the "Flemish" invention; his lunar observations and analysis; and finally, on D4r (fo. 16r) at line 12 and after, a more miscellaneous report on various observations of the fixed stars. The pages of this section contain not just typography but also seven woodcuts printed together with the typographic settings; and four etchings, separately printed under a rolling press — a process independent of and subsequent to the type-printing. One of the etchings is repeated on two pages, for a total of five impressions.

Quires E, F, G (12 leaves, 24 pages) contain Galileo's account of his observations of Jupiter and discovery of four "planets" revolving around it, concluding with an epilogue. On these pages there are, interspersed with the text proper, 65 diagrams representing Jupiter and its four satellites, as observed on different nights and times of the night. The diagrams are typographical constructs, using a large-font capital O for Jupiter, and asterisks and star-symbols of four different sizes to represent differences in luminosity or magnitude of the satellites.

Following the unnumbered title leaf, the leaves of *Sidereus Nuncius* are numbered in the right-page headlines from 2 to 28, not to 30. The irregularity comes in quire D. This was originally calculated as a conventional four-leaf quire but then was temporarily held back from production as its text and illustrations were extended to fill exactly two more leaves.[1] Meanwhile, the pages to be printed on sheet E were set and proceeded to the press, as discussed in detail below in chapter X. Leaves D1–4 were numbered 13–16 in continuation from sheet C, but D5–6 were given no numbers, since leaves E1–4 had already been printed off with numbering 17–20. For reference purposes, D5 and 6 are designated as leaves 16A and 16B.

The materials that went into making *Sidereus Nuncius* fall into two classes: those still extant – the handmade paper of all the surviving copies, and the ink that they bear – and those now lost – the metal types of several fonts, the woodcut ornamental material (a headpiece and two historiated initials), the various woodcut diagrams designed by Galileo,

1 Gingerich and Van Helden 2003, pp. 262–263: "The material intended for the skipped signature D was a patchwork of insertion upon insertions ...". The text expansions are discussed in chapters VI–IX.

and the copperplate etchings depicting the Moon. All these latter were inked and impressed onto multiple sheets of paper, leaving mirror images of their flat surfaces. These materials, and the actions of the printing craftsmen upon them, set the conditions under which the *Sidereus Nuncius* was, over a period of roughly six weeks, brought into being.

Paper

The paper stocks are of particular importance in understanding the making of *Sidereus Nuncius*. Five different stocks appear – not all in any particular copy, but overall, in the totality of all surviving copies that represent the edition. These five stocks define unambiguously three different sets of copies, hitherto not distinguished from each other. First, there is a proof copy, that is, a copy assembled from proof impressions of the type pages, made before the full press run began, but preserved and gathered into a copy by Galileo himself. This is the copy, hereafter designated as ML, in the possession of the New York book firm Martayan Lan. It contains five drawings of the Moon's phases, added on pages 8r (B4r), 9v (C1v), 10r (C2r), and 10v (C2v) in the spaces that in most copies are occupied by closely corresponding etchings. The drawings have recently been intensively studied by Horst Bredekamp, who concludes that they are Galileo's own drawings, made as models for the etchings which in turn were also probably the work of his own hand.[2] The discovery that this copy is a unique proof set brings it, from a line of evidence and reasoning independent of that pursued by Bredekamp, into an even closer connection with Galileo himself, who signed it on the title-page.[3] These proofs were printed on a different paper stock from those of the main print run, and by a different press operation (see chapter XIII).

Second, many dozens of copies, the great majority of those that survive, belong to the ordinary-paper issue. Essentially, this is the issue that was published by Tommaso Baglioni. With varying degrees of detail, relevant information has now been gathered on seventy-one copies of the ordinary issue. Its copies were printed primarily on two stocks of paper, one found exclusively in quire A, the other exclusively in quires D–F. A small number of copies contain also a third, minor stock that appears only in some examples of the half-sheet D3.4, and is one of the clues showing that this half-sheet represents a late stage of printing.

Third, a restricted number of copies, perhaps thirty originally – so far, nine survivors have been identified – were printed on finer paper, a unique stock of heavier substance, which produced richer impressions of the etchings. This issue, not previously noticed except for brief and quickly forgotten references in the antiquarian book trade, comprised Galileo's own copies for presentations; a small number of hand-corrections by Galileo appear in all nine known copies, discussed further in chapter XIII.[4]

2 Bredekamp 2007, chapters V–VIII.

3 Bredekamp 2007, pp. 186–189, identified several significant differences in setting between the ML copy and other copies, but the evidence was not enough to clarify fully the circumstances of production.

4 Two fine-paper copies were in the book trade between 1977 and 1988. The first, now at Stanford, was sold by Sotheby's Parke-Bernet in 1977 when the Franklin Institute disposed of its important collection of science books, and the catalogue described it accurately as a "superb thick paper copy"; in H. P. Kraus Catalogue 164 in 1982 it was also so described. However, when it was acquired by Stanford as part of the Barchas Collection,

Paper was made on twin moulds – wooden frames across which was stretched under tension a wire interlace, with fine wires interwoven across thicker supporting wires called chains. The chains ran from one long side of the frame to the other, and the interweaving wires ran from one short side to the other. The two moulds would share a single deckle, an open wooden frame that sat on top of the mould and created a boundary for the "stuff": a watery suspension of linen fibers placed in a vat into which the moulds were dipped. After the correct amount of stuff had been taken up, the vat man would tilt the mould back and forth to move the stuff evenly across the wire surface, and expertly shake the mould in such a way that the linen fibers would interlock. All the while, excess water would run out of the mould through the interstices of the wire interlace. The small runnels of stuff that worked their way slightly under the deckle created the irregular boundaries of the untrimmed sheet called deckle edges.

Twin moulds were used for an efficient work tempo: the vat man would dip and form one sheet, then hand over the mould to the "coucher", who would turn it out onto a piece of felt, gradually forming a stack of wet paper sheets alternating with felts to absorb more moisture. At this handover, the coucher would give back the mould that he had just laid; the vat man would clap on the deckle, which he retained, and dip and form the next sheet; and so forth.

A shaped wire figure was usually attached to the surface of the mould, which impressed into the paper an identifying symbol, visible in the sheets when held against the light: the so-called watermark. Typically the watermark was centered in one half of the mould. Very commonly, with twin moulds, one wiremark would be centered in the left half of one mould, the other in the right half of the second. The two wiremarks depicted the same symbol, but since they were individually formed, they would not be precisely identical, and it is usually an easy matter to pick out earmarks that distinguish the one wiremark from the other.[5] A ream of five hundred sheets – the common packing quantity – would contain a nearly fifty-fifty mixture of sheets from the two moulds. When both moulds have been identified, the paper stock is adequately defined. Since the late fifteenth century, in the paper mills that supplied the printing shops of Venice and the Veneto, it was frequent practice to make paper not just with watermarks centered in one half of the sheet, but also with cornermarks, usually monograms or devices of the papermaker or other simple symbols, visible in the lower corner of the opposite half of the sheet.[6] Three of the *Sidereus Nuncius* stocks have cornermarks.

the knowledge that it belonged to the extremely rare thick or fine paper issue seems to have fallen away. The second copy, now at Linda Hall Library, was sold by Sotheby's London in 1988, but was not noted as on fine or thick paper. Thus, what was known to Sotheby's New York branch and to the firm of H. P. Kraus in the 1970s and early 1980s became forgotten again by the end of the 1980s.

5 In the eighteenth century (much less commonly earlier, though a few examples even of the fifteenth century could be cited) complex watermarks were often shaped on a pin template or jig, so that the twin wiremarks are difficult to distinguish, unless perhaps one was flipped over with respect to another, and/or if they were sewn to opposite sides of their moulds.

6 Needham 1994; Briquet, p. 14. Occasionally, the cornermark was sewn to the same half of the mould as the main watermark.

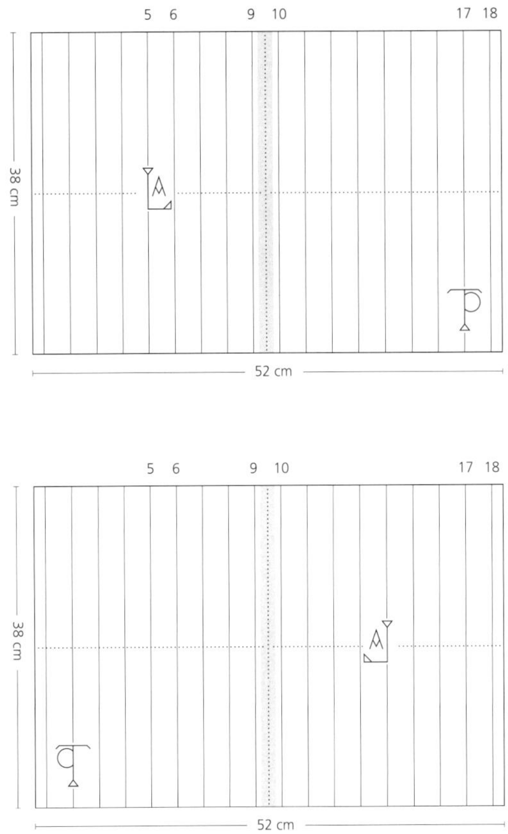

II.3: Twin Marks.

A few more points about paper need emphasis, for although they follow directly from what has just been said about the European technique of hand papermaking, they are nonetheless easily overlooked.

First, paper was made in distinct sizes, the sizes being determined by the inner dimensions of the twin moulds and their common deckle. For the fifteenth century those sizes are quite thoroughly known, and even the names used in various languages for most of the sizes are known. For the early seventeenth century, as indeed for the sixteenth, very little about paper sizes has been properly recorded. The topic has not engaged the interest of researchers, not even of those deeply involved in the study of printing at this time. An enormous body of unexploited primary evidence resides in uncut or untrimmed sheets and leaves as found in copies of many printed books (including the *Sidereus Nuncius*), manuscript books, prints and drawings, letters and documents, all across Europe. All that is needed is for someone, or a small group, to begin noticing these things: to take measurements, record their sources, and organize the results intelligently.

Second, paper is a mirror image of the wire frame of the mould on which it was made, in just the way that printing on a page is a mirror image of the flat surfaces of the inked types. Third, paper has two distinguishable sides: the mould side, which directly touched the mould and bears its impress and the slightly deeper impress of the wiremark; and the felt side, the side that first touched the felt on which the coucher laid it. It is impossible to be properly oriented when thinking about paper if the two sides are not clearly distinguished. In all that follows, the features of paper will be discussed from the viewpoint of the mould side. For instance, the watermark in a sheet of paper can be described as placed in the right (mR) or left half (mL) of the sheet, *as viewed on the mould side*. Fourth, when paper was packed into reams, it was gathered into "books" of sheets folded over once, typically twenty books of twenty-five sheets each to the ream. Thus, paper began its commercial use with a fold already in it. In a folio book, this ream fold corresponds to the fold through which sewing was done as part of the process of binding: each leaf is a half of a sheet, and its conjugate is the other half.[7] In a quarto book, this ream fold is the top edge of the folded sheet before it is cut open. Normally in bound quarto books the ream fold is removed by trimming the top edge. But in four uncut copies of *Sidereus Nuncius*, the ream folds are merely slit open, not trimmed, thus preserving the full size of the sheet except perhaps for a fraction of a millimeter or so.

The five paper stocks used in producing *Sidereus Nuncius* may be briefly described:

I. "LA" WATERMARK WITH "TP" CORNERMARK (SIZE: PRESUMABLY SUPER-MEDIAN):

II.4: SNML, A2v.

II.5: SNML, C2v.

The twin watermarks are mR and mL. This watermark type is not reproduced in the watermark albums of Briquet, Piccard, and Woodward (nor in various other albums), but many North Italian papers incorporate the letter A, and very commonly, as here, it was formed with a V-shaped crossbar. This stock was used to print all gatherings of the ML proof copy, except for one half-sheet, discussed in detail below, chapter XIV. Because of trimming and tight binding, only fractions of the main mark and cornermarks can be seen.

7 The word "folio" is inherently ambiguous, for book students have used it to refer both to a format (the fraction of a full sheet occupied by a leaf) and to a leaf itself, whatever its format. In earlier studies of the *Sidereus Nuncius*, the expression "folio leaves" sometimes appears, meaningless when applied to a quarto book, whose leaves are, self-evidently, "quarto leaves," that is, quarter-sheets. To avoid ambiguity leaves are always called leaves, never folios, although references to specific leaves are sometimes abbreviated in the traditional way as, for example, "fo. 5".

II. "LA" watermark with "Tp" cornermark (size: Super-Median):

The watermarks and cornermarks are the same types as stock I, and also with mR and mL twins, but the marks are of slightly different shapes and the spacing of the chainlines differs. That is, the stock was made on a different mould pair from stock I, but presumably at the same papermill. This stock was used to print gatherings B–G of the ordinary issue, including most copies of the half-sheet D3.4.

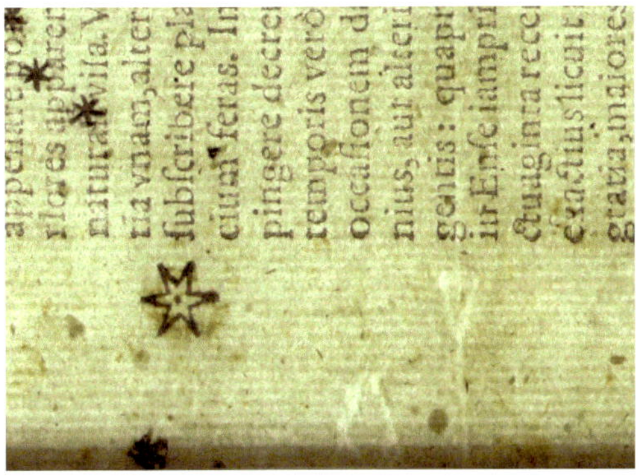

II.6: Sidereus Nuncius, D5r, Graz.

II.6: Sidereus Nuncius, B1r, Graz.

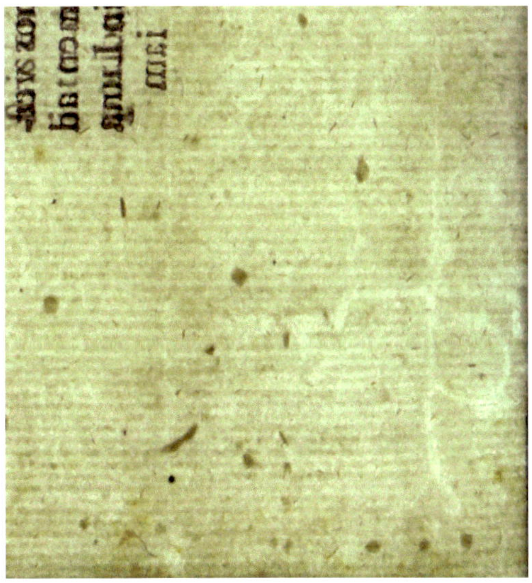

II.7: Sidereus Nuncius, D3r, Graz.

III. "Anchor in circle" watermark with a cloverleaf surmount, and no cornermark (size: Super-Median).

The twin watermarks are mR and mL. The Anchor in circle was a common watermark type. Woodward 177 almost certainly represents the same paper stock as *Sidereus Nuncius,* and see below regarding the "1609" [1610] *Fortificationi* of Buonaiuto Lorini. This stock was used to print all copies of gathering A of the ordinary issue of *Sidereus Nuncius.*

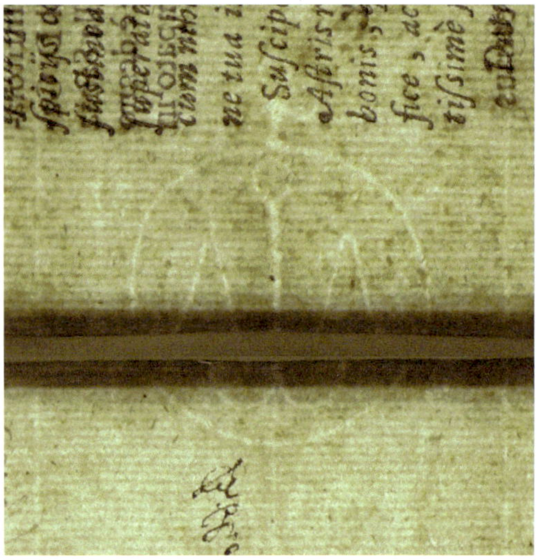

II.8: Sidereus Nuncius, A4r, Graz.

IV. Monogram mark (apparently a GZ monogram), no cornermark, twins not identified (size: Super-Median).

The mark is quite small, and being divided at the inner folds of the quarto gatherings is typically visible only fragmentarily. I have seen it really clearly only in a few copies. The twin marks seem to be mR and mL; they are not centered in their sheet halves, but are, rather, somewhat close to the ream fold. The stock was used only to print some copies of the half-sheet D3.4 – probably the last copies printed of the half-sheet. Although the size of this paper appears to be Super-Median and not Median, its moulds were slightly smaller than those of the three preceding paper stocks, as is clear from the Beinecke uncut copy.

II.9: Sidereus Nuncius, A1r, Graz.

V. "Crown with star" watermark, no cornermark (size: Median).

The twin watermarks are mR and mL. Crowned Star was a common watermark type, represented for instance in a Venetian context by Woodward 263–276, but none of these are closely similar. This stock was used to print the fine-paper issue.

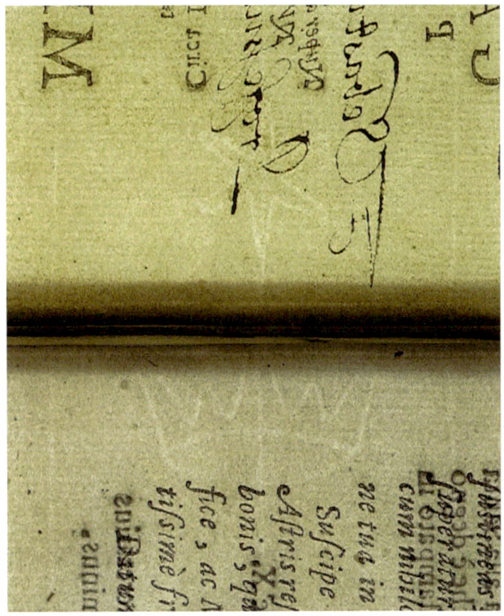

II.10: Sidereus Nuncius, A1v, Münster.

II.11: Sidereus Nuncius, A4r, Münster.

On 19 March 1610, the week after *Sidereus Nuncius* was completed, Galileo wrote to Cosimo II's secretary that 550 copies had been printed.[8] This figure probably does not include the thirty fine-paper copies that were Galileo's own property, and that he apparently refers to (although without stating that they were fine-paper) in the same letter. From these figures, we can make a reasonable estimate of how much paper was needed for the full printing job. Seven and a half sheets per copy, times 550 copies, would come to a minimum of eight and a quarter reams of 500 sheets each of the ordinary paper. Reams would often contain some unusable sheets, especially those packed at the top and the bottom; other sheets might be spoiled by printing accidents. To make sure that a full 550 clean copies could be assembled, a run of something over 550 copies per sheet would be advisable. Therefore, the amount of paper actually bought and consumed would have been something more than the minimal calculation. Most of the reams would have been stock II, but more than a single ream of stock III, Anchor, would have been required to print all 550 copies of sheet A. As for the fine-paper issue, assuming it totaled 30 copies, it would have required at a minimum 225 sheets of stock V, Crown/star: say a half-ream or so. Finally, only a small number of sheets were needed to print the proofs, whose only survivor is, so far as we know, the ML copy. A few more copies may have remained in the printing shop,

8 *Opere* X, p. 300 (Ep. 277): "… 550, che ne hanno stampati, sono già andati via tutti."

and gone to waste. It is likely that the paper of the proof or proofs was a remnant of some earlier job in the same printing shop.

It is especially noteworthy that the copies of *Sidereus Nuncius* were printed on two different sizes of paper, marked above as Median (Mezzana) and Super-Median (Mezzana grande). These two sizes and a third, which may be called Chancery (Comune), are relevant to visualizing both the *Sidereus Nuncius* as a printed edition, and Galileo's drafts for that edition. Two of the sizes, Median and Chancery, were common in the Italian paper trade by the late fourteenth century, and probably even earlier.[9] The third size, Super-Median, has a history still mostly unexplored, but it was made at least by the last two decades of the fifteenth century. Super-Median paper is mentioned explicitly in a printing contract of 13 April 1499 relating to the editio princeps of the Greek lexicon *Suidas*, printed in Milan by Johannes Bissolus and Benedictus Mangius, at the costs of Demetrius Chalcondylas, 15 November 1499. The printers undertook to produce the *Suidas* on paper "somewhat larger than Median" (*Benedictus et Johannes Magistri impressionis librorum ... obligati Sunt dictum opus imprimere in papiro in forma paulo maiori quam sit mezana*).[10] A number of Venetian editions of the later 1480s and 1490s, both in folio and quarto formats, can be identified as printed on Super-Median paper.

We know that the size continued to be used, for it is mentioned again in the 1594 contract between the Bologna naturalist Ulisse Aldrovandi and the Venice publisher Francesco de' Franceschi, which laid out plans to print various of Aldrovandi's works. Aldrovandi stipulated that these works were to be printed "on paper called mezana grande, of the quality of that used for Valgrisi's 1565 edition of the writings of Mattioli" (*di quella sorte che si chiama mezana grande, la quale habbi da essere della qualità e bontà ch'è quella dell'opera di Pietroandrea Mattioli stampato dell'anno 1565 per il Valgrisio*).[11] In fact, Super-Median seems to be a size well known in the book trade and used not infrequently in northern Italian printing, especially in folio format, in the late sixteenth and early seventeenth centuries. Unfortunately, students of printing within this period have given little or no attention to the paper of the books they study, and especially have neglected to record the leaf dimensions of the copies they examine, so that paper sizes in the book trade remain uncharted territory.

The measurements of these three paper sizes, as they came uncut into Galileo's and his printer's hands in early 1610, are known with good accuracy. The Chancery sheets – the papers on which Galileo wrote his drafts and maquettes – measured roughly 30.5 × 41 cm, so that the once-folded sheet measured roughly 30.5 × 20.5 cm. The Median sheets – the fine-paper stock (V) – measured roughly 34.5 × 48 cm (folded sheet: 34.5 × 24 cm). The Super-Median sheets, from the direct testimony of the uncut ordinary-paper copies of

9 Needham 1994a, p. 125. A Bologna statute of 1389 named these sizes as Reçute and Meçane respectively, and established minimum dimensions of roughly 32 × 45 cm and 35 × 51 cm, which are inscribed on a stone market-tablet that still survives. Later commentators have assumed too easily first, that only the four sizes on the Bologna market-tablet were ever made; and second, that the exact dimensions on that tablet defined the standard with precision, for centuries. Neither assumption is correct.

10 Motta 1893, pp. 163–165.

11 Marabini, Donati & Vai 2003, p. 118.

Sidereus Nuncius, measured roughly 38 × 52 cm (folded sheet: 38 × 26 cm). It is useful to use the dimensions of once-folded sheets, for that is the way they came out of the ream, and it is the way that their buyers looked at them.[12] In bookmaking, the dimensions of the folded sheet are those of an uncut copy of a folio printed on that paper. It is to be emphasized that Galileo, his publisher, his printer, and many of his readers knew these sizes well, from long experience. In handling the papers, they could hardly have been unaware of the different sizes and qualities. Stock III, Anchor, was used in Venice, concurrently with *Sidereus Nuncius*, to print a more substantial work undertaken by a different publisher and apparently in a different printing shop: the *Fortificationi … nuovamente ristampate* of Buonaiuto Lorini, with imprint "Venezia: Presso Francesco Rampazetto, 1609", a Super-Median folio[13]. Although the title-page date of Lorini's work is stated as 1609, it was probably actually completed in 1610. Its several dedications are dated 25 February 1609. These dates were all certainly written in the "Venetian mode", whereby the year-date changed not on New Year's day (1 January), but rather on 1 March. Galileo and many of his correspondents similarly used the "Venetian mode" in their written dates, as made clear by the Edizione Nazionale.

Special issues of printed books, it should be noted, have a history going back to the earliest printing: the Gutenberg Bible (1455) was printed both on paper and on vellum. The vellum copies – making up about a quarter of the original print run of perhaps 180 copies – would naturally have been more expensive. When the Bible was reprinted in Mainz, 1462, by Johann Fust and Peter Schöffer, more than half the copies were printed on vellum. Later in the fifteenth century, vellum issues, if they were made at all, tended with just a few exceptions (most notably, Books of Hours) to be very limited, often so limited as to suggest that they were not created for commercial sale, but rather for special presentations, and perhaps as copies for the financial investors in the edition. For example, of the *Hypnerotomachia Poliphili* printed by Aldus Manutius in Venice, December 1499, only three vellum copies are known out of more than 280 surviving. In the early sixteenth century, Aldus Manutius introduced the practice of printing some copies of his publications on large paper (and in a few cases on blue-tinted paper), either instead of or alongside, vellum copies. This practice goes back at least as far as his August 1502 Dante.[14] The most

12 In the Rawlinson manuscripts at the Bodleian Library a list is preserved of more than sixty different papers of varying sizes, qualities, and sources, that could be supplied to the Oxford press by two London paper merchants in 1673/4. The anonymous maker of the list, who endorsed it as "Severall Sortes of Paper with the number of Sheetes, Markes, Largenesse & Price of Them", recorded for each paper "How many inches Long" and "How many inches Broad." In every case, the breadth is the width of the folded, not of the open sheet. See Chapman 1927 (the merchant Chapman names "Merreatt" was the bookseller Alexander Merreall, of Bread Street, London).

13 An uncut copy of the *Fortificationi* has been recorded with leaf dimensions of 37.8 × 25.5 cm. The 1596/1597 edition was also a Super-Median folio, and an uncut copy has been measured as 37.9 × 26 cm: see Breman 2002, pp. 202, 206.

14 In the literature on Aldus, there are references both to "thick-paper" and to "large-paper" copies, but they are the same thing: most commonly, octavos printed on specially trimmed Royal paper. As "large-paper" copies, they have enormous lower margins. But many owners, including probably later collectors of Aldines, did not like the giant lower margins, and so had their binders cut them down to more conventional size. However, being on intrinsically heavier Royal paper, the copies are almost half again as thick as the copies printed on Aldus's usual "Slender Median" paper.

striking example of "issue-hierarchy" in the sixteenth century is surely that of Christopher Plantin's Polyglot Bible, Antwerp 1569–1572, which was printed in five issues: two on Royal paper of two different costs and qualities; two on Imperial paper of two different costs and qualities; and one on vellum, printed on private commission for the edition's patron, Philip II of Spain.

It is noteworthy and almost counter-intuitive that the fine-paper issue of *Sidereus Nuncius* was on the slightly smaller size, Median, and the ordinary-paper issue was on the slightly larger size, Super-Median. The uncut copies of the ordinary-paper issue have leaf dimensions of about 26 × 19 cm, height by width. None of the recorded fine-paper copies is uncut or untrimmed, but we know the dimensions of Median paper in Italy at this time from uncut copies of other books, such as the uncut copy of Galileo's *Istoria e dimostrazioni intorno alle machie solari* (Rome: Appresso Giacomo Mascardi, 1613, Median 4°) in the Beinecke Library, Yale University: its leaf dimensions are about 24 × 17 cm. The existence of a fine-paper issue is a reflection of Galileo's belief in the significance of his discoveries, and his intent to bring them, under the aegis of Cosimo II, Grand Duke of Tuscany, to the attention of all of Europe. It is to be noted that his first book, or at least the first he published under his own name, *Le operazioni del compasso geometrico et militare* (Padua: In Casa dell'Autore, 1606, Chancery F°), also had a fine-paper issue, as attested by a copy in the British Library.[15] However, Galileo's *Difesa ... Contro alle Calunnie & imposture di Baldessar Capra Milanese* (Venice: Presso Tomaso Baglioni, 1607, Median 4°) did not, at least so far as we now know, have a special-paper issue. Nor, again so far as we know, did the two very rare pamphlets relating to the Supernova of 1604 that Galileo apparently wrote under pseudonyms, the *Dialogo de Cecco di Ronchitti da Bruzene In perpuosito de la Stella Nuova* (Padua: Appresso Petro Paolo Tozzi, 1605, 4°), in Paduan dialect, and the *Considerazioni d'Alimberto Mauri ... intorno alla stella apparita 1604* (Florence: Appresso Gio. Antonio Caneo, 1606, 4°). The general inattention to paper stocks that characterizes the large majority of bibliographical studies has insured that we remain more ignorant than we need to be about which early editions had fine-paper issues. As a brief guide to the four Galilean editions just mentioned, a brief note on their paper stocks may be helpful, to encourage owners of these rare works to look at the paper of their copies through transmitted light. The ordinary-paper issue of the *Compasso* 1606 is watermarked with an open (that is, not enclosed in a circle) Crossbow, with cornermark of 3 Mounts, while the fine-paper stock is marked with a seven-spoked wheel, its inner hub made of two concentric circles. The *Difesa* 1607 has several paper stocks: one is marked with an open Crown with cornermark of Z linked to another letter, a trefoil on the link; another is

15 The *Operazioni del Compasso* is commonly (cf. Drake 1978b, p. 24) referred to as having been printed in Galileo's house. The title-page imprint reads, "In Padova, In Casa dell'Autore, Per Pietro Marinelli. M D C V I." It is doubtful if this means that Marinelli moved a printing press and other equipment to Galileo's residence. It is much more likely that, as not uncommonly in other Italian title-page imprints and colophons, both the publisher (Galileo) and the printer (Marinelli) are named. Galileo states in his preface "Ai Discreti Lettori" that he had only sixty copies printed (*Et per tal causa ne hò io fatte stampare appresso di me 60. copie sole, per presentarne insieme con lo Strumento con la somma diligenza*) – that is, to sell along with the compass itself. Earlier, he had paid a scribe, Messer Silvestro, to write out copies of his instructions on the compass (*Opere* XIX, p. 167).

marked with M M, the link between the letters with a trefoil surmount; another is marked with A M, the link with a trefoil surmount; and another marked with A P, the link with a trefoil surmount. The *Dialogo de Cecco di Ronchitti* 1605 is watermarked with an open Crossbow (different from that of the *Compasso*) with cornermark of 3 Mounts. The *Considerazioni d'Alimberto Mauri* was printed on an unwatermarked stock.

III.
Printing Materials

The *Sidereus Nuncius* is not a lengthy text: it totals something under ten thousand words. In terms of typographic composition, its size may be calculated in terms of em quads or en quads, that is, notional square types (ems) or notional types half that width (ens).[1] With allowances for the space occupied by diagrams, and omitting the blank page A1v, the typesetting can be estimated as about 61,000 en quads; or, by separate calculation, as about 76,000 individual types. Both estimates include, of course, the spaces between words, spaces at paragraph indentations, and so on, which were set with blind quadrats – pieces of type without raised characters, less tall than the types that were inked. Occasionally, these blind quadrats worked themselves up in the forme and took some ink, asserting their physical reality.

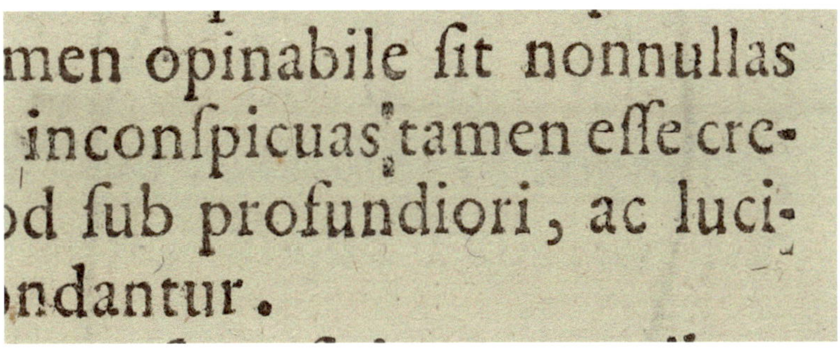
men opinabile ſit nonnullas
inconſpicuas tamen eſſe cre-
d ſub profundiori, ac luci-
ndantur.

III.1: Sidereus Nuncius, D1r, Institute for Advanced Study.

Besides the two chief fonts of a roman (text) and italic (dedication) character, additional fonts, both larger and smaller, were used in setting the title-page and the headings on fo. 2r (A2r) and 5r (B1r), which are examined more closely in the next chapter.

The late Martin Boghardt developed an elegant and fruitful concept of locating all the processes of typographic printing within the "typographic cycle," the stages by which types are drawn from their cases to set text, are printed from, and then after the printing

1 This definition of "em" and "en" as the width respectively of a square or half-square piece of type, common in the twentieth-century printing trade, is different from older English printing-house usage, as in Joseph Moxon's *Mechanick Exercises on the Whole Art of Printing* (Moxon (1683) 1962, p. 171), where an "em" is taken to be a quadrat specifically the width of the letter "m" in a particular font of type. The more recent definition has the advantage that it depends solely on the body height of a font, and so is identical for all fonts of that body height, regardless of the specific designs of their lower-case "m"s.

is complete, are broken up and returned to the cases for re-use.[2] There are four cardinal points in the cycle: composition, imposition, printing, and distribution. The imposition stage – the point at which the compositor's lines of setting are assembled into type-pages, and the type-pages are arranged for correct printing – is of critical importance in visualizing the production of a book. This is the stage at which the book is given its layout and design. Many small adjustments to the compositor's lines of setting may take place, and the decorative and illustrative materials meant to be printed directly with the types are put into their proper places. In the case of *Sidereus Nuncius*'s Moon etchings, which were printed separately, blank spaces on the relevant pages had to be calculated and laid out to accommodate their later addition.

The final stage of imposition, after each type-page had been properly laid out, was the making of formes: a tightly blocked arrangement, within a frame called a chase, of the type-pages that were to be printed on one side of a sheet. Blocks and wedges, called in the aggregate furniture, were used to apply side pressure to secure the pages firmly in place in the formes. Occasionally, the wooden blocks around the type-pages were touched by the inking ball, and left an impression on the printed page (figure III.2). If not well constructed, the formes ran the risk of collapsing, with the types falling into a heap (pie) as they were moved from the imposition table to the press bed, and hours of composition would be lost. There is one place in *Sidereus Nuncius* where the forme-locking was careless, and yet no accident resulted. In all copies examined, from the ML proof copy onward, lines 11–14 of 25v (G1v) show skewed typesetting that remained nonetheless firmly locked in place while the inner forme of sheet G was being printed (figure III.3).

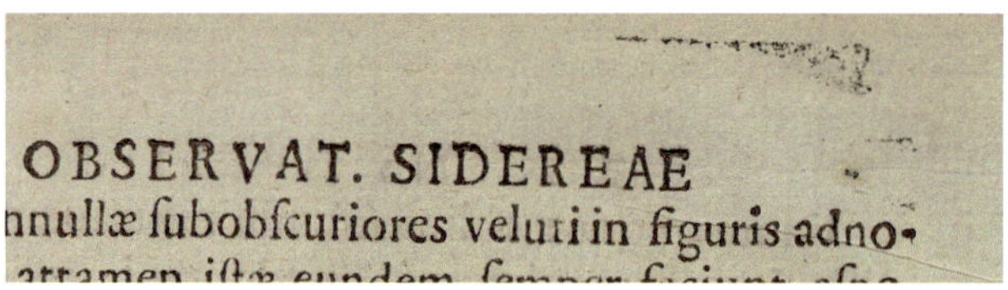
OBSERVAT. SIDEREAE
nnullæ ſubobſcuriores veluti in figuris adno-

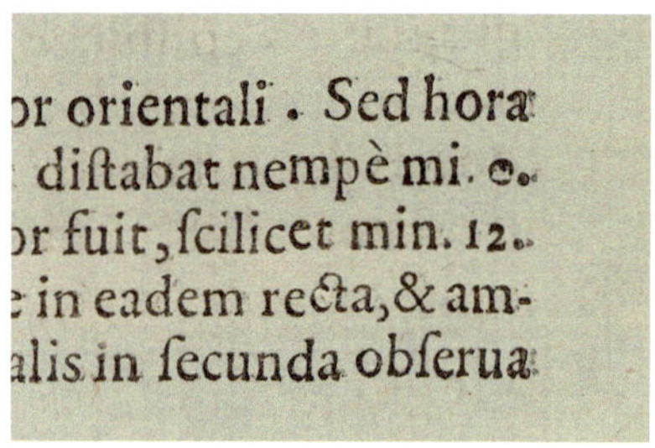
or orientali . Sed hora
diſtabat nempè mi.
or fuit, ſcilicet min. 12.
in eadem recta, & am-
alis in ſecunda obſerua

III.2: Sidereus Nuncius, C3v, Institute for Advanced Study.

III.3: Sidereus Nuncius, G1v, Institute for Advanced study.

Each sheet was printed from two formes, one for each side (figure III.4). Thus, sheet B (say) of *Sidereus Nuncius* would be printed in two successive stages. The first or outer side of the sheet would be printed over a forme containing the type-pages for 1r, 2v, 3r and 4v (outer forme); the second or inner side over a forme containing the type-pages for 1v, 2r, 3v and 4r (inner forme). If the type-pages were not put in the forme in correct order, the sheets, when folded, would not read continuously. Scattered instances of such misimposition have been noted in many printed editions, from the fifteenth century onward, but the *Sidereus Nuncius* formes were all assembled correctly.

Several points about formes, the fundamental printing units of an edition, need emphasis. It is self-evident that a particular type cannot be in two places on the same type-

2 Boghardt 1990.

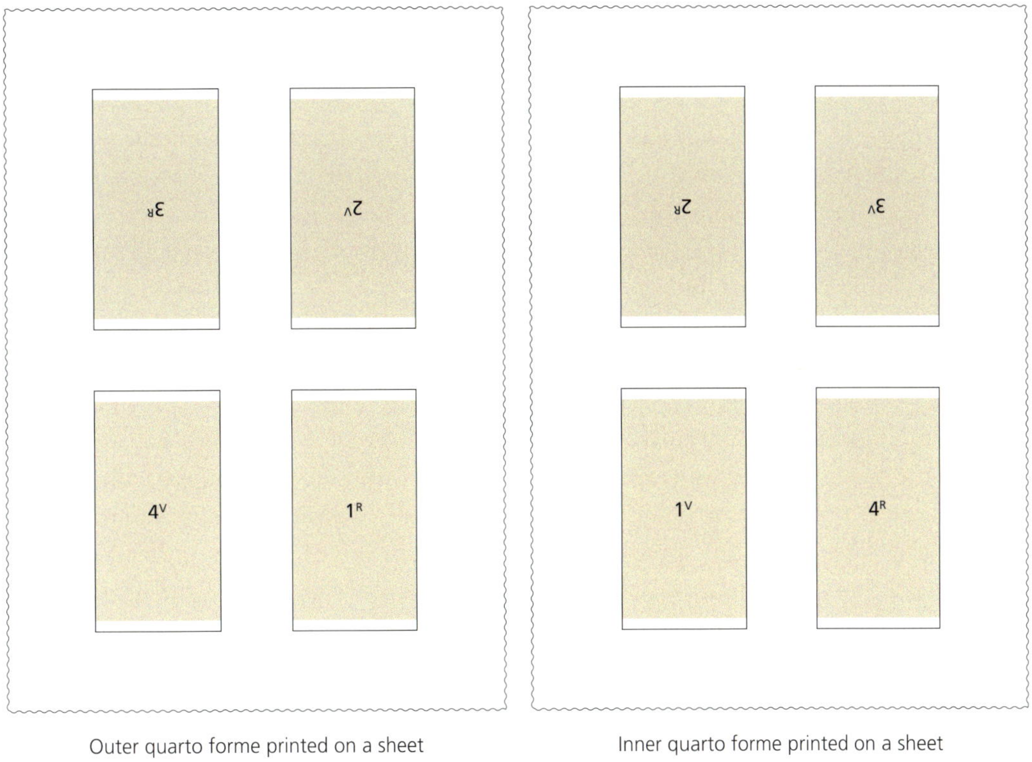

III.4

page. Less self-evident for those not accustomed to analyzing the printing process is a corollary: a type cannot be in two places in the same forme. When a printed book is studied page-by-page, as bound, one may spot a type impression with apparent distinctive damage on one page, and then again on another page. But the apparent identity must always be tested against the forme-structure of the edition. In a bound volume, the first and last pages of a quire, for example, will seem to be remote from each other, yet they will have been printed from the same forme.

However, it is in the nature of typographic printing that types will be moved from one forme to another as they proceed through the typographic cycle. Once all the required copies had been printed off, the forme would be removed from the press, scrubbed clean of caked ink, and taken apart, its types distributed back to their correct (or occasionally incorrect) boxes, to be selected again as they were needed to set the text for some page to be subsequently printed. Thus, in *Sidereus Nuncius*, we see a distinctly damaged 0 (the arabic numeral, not the letter o) in the outer forme of sheet E, and again in the inner forme of sheet G: After E's outer forme was printed off, this type returned to the 0-box, and was

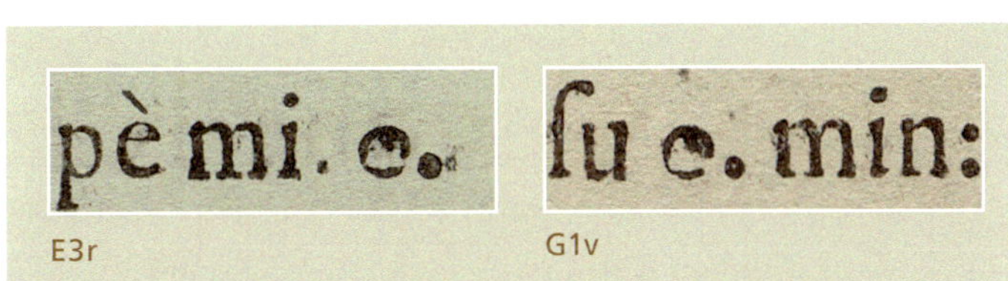

III.5: Sidereus Nuncius, Institute for Advanced Study.

drawn again when the text of quire G was being set – specifically, the second page (G1v) of the quire (figure III.5).

All the types in a forme were not always completely disassembled. In particular, when the headlines of a book did not need to be reworded, either for the entire book or for long sections of it, the headline settings were often left standing in type, so that they recur from one forme to another. In the *Sidereus Nuncius*, the common headline for each opening of quires B–G, apart from the leaf number at the right of each recto page, was a running title divided between the verso page and its facing recto: OBSERVAT. SIDEREAE | RECENS HABITAE. In the nature of things (see figure III.4), each quarto forme had the headline text in duplicate settings, for each forme would contain two recto and two verso pages. In several instances, a distinctively damaged letter shows that a headline setting was carried over intact from one forme to another, such as the "A" of OBSERVAT. that appears in the headline of, among other places, both C3v (inner forme) and F2v (outer forme; figure III.6). As discussed in more detail below, chapter XI, the headline evidence gives independent support for the argument of Gingerich and Van Helden among others that quire D, or at least part of it, was postponed at the press, and that quire E, the beginning of the Jupiter observations, was printed before quire D.

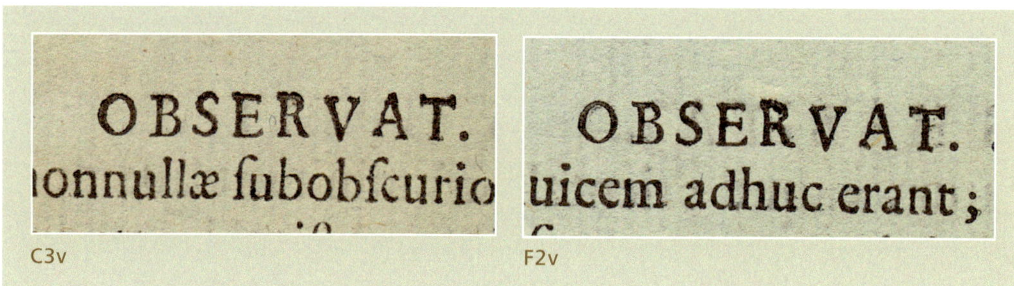

III.6: Sidereus Nuncius, Institute for Advanced Study.

The two formes for a sheet could be printed in either order: the outer forme first and the inner forme second, or vice versa. The regular pattern of *Sidereus Nuncius* was that inner formes were printed first, and outer formes second.[3] When unpressed copies of type-printed books can be examined under well-directed light, it is usually possible to determine the printing order of the formes. The close scrutiny of leaves to determine the order in which their two pages were printed is analogous in miniature to Galileo's observations of the irregular surface of the Moon: both are most conspicuous when examined under a raking light, regardless of whether that light comes from a desk lamp or flashlight at short distance, or from the Sun at a distance of tens of millions of kilometers. A printed leaf of handmade paper when closely examined has a complex surface. One side, the mould side, will have visible indentations created by the chains and the wiremark(s) of the paper mould, like swales on a broad prairie. Upon this surface, the inked types, pressed into the damp-

3 This was the order assumed as normal by Joseph Moxon: see Moxon (1683) 1962, p. 224 (plate 26), where the inner forme is the "White Paper Form", and the outer forme the "Reteration." For comments on varying practices in a sample of "forme-diagnosed" English books, see Povey 1965.

ened paper, make further indentations. The first side presses through and creates raised ridges on the second side. When the second side is printed, its types push in the opposite direction. Thus, the type indentations of the first side may, at points of overlap, be partly pushed back upward by the types of the second side: within the indentations a kind of counter-ridge can be seen.

It is to be kept in mind that the *Sidereus Nuncius*, like essentially all books of its time, was printed on a two-pull press, a form of press which was invented in the early 1470s and which a decade later had come into nearly universal use in European printing shops. The bifolium D3.4 presents a separate problem, discussed in chapter XIV. The platen of such a press, worked by lever action of a stout screw from which it was suspended, pressed a sheet of paper into the inked forme lying in the bed of the press. The platen was something under half the dimensions of the sheets it printed. Consider a book printed in folio, with two-page formes. Both pages of the forme would be inked. The platen would be brought down over one half of the forme (one type-page), then would be raised and the second half of the forme (the second type-page) would be positioned under the platen, and the platen would be brought down a second time.[4] For a quarto like the *Sidereus Nuncius*, the forme had

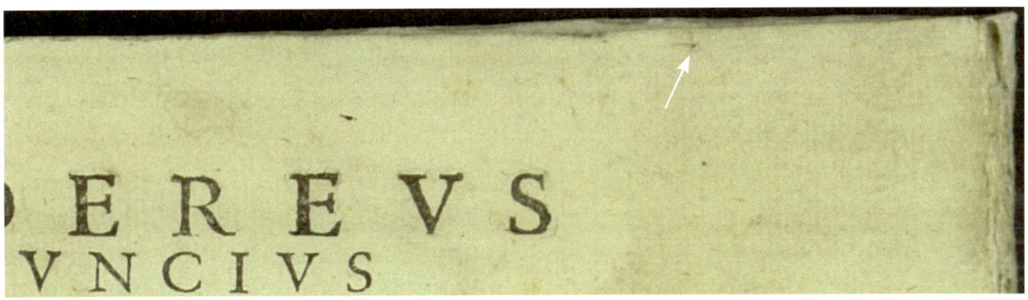

III.7: Sidereus Nuncius, A1r, Library of Congress.

four type-pages, of which two pages imposed side by side were printed at the first pull (say, 2v and 3r), and the other two pages at the second pull (say, 1v and 4r).

The sheet of paper itself was held in position by a frame called the tympan, which had two raised pins, placed in the middle of its two long sides, upon which the sheet was pressed The holes made by these pins lie on – or nearly so, varying according to the manual placement of the sheet on the tympan – the ream fold of the sheets (figure III.7). In a bound folio book, the pinholes are lost to view, for they lie in the gutter folds of the book. In a quarto book, as already noted in the preceding chapter, the ream fold, with its pinholes introduced by the printing process, goes across the top edge of the leaves. In typical bound quartos, these pinholes would be trimmed away by the binder. However, four copies of *Sidereus Nuncius* are completely uncut: the top folds of their leaves have been slit open, but not trimmed. In these copies, the remains of various of the pinholes are clearly visible: for a given sheet, one pin was placed about 4 cm in from the deckle edge at the top margin of leaves 1 and 2 (which were joined at the top before slitting), the other about 4 cm in

4 More detailed descriptions of hand typesetting and of the working of a printing press are provided in various bibliographical manuals, notably Gaskell 1972.

from the deckle edge at the top margin of leaves 3 and 4. Parenthetically, it may be noted that in these uncut copies, because of the visible pinholes, we have essentially a physical proof that the edition was printed over four-page formes. These pinholes are marks made by the printing press itself, rather than by the formes laid in the bed of the press: they tell us directly what the distance was between the pins on the tympan. The approximate dimensions of the uncut Super Median sheet are, as noted, about 38 × 52 cm. The pinholes lie about 4 cm in from the deckle edge of the ream fold, thus the distance between them, and therefore the distance between the two physical pins on the press's tympan is about 30 cm. It may be noted that this placement, equidistant from the top and bottom edges of the sheet, is different from what Moxon strongly recommended, although it corresponds to the pinhole placements found in uncut copies of Aldine octavos, such as the August 1501 Juvenal at the Morgan Library. Moxon explained at length why the pin at the lower edge of the sheet should be placed higher up – that is, at a greater distance in from the bottom edge of the chase – than the top pin. This made it more difficult, when laying on a sheet to begin printing the second forme, to position it upside-down.[5]

There is another place in various copies of *Sidereus Nuncius* where a feature of the printing press, and not of the typographic formes, has left an impression on the page. The stamping by ink balls of viscous ink onto the surface of the formes created many places where unwanted ink might end up on the sheets. To prevent this, a frisket was prepared for the formes, typically cut out of parchment, which like a picture frame would fold down over the tympan and fit closely around the type-pages, to block out any extraneous ink. When the forme itself was irregularly shaped, for example with portions to be printed that lay outside the standard rectangular page (as happens at several places in *Sidereus Nuncius*), the frisket would be cut to fit that shape. In many copies of *Sidereus Nuncius*, including all the fine-paper copies, it appears that the frisket for the inner forme of sheet D came loose and partially slipped over the type-page of D6r, preventing the full printing of two lines.

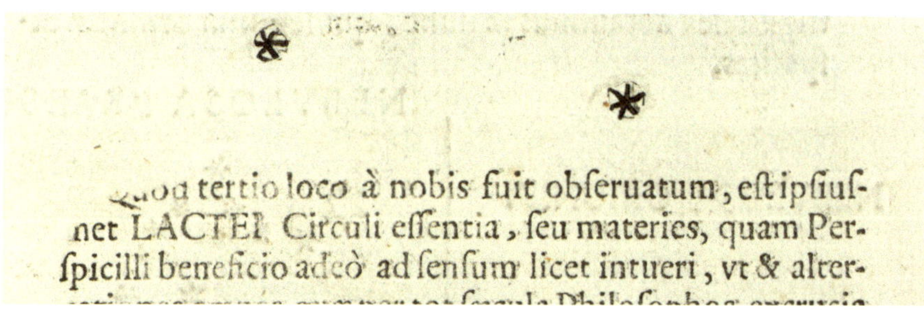
od tertio loco à nobis fuit obſeruatum, eſt ipſiuſ-
net LACTEI Circuli eſſentia, ſeu materies, quam Per-
ſpicilli beneficio adeò ad ſenſum licet intueri, vt & alter-

III.8: Sidereus Nuncius, D6r, Linda Hall Library.

5 Moxon 1683 (1962), pp. 264–266.

IV.
Design and Layout

Every printed book has a design, which is to say an organization or arrangement of the printed matter upon its pages. In principle, the design elements may be defined by a series of composition and imposition rules stating, for different textual situations, how the type-pages were to be laid out and assembled into formes, whose inked printing surfaces then create the design. It may easily happen, and often occurs, that one or more of the notional rules were not followed consistently in the printing of a given book; or were changed at a given point or points. All such changes may be looked on as potential clues to the production of the book in question; all are the result of human decisions, both conscious and unconscious. For a prose text without associated illustration, the simplest design is the equivalent of imagining the source copy to be a long string of letters and intervening spaces comprising the entire text from beginning to end, which is then cut up into individual lines. These "cuts" arrange the words into so many lines of stated length per page, perhaps with some words separated between lines. Subsidiary rules give instructions on such matters as whether the lines were to be set to uniform ("flush right") or variable ("ragged right") width – it being a given, according to the tradition of centuries of western book-writing before the invention of printing, that lines will have an even left margin unless special conventions of paragraphing or verse structure call for certain lines to be inset or, occasionally, outset from the main body of text. It is very rare to find a printed edition that contains no more arrangement than the mere breaking up of a textual string into lines, arranged into type-pages. In the early years of printing, perhaps the closest approach to this fundamental simplicity is provided by various brief, non-chaptered texts printed by, for instance, Cologne's first printer, Ulrich Zell, which use a single type font, and are almost entirely defined by just two measures, the number of lines per page, and the width of those lines. However, even these editions begin the text with a blank square of several lines' depth into which an initial letter can be supplied by a rubricator, and there is space above the first line for the rubricator to enter the title of the text. In such a book as the Gutenberg Bible, again set with a single font (and, a few pages of some copies apart, printed in a single color), the spaces provided by blank quadrats impose a complex textual hierarchy, graded by size according to whether the spaces mark the beginning of books, prefaces, chapters, or (in Psalms) verses, and according to whether spaces must also be provided for titles, chapter numbers, and a few other text items not set in type.

Most of the design elements of *Sidereus Nuncius* correspond to the widespread practices of commercial publishers in Venice and other substantial European printing towns.

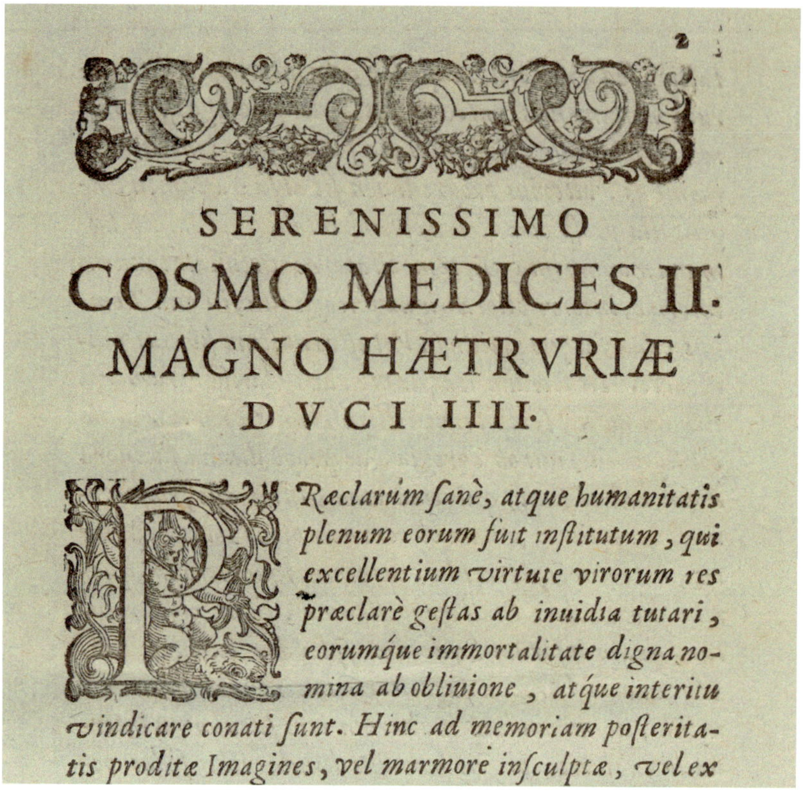

2

SERENISSIMO
COSMO MEDICES II.
MAGNO HÆTRVRIÆ
DVCI IIII.

Præclarum ſanè, atque humanitatis plenum eorum fuit inſtitutum, qui excellentium virtute virorum res præclarè geſtas ab inuidia tutari, eorumque immortalitate digna nomina ab obliuione, atque interitu vindicare conati ſunt. Hinc ad memoriam poſteritatis proditæ Imagines, vel marmore inſculptæ, vel ex

IV.1: Sidereus Nuncius, A2r, Institute for Advanced Study.

None of the elements is obviously unorthodox: all have numerous analogues in other contemporary printing in Italy and elsewhere, and all have roots in medieval scribal practices and traditions. The two chief sections of *Sidereus Nuncius*, the dedication and the text proper, both open with a woodcut decorative headpiece, a dropped heading of centered lines set in several grades of type, and a woodcut decorative initial. This same combination of elements appears on the dedication pages of *Discorso ... intorno all'opere di messer Gioseffo Zarlino da Chioggia*, by Galileo's father Vincenzo, printed in Florence, 1589; of the Paduan-dialect *Dialogo di Cecco di Ronchitti da Bruzene*, printed in Padua, 1605, which Favaro revealed to be a pseudonymous work of Galileo's; and of Baldassare Capra's *Usus et fabrica circini*, printed in Padua, 1607, the plagiarizing work which aroused a few months later Galileo's scathing *Difesa*, Venice, Baglioni, 1607, bitterly attacking Capra's "impostures and calumnies". Decorative headpieces were part of the "ornament stock" of hundreds of European printing shops of the sixteenth and seventeenth centuries. The history of their use goes back to 1495, to Aldus Manutius's first Greek printing in Venice, including the *Opera* of Aristotle and Theophrastus in Greek. The ornamental woodcut headpieces of Aldus were "translated" from a layout tradition of Byzantine manuscripts that can be traced as early as the tenth century.[1] The high reputation of Aldus must have

1 Compare Wilson 1973, nos. 21, 25, 28, 34, 40, 43, 49, 54, 55, 58, 77, 84. The last of these is a manuscript written by Zacharias Callierges of Crete, who in 1499–1500 produced in Venice four magnificent Greek-language incunables, all supplied with beautiful headpieces.

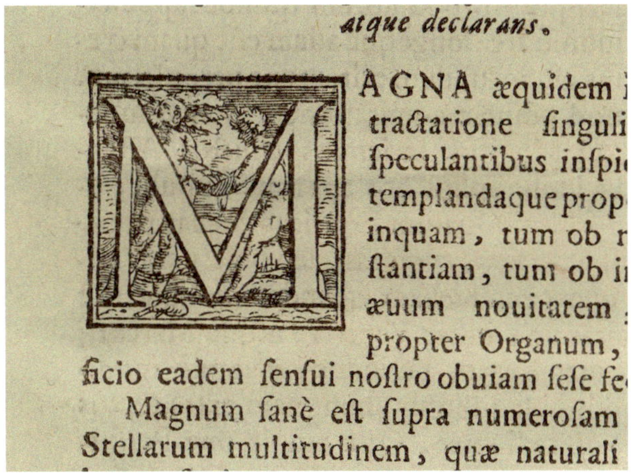
atque declarans.

MAGNA æquidem i
tractatione ſinguli
ſpeculantibus inſpi
templandaque prop
inquam, tum ob r
ſtantiam, tum ob i
æuum nouitatem
propter Organum,
ficio eadem ſenſui noſtro obuiam ſeſe fe
Magnum ſanè eſt ſupra numeroſam
Stellarum multitudinem, quæ naturali

IV.2: Sidereus Nuncius, B1r, Institute for Advanced Study.

played a role in the process by which other printing shops, in following decades, increasingly acquired woodcut headpieces – or, not infrequently, constructed headpieces out of assemblages of printer's ornaments. The pattern of headpiece + dropped title-heading + decorated initial became throughly domesticated in western European printing shops. No printer of the later sixteenth or early seventeenth century, when laying out an opening text page to this pattern, would have had the least notion that he was thereby "byzantinizing" his book.

Concerning the headpiece and decorative initials of *Sidereus Nuncius*, a further remark is necessary. The book's publisher, Tommaso Baglioni, was only a publisher, not a printer. Not until 1615 did he purchase a printing shop. In 1610, he would have commissioned some printing shop to produce *Sidereus Nuncius* and other books that he published. The shop within which the *Sidereus Nuncius* was actually composed and printed remains unidentified. The best clues to its eventual identification are the two woodcut historiated initials and woodcut ornamental headpiece used in *Sidereus Nuncius*. They do not appear in a variety of other books published by Baglioni in the years 1609–1611, but well over six hundred different editions were produced in Venice in those years. It would be surprising if the decorative materials of *Sidereus Nuncius* turned out to be unique to that book. Both initials, with putti in their backgrounds – as found in other Venetian initials – are likely to belong to alphabetical sets. Presumably, they were not, at least originally, part of a single set, as the M has a thick-line border, while the P is open.[2]

The text of *Sidereus Nuncius*, quires B–G, was set with a roman font of which twenty lines measure about 94 mm. This corresponds closely to a size that in English and French printing trade nomenclature of the time was called English or Saint-Augustin. According to the modern measure of typographic points, with 1 point = 1/72 inch (or 0.353 mm), this

2 I have left this paragraph as originally drafted, as an implicit prediction. Subsequently, Dr. Nicholas Wilding has found at least four Venice editions from 1600 to 1607 that contain one or more of the two initials and the headpiece, and shown that the printing shop of Nicolo Polo is at least a solid candidate as the place where all were printed. His detailed study is forthcoming, and I am grateful for his generosity in giving me advance notice of it.

is just over 13 points. The standard page was set to 34 lines of 95-mm width. Since twenty lines of the type measure 94 mm, this line width is very nearly twenty em quads. Above the text proper is a headline, set in the same type without added spacing below its line, which contains as noted in the preceding chapter a running title divided between the verso and recto of an opening, and the leaf (folio) number, which was printed on the rectos. A direction line at the foot of the page in the same type contains the catchword (the first word or portion of word of the following page) and the quire signature, set only on the recto. Quire A, with Galileo's dedication letter, was set in a larger italic font, measuring about 113 mm over twenty lines. This is slightly smaller than the size called in English Great Primer and in French Gros Romain, and corresponds in modern type measurement to 16 points. The dedication was set to 29 lines of 101-mm width, or about 18 em quads. Its headline has no running title, only the foliation 2, 3, 4.

The standard type-page of *Sidereus Nuncius* is best shown as a two-page opening, for this is the way readers see books. When, as here, the opening of an uncut copy is available, we see the original proportions of margin to printing (figure IV.3). As to the text proper, the only break from solid lines of setting is the paragraph indentation on the first line of the verso. A few words should be said about the history of the "paratextual" elements above and below the text: the running titles, leaf numbers (foliation), signatures, and catchwords. All these were imported into printed books from medieval manuscript traditions; each has its own history or pattern in the early printing shops.

Running titles are common in many Gothic manuscripts, according to the nature of the text or texts. A very common class of text typically given running titles is the Bible, in Latin and many other languages, where the different books of the Bible need to be identified. Running titles began to appear in printed books in the early 1470s, that is some twenty years after the invention of printing, and in the earliest examples they were stamped in by hand. In *Sidereus Nuncius* the running titles are centered above their respective pages: that is, they follow the same principle of centering – a very widespread convention that lives on today – that governs equally the layouts of the title-page, the dropped heading of the dedication, and the dropped heading of the first text page, B1r (fo. 5).

Two paratextual elements of the two-page openings of *Sidereus Nuncius* descend from medieval manuscript systems for guiding or verifying the assemblage of written leaves in correct textual order, and thus correct order for binding.

Signatures: The printed signatures of *Sidereus Nuncius* indicate the correct order of the gatherings and also of the bifolia within the gatherings. Signatures of identical pattern and purpose are common in manuscripts of the fourteenth and fifteenth centuries but, being usually written in the lower right corner of the rectos of leaves, they have often been trimmed away by binders.[3] Written signatures were added – again, subject to trimming by binders – to many copies of books from the earliest European printing shops, which did not attempt to print signatures. Several incunabulists have stated more or less dogmatically that such written signatures were entered in the printing shop, before copies were

3 Derolez 2003, p. 34.

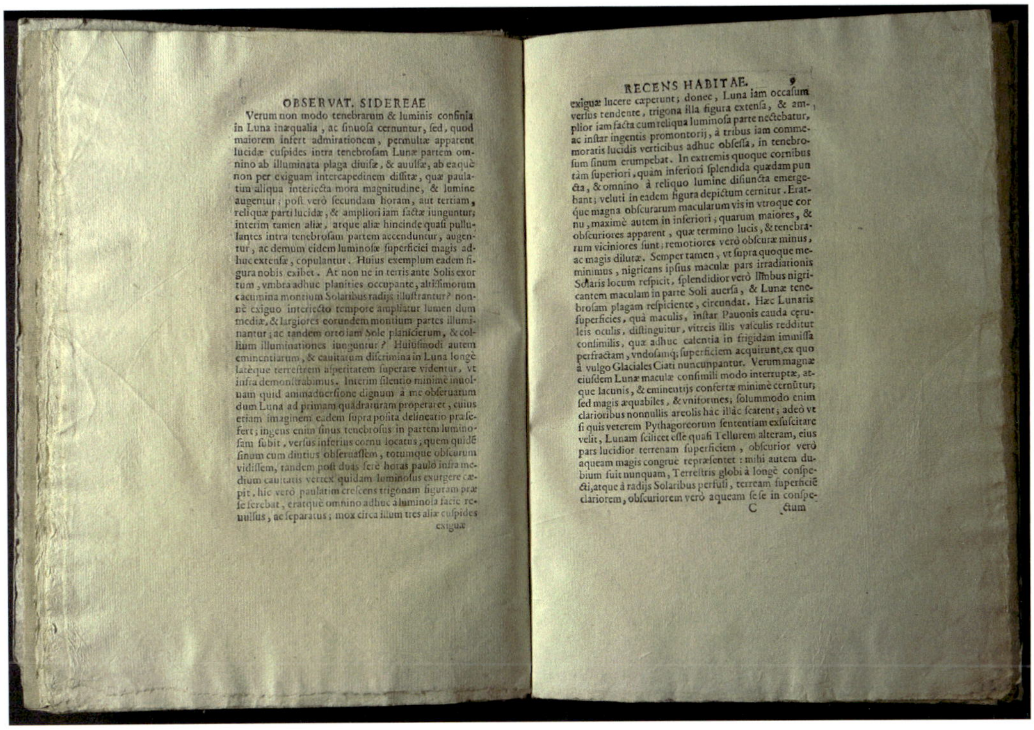

IV.3: Sidereus Nuncius, B4v–C1r, Library of Congress.

put on the market, but it is far from certain whether this was a regular practice, and there is a dearth of reliable evidence.

Printed signatures, typically placed at the right side of the direction line, began to be used in printing shops in the early 1470s. In several shops, especially in Italy, such signatures were at first not printed under the press, but stamped in, irregularly positioned, by hand. No systematic conspectus of the spread of printed signatures seems to have been carried out, but we have more or less comprehensive information on many and even most of the fifteenth-century printing shops. For example, the Venice printer Nicolaus Jenson did not set printed signatures in his editions of 1470–1475, but used them regularly in 1476 and after. They were inaugurated at essentially the same time in books printed by his Venice contemporaries Franciscus Renner, Gabriele di Pietro, Jacobus Rubeus, and others. By the mid-1480s it is unusual to find books without printed signatures.[4] Signatures in sixteenth-century books almost invariably used alphabetical (whether uppercase or lowercase) rather than numerical signatures, whereas in the fifteenth century a notable, though still small, group of editions contain arabic-numbered signatures.

Sidereus Nuncius follows a long-established practice of "signing to the fold". That is, for a four leaf gathering, the first two leaves are signed, in the form B, B2. Gathering D being of six leaves, the first three are signed: D, D2, D3. For folio books this served a nec-

4 Smith 1988 p. 58 presents a chart which seems to indicate that within a sample of almost 4,200 fifteenth-century editions, about 30% of those datable from 1475 through 1479 have printed signatures; almost 60% of those from 1480 through 1484; about 80% of those from 1485 through 1489; and about 90% of those from 1490 through 1494.

essary purpose, for each leaf of the first half of a quire belongs to a separate sheet, which must be "sandwiched" in the correct order. For quartos printed on full sheets of paper, the secondary signatures B2, C2 et cetera are more or less redundant. Given a sheet with four pages to a side, and the first page marked B (let us say) on its recto, it is virtually impossible to fold the sheet wrongly. There was no real need to refer to the second signature, but it may be that printers generally wanted a signature on both sides of the printed sheet, that is, at least one set in the outer (1r) and one in the inner (2r) forme. To a modern eye, the use of arabic numbers to mark the second leaf (or more) of a gathering looks entirely conventional, and this was the practice also of much sixteenth-century printing, but it has been noted that in Paris books of the sixteenth century and well beyond, the convention was to use roman numerals (e.g., b ii, or b ij).[5]

Catchwords: In medieval manuscript book culture, catchwords performed part of the same function as signatures: they were written on the last page of a gathering, to indicate the first word of the following gathering. As such, they defined only the outer bifolium of a gathering; if inner bifolia were out of order, the catchword would not indicate it. Catchwords are found in some manuscripts as early as the tenth century, and are common in the thirteenth century onward.[6] The first dated printed book with catchwords is an edition of Martial completed in Ferrara, 2 July 1471: catchwords appear only on the final page of quires 13–18 (out of 19 quires in all). The white space between the catchwords and the bottom line of text is greater than a single line and some catchwords are slightly skewed, yet it seems clear on close examination that they were printed as part of the type-page, not hand-stamped. In three editions printed in Bologna, 1472, the catchwords are vertical (a catchword orientation found in many Bologna manuscripts), to the right of the last lines of the text block. Here too close examination shows that they were incorporated into the formes of the pages, and not hand-stamped.[7]

The scattered printed catchwords of the 1470s are found almost entirely in northern Italian printing (Perugia the southernmost of fourteen towns) except for experiments in Cologne and Bruges. They may well reflect, in large measure, the influence of the exemplars from which the editions were set in type. In the early 1480s printed catchwords almost died out, except in a few printing shops. Then in the late 1480s and 1490s catchwords were revived again, especially in the major printing shops of Venice, and there in particular they became common although not standard. In many of these later editions, including various works printed by Aldus Manutius, the catchwords were set on a quite different basis from the manuscript tradition of placing catchwords only on the last page of quires. One common Venetian pattern was to set catchwords on the verso of every page up to the quire fold, and then one more on the last page of the quire. Used thus, they served as an

5 Sayce 1966 (1979), pp. 21–22. The Aldine catalogue Ahmanson-Murphy 2001 consistently indicates whether the leaf numbers of signatures are in roman or arabic numerals, a detail overlooked by most bibliographical catalogues.

6 Derolez 2003, p. 34. Derolez 1984, I pp. 59–63, makes a detailed analysis of variant catchword styles within a corpus of 1,200 Italian fifteenth-century humanist manuscripts on vellum.

7 Sorbelli 1925 lists a number of Bologna manuscripts with similar vertical catchwords; Derolez 1984, I p. 61 cites eighteen Bologna-localized fifteenth-century manuscripts with vertical catchwords.

independent check on the printed signatures: both systems indicated the correct order of every bifolium.

In the sixteenth century, according to a pattern of dispersed influence that has never been studied in detail, it became common to set catchwords on every page, the pattern of *Sidereus Nuncius*. One writer has noted that "the practice is all but universal" in Italy from the mid-sixteenth century, and in the seventeenth cenury, in many parts of Europe, it was standard practice.[8] Thus the English printer Joseph Moxon, in his printing manual of 1683–84, described the setting of "directions" (catchwords) as a regular part of the compositor's work: "he *Sets* on till his *Page* is *Out*, Remembring after the last *Line* of every *Page* to set a *Direction*: That is, he *Sets* a *Line* of *Quadrats* and at the end of it the first word of the next *Page* ...".[9] The roots of the page-catchword convention reach back to the last decade of the fifteenth century, where they are found in several Venice incunables: for example, a 1495 multi-volume edition of the Vulgate Bible printed by Paganinus de Paganinis, with surrounding columns of the text of the Glossa ordinaria and of the Postilla of Nicholas of Lyra. When this work was reprinted in Basel in 1498 by Johann Froben and Johann Petri (with added matter), the printers set multiple catchwords on the pages, to guide or link not just from column to column, but also to the correct commentary continuations, which were intermixed within the columns. In this work, the catchwords were obviously intended as a reading aid, rather than an aid to binders.[10]

This small hint of a transfer of Venetian catchword practices to Basel, with further development there, deserves closer study. On present and very incomplete knowledge, the first printer to set catchwords on every page with some regularity was Johann Froben. Page catchwords appear, for instance, in his March 1515 quarto edition of Erasmus's *Moriae encomium*, and in his February 1516 folio edition of the Greek New Testament with Erasmus's Latin version in parallel columns. By contrast, the full Greek Bible published by Aldus's Heirs in Venice, folio, February 1518 (its New Testament portion dedicated to Erasmus), has only quire-catchwords. At any rate, by the time *Sidereus Nuncius* was produced, page-catchwords were set as a matter of course in many printing shops all across Europe, and it is doubtful that the printers ever stopped to examine why they set them: catchwords were simply part of the compositor's work pattern.

Foliation versus pagination: A typographic element of *Sidereus Nuncius* that could conceivably be seen as slightly archaic is its foliation, with numbers printed on recto pages only, as opposed to pagination, the now universal practice.

Foliation in medieval manuscripts has been little studied, but a scholar of exceptionally wide experience has stated that it "remained rare till the end of the Middle Ages".[11] That is, foliation supplied as part of the original layout of a manuscript is rare. On the other hand, early readers might add foliation informally for reference purposes, and similarly there are numerous examples of fifteenth-century printed books whose readers added foliation. The first printed foliation appears in Cologne, 1470, in two tracts by the

8 Sayce 1966 (1979) p. 30.

9 Moxon 1683 (1962), p. 210.

10 I am grateful to Eric White for examining and sending images of the 1498 Basel Bible, from the copy at The Bridwell Library, Southern Methodist University.

11 Derolez 2003, p. 34.

Cologne Carthusian Werner Rolewinck († 1502), and the first printed pagination appears in a 1474 Cologne edition of the same author's world chronicle, *Fasciculus temporum*. In the *Fasciculus temporum*, the pagination allowed reference to a detailed printed index, which for each entry identified the page (*lamium*) on which it was discussed.[12] The Cologne experiments did not take root even in Cologne, where printed foliation remained rare in the 1480s and '90s, and beyond. Roughly speaking, printed foliation was very rare before the early- to mid-1480s, and although the practice increased thereafter, it was not the invariable practice of any major printing shop.[13]

The city producing the largest number of books with printed foliation was Venice, and here a sample based on the holdings of the British Library allows a degree of quantification: from 1481 to 1485, about 6% are foliated (of a sample of 277 editions); 1486 to 1490, about 20% (sample of 260); 1491 to 1495, about 40% (sample of 443); and 1496–1500, about 45% (sample of 444). The printers of the 1490s included, of course, Aldus Manutius. Most of his editions were not given foliation (foliation is commonest, probably for reference purposes, in Venice law books, which Aldus did not print), but he did foliate his massive editio princeps of Aristotle in Greek, in its second volume and after, and two other works; and he paginated his edition of the lexicographic commentary on Martial by the humanist Niccolò Perotti, the *Cornucopia* – the second and last fifteenth-century instance of printed pagination. In the sixteenth century, until his death in early 1515, Aldus paginated some sixteen editions, and foliated about twenty. In the early years 1501 to 1504 most of his editions, particularly the innovative octavo editions of classical texts, had no numeration. After his death, until the 1550s, Aldus's heirs and successors used foliation almost exclusively, in hundreds of editions. In the years 1551 to 1565 Aldus's son Paulus Manutius began paginating a distinct minority of editions, some ten percent or less of his output. Then in the last years of his life, 1566–1573, pagination began to dominate, and Paulus's son and successor Aldus the Younger paginated almost exclusively, 1574–1597.

It is clear that the eventual wide spread of pagination did not come from Venice. Rather, there is reason to suppose that pagination first became prominent in the editions of the Basel printer Johann Froben, and that Froben's high reputation, drawn in part from his widely known connection with Erasmus, may have influenced other printing shops to copy the practice. Of a group of sixty-one Erasmus editions that Froben printed 1513 to 1520, five have no printed numbering, four are foliated, and fifty-two are paginated. This contrasts with a group of fifty-eight Erasmus editions printed in Strasbourg or Sélestat 1509 to 1520 by Thomas Schürer, who also had a personal acquaintance with Erasmus. In the Schürer group, twenty-one have no printed numbering, nine have pagination. and the remaining twenty-eight have foliation. Among the nine with pagination, two were preceded by Froben's paginated editions of the same texts.[14] Finally, consider another printer very

12 For the same purpose, an edition of the *Mammotrectus* of Johannes Marchesinus printed in the small Swiss town of Beromünster, on the "Vigil of St Martin 1470", used a complex system of printing sheet numbers and column letters, keyed to its index. The date of this edition is not entirely certain, for both its feast-day date formula and other phrases in its colophon appear to be borrowed from Peter Schöffer's Mainz edition of the same work, with completion also on the Vigil of St Martin 1470.

13 Smith 1988 provides a useful summary.

14 Data drawn from Bezzel 1979, a carefully made catalogue of Erasmus editions in Bavarian libraries.

closely connected with Erasmus, Thierry Martens of Louvain. A comprehensive list of 169 editions he printed in Louvain 1501–1529 includes only three with pagination and ten with foliation; the remainder are all without printed numbering.[15]

Froben's practice and influence is also hinted at in a group of twenty editions of Terence, with and without commentaries, printed between 1511 and 1530 and preserved in Cambridge libraries.[16] Froben's octavo edition of March 1519 has pagination; the remaining nineteen editions, printed in nine towns including Paris, Lyons, Strassburg, Leipzig and Venice, either have no numbering (6) or have foliation (13). In 1532 Froben's heirs printed another edition with pagination, and in the same decade pagination appears for the first time outside Basel, in an octavo edition printed by Gryphius in Lyons, 1531, and in octavo and folio editions printed by Robert Estienne in Paris, 1536. A broader sample from the holdings of Cambridge libraries suggests the same overall pattern: pagination begins in Basel in the fifteen-teens, in Paris and Lyons in the fifteen-thirties, and in Venice – as also in Florence and Rome – in the fifteen-fifties and onward.

The proportion of foliated to paginated printed books from Venice shops in the first decade of the seventeenth century is not known: the census of seventeenth-century Venetian printing is rigorously brief, and provides no information on the question. However, there is some reason to suppose that printed foliation continued to be common in Venetian books in the late sixteenth century, more so than in many other of the major printing towns. Among Galileo's own publications, those that precede *Sidereus Nuncius* likewise have foliation: the *Compasso* (Padua, 1606), *Difesa* (Venice, 1607) and – if it is his composition – the *Considerazioni d'Alimberto Mauri* (Florence, 1606).[17] But his subsequent book publications, beginning with the *Cose che stanno in sù l'acqua* (Florence, 1612), are all paginated. There may be a hitherto unexplored difference in custom in this matter between, roughly, Italian and transalpine printing shops. The Frankfurt octavo reprint of *Sidereus Nuncius*, (autumn) 1610, is paginated. Conversely, the first edition of Kepler's response to the *Sidereus Nuncius*, his *Dissertatio cum nuncio sidereo*, Prague, [May] 1610, is paginated, while the Florence reprint of the same year is foliated. From a broad historical standpoint, the foliation of *Sidereus Nuncius* is a small but noteworthy feature. To contemporary readers, it was certainly a matter of indifference.

Another small detail in the two-page layout of *Sidereus Nuncius* also seems to reflect an overall difference in custom between Italian and German (and perhaps, more generally, transalpine) printing practice: the placement of the catchwords and foliation. Consistently in *Sidereus Nuncius*, and in much other Venetian printing of the time, catchwords and leaf numbers are inset from the right margin of the type-page proper, while in much northern European printing, including the Frankfurt 1610 *Sidereus Nuncius* and the Prague *Dissertatio cum nuncio sidereo* just mentioned, these elements are set "flush right" in alignment. It should be kept in mind that the placements of catchwords and leaf numbers belong to two different stages of production: the compositor set the direction line with catchwords

15 Data drawn from NK.

16 Adams T303–T323.

17 The *Dialogo de Cecco di Ronchitti da Bruzene* (Padua, 1605) of eighteen leaves has no numbering, which was often the case for brief pamphlets.

and signatures, but the leaf numbers belong to the headline. This was not part of the compositor's page-setting, but was added at the time of imposition of pages into formes.

The most singular feature of *Sidereus Nuncius*'s layout is also the least obvious, for the binding history of most copies has thoroughly submerged it: the designedly small type-page with respect to the size of the quarto leaf. This ratio, a fundamental measure in book-making, can be compared against that of various other contemporary editions, including works by Galileo. The one complication with *Sidereus Nuncius* is that copies divide into two slightly different sizes, as already noted in chapter III: Super-Median quarto for the ordinary-paper copies, and Median quarto for the fine-paper. Calculations are best made

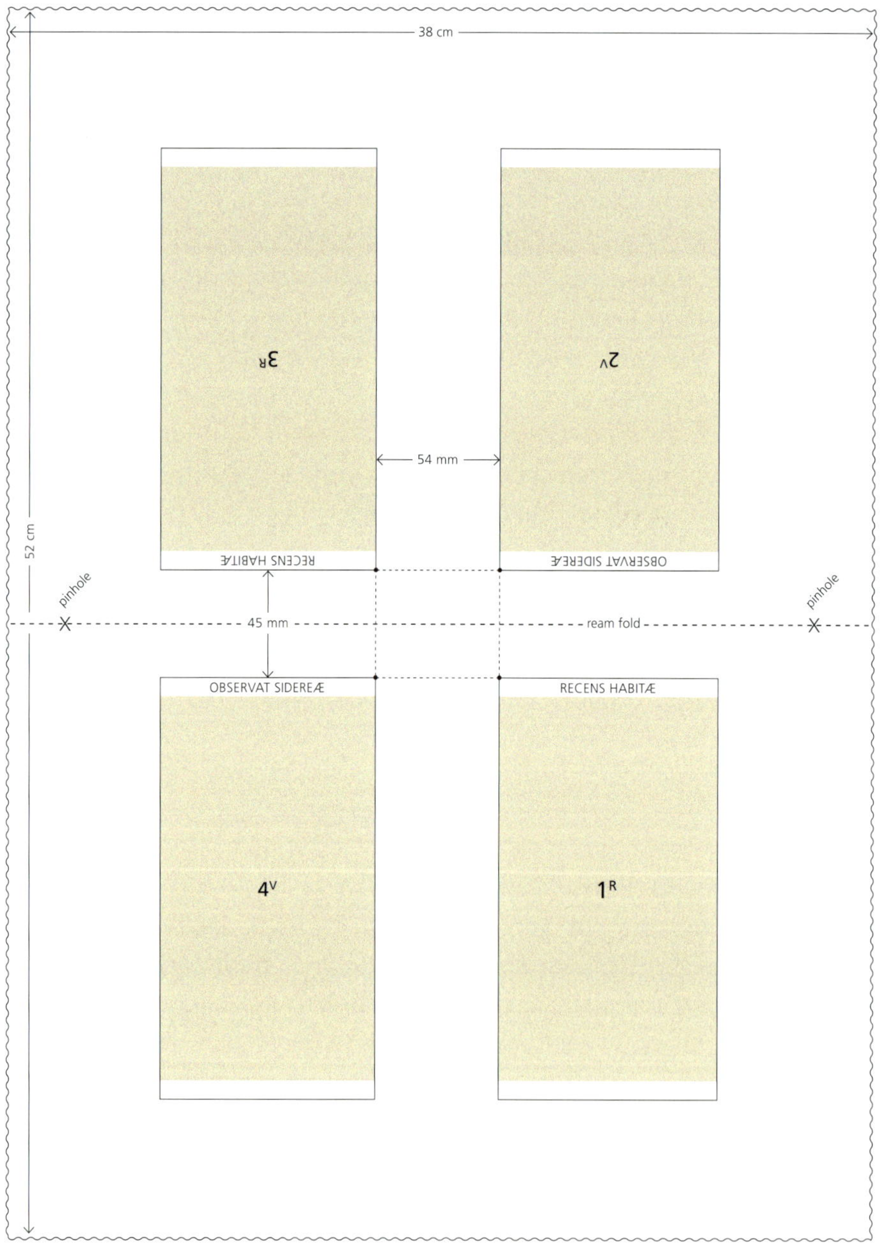

IV.4: placement of type-pages on the unfolded sheet.

on the basis of the ordinary-paper copies, for these comprise by far the greatest share of the edition. The Super-Median quarto leaf, uncut, measures about 26 × 19 cm, or about 494 cm^2. The *Sidereus Nuncius* 34-line type-page measures about 16.3 × 9.5 cm, or 155 cm^2. Thus, the "black" area of the type page takes up roughly 0.31 of the page surface, and the white margins of the page takes up about 0.69. Similar calculations based on uncut or nearly uncut copies of other Galileo editions show a considerably higher text block-to-white margin ratio: for the *Operazioni del compasso geometrico* (Padua, 1606, Chancery F°) the ratio is about 0.475 to 0.525, as also for the *Macchie solari* (Rome, 1613, Median 4°). For the *Difesa* (Venice, 1607, Super-Median 4°) it is 0.45 to 0.55, as also for the *Cose che stanno in su l'acqua* (Florence, 1612, Chancery 4°). As for the *Dialogo* (Florence, 1632, Median 4°), its type-page dimensions are essentially equal to those of *Sidereus Nuncius*, so that its text block-to-margin ratio is nearly that of the fine-paper (Median-sheet) copies of *Sidereus Nuncius*: about 0.375 to 0.625. However, its broad outer margins contain frequent printed marginal notes of 20-mm width: that is, the margins were planned for a specific purpose of textual presentation, and not simply to stand out owing to their amplitude. In short: *Sidereus Nuncius* was planned from the beginning to have conspicuously large, one could even say enormous margins. The margins were themselves a bold visual statement. A publisher, fully aware of paper costs, would never for ordinary commercial purposes have laid out a book in this "wasteful" way. We must suppose that the margins express

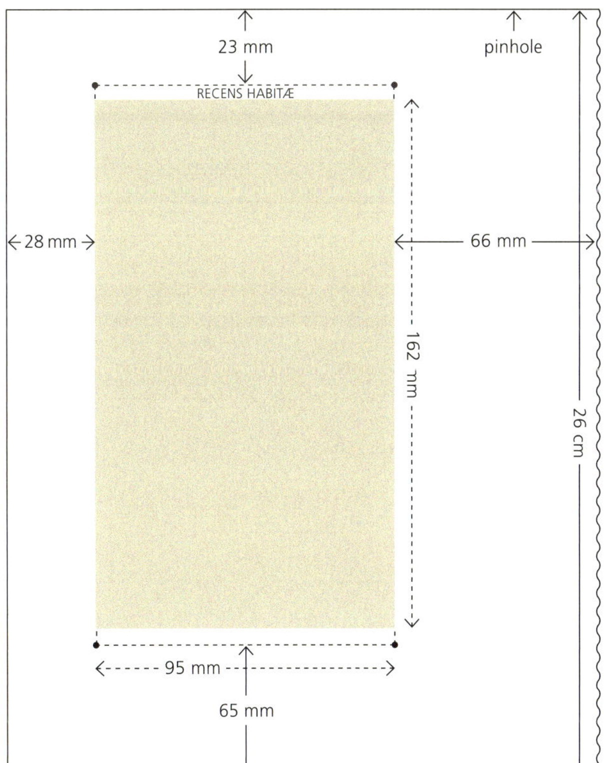

IV.5: type-page and margins of a single uncut page.

Galileo's own wish: to create, however rapidly, a work whose physical appearance would be one of imposing luxury. In his correspondence Galileo referred to the *Sidereus Nuncius* as an "avviso," that is, essentially a news report. But he did not want it to look like an avviso.

Because uncut copies of *Sidereus Nuncius* survive, we can calculate not only the text block-to-white margin ratio, but also the exact placement of the text-block on the full page, which is to say, the placement of all four text-blocks on the four pages of one side of the sheet (figure IV.4). For a quarto forme, the margin layout is defined by the interaction of two measures: the width and depth of the type-pages proper, and the dimensions of the central rectangle where the inner corners of the pages come closest. Each type-page may be thought of as suspended from one of the corners of this rectangle. Knowing that the dimensions of the rectangle are about 45 × 55 mm; that the type-pages measure 163 × 95 mm with a headline above of the same font; and that the sheets used are Super-Median in size, the dimensions of each margin are automatically set: there is a top margin of about 23 mm (to top of the headline), an inner margin of 27 or 28 mm, an outer margin of 66 mm, and a bottom margin (from the foot of the direction line) of 65 mm. Small irregularities in folding the sheet can change all these dimensions slightly, leaf by leaf and copy by copy (figure IV.5).

In bound copies all these measurements, except that of the inner margin, are typically cut into, but once we have the measures of an uncut copy, we can also closely calculate, for each bound copy, just how much was trimmed from each of the other margins. The chief reason so many copies of *Sidereus Nuncius* have been cut down below their expected norm is that, as already noted, early owners often had their copies bound with other more or less related quarto books printed on considerably smaller Chancery paper, and so the binder naturally trimmed according to the smaller works. Only a handful of the surviving copies of *Sidereus Nuncius* still preserve what may be called their natural margins: the four uncut copies, and a few other copies in seventeenth-century bindings, such as that at Wellesley College, that were not cut down to conform with other books put into the same bindings.

V.
The Sidereus Nuncius Autograph Dossier

The extensive autograph papers documenting Galileo's preparation of the text and illustrations of *Sidereus Nuncius*, mentioned briefly in the Introduction and chapter I, need closer attention as a group. Their survival is not a random accident, the chief actors in their preservation being Galileo himself, and Vincenzo Viviani (1622–1703), who served as Galileo's secretary at Arcetri in the last two years of the scientist's life, and thereafter proudly titled himself "Galileo's last disciple."[1] One scholar has made the attractive suggestion that it was the publication of the *Sidereus Nuncius* itself that awakened Galileo to the importance of preserving his own writings; we do not have similarly preserved "working archives" of the other Italian natural scientists who were his contemporaries.[2] With regard to Galileo's awareness of the epochal significance of his discoveries, and of the importance of preserving their records, we may draw a connection with a passage in the draft of his letter to Belisario Vinta, 19 March 1610, to accompany the dedication copy for Cosimo II of *Sidereus Nuncius*: "I send this to His Highness together with the very telescope with which I discovered the [Medicean] planets and made all the other observations. I send it unadorned and uncleaned, just as I used it; and seeing that it was the instrument by which so great a discovery was made, I beg that it be left in this same state. It would not be appropriate to remove any of its old elements, and to honor it with new ones which did not form part of the late nights and fatigues of the discoveries."[3]

Immediately after Galileo's death, his son and heir Vincenzo wrote to Viviani anxiously, fearing that the inquisitors might try to seize his father's papers, and authorizing Viviani to remove them to a place of safety if that seemed advisable.[4] From that time on-

1 As in his *Quinto libro degli elementi d'Euclide ... spiegata colla dottrina del Galileo ... per la prima volta pubblicata da Vincenzio Viviani ultimo suo discepolo*, Florence: Alla Condotta, 1674, 4°. See also Bonelli 1972.

2 Bucciantini 1998, p. 23.

3 Opere X, pp. 297–298 (Ep. 277): "... la quale mando a S.A.S. insieme con quello stesso occhiale col quale ho ritrovati i pianeti et fatte tutte le altre osservazioni, et lo mando così inornato et mal pulito quale me l'havevo fatto per mio uso; ma da poi che è stato strumento a sì grande scoprimento, desidero che sia lasciato nel suo primo stato, non convenendo che si rimuova cosa alcuna delle vecchie per onorarne delle nuove che non sono state a parte nelle vigilie et fatiche delle osservationi." It appears that Galileo decided not to send his working telescope, for in the letter as sent he speaks only of "a quite good telescope": *un occhiale assai buono*.

4 Favaro 1885 provides the indispensable fully documented account of the movement of the Galileo papers from their author's death to their incorporation into the Biblioteca Nazionale of Florence. Further analysis is in Camerota 1992, pp. 20–23.

ward Viviani became the active figure in safeguarding the papers. As decades passed, and as Galileo's children and grandchildren predeceased Viviani, the latter became effective owner of the papers, which through contacts with other scientists he labored assiduously at enlarging. His grand project, never fulfilled, was to publish a stately folio collection of Galileo's writings, the Italian texts set in parallel with Latin translations, to include even the banned 1632 *Dialogo* and other writings that had been condemned by the Inquisition. At Viviani's death the papers passed to his nephew, the mathematician Jacopo Panzanini, under whose care they were made available to Tommaso Buonaventuri, editor of the 1718 Florence edition of Galileo's *Opera*.[5] At Panzanini's death in 1733 the papers were inherited by his nephews, and here the chain of pride and piety was broken. Galileo's papers (and many other scientific papers assembled by Viviani, notably his own and Torricelli's) remained in a grain bin in the Casa dei Cartelloni where Viviani had lived, but became subject to commerce and contempt. The nephews quietly sold off individual items or groups to Florentine collectors including Alessandro Tosi-Galileo, Giovanni Targioni-Tozzetti, Dr. Antonio Cocchi, and Marchese Capponi.

A critical stage in the preservation of the papers came in the summer of 1750, when the Florentine senator Giovanni Battista Clemente de Nelli bought sausages from a pork butcher, and noticed that they were wrapped in a Galileo autograph. By successive inquiries he acquired the remaining "waste paper" from the butcher, then tracked their source to the Panzanini nephews. From the negligent and ignorant nephews he purchased for a very low sum all the papers that remained, of Galileo, Viviani, Torricelli and Borelli, and many of Viviani's mathematical instruments.[6] Nelli later compared the Panzanini nephews to the Goths and Vandals.[7] The massive collection of papers stayed with Nelli until his death in 1793, when they were inherited by his sons. After various vicissitudes, the papers were finally purchased by Grand Duke Ferdinand III in 1818, and joined the grand-ducal collections, the Biblioteca Palatina, housed in the Pitti Palace. The *Sidereus Nuncius* papers, perhaps already partly disunified, lay within this large and as yet unorganized collection.

Ferdinand's son and successor Leopold (1797–1870), who succeeded as Leopold II in 1824, was especially enthusiastic about the acquisition, and asked permission from his father to begin arranging Galileo's papers, a task in which his friend Vincenzo Antinori (1792–1865), whom he subsequently appointed as director of the Museo Fiorentina della Specola, was his faithful companion.[8] Antinori was chiefly responsible for organizing the

5 Favaro 1885, p. 157 (doc. LXVIII) and after.

6 Favaro 1885, p. 163 (doc. LXXV): we have a minute of the purchase written by Nelli 18 July 1750, and a letter he wrote to Giovanni Lami, librarian of the Riccardiana, 15 September 1750, reporting particularly on the Torricelli papers, and promising to write more about the Galileo and Viviani papers.

7 Nelli 1793, II p. 765: "Questo fu il funesto avvenimento, e l'infelice sorte de' MSS. … del sommo Filosofo della Toscana, che furono trattate inumanamente, e che una simile persecuzione non potevano essi aspettarla, che da' Goti, e da' Vandali." As noted by Favaro and others, Nelli's monograph, dated "Losanna 1793" on its title-pages, was not actually printed and published until 1818 or after: see Camerota 1992, p. 23 note 15.

8 Pesendorfer 1987, p. 29: "In quel tempo [c. 1819] sepi essere stati veduti in vendita scritti originali del Galileo e dei suoi scolari, e mi operai presso il padre perché di quei preziosi manoscritti facesse acquisto per la Libreria [the Biblioteca Palatina]. Chiesi volesse permettermi di ordinarli, e m'associai il giovine cavaliere Vincenzo Antinori, persona profondamente culta nelle scienze e per semplici costumi ed auree maniere apprezzabilissimo, che divenne amico a me sincero e costante."

great mass of papers, and for gathering them into distinct bindings, each volume being given a handsomely calligraphed title-page with its class number, and a table of contents. Antinori placed Galileo's astronomical papers in his third group (the first group being Galileo's correspondence, and the second, his writings on fortifications, the calculating compass, and the balance), and gathered what he recognized as *Sidereus Nuncius* materials into the third volume of this group, currently classed as Ms. Gal. 48: "Opere di Galileo Galilei } Parte 3a Tomo 3 Astronomia." Volumes four and five of this class (Mss. Gal. 49–50) are much more miscellaneous, being essentially unorganized notes, most but not all in Galileo's hand, and most but not all relating to the Medicean planets, that is, the moons of Jupiter.

It is to be kept in mind that Antinori's classification of the Galileo papers was not an attempt to reconstruct Galileo's working archive or to preserve it in its original form – the very concept of "respect du fonds" was not fully developed until the end of the nineteenth century – but rather to collect in some rational order all of Galileo's writings and correspondence regardless of source, with frequent recourse to copies of materials that were searched out in other collections, both in Florence and outside. The *Sidereus Nuncius* volume itself, Ms. Gal. 48, demonstrates well Antinori's thematic approach. Its first leaves are not Galileo autographs, but rather an introductory group of passages, written carefully in a nineteenth-century hand on ruled paper, concerning Galileo's and other telescopes in Venice, extracted from letters sent to Vinta in the autumn of 1609 by the secretary of the Florentine consul in Venice. These letters were preserved in the Tuscan state archives. On the same thematic principles, a number of the volumes in the Galileo series consist not of manuscripts but of printed books: for example, Parte 2a Tomo 10 (Ms. Gal. 39) is a copy of the printed edition of Galileo's 1606 *Compasso*, while Tomo 11 (Ms. Gal. 40) is Galileo's annotated copy of Baldassare Capra's plagiarized 1607 *Usus et fabrica circini*.

Leopold II continued to be active in adding to the Palatina collection of Galileana. Galileo materials in the already mentioned collections of Cocchi and Targioni-Tozzetti, which after their owners' deaths (1758, 1783) had gone to the Magliabechiana, were removed from there and brought to the Pitti Palace. Leopold, like his father and grandfather, was a learned book collector, but his very interest in literary matters caused him to intervene, sometimes capriciously, in the workings of Florence's great libraries and the state archives. Galileo letters preserved in the archives of the Tuscan secretariat of state were removed and brought to the Palatina shelves. These letters typically bear file and item class-numbers from their original context. Antinori included all these external acquisitions, some part of Galileo's original archive and some not, in his classification and binding scheme.

In studying the *Sidereus Nuncius* dossier, it is fundamental to remove them – notionally – from the bindings in which they were placed in the first half of the nineteenth century, and to return them – notionally – to their original state of unbound but organized loose papers, just as they were under Galileo's own hands when he created them. In the case of their main locus, Ms. Gal. 48, we must keep in mind that Antinori put them there, and that it is possible (and in fact certain, as will be argued subsequently) that he made errors both in omitting dossier items from this volume, and in placing non-dossier items within the volume.

It is useful to picture Galileo as producing two kinds of archive, the first his own retained papers plus the papers he received and kept from others, and the second a kind of "reverse" or "outward" archive: the papers he sent out.[9] Both reveal and reflect his several capacities as a professor, as a scientist, and as the head of a family in which he took pride and toward which he had responsibilities. As to pride in family, it may be noted that Galileo preserved many of the music and music theory manuscripts of his father, which Antinori arranged and bound under the rubric "Anteriori" (Mss. Gal. 1–8).

The difference between the retained and the "outward" archives is most easily seen in Galileo's correspondence. He kept the letters he received from his many correspondents, and so they were in his possession at his death. The letters he wrote and sent entered many other archives, personal (mostly broken up over time) and official. His letters to Belisario Vinta were naturally treated by the latter as "official" and not personal, and so eventually joined the other papers of the secretarial office in the Tuscan archives until, as just mentioned, Leopold II had them removed to his personal collection, the Palatina library. Thus, Galileo's letter to Vinta of 20 November 1609, now Ms. Gal. 14, fos. 41–42, bears an early archival annotation, "Filza 301, 239." However, Galileo made copies – drafts and minutes – of a number of his sent letters, and these copies are thus part of his retained archive. The distinction between a retained and an outward archive is relevant even when studying not Galileo's autographs, but later copies. In chapter I, Galileo's crucial letter of 7 January 1610 was mentioned, which survives in two later copies, with significant textual variations. Of the dated copy found in the Vatican and then lost again, nothing positive can be said, although it may be conjectured that it derived from the archive of the unknown recipient. But the undated copy still in Florence is almost certainly based on a now-lost original draft or minute by Galileo; it was bound by Antinori in the miscellaneous collection classed as Parte 3a Tomo 7(1), now Ms. Gal. 53, fos. 58–61.

Galileo's retained writings, including the *Sidereus Nuncius* dossier as his chief written work in the months January-March 1610, and his outgoing letters of the same general period, come together again in the paper on which they were written. We may even say that they come together naturally, for all were his written products over this interval of time. The letters, being fully dated, provide a chronological framework or matrix that can be used to gauge the probable dates of Galileo's undated writings. At the least, if such physical evidence is not brought into play, it is premature to say that an undated leaf of Galileo's, for instance some notes on astronomy or mechanics, is undatable. The paper evidence also serves as a salutary test of suggested dates derived from more theoretical or abstract reasonings made without consideration of the physical materials being dated. The brilliant pioneer of this approach to Galileo's chronology was Stillman Drake, who investigated briefly, but with his characteristic sharp intelligence, the watermarks of a substantial group of Galileo's autograph letters.[10] His interest was in enlisting their aid to infer the probable order of the large group of Galileo's notes on motion, written over many years, and brought together by Antinori within the volume "Parte 5 Tomo 2, Meccanica" (Ms. Gal. 72).

9 Further to the general concept of dual or mirror archives see Needham 2000.
10 Drake 1972, pp. 62–65; Drake 1979, pp. XXIV sqq.

As to the *Sidereus Nuncius* dossier it may, at first consideration, appear to contain little chronological mystery, for we have, in particular, precise knowledge of the dates on which he made his Jupiter observations. And yet, the paper stocks of the dossier do yield further clues and verifications that could arise from no other channel. The paper stocks (all Chancery paper) of nine autograph letters bracketing the period of writing and production of *Sidereus Nuncius* may be defined, with a tenth for comparison, as follows:

1. 30 October 1609, Padua, Galileo to Belisario Vinta (Opere X, p. 262, Ep. 247): BNCF Ms. Gal. 14, fos. 39–40
2. 20 November 1609, Padua, Galileo to Belisario Vinta (Opere X, p. 268, Ep. 253): BNCF Ms. Gal. 14, fos. 41–42
3. 4 December 1609, Padua, Galileo to Michelangelo Buonarroti (Opere X, p. 271, Ep. 257): British Library Ms. Add. 223139, fos. 39–40
4. 30 January 1610, Venice, Galileo to Belisario Vinta (Opere X, p. 280, Ep. 262): BNCF Ms. Gal. 86, fos. 22–23
5. 13 February 1610, Padua, Galileo to Belisario Vinta (Opere X, p. 282, Ep. 265): BNCF Ms. Gal. 86, fos. 24, 26
6. 13 March 1610, Venice, Galileo to Belisario Vinta (Opere X, p. 288, Ep. 271): BNCF Ms. Gal. 86, fos. 27–28
7. 19 March 1610, Venice, Galileo to Cosimo II (Opere X, p. 297, Ep. 276]: BNCF Ms. Gal. 14, fo. 45
8. 19 March 1610, Venice, Galileo to Belisario Vinta (Opere X, p. 297, Ep. 277(A): draft]: BNCF Ms. Gal. 86, fos. 29–30
9. 19 March 1610, Venice, Galileo to Belisario Vinta (Opere X, p. 297, Ep. 277(B): sent]: BNCF Ms. Gal. 86, fos. 31–32
10. 7 May 1610, Padua, Galileo to Belisario Vinta (Opere X, p. 348, Ep. 307): BNCF Ms. Gal. 86, fos. 34–37

Four paper stocks appear in letters 1–9, while letter 10 was written on a fifth:

A. Unicorn, cornermark of 3 Mounts, circle surmount: Letters 1–3, Padua, 30 October–4 December 1609.

B. Rampant Lion, cornermark monogram ZFM (the F as a surmount of the line linking Z and M): Letter 4, Venice, 30 January 1610.

C. Armorial of Crossed Keys, cornermark of 3 Mounts, circle surmount with a St Andrew cross: Letter 5, Padua, 13 February 1610.

D. Crossbow in circle, cornermark monogram RA with trefoil on the link: Letters 6–9, Venice, 13–19 March 1610.

E. Crossbow (open), cornermark 3 Mounts: Letter 10, Padua, 7 May 1610

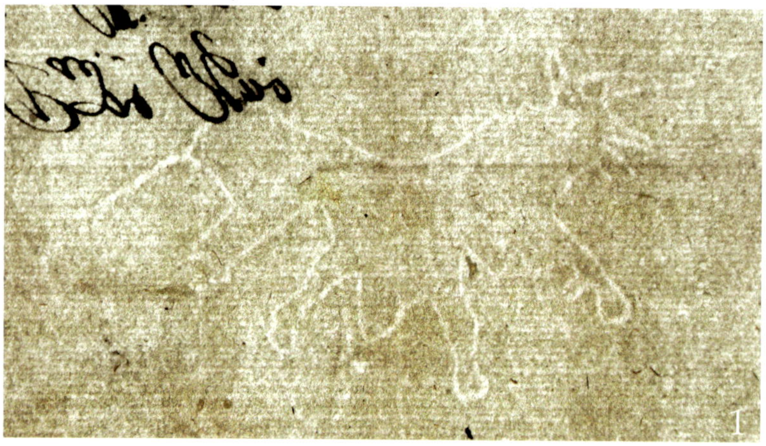

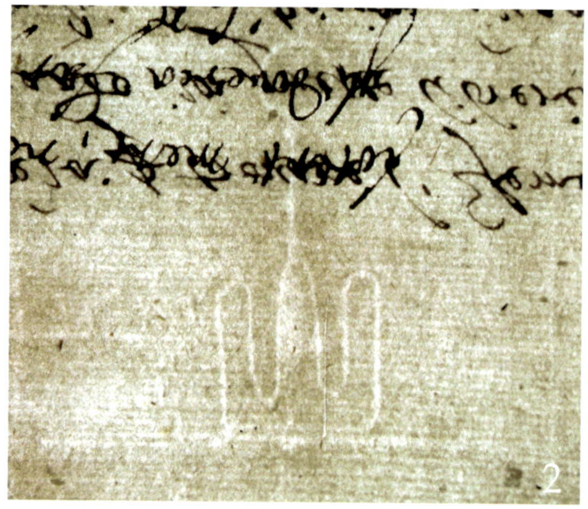

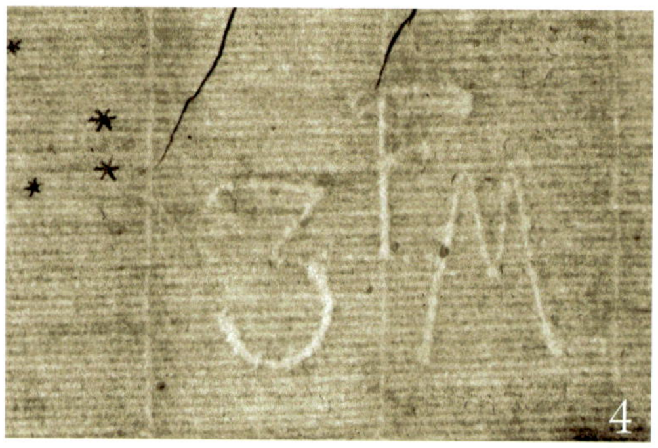

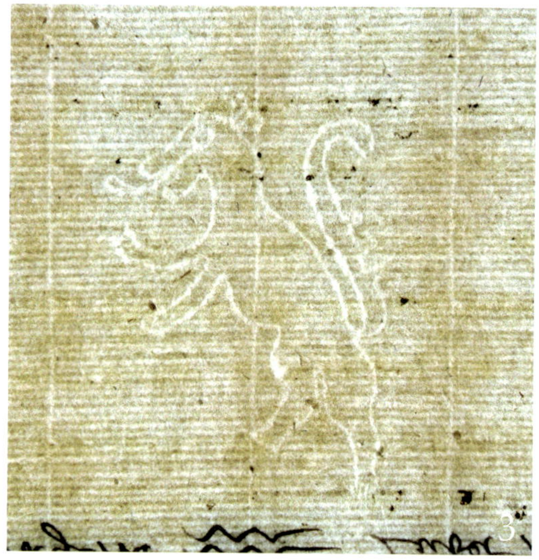

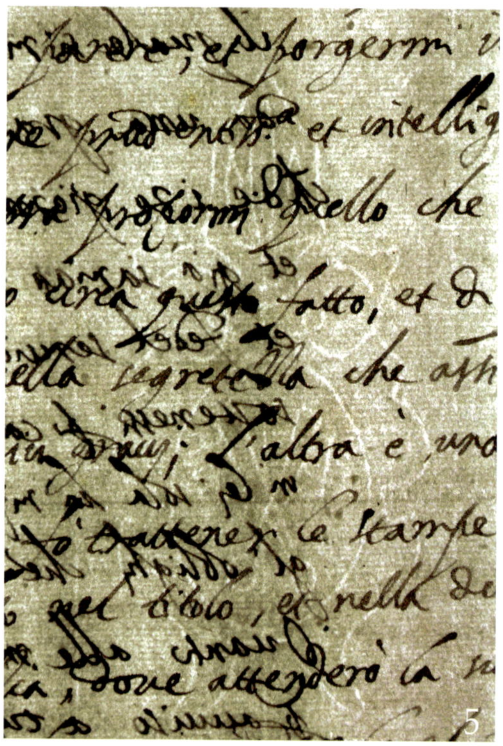

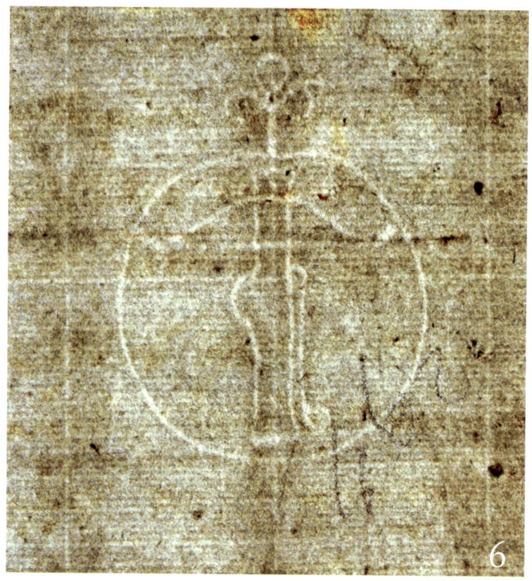

V.1: Stock A.

V.2: Stock A, cornermark

V.3: Stock B.

V.4: Stock B, cornermark

V.5: Stock C

V.6: Stock D

The four watermarks A–D, and no other, define all the paper stocks used in the *Sidereus Nuncius* dossier. Put another way: over the months October 1609 through March 1610, all surviving Galileo autographs, both retained and sent, were apparently written only on these four stocks.[11] The possible exceptions are a handful of smaller leaves that happen to bear no traces of either main mark or cornermark. By May 1610, Galileo was using a new paper stock, Open Crossbow, unrelated to the *Sidereus Nuncius* dossier, and autograph letters he wrote in Padua in June and July 1610 are on yet other papers, again different from those of the *Sidereus Nuncius* dossier.

To reconstruct the original state of the *Sidereus Nuncius* dossier we need to know not only what papers Galileo used, but how he used them. For instance, he might have written in folio notebooks, that is fascicles of two or more single-folded sheets sandwiched together. He might have written on loose individual folio leaves, that is half-sheets. He might have written on individual quarto leaves, that is, leaves that represent a quarter of a sheet. A quarto leaf might be used in upright orientation, with horizontal chainlines, or oblong orientation, with vertical chainlines. In fact, all of these variations, and others, are represented in the *Sidereus Nuncius* dossier, but none of the necessary information can be elicited from the facsimile reproductions of much but not all of the dossier in the Edizione Nazionale, nor even from the online digital facsimile of Ms. Gal. 48 maintained by the Istituto e Museo di Storia della Scienza in Florence, recently renamed the Museo Galileo.[12] Only from the originals can the chainlines and watermarks of the paper be identified, the mould sides and felt sides of the leaves be determined, and the sewing structure of Ms. Gal. 48 itself be examined – the necessary steps to reconstructing the original structure of the various constituents of the dossier.

The dossier as preserved at the Biblioteca Nazionale Centrale of Florence can be divided into four components: the full Draft, the incomplete Fair Copy, the original Jupiter Observations, and three Maquettes, that is precise models for two of the edition's asterism woodcuts and for the typographic layout of the Jupiter-moon diagrams. The full Draft is the core component. It divides into two parts, corresponding to the two divisions of the printed edition: (I) Moon and stars, (II) Jupiter and its moons. The structure of the Draft and of the Jupiter Observations is shown in figure V.7.

Part I was written on eleven folio leaves of Chancery paper. These comprise five distinct writing units, each independent of the others, each containing its own topic or pair of topics. Placed in final textual order (as Antinori arranged them), the units are: a single sheet (fos. 1–2), a fascicle of two sheets (fos. 3–6), another single sheet (fos. 7–8), a single leaf (fos. 9), and a final single sheet (fos. 10–11). Three of the units were written on the Unicorn paper stock, and two on the Rampant Lion stock. As will be argued more fully in the next two chapters, there is a good presumption that the three units on the Unicorn paper were written before the two units on Rampant Lion paper, for Galileo had the Unicorn paper in his possession in Padua by October 1609, whereas there is no positive evidence that he still had the supply in hand by late January, when he had moved to Venice to

11 This statement can eventually be tested by a Galileo autograph letter, Padua 4 November 1609, in the Archivio di Stato in Venice, that I have not had the opportunity to examine: *Opere* X, p. 264 (Ep. 249*). It may be predicted that this letter was written on the Unicorn paper stock.

12 http://fermi.imss.fi.it/rd/bd?collezioni=galileiana.

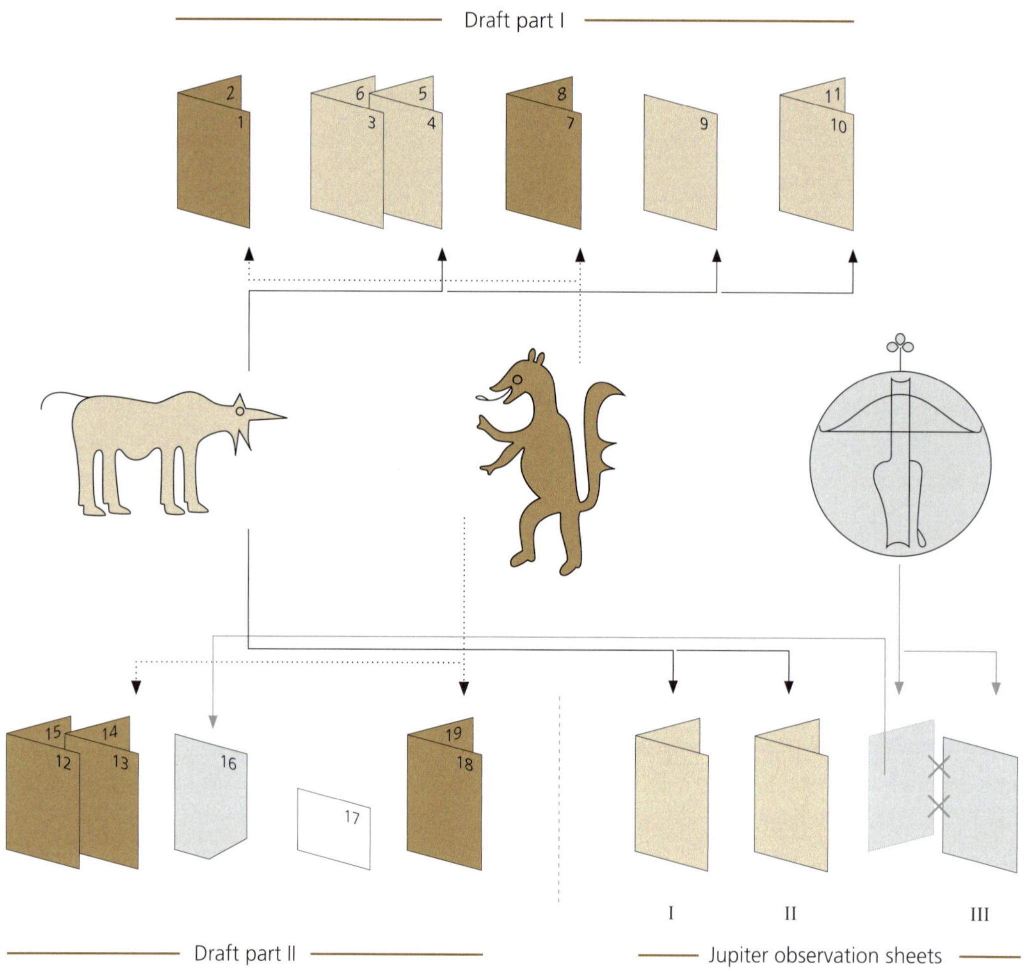

V.7: Structure of Draft.

oversee production of *Sidereus Nuncius*. The Rampant Lion paper was in his possession in Venice on 30 January, when he used it to write a letter to Vinta.

Part II of the Draft is generally simpler in structure than part I, all except two single leaves being on the Rampant Lion paper. It is noteworthy that the main section of part II, the Jupiter observations of 9 January through 16 February, was written in a single fascicle of four leaves (fos. 12–15), whereas the conclusion is on a separate sheet (fos. 18–19) Therefore, the conclusion could have been written before the Jupiter observations were finished. Textual revisions on this sheet, examined in chapter X, suggest that this was the case.

The original Jupiter Observations, recopied and revised in the Draft, were written not in a fascicle, but on three successive folded sheets. These sheets reinforce the suggestion that the Unicorn paper was distinctive of the Padua phase of Galileo's work on *Sidereus Nuncius*: the first two sheets are that Unicorn paper, the second of them being taken in hand when Galileo began to record his observation of 24 January, a time when he apparently had not yet left Padua for Venice. We may suppose that he took the sheets with him, and continued adding entries to the second sheet while in Venice. Then the third sheet, beginning with his observation of 17 February 1610, needed new paper: the Crossbow paper that he still had with him in Venice when he wrote letters on 13 and 19 March 1610.

The Maquettes, and a small group of more miscellaneous leaves of asterism sketches associated with them, are divided between Unicorn (Padua) and Rampant Lion (Venice) paper. It must be emphasized that naming the Unicorn paper as Paduan and the Rampant Lion paper as Venetian implies only that these were the places where Galileo acquired his supplies, not that the papers were made in those places. There were a great many paper mills in the Veneto, but little specific is known about them. Paduan and Venetian stationers would have had essentially the same sources for the papers they sold; it is very unlikely that some mills sold their papers only to Padua, others only to Venice.

Two rough sketches of the Pleiades asterism are drawn, the smaller above the larger, on an oblong folio leaf of the Unicorn paper (Ms. Gal. 70, fo. 3). This leaf has a vertical fold across its center, which probably goes back to Galileo's handling of it: that is, it appears that he made his sketches, then folded the leaf in half to make a quarto bifolium with the sketches preserved on the inside. The leaf shows a variant of the Unicorn stock in which the 3 Mounts cornermark was sewn to the same side of the sheet as the main mark, rather than to the opposite side. The finished Pleiades Maquette which served as model for the blockcutter (Ms. Gal. 70, fo. 12b) is an oblong octavo leaf, with no watermark visible.

There are two other quarto bifolia on Unicorn paper, one bearing an asterism sketch focused on Sirius, and titled by Galileo "Canicula" (Ms. Gal. 70, fos. 28–29), and the other bearing a preliminary asterism sketch of Orion, titled by Galileo "Circa cingulum orionis" (Ms. Gal. 70, fos. 12–13).[13] The finished Orion Maquette (Ms. Gal. 70, fo. 18) is a cut-down folio leaf of (Venice) Rampant Lion paper.

The Maquettes for the Jupiter-moons diagrams are three folio leaves of Rampant Lion paper (Ms. Gal. 50, fos. 70–72). When these three leaves are carefully noted as to whether their rectos are mould side or felt side, and as to their possession (and placement) of main mark and cornermark, it can be determined that fos. 70 and 72 were probably originally a single sheet, but that Galileo tore the sheet in half before he began to write on it. On the verso of the third leaf, Ms. Gal. 50 fo. 72, Galileo scribbled brief notes on the maximum elongation of the outermost moon from Jupiter, to east and west alternately, at eight-day intervals, ending with his observation of 27 February. A somewhat later hand, possibly Viviani's, added a caption: "Fragmenti di Osseruazioni intorno alle Medicee." The nondescript identification suggests that even in the seventeenth century the *Sidereus Nuncius* dossier components had gotten separated from each other, and that Viviani did not recognize that these three leaves belonged specifically to *Sidereus Nuncius*. The leaf following the three Jupiter-moon Maquettes, Ms. Gal. 50, fo. 73, includes Galileo's sketch of a series of horizontal lines used as a scale on which he indicated, roughly, the maximum elongations of the moons from Jupiter. It might be surmised that these notes belong with the Maquettes, but the paper is watermarked with a Bird in a circle which appears in several other leaves of calculations on the Medicean planets, some of which are dated January through March 1612. Therefore, this leaf is later, and is not part of the *Sidereus Nuncius* dossier. The Maquettes had been variously recognized as relating to *Sidereus Nuncius* by

13 The sketches of the bifolium now bound as fos. 28–29 are reproduced in E.N. III pt. 2, pp. 963–964, but without any indication of leaf edges or of the orientation of the sketches with respect to the tops of their leaves. On what is now fo. 29v, Galileo wrote (expanded) "Illustrissimo Signore Osservatissimo", as if beginning a letter.

Favaro (Orion, and its preliminary sketch), Drake (Jupiter-moon diagrams) and Pantin (Pleiades). These identifications are strengthened, and in fact made certain, by the evidence of their paper stocks.

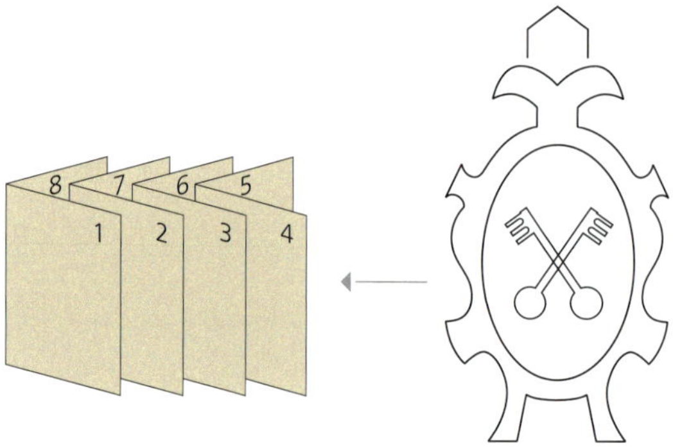

V.8: Fair Copy fascicle.

The partial Fair Copy is physically the least complex component of the *Sidereus Nuncius* dossier. It consists of a single folio fascicle of 4 sheets or 8 leaves, written on a single paper stock, the Crossed Keys coat of arms that Galileo used also for a letter dated from Padua 13 February 1610. In that letter, addressed to Belisario Vinta, Galileo wrote that he had returned the day before from Venice. It is just possible, therefore, that Galileo did not already have this paper at his lodgings in Padua, but had brought it with him from Venice. This has some bearing on the question of when Galileo began to write out the Fair Copy. The Crossed Keys paper appears to be both slightly larger, and of higher quality, than the other stocks in the dossier.

It is worthwhile to consider whether there are other materials in Galileo's autograph that belong to the *Sidereus Nuncius* dossier? There are three strong candidates. The first is the "Michigan Leaf" illustrated in chapter I. These notes on the first days of Galileo's Jupiter observations, 7 to 15 January 1610, are written on an untrimmed Chancery folio leaf, immediately below a draft or outline of the letter that accompanied Galileo's official presentation of an improved telescope to the Doge of Venice, an event that took place on 24 August 1609.[14] Thus, Galileo made his Jupiter notes on what was essentially scrap paper, a common practice of Galileo's well attested by various other of his autographs. This leaf was almost certainly among Galileo's own papers at the time of his death, but we have no knowledge of when it strayed from the authorial archive. It first came to public notice in May 1934 when it was sold from the library of the late Roderick Terry, a wealthy clergyman of New York City and Newport, Rhode Island, and collector of high spots.[15] The purchaser was Tracy McGregor, a Detroit businessman who retired to Washington, D.C., and collected heavily in the last years of his life. After his death in 1936 his executors donated most of McGregor's collection to the University of Virginia, but the beautiful Galileo leaf to the University of Michigan in honor of the longtime professor of astronomy there, Heber D. Curtis. It had come out of Italy relatively recently, for it was examined and au-

14 The finished letter, preserved in the Archivio di Stato of Venice, is printed in *Opere* X, pp. 250–251 (Ep. 228).

15 American Art Anderson Galleries, The Library of the Late Rev. Dr. Roderick Terry ..., Part I, 2 May 1934, lot 92, illustrated.

thenticated, probably sometime in the nineteen-twenties, by Cardinal Pietro Maffi, Archbishop of Pisa, whose admiration of Galileo was boundless, and who compared the leaf with a Galileo autograph letter in his own possession.[16] The Michigan leaf is watermarked with the initials A S in a circle, the initials B M O below. The stock is different from any he is known to have used in Padua in 1609. It may be part of a sheet that Galileo picked up in Venice in late August 1609, and should eventually be compared with the original of Galileo's presentation letter to the Doge in the Archivio di Stato, Venice.[17]

The second candidate has been identified recently by Nicholas Wilding, who will discuss it in a forthcoming publication. This is BNCF Ms. 70, fo. 4, a Galileo autograph including Greek and Latin variants for the word *nuncius*, and transcriptions from Pliny's Natural History, and Suetonius's Life of the Caesars. As Dr. Wilding points out, these notes must be preliminary to Galileo's drafting of the dedication to Cosimo II, a text for which we possess otherwise no dossier materials. This is a quarto leaf in oblong format, and bears no watermark.[18]

The third candidate is the "ML" copy of *Sidereus Nuncius* itself, briefly mentioned in chapter II: a copy, containing in the spaces provided for the Moon etchings, careful pen and wash drawings of the relevant Moons. By Bredekamp's argument these drawings are from Galileo's hand, and served as Maquettes for the etchings; and by my argument the copy itself is gathered from proofs, that is, bifolia that were printed before the main print runs for each forme of the book began. Our respective arguments, and the arguments of another scholar that the ML drawings are modern forgeries, are discussed in detail in chapters VII, VIII, and XIII.

This survey of the *Sidereus Nuncius* dossier has been focused almost exclusively on its material aspect. In essence, it is an account of the papers with their writing stripped off. Nonetheless, this approach reveals characteristics of and differences among the several constituents of the dossier that are in themselves telling. One can say that each part of the dossier has something like its own personality, each reflecting different aspects of Galileo as a writer. For example, consider the eleven leaves of part I of the Draft (figure 7 above). Even without knowing specifically which leaves contain what text, we know that this was a complex composition, apparently begun in one place (Padua) and finished in another (Venice), and drafted in an order quite different from that of the printed edition. For the eight leaves of part II of the Draft (Jupiter and its moons), we see a simpler fundamental structure: a fascicle of four leaves for the main body, and a sheet in two leaves for the conclusion, all using the same paper stock. Yet here too we see that the conclusion formed a writing unit distinct from the main portion of the text. The additional complexities of part II of the Draft – the interpolation of one leaf of the Observation sheets, followed by a second re-used "scrap" leaf – are analyzed in chapter X. The Observation sheets are just that: successive sheets, thus neither a single fascicle, nor a collection of single leaves. The

16 Maffi's authentication is mentioned in the Terry sale catalogue; it has gone missing since the sale. On Maffi see DBI 67, 271 and Vergara Caffarelli 2004, pp. 97–120.

17 I am grateful to Peggy Daub of the Special Collections Department, University of Michigan, for photographs of the watermark, and much other information on the Michigan Leaf.

18 The average distance between chainlines of this leaf, 26–27 mm, suggests that it is most likely to be the (Venice) Rampant Lion stock, but this is well short of certainty.

Maquette for the Jupiter-moons diagrams consists of three single leaves, each disconnected from the others, while a fourth leaf, bound immediately after them in Antinori's time, and reproduced sequentially after them in the Supplement (Volume III, Parte Seconda) of the Edizione Nazionale, postdates *Sidereus Nuncius* by about two years. As noted above the most regular structure in the dossier is that of the uncompleted Fair Copy: an eight-leaf folio quire of uniform paper. This is the dossier component that corresponds most closely to the structure of a conventional book, whether handwritten or printed, and it is significant that it is the one where Galileo's own writing most closely approaches a formal bookhand.

VI.
Quire B

In his Saturday, 30 January 1610 letter to Belisario Vinta, Galileo wrote, "I find myself currently in Venice, bringing into print certain observations which I have made among the celestial bodies by means of my telescope" (*Io mi trovo al presente in Venezia per fare stampare alcune osservazioni le quali col mezo di uno mio occhiale ho fatte ne i corpi celesti*).[1] For reasons reviewed in detail in chapter XI, it is likely that on this date Galileo had just arrived in Venice, and that the actual printing of *Sidereus Nuncius* did not begin for another week or more. On the next evening, Sunday 31 January, Galileo sketched into his second Jupiter Observation sheet, which he had begun a week earlier in Padua, not just Jupiter and its moons, but also the stars of the Pleiades, within Taurus – the many "new", smaller stars, that is, that the telescope brought out among its six conspicuous stars. He drew a box around the sketch, and labelled it the "requisita descriptio": the "required image." Higher up on the page, he also made a sketch of the Hyades star cluster in Taurus, which he labelled as the "succulae", or suckling pigs (figure VI.1).

The woodcut-printed diagram representing the Pleiades appears in *Sidereus Nuncius* near the end of quire D, suggesting that by January 31st, Galileo's general plan for the text and illustrations of the first part of *Sidereus Nuncius* had been settled: it would discuss both the Moon and the fixed stars.

As we try to visualize the processes that created *Sidereus Nuncius*, we must steadily keep in mind the two composition stages that precede the text we find in the printed edition, and that are preserved in Galileo's handwriting: that is the full Draft and the partial Fair Copy. Between these two texts and the printed edition came an autograph that does not survive: the setting copy that Galileo brought to the printing shop. The Draft and the Fair Copy stayed with Galileo and were preserved in his working archive. The setting copy presumably remained in the printing shop, and went to oblivion.

As noted in the preceding chapter, the first portion of the Draft, dealing with the Moon and the stars, was written on eleven folio leaves of Chancery paper, which divide into five writing units, each independent of the others: a single sheet (fos. 1–2), a fascicle of two sheets (fos. 3–6), another single sheet (fos. 7–8), a single leaf (fo. 9), and a final single sheet (fos. 10–11). The second, fourth and fifth of the units (as put into their final sequence

1 I translate Galileo's Italian *occhiale* and Latin *perspicillum* simply as "telescope". Van Helden translates as "spyglass," on the grounds that "the word telescope was unveiled only in 1611"; yet by that line of reasoning spyglass is even more anachronistic: the *Oxford English Dictionary*'s first citation for spyglass dates to 1706. Both Galileo's Italian and Latin words, in plural, were used long before him to mean glasses, i.e. spectacles.

of order) were written on the (Padua) Unicorn paper stock, and the first and third units on the (Venice) Rampant Lion stock (see figure V.7)

The broken-up and disparate physical structure of part I of the Draft mirrors the complexity of Galileo's composition of the text. As will be seen, further complexities reside within the text itself: that is, to speak in physical terms, in the ink and pencil marks on the paper, not in the paper itself. The order in which Galileo wrote and assembled his words for the first part of *Sidereus Nuncius* is quite different from the order in which they were printed as gatherings B–D of the completed work. *Sidereus Nuncius* begins with two introductory sections, a general introduction (5r: *Magna aequidem in hac exigua tractatione* …), and a brief discussion of the history of the telescope and of his improvements to it (6r line 3: *Mensibus abhinc decem ferè* …). Galileo wrote these in two stages on the first sheet of the Draft as bound (fos. 1–2), but this sheet was written later than the second unit, the four-leaf fascicle (fos. 3–6), which contains the central discussion of the lunar observations.

Thus, we may suppose that Galileo began drafting the lunar observations of *Sidereus Nuncius* in roughly mid-January 1610, while he was still in Padua. As already noted in chapter I, by midnight of Friday, 15 January, Galileo had decided that his observations were to be printed. Assuming that after the observations he went to sleep, he could have begun writing his Draft, starting with fo. 3, as early as Saturday the 16th. As for the two introductory lead-ins to the lunar observations, it is likely that Galileo drafted them in Venice at the end of January or early in February, using the "new" Rampant Lion paper that he acquired there, and that he used also for his 30 January letter to Vinta. But the difference in time between these two portions of the Draft would have been relatively slight. In the general introduction (Draft, fo. 1r), Galileo wrote that his discovery of "four erratic stars" revolving around an as yet unnamed "certain star among those that are known" (*circa stellam quandam e numero cognitarum*) had been made by divine grace "just a few days ago" (*Diuina prius illuminante gratia, paucis abhinc diebus reperta* …). By the time *Sidereus Nuncius* was published, in mid-March 1610, those words *paucis diebus* had, unnoticed, become rather anachronistic: the few days had lengthened to a month and a half.

The Draft is the original document of composition, written in a cursive hand to the very edge of the outer margins of the recto leaves and to the inner margins of the verso leaves, with numerous deletions, changes, and expansions of text variously incorporated within the lines, between the lines, and in the margins, sometimes displaying a thinner ink and a pen with a narrower nib. That is, the revisions themselves embody more than a single editing stage: structurally speaking, we may say that they occupy more than one stratum. The Draft is a precious survival, for it brings vividly before our eyes a picture of Galileo as a "writer at work". Consider just one small variation and how Galileo got to it. At one place (Draft, fo. 4r), Galileo discussed the interpretation of certain small points of light that were visible within the darkness beyond the Moon's horns at a certain stage of the lunar cycle. These points of light, he wrote, were seen to be entirely separated from the broader lighted surface of the Moon: *quędam puncta, … omnino à reliquo lumine disiuncta conspiciebantur.* On rereading the passage, he must have been dissatisfied with the word *conspiciebantur.* He underlined it, and added in a smaller hand, in the narrow inner margin, six alternatives: *cernebantur – perspiciebantur – se in conspectum dabant – obijciebant – apparebant – emergebant* (see figure VII.7). For the Fair Copy (fo. 5v), he settled

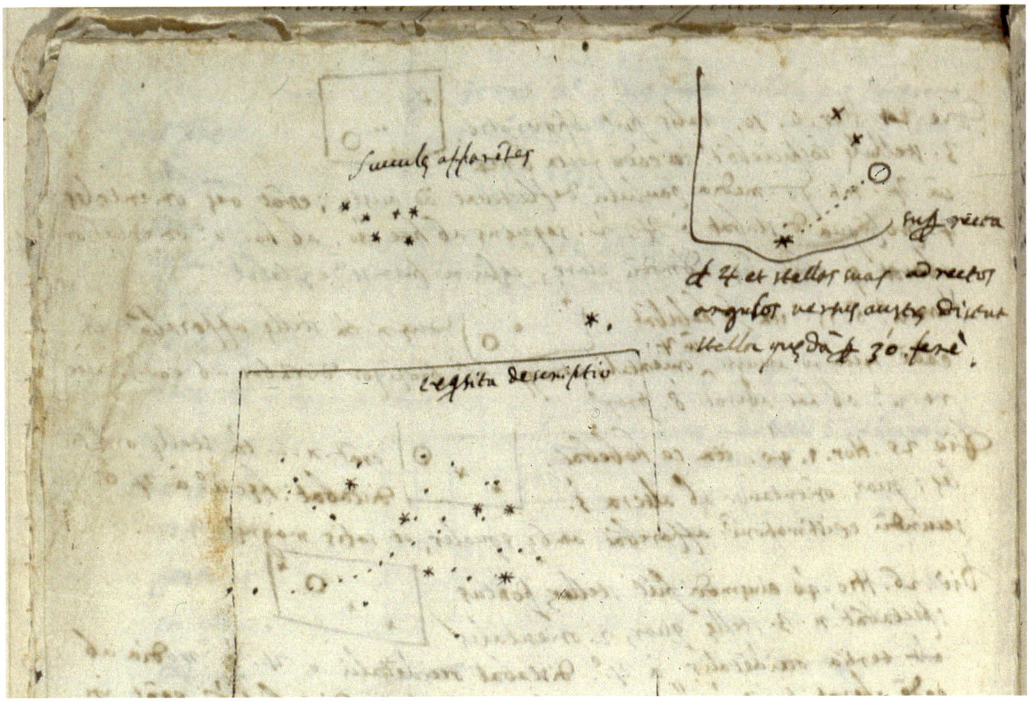

VI.1: BNCF, Ms. Gal. 48, 32v.

on the last of these, and that is the reading that went into type in *Sidereus Nuncius* (9r line 7): *quaedam puncta, … omnino à reliquo lumine disiuncta emergebant.*

The Fair Copy is, as noted in the preceding chapter, a folio fascicle of eight leaves or four sheets, carefully written in a larger and more formal hand, with narrower and fewer lines than the Draft pages, leaving ample margins. The leaves are scored with two vertical blind lines to mark the right and left boundaries of the written space. These boundary lines are another signal that the Fair Copy was written more or less as if it were a book. The last ten pages of the Fair Copy were written regularly with 25 lines per page (in the preceding pages there is slight variation), whereas the Draft fluctuates from 30 to 33 lines per page. The Fair Copy breaks off in mid-sentence at the foot of 8v, at a place corresponding to *Sidereus Nuncius* 12r (C4r) line 28. Galileo added catchwords at the foot of 8v of the Fair Copy, suggesting that he continued – or intended to continue – it in a second fascicle that, if it was written, does not survive (figure VI.2).

There is a tenuous hint in its paper stock that the Fair Copy was written, or at least begun, while Galileo was still in Padua, that is, in the latter half of January 1610. As noted in chapter V, it uses a paper stock different from the Draft, watermarked with an armorial of Crossed Keys. During these months that paper appears in only one other surviving autograph: the letter Galileo wrote to Vinta from Padua on 13 February, having returned to there from Venice the day before. Perhaps, if only for reasons of economy of hypothesis, we may suppose that this is a paper he had already in his Padua lodgings, rather than one that he carried with him from Venice.

A multitude of readings, several of which are discussed in the following chapters, verify that the Fair Copy stands directly between the Draft and the printed edition. By comparison with the Draft, the Fair Copy is indeed a "fair copy," and yet it is not in the stricter sense a clean copy, such as could have been intended, for instance, for a presentation, for at several places it contains words struck through and interlinear changes. The Fair Copy is presumably close to the now-lost setting copy, and it may be seen, probably, as Galileo's retained copy of the text at nearly settled state, just before he had his setting copy ready to take to the printing shop. However, there are still some dozens of variations in text between the Fair Copy and the printed *Sidereus Nuncius*. In most of these cases we cannot be certain whether the variants were already entered in the setting copy, or were introduced by him at a later proofing stage, after typographic composition had begin. A striking example comes on the second page of the text of *Sidereus Nuncius* (5v), within the introductory paragraphs where Galileo laid out his topics. He first wrote, in Draft and Fair Copy:

> *Altercationes insuper de Galaxya, seu de Lacteo circulo substulisse, eiusque essentiam sensui, nedum intellectui manifestasse, parui momenti existimandum minimè videtur.*
> (To have laid to rest the controversies about the Galaxy or Milky Way, revealing its true essence to the senses and the intellect, can hardly be seen as a trivial matter.)

But after writing the Fair Copy, Galileo must have recalled that he had silently passed over the last of his topics, the nature of the nebulous stars: a topic explicitly if briefly mentioned in the drop-title printed on the preceding page, 5r. He therefore added a clause present in neither the Draft nor the Fair Copy:

> *insuperquè substantiam Stellarum, quas Nebulosas huc vsquè Astronomorum quilibet appellauit digito demonstrare longèque aliam esse quam creditum hactenus est, iocundum erit, atque perpulcrum.*
> (and moreover to demonstrate that the substance of the stars that astronomers have called nebulous is wholly different from what has hitherto been believed will be pleasant and indeed most elegant.)

It must be emphasized that this discussion of the nebulous stars was not itself an afterthought in Galileo's plans for the text; he forgot only to refer to it in his introduction. In the Draft, on fo. 11r – in one of the units, it will be recalled, written on the (Paduan) Unicorn paper – the discussions of the Milky Way and of the nebulous stars were written continuously in a single stint, without pause or hiatus.

A small sample of variants from Draft to Fair Copy to *Sidereus Nuncius* provides a practical sense of how the texts of the first two led to the printed text. A number of small changes were aimed at creating a higher style. Thus, at one place (*Sidereus Nuncius* 7r 13), Galileo wrote *interstitia* in the Draft, referring to the intervals between the stars on the celestial sphere. In the Fair Copy, he changed this to *intercapedines*, and so it appeared in the printed edition; and at other places, too, the poetic word *intercapedines* served as a useful alternative to the more prosaic *interstitia*. Similarly, the poetic word *Tellus* was used a number of times as an alternative to *Terra*. At one of these places (*Sidereus Nuncius* 9r

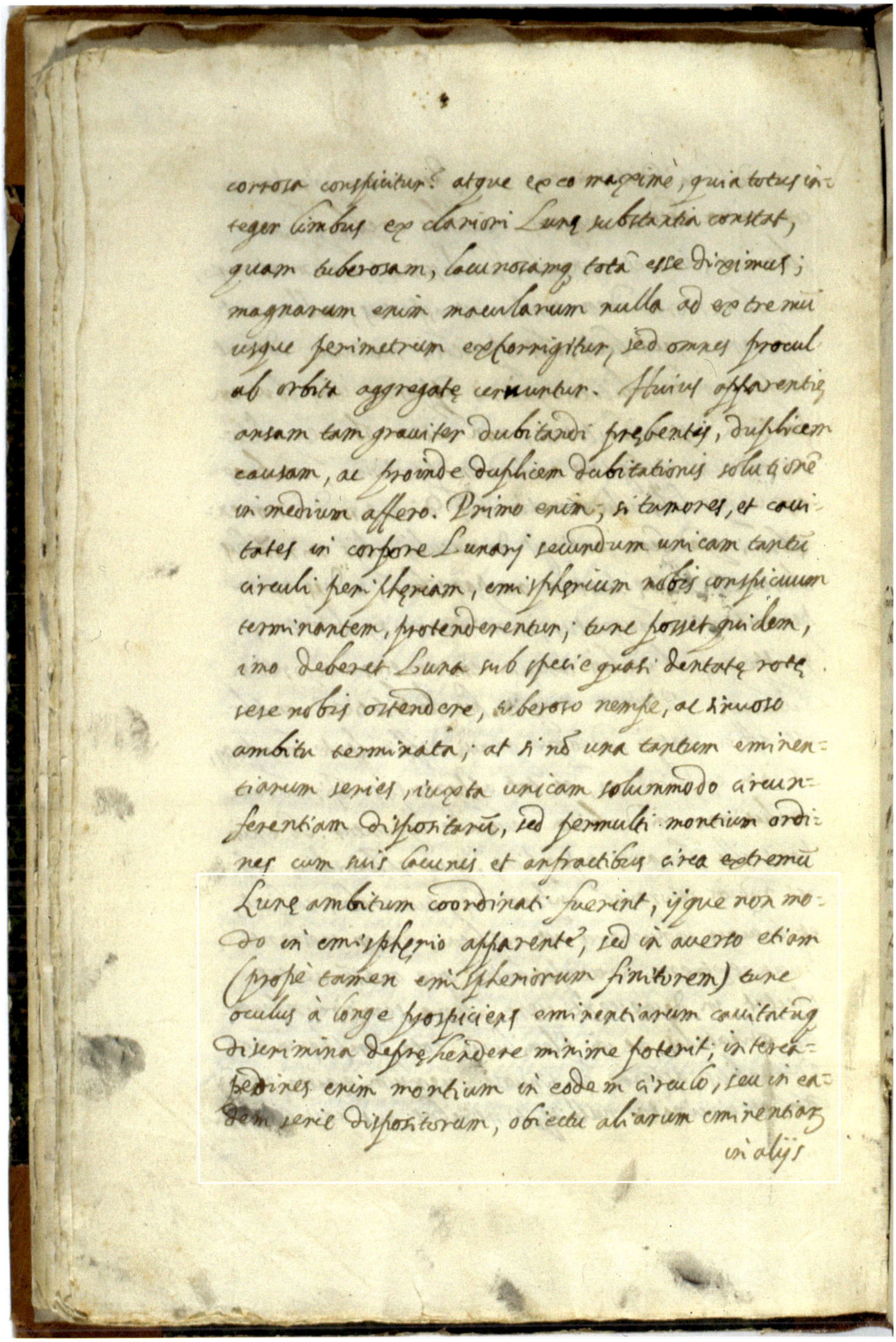
corrosa conspicitur: atque ex eo maxime, quia totus in-
teger limbus ex clariori Lunę substantia constat,
quam tuberosam, lacunosamque totã esse diximus;
magnarum enim macularum nulla ad extremũ
usque perimetrum exporrigitur, sed omnes procul
ab orbita aggregatę cernuntur. Huius apparentię
ansam tam graviter dubitandi prębentis, duplicem
causam, ac proinde duplicem dubitationis solutionẽ
in medium affero. Primo enim, si tumores, et cavi-
tates in corpore Lunari secundum unicam tantũ
circuli periferiam, emisphęrium nobis conspicuum
terminantem, protenderentur; tunc posset quidem,
imo deberet Luna sub specie quasi dentatę rotę
sese nobis ostendere, tuberoso nempe, ac sinuoso
ambitu terminata; at si nõ una tantum eminen-
tiarum series, iuxta unicam solummodo circun-
ferentiam dispositarũ, sed permulti montium ordi-
nes cum suis lacunis et anfractibus circa extremũ
Lunę ambitum coordinati fuerint, iique non mo-
do in emisphęrio apparente, sed in averso etiam
(prope tamen emisphęriorum finitorem) tunc
oculus à longe prospiciens eminentiarum cavitatumque
discrimina deprehendere minime poterit; interca-
pedines enim montium in eodem circulo, seu in ea-
dem serie dispositorum, obiectu aliarum eminentiarũ
in aliis

VI.2: BNCF, Ms. Gal. 48, 45v.

line 29), Galileo originally wrote *Terram* in the Draft, then modified it to *Tellurem*, and so it went to Fair Copy, to the presumed setting copy, and thence into the printed work. At four other places, later in the text than the Fair Copy extends, *Terra* in the Draft was transformed to *Tellus* in print. At the beginning of the introduction, the Draft's *inerrantium stellarum* became *Stellarum inerrantium* in the Fair Copy, then returned to *Inerrantium Stellarum* in *Sidereus Nuncius* (5r line 10). Similarly at another place Galileo wrote *hic uero paulatim* in the Draft, changed to *hic paulatim* in the Fair Copy, and returned to *hic vero paulatim* in *Sidereus Nuncius* (8v line 32). At more than one place, Galileo heightened the language by anastrophe, that is, selection of a less direct word order, a device to which Latin is well suited. At one place, *nullo labori* in Draft and Fair Copy become *labori nullo* in *Sidereus Nuncius* (6r line 25); at another, *luna iam uersus occasum tendente* in Draft and Fair Copy became *Luna iam occasum versus tendente* (9r line 1).

One change, often commented on, must reflect Galileo's recognition that the production of the printed edition would occupy some weeks. In the Draft, in his account of first learning of the invention of a telescope by a certain "Belgian", Galileo wrote *Mensibus abhinc 8. ferè ad aures nostras increpuit rumor à quodam Belga perspicillum elaboratum fuisse*. The Fair Copy keeps this wording, apart from spelling out "octo". In *Sidereus Nuncius* (6r line 3) the interval became not eight months ago, but ten (*Mensibus abhinc decem ferè rumor ad aures nostras increpuit* ...). For Galileo writing in January, "almost eight months ago" – reaching back to May 1609 or possibly the beginning of June – was correct. As he states, that is when he received a positive report, confirming the rumor, that Jacques Badover, a former pupil, sent from Paris.[2] It is noteworthy that this is the second time that Badover's name appeared in one of Galileo's published writings. In his 1607 *Difesa* against Baldassare Capra, Galileo printed the autograph statement, dated from Padua, 13 May 1607, of "Io Giacomo Badouere Francese," who attested that he had first acquired one of Galileo's calculating compasses nine years earlier.[3]

In preparing the setting copy, or possibly when examining a proof, Galileo increased the number of months to ten, so that the statement would be correct not for January but for March, when the projected book was to appear on the market. The same passage provides an excellent example of the heightening of style that Galileo brought to the published text. In the Draft, he wrote about the newly invented telescope: "by its aid visible objects, though far removed from the eye of the observer, were distinctly seen as though very close; and many reports circulated of this assuredly amazing effect" (*cuius beneficio obiecta uisibilia, licet ab oculo inspicientis longe dissita, ueluti propinqua distincte*

2 Van Helden 1977, p. 44; Pantin p. XVI n. 21, p. 58 note 11. For more detailed information on Badover and his relations with Galileo see Reeves 2008, pp. 125–138. Various commentators including Van Helden state as a positive fact that Galileo first heard about the telescope in the "summer" of 1609, this being inferred from his statements in *Il Saggiatore*, 1623. Galileo's direct remark of January 1610 is, however, surely preferable to his retrospective account many years later, made in the context of defending himself against what he believed to be an unfair accusation negating his role in the telescope's development.

3 Galileo Galilei, *Difesa ... Contro alle Calunnie & imposture di Baldessar Capra Milanese* (Venice: Tommaso Baglioni, 1607, 4°), fo. 10r, Badover stated: "...che sendo io già noue anni passati alloggiato nella propria casa [of Galileo] ... non pure viddi diuersi de' suoi Campassi Geometrici, & Militari, mà ne fui gratificato di vno, & di più della sua dichiarazione, mostrandomi in oltre le regole, che teneua intorno al modo del comparlo ...".

cernebantur, ac huius profecto admirabilis effectus nonnullę experientię circumferebantur). In the printed text, he tacked on a chiastic flourish which may be roughly englished as "credited by some, while others had doubts" (*quibus fidem alij praebebant, negabant alij*). Here, indeed, we can move through four layers in Galileo's formulation, the chiasmus entering as the final refinement:

> [Draft, fo. 1v, state A]:
> *... cuius beneficio obiecta uisibilia, licet ab oculo inspicientis longè dissita, ueluti propinqua distinctè cernebantur, ac huius certe admirabilis effectus nonnullę experientię circunferebantur. idem paucis post diebus mihi per literas à nobili Gallo Jacobo Badouere ex Lutetia confirmatum fuit ...*
>
> [Draft, fo. 1v, state B, changes noted in small capitals]:
> *... cuius beneficio obiecta uisibilia, licet ab oculo inspicientis longè dissita, ueluti propinqua distinctè cernebantur, ac huius* PROFECTO *admirabilis effectus nonnullę experientię circunferebantur. idem* PAUCOS POST DIES *mihi per literas à nobili Gallo Jacobo Badouere ex Lutetia confirmatum* EST ...
>
> [Fair copy, Gal. 48, fo. 2r, changes from Draft state B noted in small capitals:]
> *... cuius beneficio obiecta uisibilia, licet ab oculo inspicientis longe dissita, ueluti propinqua distinctè cernebantur; ac huius profecto admirabilis effectus nonnullę experientię circunferebantur,* QUIBUS ALIJ FIDEM PRĘBEBANT, ALIJ NEGABANT; *Idem paucos post dies mihi per literas à nobili Gallo Jacobo Badouere ex Lutetia confirmatum est ...*
>
> [*Sidereus Nuncius*, Venice, 1610, fo. 6r, changes from "Fair copy" noted in small capitals]:
> *... cuius beneficio obiecta visibilia, licet ab oculo inspicientis longè dissita, veluti propinqua distinctè cernebantur; ac huius profectò admirabilis effectus nonnullæ experientiæ circumferebantur, quibus* FIDEM ALIJ *præbebant,* NEGABANT ALIJ. *Idem paucos post dies mihi per literas à nobili Gallo Iacobo Badouere ex Lutetia confirmatum est ...*

A few sentences later, perhaps in the same round of editing, Galileo again heightened his prose with chiasmus. He wrote (*Sidereus Nuncius*, 6r line 25) of developing finally a thirty-power telescope, "sparing neither labor nor expense in doing so" (*Tandem labori nullo, nullisquè sumptibus parcens ...*). In both the Draft and Fair copy, Galileo originally wrote *Tandem nullo laborj, nullisque sumptibus parcens.*

Two other changes in the account of Galileo's involvement with telescopes are noteworthy. In *Sidereus Nuncius*, as in the Fair Copy, news of the Dutch discovery was anonymous, the story was simply "noised about" (*rumor ... increpuit*). But in the Draft, Galileo began to sketch out details of the circumstances: the rumor was that a certain Dutchman had invented a telescope "and ... presented it to Count Maurice" (*ac ... Comiti Mauritio dono datum*) – that is, Galileo himself wrote in his draft the marks of elision, as if he was considering the idea of writing more circumstantially, making an elision that he would eventually fill in with details. The story of Hans Lipperhey's presentation of his invention to Maurice and to the Dutch States General is documented in official records, and was also reported as an addendum to a rare little newsletter, the *Ambassades du Roy de Siam … arriué à la Haye le 10. Septemb. 1608.*[4] It may be noted that although this news did not reach Galileo until late May 1609, Paolo Sarpi in Venice, with whom Galileo was well acquainted, had already been sent a copy of the Hague newsletter in early November 1608, as he reported to a correspondent in Paris, Francesco Castrino, in a letter dated 9 December 1608.[5]

In the Draft Galileo at first expatiated on the excitement that led him to turn his energies to creating a telescope of his own: he was captivated by the beauty of the idea (… *eiusque pulcritudinis cupidine captus* …). He dropped this phrase when writing out the Fair Copy. In removing these elements, Galileo moved more quickly to what was for him the central point: yes, he first heard news from abroad of a telescope, but it was his own effort, initiative and imagination that enabled him soon to make a greatly improved instrument, and to train it on the night skies. He omitted entirely the summer months of 1609, when he demonstrated his nine-power telescope to the state of Venice, and received in reward an ample increase in his professorial salary. In a work dedicated to the Grand Duke of Tuscany it would have violated all decorum to refer to this earlier reward. In *Sidereus Nuncius* there remains the merest hint of his transactions with the Republic: he wrote that it would be "superfluous to list" all the advantages of his telescope in matters of land and sea (*superuacaneum foret enumerare*); and then, "these matters left aside, I turned my thoughts to the heavens" (*missis terrenis, ad Cœlestium speculationes me contuli*).

Galileo's Draft and Fair Copy provide evidence for orthographic changes introduced by the compositor (or compositors) of *Sidereus Nuncius*, and in so doing reveal the cause for several typographical errors. In Draft and Fair Copy Galileo always wrote out the word "et"; in *Sidereus Nuncius*, this was consistently changed to the ampersand, "&". Galileo also consistently wrote a caudate e (ę), a letter form many centuries old, for both diphthongs "ae" and "oe". The compositor generally, but not always, set these with the ligatured single types "æ" and "œ". But the roman text font of *Sidereus Nuncius* also possessed ę types, as did many or most of the roman fonts of this time, and so there are not infrequent instances where the compositor incorrectly followed Galileo's use of caudate ę, for instance fo. 8r line 9 *pęnitus*, the correct spelling being *penitus*. The second word of the first sentence of *Sidereus Nuncius*, *æquidem* (equidem) is likewise misspelled because the compositor responded to Galileo's incorrect form *ęquidem*. Galileo regularly misspelled the word *peripheria* as *periphęria*. In *Sidereus Nuncius* it is spelled five times as *periphæria* against two correct spellings of *peripheria*. These latter two settings are doubtfully to be

4 Van Helden 1977, pp. 25–26, 35–44; Drake 1976, including a facsimile of the newsletter, a six-leaf 12mo; a copy is preserved at the Universiteitsbibliotheek Gent (Bib. Meul. 1120). The newsletter was reprinted as an octavo in Lyons, with the date of the ambassadors' arrival changed to 12 November 1608; a copy is preserved at the British Library (C.32.d.3).

5 Sarpi 1931, p. 15.

seen as conscious corrections; it is more likely that the spellings are inadvertently correct because the compositor overlooked Galileo's "ę". Likewise *Sidereus Nuncius*'s misspellings *sphære* for *sphæræ* (12v line 28) and *spheram* for *sphæram* (28v line 28) presumably resulted from failures to notice the caudate e in Galileo's writing. Galileo spelled at least two words with "ci" for "ti": *planicies* and *negocium*. In *Sidereus Nuncius*, the spelling of the second word was corrected to *negotium* (6v line 17, 17r line 3), but on 8v the compositor corrected to *planities* on line 15, then moments later set *planicierum* on line 19.

The other notable orthographic feature of Galileo's Latin writing was his frequent omission of "h" at the beginning of syllables, for example *anteac* for *antehac* (in this case, *anteac* in the Draft became *antehac* in both Fair Copy and *Sidereus Nuncius*). The compositor of *Sidereus Nuncius* regularized many of these spellings, but there are also oversights, such as *coors* for *cohors* (7v line 17), *exibet* for *exhibet* (8v line 14), and *nocturnus orror* for *nocturnus horror* (14v line 4).[6] At one place (11v line 11), where Gaileo discussed the shadows cast by the Sun across the floor of a great crater, the compositor made a false correction, changing Galileo's *ora* – edge or boundary – to the confusing *hora*, hour (*vt extrema hora tenebrosæ Lunæ parti contermina Solis lumine perfusa spectetur*). Galileo consistently omitted initial h from *hemisphaerium*, and the *Sidereus Nuncius* compositor never supplied it. In the third syllable, the compositor generally transmuted Galileo's "ę" to "æ", but there is one spelling *emispherij* (15v line 14), while on 12r the compositor set *emisphærio* on one line, and *emisphęriorum* on the next (lines 22–23). In essence, for the diphthong "ae" the compositor was at liberty to reach either for the æ or the ę box. It was not an error to select the latter type.

The outer forme of sheet B contains the first page set of *Sidereus Nuncius*, with the beginning of the text placed below a lengthy drop-title summarizing the contents (figure VI.3). This page has already been discussed in chapter IV with regard to its "Byzantine" layout of headpiece + heading + decorative initial. Two other features should be briefly noted. First, the text of the drop-title is centered, line by line, and laid out in the pattern of a series of inverted pyramids. This style of text arrangement for titles and headings in books became extremely common in the sixteenth century; in printed books it can be traced back at least to the 1490s in Venice.[7] The other feature of the drop-title can be called typographic modulation. The drop-title makes use of what may be seen as four grades or modes of type: two of the lines are in roman capitals, but in different body sizes. And although the italic typesetting is of a single font, it has two grades: five lines are in the upper- and lowercase of ordinary reading, and two others are set in full capitals. This use

6 "Nocturnus horror" is one of the most vivid expressions in *Sidereus Nuncius*. In the relevant passage Galileo discusses the phenomenon of the "ashen light" visible on the Moon's dark surface near the time of the New Moon, especially when the "nocturnal horror" has increased due to the Sun's absence (*potissimum, si iam nocturnus* [h]*orror ob solis absentiam increuerit*). One commentator (Lambert 2002, pp. 43–44) has interpreted this literally as the terrors of the night. It seems more likely that what Galileo had in mind was the increasing cold as the Sun falls below the horizon, a time of "shivering", as well expressed in Van Helden's translation, "when the chill of the night has already increased through the absence of the Sun." It will be recalled that Galileo's observations of the Moon were made from late November to early January.

7 Smith 2000, pp. 60–61 ("half-diamond indentation"), figs. 6.3 (Venice, 1492) and 6.7 (Venice, 1499). The latter is Aldus Manutius's edition of *Hypnerotomachia Poliphili*. See also Barberi 1969, I p. 102: "Un titolo lungo e articolato viene spesso ripartito in due, tre, perfino quattro triangoli o coni tronchi … sottoposto l'uno all'altro …."

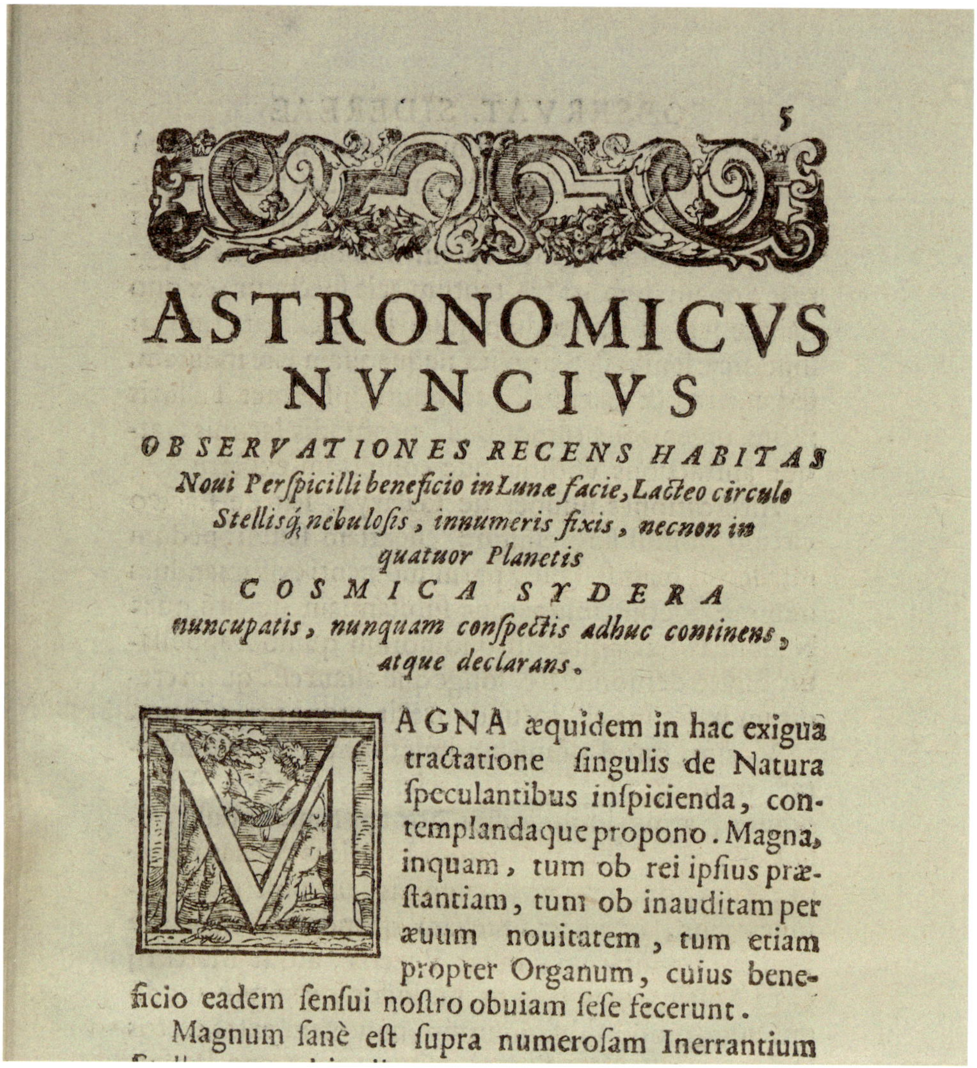
5

ASTRONOMICVS NVNCIVS

OBSERVATIONES RECENS HABITAS Noui Perſpicilli beneficio in Lunæ facie, Lacteo circulo Stellisq́; nebuloſis, innumeris fixis, necnon in quatuor Planetis *COSMICA SYDERA* *nuncupatis, nunquam conſpectis adhuc continens, atque declarans.*

MAGNA æquidem in hac exigua tractatione ſingulis de Natura ſpeculantibus inſpicienda, contemplandaque propono. Magna, inquam, tum ob rei ipſius præſtantiam, tum ob inauditam per æuum nouitatem, tum etiam propter Organum, cuius beneficio eadem ſenſui noſtro obuiam ſeſe fecerunt.

Magnum ſanè eſt ſupra numeroſam Inerrantium

VI.3: Sidereus Nuncius, B1r, Institute for Advanced Study.

of a wide variety of typographic "modes" for setting titles and headings is a sixteenth-century phenomenon, with a long continuation reaching to the present, that has not been closely studied.[8] When the Münster antiquary Bernard von Mallinckrodt published in 1640 the first monograph on fifteenth-century printing, the small number of fonts used in setting early printed books – including many hundreds of editions set with a single font throughout – struck him as a crude oddity. To his eye, the diversity of types in the printing of his own day gave elegance and beauty, and reduced the tedium of the reader by

8 A broad conspectus of sixteenth-century French and Italian books with both inverted-pyramid layouts and varying degrees of typographic modulation is presented by the illustrations of Mortimer 1964 and 1974. Barberi 1969 I, p. 104 briefly discusses typographic modulation, noting that "nel tardo Cinquecento gli esempi si fanno sempre più numerosi."

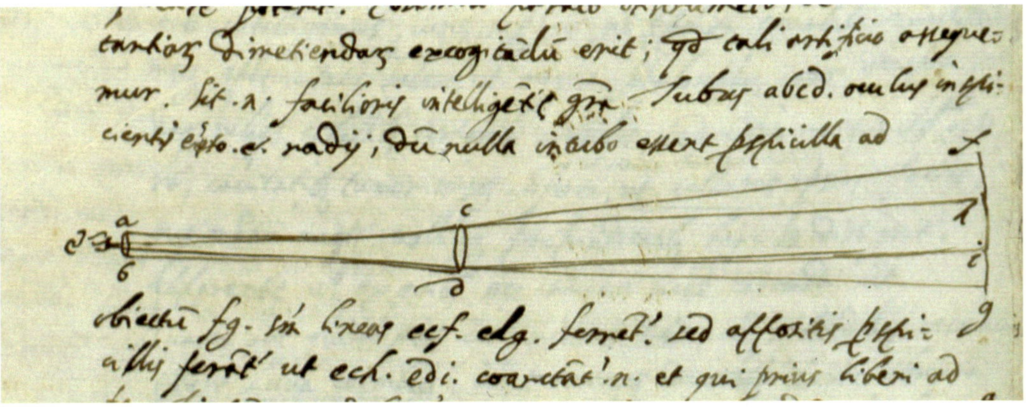

VI.4: BNCF, Ms. Gal. 48, 9r.

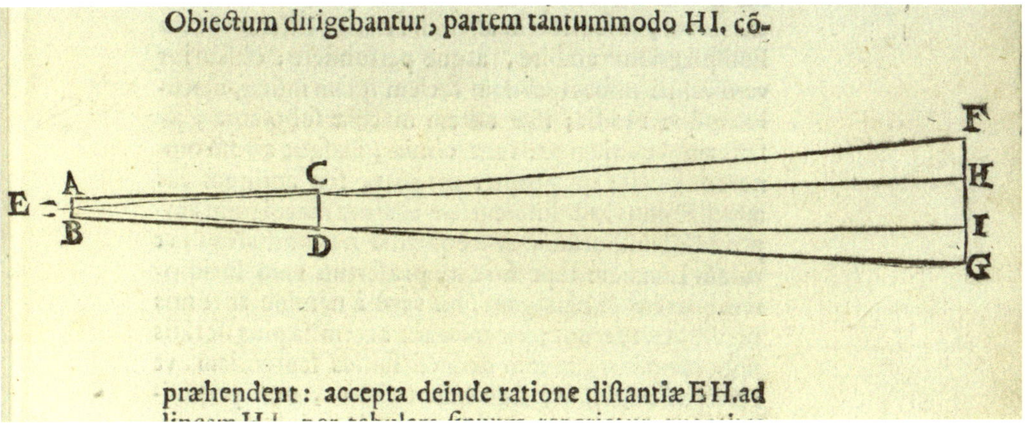

VI.5: Sidereus Nuncius, B3r, Linda Hall Library.

conspicuously marking distinctions in the text; in short, to him it was an unalloyed improvement.[9]

We see in the drop-title Galileo's first idea for the title of his work: *Astronomicus nuncius,* not *Sidereus Nuncius.* The change was yet another instance of preferring the poetic to the technical word. In fact, at one point after another the drop-title and the title-page, whose text was formulated by Galileo and put into type some weeks later, contrast with each other. The drop-title emphasizes that Galileo's observations are of the freshest – *observationes recens habitae* – a point that fell out of emphasis on the title-page, but was the central point of the running title: *Observationes sidereae recens habitae.* In the drop-title the telescope is simply a new instrument (*Noui Perspicilli*), whereas the title-page features Galileo's development of his own superior instrument (*Perspicilli nuper à se* [Galileo] *reperti*). Most notable are the differences between drop-title and title-page in referring to the most exciting of Galileo's discoveries, the four new planets. Their often-discussed renaming on the title-page as "Medicean" and not, as originally, "Cosmic" plan-

9 Mallinckrodt 1640, p. 105: "Sed & multiplex typorum diuersitas & varietas, hodie Typographiam mirificè exornat … cum vna saepè pagina eius generis decem, duodecim aut amplius conspiciendas exhibeat diuersas species." To the book scholar A. W. Pollard, who could not shake off his aesthetic senses, the seventeenth century was an age of decadence in book design, marked by "an inane ambition on the part of the printer to show off the multiplicity of his types" (Pollard 1891, p. 33)

ets is explored below in chapters XI and XII. Of equal significance is that in the drop-title the four planets are so to speak "lost in space," of unstated location (*quatuor Planetis Cosmica Sydera nuncupatis, nunquam conspectis adhuc*), whereas the title-page locates them precisely as satellites of Jupiter, revolving about it with wondrous speed (*quatuor planetis circa Iovis Stellam disparibus interuallis, atque periodis, celeritate mirabili circumuolatis*). The reason for Galileo's reticence in sheet B is obvious: once that sheet had been printed in 550 or more copies, the greatest surprise of the *Sidereus Nuncius* was already running a strong risk of going outside Galileo's control. The power of the printing press to broadcast the word brought its own dangers.

The outer forme of sheet B contains (B3r: fo. 7r) one item of printing material that would certainly not have been at hand in the printing shop's stock, as it did not yet exist as an illustration: the diagram of a telescope, corresponding to the sketch on fo. 2r of Galileo's Draft (figures VI.4, 5). This was a professionally made woodcut, with minimum block dimension (for the blocks are unbordered, and could have extended beyond their printed lines) of 28 × 165 mm. The space within which the block was placed was equivalent to twelve lines of typesetting. The block extends well into the inner and outer margins of its page – the line width, it will be recalled, is about 95 mm. At the inner margin, it looks in fact as if the block directly abutted the longer of the two crossbars that presumably divided the chase into quarters, one for each page of the forme. In various tightly bound copies, the letter E at the left side of the block is obscured or entirely lost in the gutter.

The inner forme of sheet B contains (B4r: fo. 8r) the first of the large spaces left blank for the eventual supply of an etching or engraving of the Moon. The last nineteen lines of the 34-line page were left blank, but with a catchword set at the foot of the page. That nineteen-line space would have been filled in the forme by blocking material of less than type height. The frisket for this page would have been cut to cover over the blank expanse, to prevent the sheets from taking extraneous ink that might have been left on the blocking material by the inking ball.

As to the Moon illustrations within *Sidereus Nuncius*, Galileo's Draft refers to various illustrations of the Moon, but neither depicts them nor indicates them by leaving blank spaces. At the place in the text referring to this first Moon illustration, the Draft reads, "as the present first figure represents it" (*ueluti in pręsens prima figura repręsentat*). In *Sidereus Nuncius*, since the words are printed on 7v, the wording was changed to "as the figure nearby represents it" (*veluti apposita figura repræsentat*), to refer to the (not yet printed) illustration on the facing page, 8r. This change of wording could presumably not have been set in type until the layout of sheet B had been established. A further implication is that all the pages of quire B were in type at the same time. Only when both formes of sheet B were put together could the relative positions of the first reference to the illustration (7v, inner forme), and of the illustration itself (8r, inner forme) be brought into coordination. At the time of writing his Draft, Galileo would have had no certain knowledge of the layout of the printed pages, or of the formes to which each type-page would belong.

In particular it must be kept in mind that when quires B and C were printed, the illustrations of the Moon did not yet exist, except in Galileo's mind. The etchings, one in sheet B and four in sheet C, were made and printed later, in a separate operation from that of the

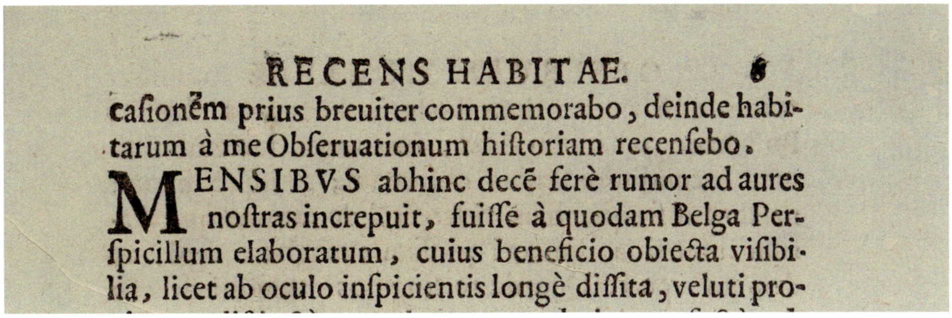
RECENS HABITAE.
casionẽm prius breuiter commemorabo, deinde habi-
tarum à me Obſeruationum hiſtoriam recenſebo.
MENSIBVS abhinc decẽ ferè rumor ad aures
noſtras increpuit, fuiſſe à quodam Belga Per-
ſpicillum elaboratum, cuius beneficio obiecta viſibi-
lia, licet ab oculo inſpicientis longè diſſita, veluti pro-

VI.6: Sidereus Nuncius, B2r, Institute for Advanced Study.

printing shop, using a rolling press. As noted in more detail in chapter XII, in a letter of 19 March 1610, six days after printing of the letterpress was finished, Galileo stated – complained – that a substantial number of copies still had not received their copperplate illustrations. This work would have been carried out by taking all the copies of sheets B and C to a printseller's shop.

Two small layout details of sheet B are variant from the remainder of the book, reflecting its temporal position as the first sheet to be printed, while adjustments continued to be made. First, B1v (5v), the first page of full text setting (almost half of B1r being occupied by the headpiece and drop-title) was set to 33 lines, whereas all the following pages, except where interrupted by diagrams, were set to 34 lines. Second, on the following page B2r (6r), Galileo's account of how he first heard about the telescope is demarcated as a kind of sub-chapter, with a 2-line capital M set as an initial, and the first word set in full capitals (figure VI.6). This implies that when printing began, the intent was to articulate the text by conspicuously marking out the several thematic sections with initials. Probably because of the increasing press of time, this idea was soon dropped. Even on E1r (17r), where the second section of *Sidereus Nuncius* – the discussion of Jupiter's satellites – begins, the first line is only indented as a paragraph. A two-line initial would have been very natural here, for this is the major division of the text: Moon and Stars in part I, Jupiter in part II.

The increase of the page depth from 33 to 34 lines may have been the first of the various steps taken to accommodate the changes of text and text placement that affected the entire typographical treatment of quires B, C, and D. The tension between growing text and limited pages may also be reflected in a small way by the paucity of paragraphing in long sections of these three quires, paragraphing that would have been useful for the reader in breaking up lengthy passages. The paragraph (*Verum non modo tenebrarum …*) that begins at the top of 8v (B4v) extends for 82 lines; the paragraph (*Verum magna hic dubitatione …*) beginning on line 26 of 11v (C3v) for 84 lines; and the paragraph (*Dum Luna tum ante …*) beginnning on line 15 of 14r (D2r) for 133 lines. Galileo's rapid composition of his Draft, which in its first portion contains only three or four clear paragraph indications, became unintentionally reflected in the printed edition.

VII.
Quire C

Quires C and D combined contain the remaining text for the first part of *Sidereus Nuncius*, as summarized in the drop-title: telescopic "observations of the face of the Moon, the Milky Way, nebulae, and the countless fixed stars" (*observationes … in Lunae facie, Lacteo circulo Stellisque nebulosis, innumeris fixis …*). As noted already, *Sidereus Nuncius* contains a powerful clue that the first part of the work was originally planned to fill 24 pages, that is the 12 leaves of B–D^4. By the printed foliation, these would have been leaves 5 to 16, so that quire E, the beginning of the Jupiter report, would start, as it did, with numbered leaf 17. Galileo subsequently added more text and illustration, the equivalent of four pages, to the first part. This was eventually accommodated by inserting a half sheet of paper, D3.4, into the center of sheet D, that is between D2 and D5. The first four leaves of quire D were foliated 13–16, leaves 15 and 16 being the two leaves of the half sheet. The last two leaves D5–6, that is, the second half of the full sheet, have no foliation. Evidently, this is because by the time quire D was printed, quire E, with its "frozen" foliation 17–20, had already been set and printed off. As will be seen in chapter XI, there is

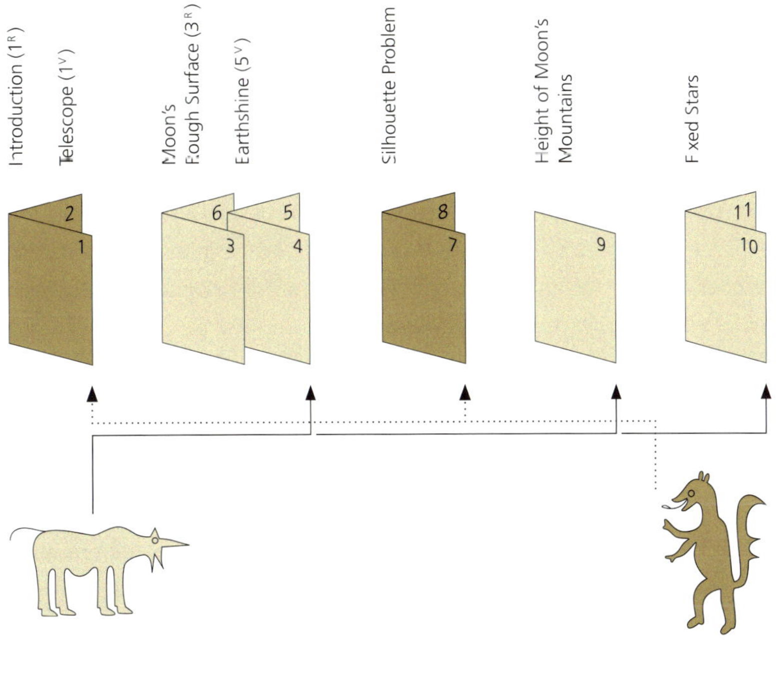

VII.1: *Draft pt. I text units, with their topics.*

reason to suppose, indeed, that sheets E to G had all been printed off before quire D, or at least its half-sheet insert, went to press.

Comparison of the printed quires C and D with Galileo's Draft brings to light substantial changes to the text as originally planned. They were introduced by Galileo in several stages in the last days of January and first half of February. These changes are the reason quire D eventually required six rather than four leaves. The contents of part I, quires B–D, can be separated out into seven discrete topics, all of which are addressed in the first eleven leaves of the autograph Draft.

In the order eventually printed in *Sidereus Nuncius*, these topics are:

I. Introduction: 5r–6r li. 2 (B1–2)
II. The Telescope: 6r li. 3–7r li. 33 (B2–3)
III. The Moon's Rough Surface: 7r li. 34–11v li. 25 (B3–C3)
IV. The Silhouette Problem: why do we not see the roughness of the Moon's mountains at its lighted perimeter? 11v li. 26–13r li. 14 (C3–D1)
V. The Height of the Moon's Mountains: 13r li. 15–14r li. 4 (D1–2)
VI. Earthshine on the Moon: 14r li. 5–16r li. 11 (D2–4)
VII. Fixed Stars: Constellations and Nebulae: 16r li. 12–16Bv (D4–6)

But this was neither the order of Galileo's original writing, nor of his original plan for the printed edition. As noted in chapter VI, he wrote the first two topics, Introduction and Telescope, on what became the first sheet of the Draft, fos. 1–2, this being (Venetian) Rampant Lion stock. On this sheet there is even a hint that the topics may have been drafted in reverse order, that is, that Galileo began by writing the Telescope section at the top of 1v, reserving the blank 1r for the eventual Introduction, estimated to occupy no more than a page. The Introduction draft on 1r contains 32 lines, whereas 1v has 30 lines in the same vertical space, suggesting that Galileo may have crowded the Introduction slightly, because he could not carry it over to the already-written following page. The next two sheets of the Draft, the fascicle consisting of fos. 3–6, on the (Paduan) Unicorn stock, were begun before the eventual first sheet was taken in hand. This fascicle contains Topic III, the Moon's Rough Surface, followed immediately by Topic VI, Earthshine. Galileo originally finished his draft of Topic III about a third of the way down 5r, ending at the beginning of line 11. Below this, he left a space of about a line, then began to write out Topic VI, Earthshine, which continued to the end of the fascicle, 6v.

As for Topics IV and V, the problem of the apparent smoothness of the Moon's perimeter and a calculation of the height of the Moon's mountains, these are later interpolations, each written separately. It is noteworthy and indicative that neither topic was broached in the 7 January 1610 letter that provided most of the framework for Topic III. Thus, Galileo's original plan was that the first part of *Sidereus Nuncius* should take in five rather than seven topics, in the order: I, Introduction; II, Telescope; III, Moon's Rough Surface, VI, Earthshine; VII, Fixed Stars. Allowing for the inserted illustrations of the Moon necessary to Topic III, analyzed in greater detail below, these topics would indeed have fit in the originally planned twelve printed leaves, that is, the quarto sheets B, C, and D. It was Galileo's decision to add Topics IV and V that made it necessary to add a bifolium to gathering D.

Within Topic III, Moon's Rough Surface, several stages of composition have left physical traces in the Draft. The two main changes are an expansion of its conclusion, and – entailing considerably more complication and calculation – a major revision of the original plans for accompanying illustrations. Topic III in its original form concludes by re-emphasizing a fundamental difference between the two chief forms of light and dark effects on the Moon that the telescope reveals. There are the ever-changing sharp contrasts of shadows and lighted portions caused by variations in the Sun's raking light across the irregular surface during the course of the lunar cycle: all the evidence, in short, that led Galileo to conclude that the Moon's surface was not uniformly smooth, but rough and jagged, like the mountainous regions of the Earth. Then there are the Moon's "large or ancient" spots (7v line 8: [maculas] *magnas seu antiquas*), such as had been observed and recorded from ancient times by naked eye observation: what we now refer to as the seas of the Moon. These evidently lie lower than the brightly lit upper regions and have a smoother surface, corresponding to the seas and oceans of the Earth, an analogy that Galileo drew out: he had never, he wrote, doubted that if the Earth could be viewed from on high, the dry land would look brighter, the seas darker (9r lines 31 sqq.: … *mihi autem dubium fuit nunquam, Terrestris globi à longè conspecti, atque à radijs Solaribus perfusi, terream superficiem clariorem, obscuriorem verò aqueam sese in conspectum daturam*).[1] Within these large spots certain darker sections (*areolae*) can be seen, but these dark sections do not fundamentally vary, apart from changes in the incidence of the Sun's rays that may make them a little more or a little less conspicuous. And, they merge gradually from one to another. All this, Galileo concludes, contrasts decisively with the abrupt, rugged and ever-changing patterns that develop on the Moon's bright surface as the Sun moves across it, always with a sharp boundary line between light and shade (… *umbrarum luminumque rudibus taglienti* [sic] *discriminibus disterminantur*).[2]

Later, after he had already filled the remainder of Draft 5r with the beginning of Topic VI, Earthshine (*Lubet hoc loco* …), Galileo decided that the conclusion of Topic III needed a final clarification, making explicit that on the "large spots," besides the already discussed darker sections, certain other lighter sections, some of them even "lucidissimae", can also be seen. Like the darker sections, these do not change in size and shape, thus there must be some real difference in substance between the darker and lighter shapes on the large spots. Galileo began to draft this addendum (*Apparent insuper intra easdem magnas maculas areolę quedam clariores* …) in the space he had left between the original end of Topic III and the beginning of Topic VI, squeezing in three lines written small with a fine nib. Then, having run out of space, he continued the addendum to its end – about eighty words in all – on a small loose slip of paper. To link the two, he drew a reference symbol, the sign of Mars (♂), at the end of the interpolation on Draft 5r, and again at the

1 The remark closely parallels what he wrote in his Italian letter of 7 January 1610: *Opere*, X, Ep. 259, p. 276 at lines 97–103: *io ho veramente ancora per avanti hauto sempre opinione, che il globo terrestre veduto da grandissima lontananza del sole, più lucido aspetto faria nella parte terrena, et meno risplendente apparirebbe il mare* …

2 Galileo apparently put the Italian word "taglienti" (sharp) in his Draft as a kind of placeholder, until he could think of an adequate Latin expression. In the Fair Copy, and thence in *Sidereus Nuncius*, the word was replaced by "ad lineam".

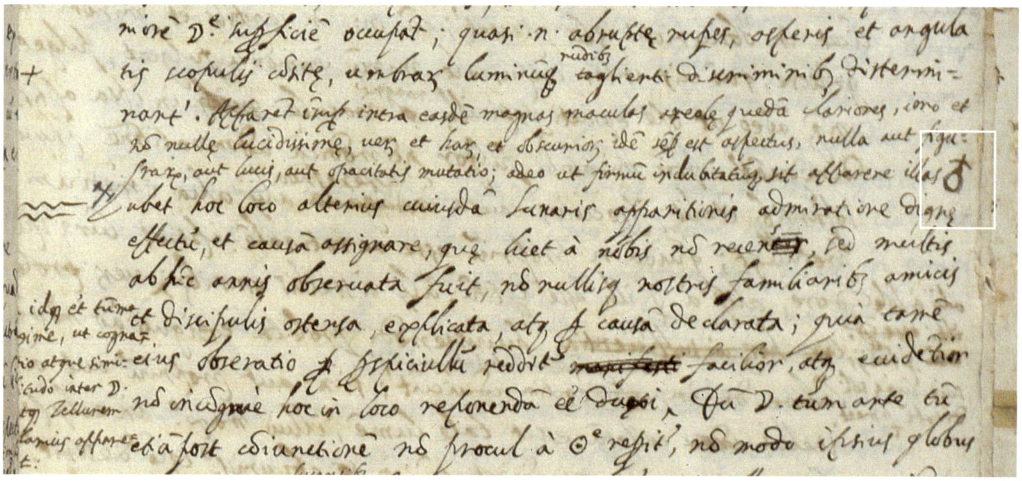

VII.2: BNCF, Ms. Gal. 48, 12r.

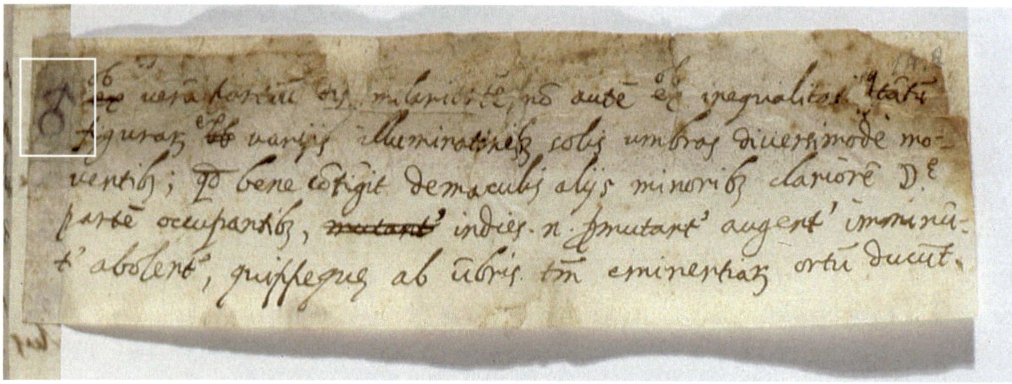

VII.3: BNCF, Ms. Gal. 48, 13bis.

beginning of the loose slip.[3] This expanded conclusion to Topic III added 12 lines to the printed text (figure VII.2, 3).

The other major change in Topic III was that of variously adjusting the plans for illustrations of the Moon. Originally Galileo planned seven illustrations, and explicitly numbered each within the draft. Though he did not include sketches, the intended purpose of each illustration is clear from his descriptions of the phenomena they were to represent. In the printed edition these became reduced to four Moon illustrations: one in quire B, and three (one repeated, for four impressions in all) in quire C. These may be referred to as Moons I, II, III, IV. The detail features originally intended for illustration in the three cancelled images were instead incorporated into Moons I, II, and III.

There must have been several interrelated rationales for this reduction. Collapsing the seven illustrations into five (counting the repeated etching) reduced by about a page the amount of space that would be taken up by Topic III as printed, and simplified the layout problems of fitting the letterpress around the spaces needed for the illustrations. It also reduced the time needed to make the etchings. These changes in the illustration scheme

3 In Ms. Gal. 48, this slip is glued at the top of fo. 14r (that is, fo. 7r of the Draft), where it does not belong. In *Opere* III pt. i, Favaro reproduced it on pl. 17, that is, it looks as if it was written at the foot of fo. 6v of the Draft. Both locations, virtual and actual, are highly misleading.

had, of course, to be finalized by the time quire C was being composed, laid out for imposition, and printed.

Galileo's original scheme for illustrations is revealed in his Draft text as follows.

Illustration 1. (Draft 3r, lines 23–28): *Quarta aut quinta post coniunctionem die, cum splendidis sese nobis offert Luna cornibus, iam terminus partem obscuram a luminosa diuidens non ęquabiliter secundum oualem lineam extenditur, ueluti in solido perfectè sphęrico contigeret, sed inęquabili, aspera, et admodum sinuosa linea designatur,* UELUTI PRĘSENS PRIMA FIGURA REPRĘSENTAT ...

Illustration 2. (Draft 3v, lines 27–29): *interim silentio minime inuoluam notabile quoddam à me obseruatum dum Luna ad primam quadraturam properaret,* CUIUS ETIAM IMAGINEM 2. DELINEATIO DEMONSTRAT; *ingens enim sinus tenebrosus in partem luminosam irrumpit* ... [emended to: ... IMAGINEM EADEM DELINEATIO ...]

Illustrations 3, 4. (Draft 4v, line 7): ... *tam supra illam quam infra ingentes quędam eminentię,* UELUTI APPOSITE PRE SE FERUNT DELINEATIONES 3. 4. *hęc eadem macula* ...

Illustrations 5, 4. (Draft 4v, lines 10–13): ... *qua uero solem respiciunt lucidiores extant* UT .5. EXPRIMIT DESCRIPTIO; *diminuta deinde parte luminosa, cum primum tota fermè macula tenebris est obducta, clariora montium dorsa eminenter tenebras scandunt,* UT 4. DELINEATIO FIGURAT.

Illustrations 6, 7. (Draft 4v, lines 15–18): ... *mediam quasi Lunę partem occupatam esse à cauitate quadam reliquis omnibus maiori, ac figura perfectę rotunditatis; hanc prope utramque quadraturam conspeximus,* EANDEMQUE IN 6. ET IN 7. FIGURA QUANTUM LICUIT IMITATI SUMUS.

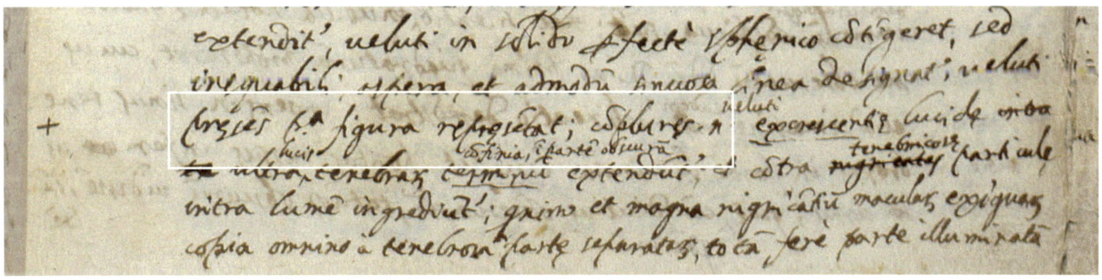

VII.4: BNCF, Ms. Gal. 48, 10r.

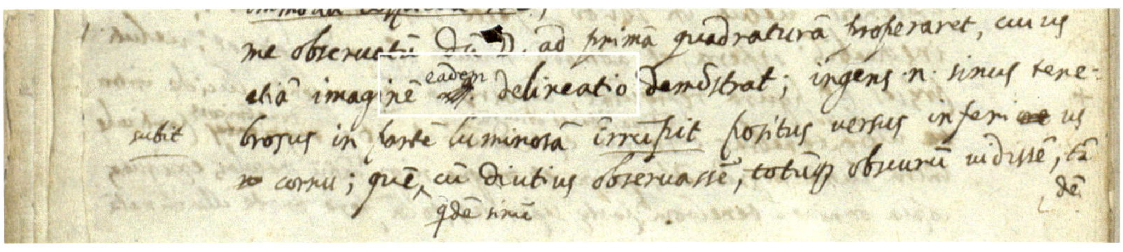

VII.5: BNCF, Ms. Gal. 48, 10v.

The first change Galileo made was to conflate Delineations 1 and 2 into a single image, that is, what became Moon I. Delineation 1, by itself, was to represent the Moon on the fourth or fifth day after the New Moon, when its horns were prominent. At this time "the

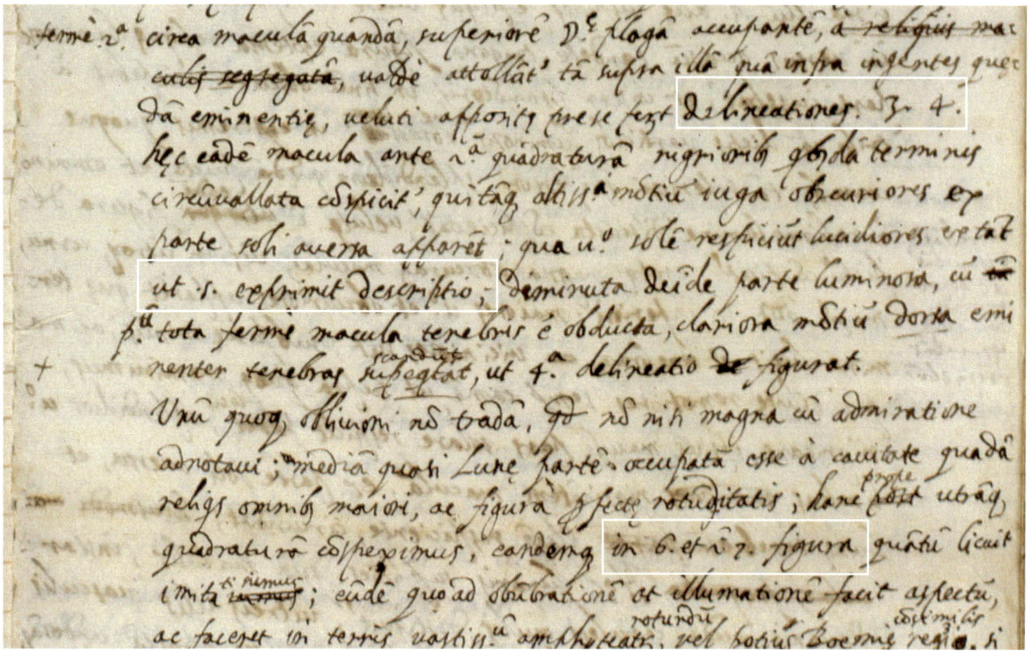

VII.6: BNCF, Ms. Gal. 48, 11v.

terminator dividing the dark part from the lighted does not extend in a smooth oval line, such as would fall on a perfectly spherical solid, but is irregular, rough, and quite wavy, as the present first figure represents it." This marked the beginning of Galileo's specific lunar observations (*Sidereus Nuncius* 7v [B3v] lines 24 and after). Galileo's text closely follows that of his 7 January 1610 letter to an unidentified correspondent (*cominciando … 4 o 5 giorni dopo il novilunio, vedesi il confine che è tra la parte illuminata et il resto del corpo tenebroso, esser non una parte di linea ovale pulitamente segnata, ma un termine molto confuso, anfrattuoso et aspro …*).[4]

Delineation 2 was intended to exemplify a quite different observation, made a few days later "when the Moon was approaching its first quarter": the intrusion of a giant shadowy bay into the lower horn (*Sidereus Nuncius* 8v [B4v] lines 23 and following). Here too the Draft closely follows a passage from Galileo's 7 January 1610 letter, quoted in the following chapter. The letter at this place includes a small illustration of the phenomenon of the shadowy bay.

We have a good idea of when Galileo made these observations. Whoever the addressee of the 7 January letter, the introductory words reveal that Galileo had written earlier to him with a briefer account of his observations, and that the addressee wrote back urging more details. Galileo's preceding lost letter to this correspondent must therefore date to roughly mid-December. The two observations would then have been made after the New Moon of 26 November 1609: at Padua's longitude the conjunction of Sun and Moon would have been at about one o'clock in the afternoon on that day. The "fourth to fifth day" after conjunction would then have been the evening of 30 November; and the time when the Moon was approaching first quarter would have been the evening of 2 December, for on

4 *Opere*, X, pp. 273–274 (Ep. 259, lines 18–21).

the evening of the next day, the Moon would have been fully at first quarter. This time frame is indicated likewise by the final sentence of Topic II, The Telescope (*Sidereus Nuncius*, 7r lines 19–20), written as a transition to Topic III, The Moon's Rough Surface: "Now let us review the observations we have made over the last two months ..." (*Nunc obseruationes à nobis duobus proximè elapsis mensibus habitas recenseamus ...*). This time period is not to be interpreted as the two months before *Sidereus Nuncius* was published in mid-March 1610, but as the two months before he wrote this part of the Draft: that is, two months back from the end of January 1610.

The passage about the shadowy bay marks a crux in the transition from written draft to printed page. As printed, the passage falls on 8v (B4v), the last page of its gathering, and near the foot of that page. The continued description of the phenomenon extends well into 9r (C1r), the first page of the next gathering. Thus, the planned second illustration would also have to be pushed forward and allotted its space in gathering C, say in the lower half of C1r. It presumably became clear to Galileo at this point, perhaps in consultation with the printer, that the close calculation of letterpress space required for the written text of part I as it then existed – estimated at 24 pages or three quarto sheets – would be thrown off entirely by the amount of space required for all seven illustrations. Galileo returned to the Draft, crossed out the "2" and substituted "eadem": "as the same delineation demonstrates", that is, Delineation 1. The textual change was brought into the Fair Copy, and from there to the setting copy, and on into *Sidereus Nuncius*. And indeed, the depiction of the "shadowy bay" was incorporated into Moon I. In the Fair Copy, the location of the illustration was specified even more precisely, from "the same delineation" to "the same delineation placed above" (*cuius etiam imaginem eadem* SUPRA POSITA *delineatio prę se fert*).

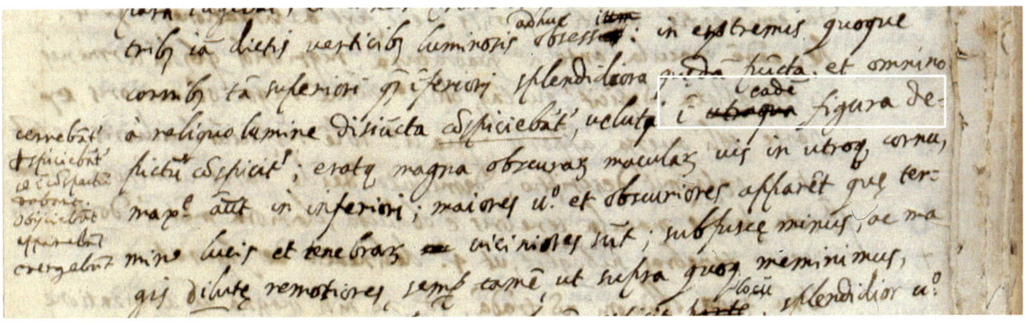

VII.7: BNCF, Ms. Gal. 48, 11r.

Eleven lines further on in the Draft (4v line 9), writing about his observations later on the same night, Galileo once more had to modify his text to reflect this decision to conflate two illustrations. He originally wrote that as the Moon was about to set (let us say, roughly at midnight), more bright points of light could be seen in the dark areas beyond the cusps of both horns, "as is depicted in both figures" (*ueluti in utraque figura depictum cernitur*). He subsequently struck through "utraque" and substituted "eadem": as is depicted in the same figure (figure VII.7).

Textually the changes were simple: two word substitutions. However, as will be explored in the next chapter, they quietly created a significant problem as to the accuracy of Galileo's observations.

As the type-pages of gathering C continued to be set and laid out, with additional spaces reserved for illustrations, further problems arose, which are reflected not in the Draft, but rather in the next writing stage, the Fair Copy. The essence of the problem was that five further illustrations were called for (Delineations 3–7) within a relatively brief stretch of text. That is, even if the space needed for the illustrations were not of itself a problem, there would still be difficulty in determining the exact locations for these not-yet-created illustrations so that they would all sit in useful proximity to the texts they were to exemplify. As printed, the sequence of text encompassing the Draft's proposed Delineations 3–7 occupies 52 full lines, that is a page and a half of solid setting. At the

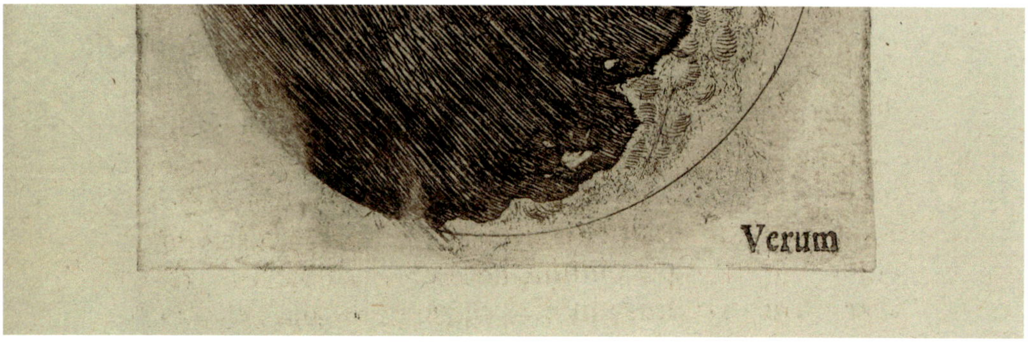

VII.8: Sidereus Nuncius, B4r, Münster ULB.

imposition stage these lines were laid out over the three pages 9v, 10r and 11r (C1v, 2r, 3r), interspersed with blank spaces on pages 9v, 10r, and 10v (this last page without letterpress) that were reserved for four illustrations. Each Moon etching required a blank space of somewhat more than a half page. As noted in chapter VI, Moon I had to be fit within a space equivalent to nineteen lines: from the foot of printed line 15 to the top of the catchword. This turned out to be quite a tight fit; a twenty-line space would have been preferable. In most if not all copies of *Sidereus Nuncius*, the impression of the plate edge of the etching on 8r lies partly on top of the earlier-printed letterpress, either above, or below on top of the catchword (Figure VII.8).

Galileo partly solved the space problem by cancelling the separate Delineations 6 and 7. These were to have illustrated his remarks on a particular perfectly round crater (*cavitas*) that he had observed, as discussed on *Sidereus Nuncius* 11r (C3r) lines 1–27 (*Vnum quoque obliuioni minimè tradam ... ab eminentiarum candore discrepet.*). The purpose of Delineations 6 and 7 was to feature the shadows that the Sun cast across the crater, both in the first quarter and with opposite direction in the last quarter. To substitute for these two detailed illustrations of the crater, Galileo decided to depict the crater directly on his two illustrations of the entire Moon at (roughly) first and last quarters. In essence, those two illustrations, Moons II and III, provided both a view of the Moon in full size, and, so to speak, an out-of-scale "pasted-on" image of the crater – an awkwardness or disparity that readers have remarked on from the seventeenth century to the present. We understand this somewhat confusing compromise better when we know its genesis. It was not part of Galileo's original plan, but was rather a space-saving makeshift, undertaken as quire C was about to be set.

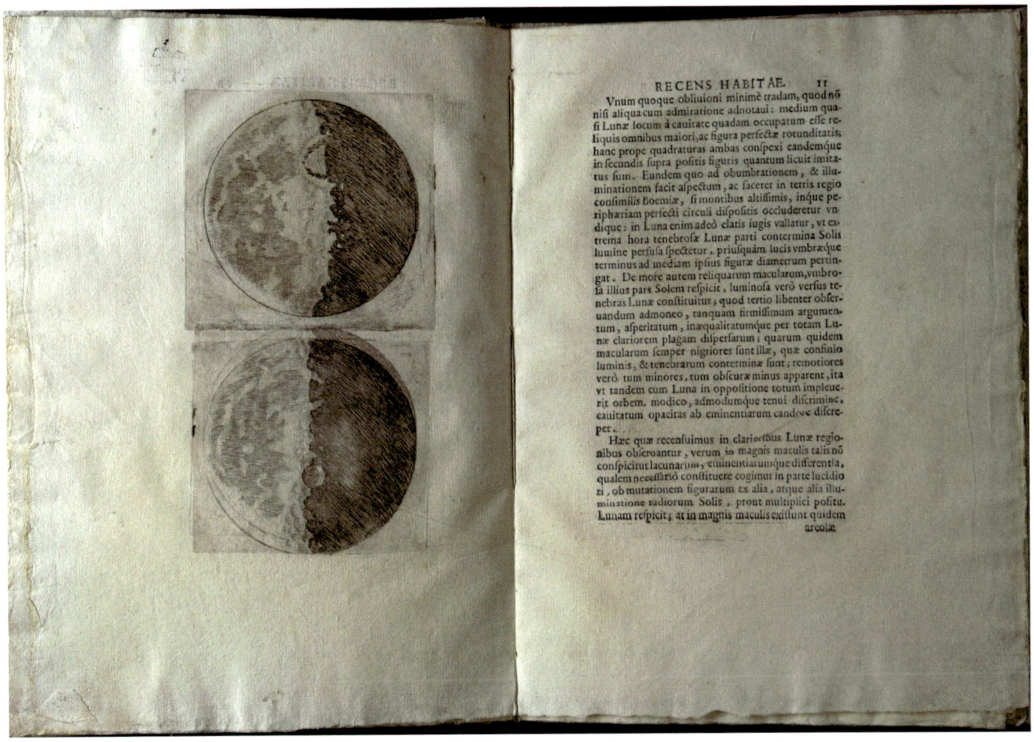
RECENS HABITAE. 11

Vnum quoque obliuioni minimè tradam, quod nō nisi aliqua cum admiratione adnotaui: medium quasi Lunæ locum à cauitate quadam occupatum esse reliquis omnibus maiori, ac figura perfectæ rotunditatis; hanc prope quadraturas ambas conspexi eandemq́ue in secundis supra positis figuris quantum licuit imitatus sum. Eundem quo ad obumbrationem, & illuminationem facit aspectum, ac faceret in terris regio consimilis Boemiæ, si montibus altissimis, inq́ue peripheriam perfecti circuli dispositis occluderetur vndique: in Luna enim adeò elatis iugis vallatur, vt extrema hora tenebrosæ Lunæ parti contermina Solis lumine perfusa spectetur, priusquàm lucis vmbræq́ue terminus ad mediam ipsius figuræ diametrum pertingat. De more autem reliquarum macularum, vmbrosa illius pars Solem respicit, luminosa verò versus tenebras Lunæ constituitur; quod tertio libenter obseruandum admoneo, tanquam firmissimum argumentum, asperitatum, inæqualitatumq́ue per totam Lunæ clariorem plagam dispersarum; quarum quidem macularum semper nigriores sunt illæ, quæ confinio luminis, & tenebrarum conterminæ sunt; remotiores verò tum minores, tum obscuræ minus apparent, ita vt tandem cum Luna in oppositione totum impleuerit orbem, modico, admodumq́ue tenui discrimine, cauitatum opacitas ab eminentiarum candore discrepet.

Hæc quæ recensuimus in clarioribus Lunæ regionibus obseruantur, verum in magnis maculis talis nō conspicitur lacunarum, eminentiarumq́ue differentia, qualem necessariò constituere cogimur in parte lucidiori, ob mutationem figurarum ex alia, atque alia illuminatione radiorum Solis, prout multiplici positu Lunam respicit; at in magnis maculis existunt quidem

areolæ

VII.9: Sidereus Nuncius, C2v.3r, Library of Congress.

In his Draft, Galileo contemplated that the two detail-images of the crater would be placed directly within his discussion: "We saw this crater at about the time of both quarters, and have represented it in delineations 6 and 7 to the best of our ability." In the printed edition, where the discussion of the crater appears on 11r, the citation of the images had to be changed to (small capitals): "I saw this crater at about the time of both quarters, and have represented it IN THE SECOND FIGURES PLACED ABOVE to the best of my ability" (*... hanc prope quadraturas ambas conspexi eandemque* IN SECUNDIS SUPRA POSITIS FIGURIS *quantum licuit imitatus sum.*). In the printed edition this reference, clear enough word by word, becomes confusing. Facing 11r is 10v, a page left blank by the printer as the location for two of the Moon etchings, Moon IV above Moon III. Yet only the lower of these etchings, Moon III, represents the crater. There seems to be a conflict between Galileo's words, referring to two images of the crater, and the facing etchings, only one of which shows the crater. Moreover, *supra positis* would better have been rendered as *appositis,* the word Galileo used for a similar reference on 9v (see above, Delineations 3, 4). (Figure VII.9)

This visual-textual dissonance highlights the essential problem of this stage of production: deciding how best, within a limited space, to bring the Moon images into relation with the text, and how, in the text, to refer to the images, which do not have printed captions. The final result contains silent compromises. In the case of the opening 10v. 11r (C2v.3r), the two Moon etchings on 10v facing the text of 11r do not relate to that text at all. It is only a coincidence that one of the two depicts the large crater. The etchings on 10v are, rather, the illustrations for Galileo's text on the preceding page, 10r, where he analyzes the shadows and light seen at various times around the large dark spot in the northern part of the Moon's face, and explains that this spot must be lower than the land around it, just as on Earth the seas are lower than the land. In particular, Galileo wanted to demonstrate that not long before the last quarter, a shadowed projection into the lighted side of the Moon indicates that a mountain ridge serves as the large spot's boundary. The side of its slopes facing the Sun at this time is brightly lit, the side turned away is in shadow. He represented this on Moon IV, then below it he placed for comparison a repetition of Moon III, which represents the face of the Moon a day later, just at the last quarter (figures VII.10, 11).

As for the Moon's conspicuous crater discussed on 11r, Galileo's reference to its depictions on the "second figures placed above" (*in secundis supra positis figuris*) refers not to the two facing etchings on 10v (Moons IV and III), but rather to the two on the preceding opening 9v (C1v) and 10r (C2r), Moons II and III. These had already done service to illustrate Galileo's discussion of the large spot in the northern hemisphere of the Moon. Readers were now instructed to turn back and examine them again with regard to a quite different feature: the large round crater in their southern hemisphere (see figures VII.12, 13). In Galileo's original plan these would have had their own illustrations immediately alongside the discussion of the crater. Why did Galileo refer to these as the "second figures"? They are second in the sense that they follow the first illustration, Moon I, printed on 8r (B4r). It is easily imaginable that although this form of reference was clear to Galileo, it would not have been self-evident to his readers, who simply saw the various etchings on various pages with nothing to label them as first, second, third, and so on.[5] But as will be seen shortly, the wording of the Fair Copy clearly indicates that this is what he meant by "second figures": the figures that were eventually created as Moons II and III.

Quire C shows the beginnings of other large changes in Galileo's plan for *Sidereus Nuncius*. Topic III, The Moon's Rough Surface, ends on 11v (C3v) at line 25. By the original indications of the Draft, the next topic should have been VI, Earthshine: the argument that the long-observed "ashen light" visible on the Moon at the beginning and end of the lunar cycle is reflected from the Earth, and has no other possible source. However, line 26 of 11v instead begins Topic IV, The Silhouette Problem, opening with the words *Verum magna hic dubitatione complures affici sentio …*: "Here, indeed, I realize, many readers may be struck by a grave difficulty, to such an extent that they may be forced to doubt the conclusions reached so far …". The difficulty was: if the surface of the Moon is as rough as Galileo has just argued, with mountains higher even than those on Earth, why does the lighted perimeter or silhouette of the Moon look entirely smooth, and not jagged? As briefly stated above, originally Galileo had not intended to deal with this problem, and we may suggest that until he had already written out the greater part of the Draft, it had not even occurred to him.

Thus, the text on C3v reflects a change of plan that arose after most of the first part of the Draft had been written, and possibly after sheet B of *Sidereus Nuncius* had already

5 Van Helden's translation (Van Helden 1989, p. 47) renders *in secundis supra positis figuris* in the singular, "in the second figure above", apparently to make the words "fit" to the second of the facing illustrations, p. 46, which includes the crater. But Galileo's entire point is that he was supplying two depictions, corresponding to the crater as seen "in *both* quarters". Drake 1983, p. 30, met the difficulty by omitting "second" entirely: "in the diagrams above." The translation of Pantin 1992, p. 14, is accurate: "dans la séconde série de figures ci-dessus."

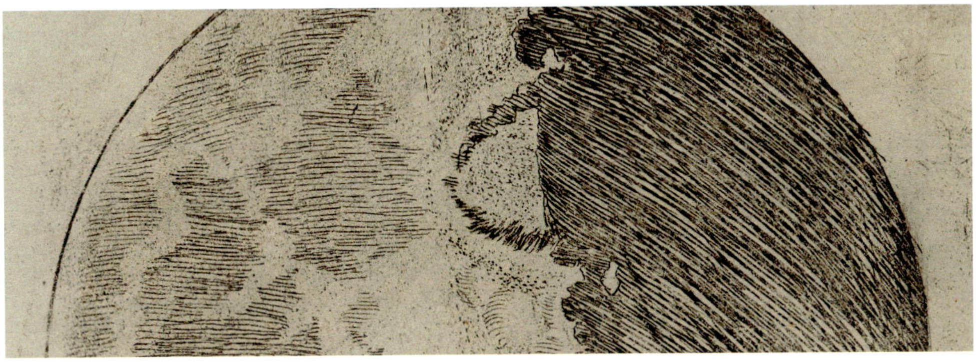

VII.10: Sidereus Nuncius, C2v, Institute for Advanced Study.

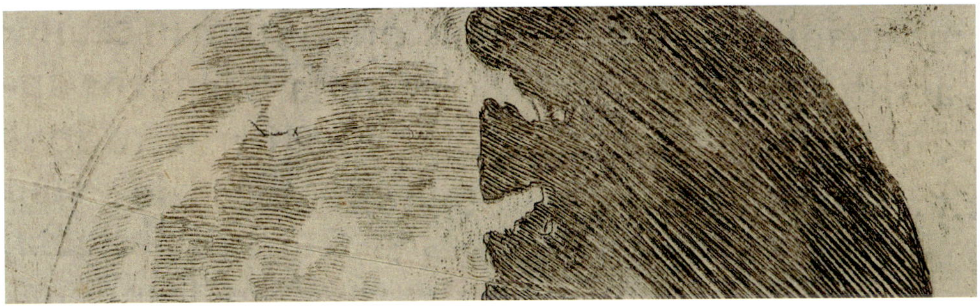

VII.11: Sidereus Nuncius, C2v, Institute for Advanced Study.

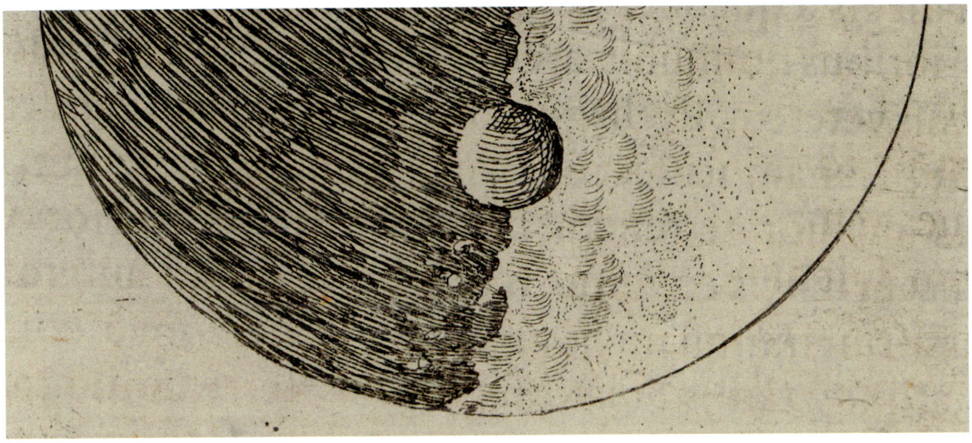

VII.12: Sidereus Nuncius, C1v, Institute for Advanced Study.

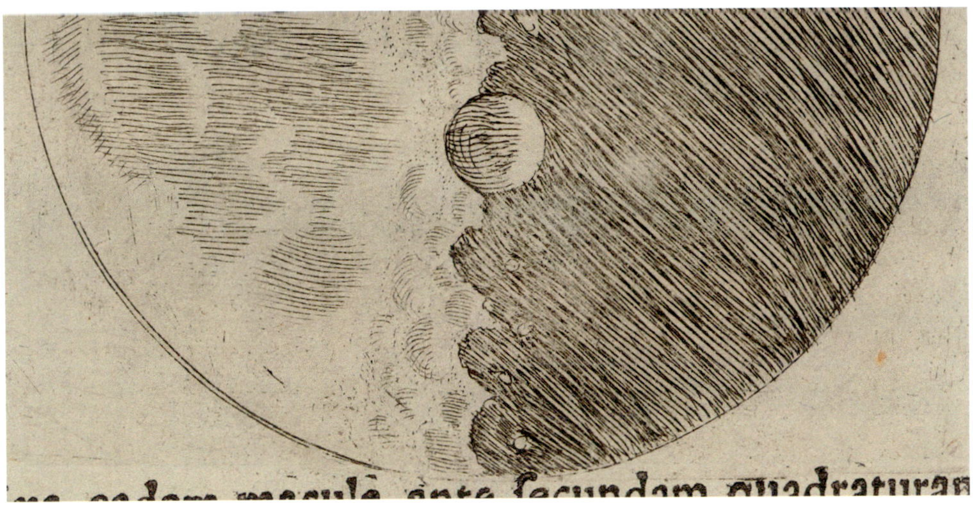

VII.13: Sidereus Nuncius, C2r, Institute for Advanced Study.

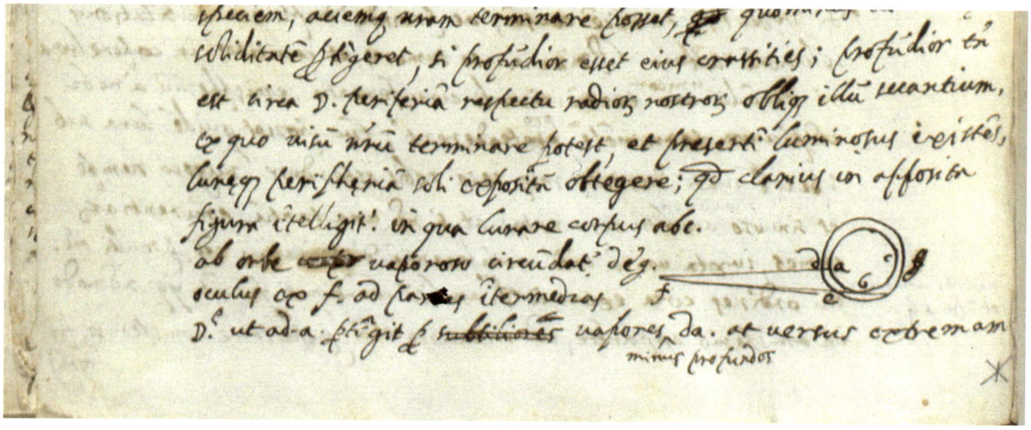

VII.14: BNCF, Ms. Gal. 48, 14v.

been printed off. Having now to deal with the afterthought of the Silhouette Problem, Galileo postponed his already-written topic of Earthshine, and brought forward the new problem, which he wrote out on a separate sheet of the Draft (fos. 7.8), using (Venetian) Rampant Lion paper. The placement was rhetorically prophylactic: it immediately followed The Moon's Rough Surface to disarm the hypothesized reader who might conceivably think of it himself as he was reading, and might wonder whether Galileo had an adequate answer, or had overlooked or even, in bad faith, evaded the difficulty?[6]

But in terms of the chronology of the Draft, even this is not quite right. When we cross-compare the Draft, the Fair Copy, and quires C and D of *Sidereus Nuncius*, we discern the traces of four distinct arrangements that Galileo considered, one after the other, for bringing together the assortment of texts or topics that make up part I. These arrangements are:

Plan A. Topics I-II-III-VI (Introduction – The Telescope – The Moon's Rough Surface – Earthshine), then VII (Fixed Stars)

Plan B. Topics I-II-III-VI-V (As A, but adding in conclusion Topic V, The Height of the Moon's Mountains), then VII (Fixed Stars).

Plan C: Topics I-II-III-IV-VI-V (Introduction – The Telescope – The Moon's Rough Surface – The Silhouette Problem – Earthshine – The Height of the Moon's Mountains), then VII (Fixed Stars)

Plan D (as printed): Topics I-II-III-IV-V-VI (As C, but reversing the last two topics to: The Height of the Moon's Mountains – Earthshine), then VII (Fixed Stars).

This final arrangement of topics added clarity, for it was natural to move directly from the Silhouette Problem in general to the specific question of calculating the height of the Moon's mountains. These changes in plan are reviewed more closely in chapter IX, in the close analysis of quire D.

6 I interpret this somewhat differently from Wootton 2009, p. 136, who writes, "Evidently Galileo had shown his manuscript to some friends or colleagues; they had raised an objection which had not occurred to him; and he thought it absolutely essential to reply … even though printing was already underway." We have no evidence that Galileo consulted others as he wrote.

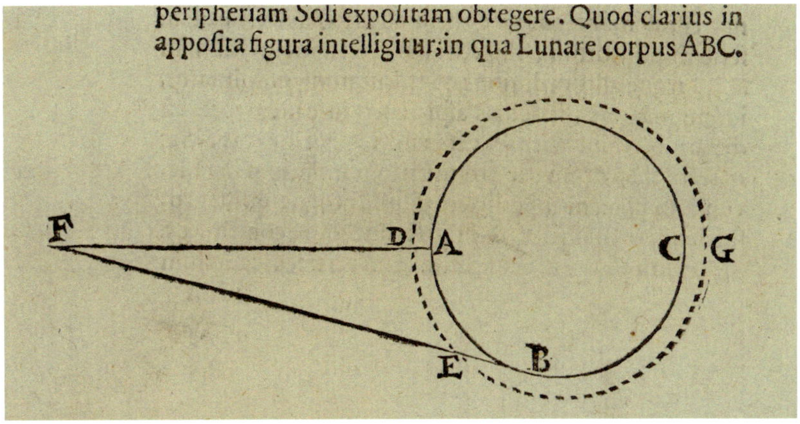
peripheriam Soli expolitam obtegere. Quod clarius in apposita figura intelligitur; in qua Lunare corpus ABC.

VII.15: *Sidereus Nuncius, C4v, Institute for Advanced Study.*

Galileo's discussion of the Silhouette Problem continued from the end of quire C onto the first page of quire D (13r, lines 1–14). As printed in the *Sidereus Nuncius,* the entire discussion occupies 84 lines of typesetting, plus a woodcut diagram equivalent in height to about 12 lines of type, which had to be made according to Galileo's sketch. This diagram exemplified his second argument on the question, which he abandoned in later years: that a vapor or atmosphere surrounds the Moon as it does the Earth, and partly obscures any irregularities in the Moon's outline (figures VII.14, 15). Including the woodcut, the Silhouette Problem takes up almost three pages of *Sidereus Nuncius,* and of itself is a major part of the reason why it would have become evident, as quire C was being composed, that the

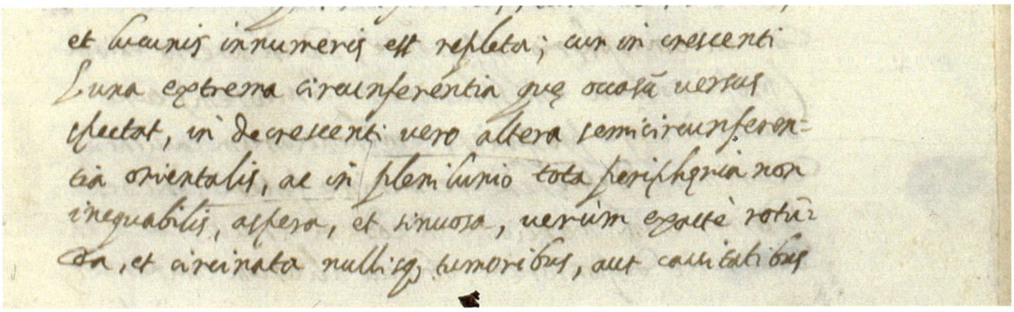
et lacunis innumeris est repleta; cur in crescenti
Luna extrema circunferentia que occasu versus
spectat, in decrescenti vero altera semicircunferen-
tia orientalis, ac in plenilunio tota periphaeria non
inequabilis, aspera, et sinuosa, verum exacte rotun-
da, et circinata nullisque tumoribus, aut cavitatibus

VII.16: *BNCF, Ms. Gal. 48, 45r.*

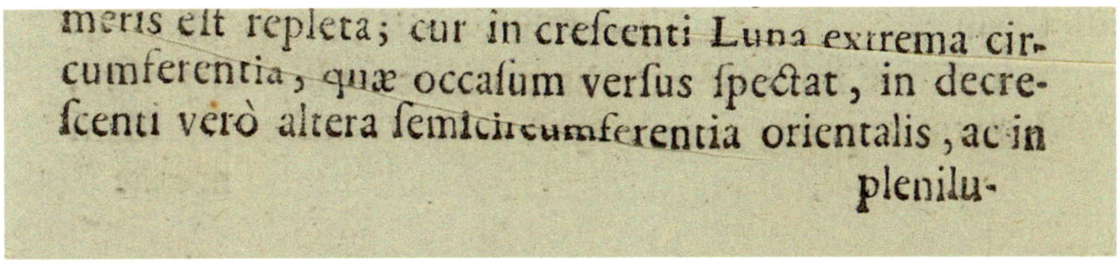
meris est repleta; cur in crescenti Luna extrema circumferentia, quæ occasum versus spectat, in decrescenti verò altera semicircumferentia orientalis, ac in plenilu-

VII.17: *Sidereus Nuncius, C3v, Institute for Advanced Study.*

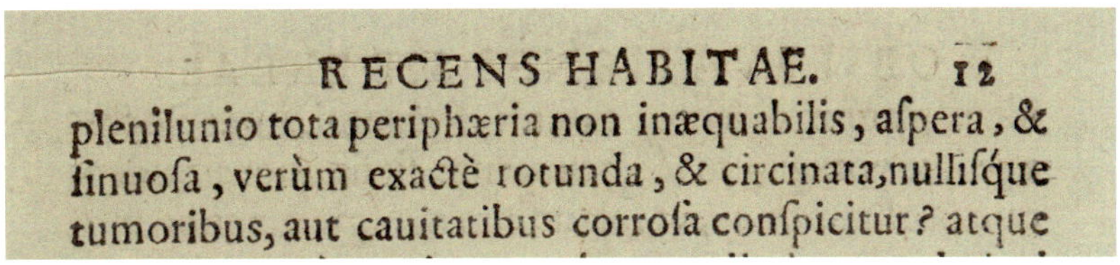
RECENS HABITAE. 12
plenilunio tota periphæria non inæquabilis, aspera, & sinuosa, verùm exactè rotunda, & circinata, nullisque tumoribus, aut cauitatibus corrosa conspicitur? atque

VII.18: *Sidereus Nuncius, C4r, Institute for Advanced Study.*

original plan of printing the first part of the text on three sheets of twenty-four pages could no longer be maintained.

As concerns the make-up of the formes of quire C, it should be noted that the outer forme (C1r-2v-3r-4v) includes one entirely blank page, C2v. This page lacks even the expected running title "OBSERVAT. SIDEREAE". As for the inner forme, its pages were imposed with two substantial blank spaces for etchings, at the foot of C1v and top of C2r. On C1v the space was ample, with twenty-one full lines between the last line of text and the catchword. C2r was not well laid-out in this respect, with only seventeen lines left for the etching between the foot of the headline and the first line of text, and this caused a real difficulty that had to be handled when it came time to print the etchings: see chapter XIV.

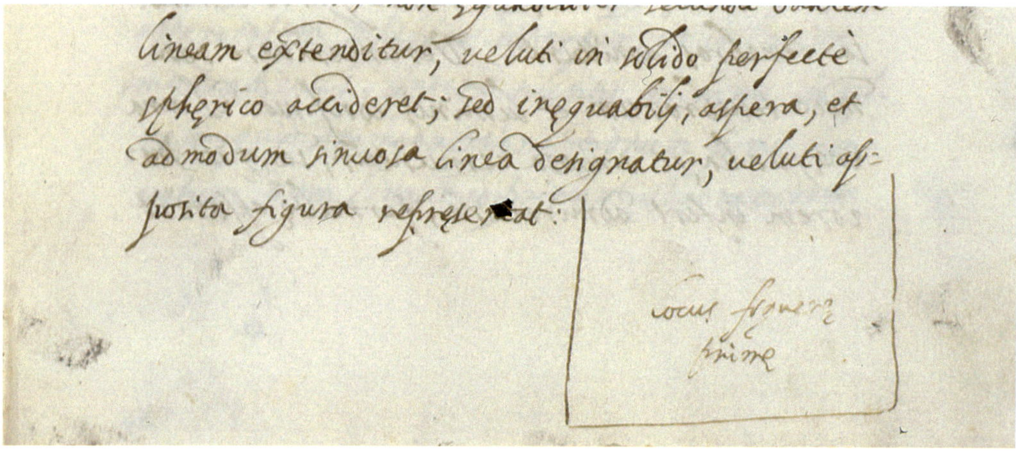

VII.19: BNCF, Ms. Gal. 48, 41r.

There are several indications that Galileo put his Fair Copy to practical use as quire C was being printed, partly to give him a clean reading of the topics in their revised order, and partly to indicate how the eventual Moon illustrations would be positioned after their number had been reduced from seven to four. Related to this is the possibility that the regular writing and lineation of the Fair Copy was a help to Galileo in calculating the space requirements of the book that was coming into print: a page of his writing of the Fair Copy corresponds with only slight irregularities to twenty-four lines of typesetting, and so the Fair Copy could be used as a gauge. On Fair Copy 8r, Galileo added a mark of separation to demarcate the end of C3v (11v) of *Sidereus Nuncius* from the beginning of C4r (12r). He made this mark with four successive deeply scratched strokes of an almost empty quill (figures VII.16, 17, 18). This can be interpreted to indicate that Galileo had the Fair Copy directly in hand at the time he was examining proofs of *Sidereus Nuncius*, an interpretation strengthened by the fact that at several places on the Fair Copy, most notably on 8v, there are inked fingerprints (figure VI.2).

The Fair Copy also enlightens the question raised earlier: how or why did Galileo, when writing about the Moon's most conspicuous crater, come to refer to its depictions as occurring "in the second figures placed above" (*in secundis supra positis figuris*)? This is specifically the nomenclature of the Fair Copy, and originally he must have written it in-

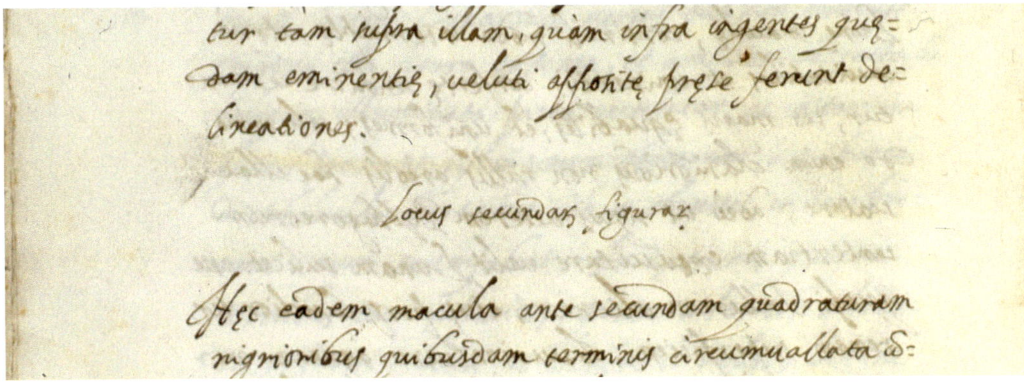
tur tam supra illam, quam infra ingentes quę-
dam eminentię, veluti apposite pręse ferunt de-
lineationes.

Locus secundarũ figurarũ.

Hęc eadem macula ante secundam quadraturam
nigrioribus quibusdam terminis circumvallata ob-

VII.20: BNCF, Ms. Gal. 48, 43v.

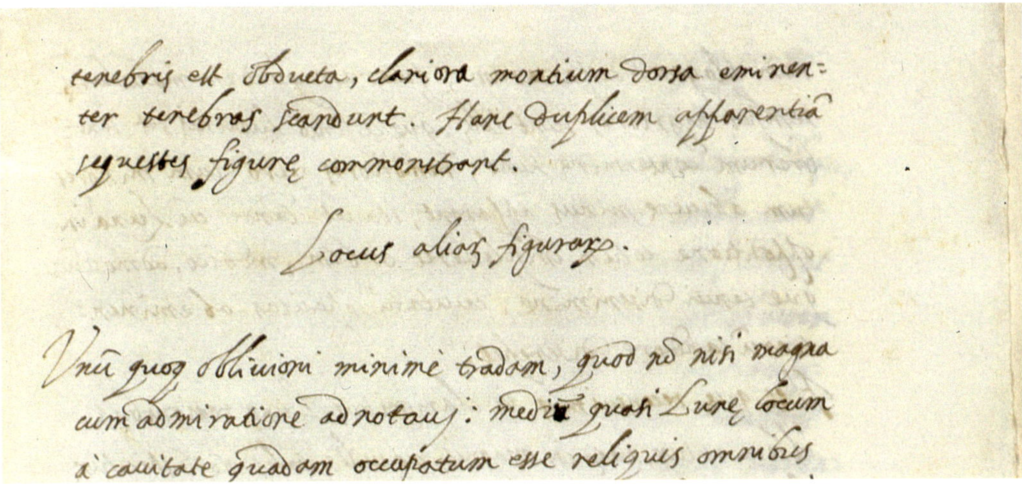
tenebris est obducta, clariora montium dorsa eminen-
ter tenebras scandunt. Hanc duplicem apparentiã
sequentes figurę commonstrant.

Locus aliarũ figurarũ.

Unũ quoque oblivioni minime tradam, quod nõ nisi magna
cum admiratione adnotaui: medium quasi Lunę locum
a cavitate quadam occupatum esse reliquis omnibus

VII.21: BNCF, Ms. Gal. 48, 44r.

formally, so to speak, for his own understanding. The Fair Copy contains no illustrations, but it leaves various blank spaces to indicate roughly where the figures should go with relation to the text. There is a three-line blank space at the place where the first diagram, of the telescope, was to be positioned (Fair Copy, 3r). At the place in the text where the first depiction of the Moon was to appear, Galileo drew a small box at the foot of the page and captioned it *locus figurę primę* (figure VII.19). At the place in the text where the next two Moon depictions belonged (Fair Copy, 6v), Galileo left another three-line space captioned *locus secundarum figurarum,* and at the place where the final two Moon depictions belonged (Fair Copy, 7r), Galileo captioned another three-line space as *locus aliarum figurarum* (figures VII.20, 21). Finally, at the place on 7r where Galileo referred to the crater, he used the same words as appear in the printed *Sidereus Nuncius* (small capitals): *hanc [cavitatem] prope quadraturas utrasque conspexi,* EANDEMQUE IN SECUNDIS SUPRA POSITIS FIGURIS *quantum licuit imitatus sum.* Thus, on these pages the Fair Copy clearly preserves a later stage of composition than the Draft, belonging to the time when Galileo had decided not to try to fit Delineations 6 and 7 into *Sidereus Nuncius* as separate detail-images, but rather to incorporate them into two of his images of the full face of the Moon – his *secundae figurae.*

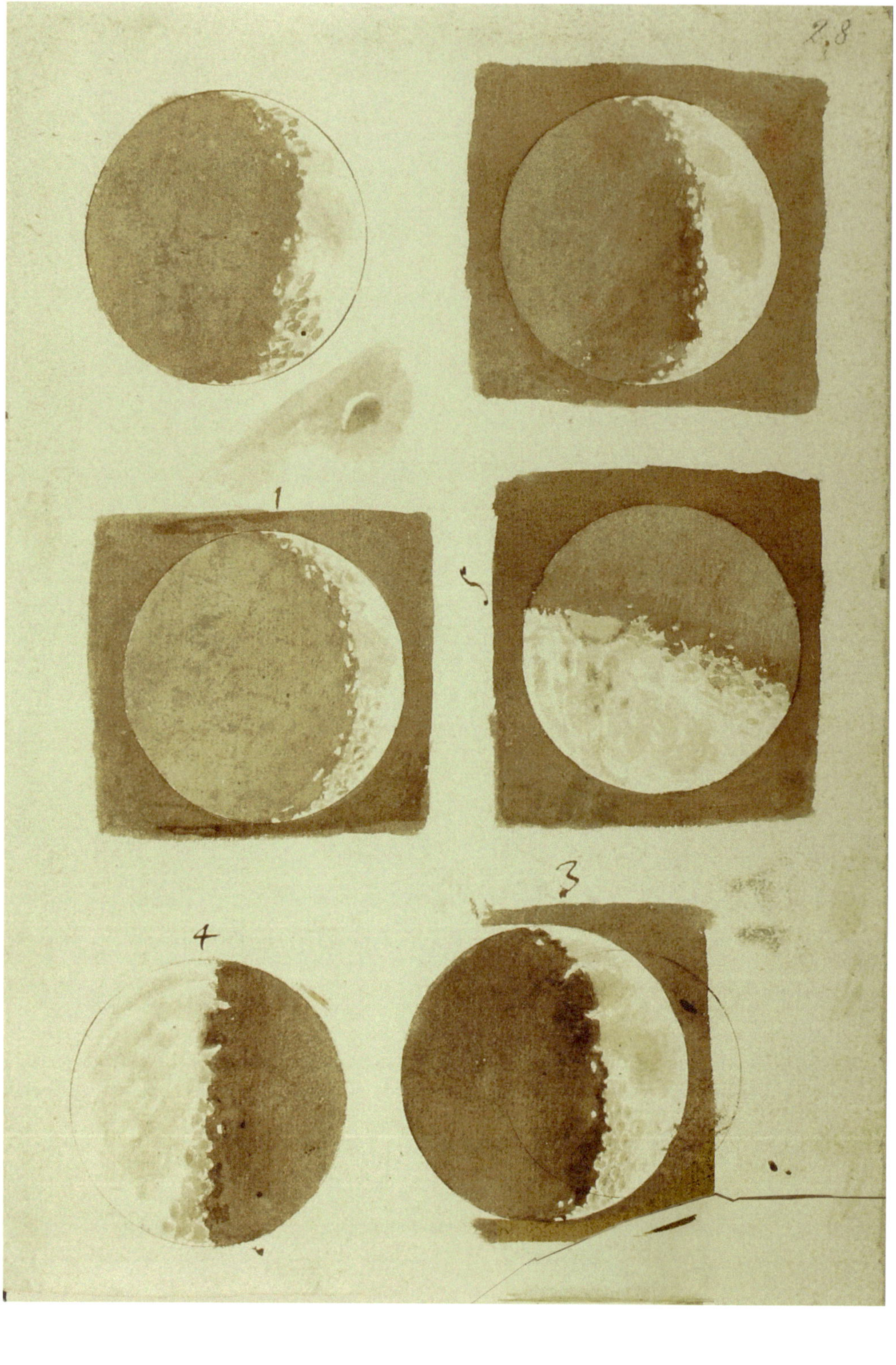

VIII.1: BNCF, Ms. 48, 28r (Seven-Moon bifolium, 1r).

VIII.
The Moon Etchings

The preceding review of Galileo's plans for the lunar illustrations as manifested at four stages – the 7 January 1610 Letter, the Draft, the Fair Copy, and the printed edition – may appear to be fixated on small matters, a mere "fine tuning" of Galileo's verbal references to the illustrations. But those references attach to a larger question about the Moon etchings (until recently called engravings). Among the Galileo papers in Florence is a leaf bearing on its outer side seven highly accomplished ink-and-wash drawings of the Moon in different aspects and also a draft of a horoscope for Cosimo II de Medici, whose birthdate was 2 May 1590. Two of the Moon sketches are unnumbered, the others are numbered, perhaps subsequently to their making, as 1, 3, 4, 5, and 8. The inner side of the leaf is blank except for an earlier sketch of the same horoscope. Galileo used this leaf – a half-sheet of Royal paper – as a bifolium, that is, he folded it to make two leaves.[1] For convenience, it may be referred to as the Seven-Moon Bifolium. Six of the Moon drawings are on 1r of the bifolium, arranged in two columns of three rows each, including the Moons numbered as 1, 3, 4, and 5. (figure VIII.1) Fo. 1v is blank; 2r contains a horoscope diagram in its lower right corner; 2v contains in the upper left corner a single Moon drawing numbered 8: this last is a waning Moon with a fixed star indicated just beyond its eastern rim. In the lower right corner of this page is another horoscope diagram. This bifolium has been steadily associated with *Sidereus Nuncius* since roughly 1820–1830, when Antinori had it bound into the volume he titled "Parte 3ª, Tomo 3, Astronomia", what is now Ms. Gal. 48. Antonio Favaro accepted its connection with *Sidereus Nuncius* and reproduced it, omitting the horoscopes and without indicating the bifolium's pages, in *Opere* III as pl. 48. Until recently, its status as one of Galileo's working papers for *Sidereus Nuncius* has not been doubted. For that matter, the question of whether the drawings on the Seven-Moon Bifolium were indeed from Galileo's hand was never raised. Is it necessary to point out that a binding decision taken two centuries after Galileo's death tells us nothing at all about the Seven-Moon Bifolium's connection or lack of connection with *Sidereus Nuncius*?

Various studies by Righini, Drake, Whitaker, Gingerich and Van Helden have discussed, disputed and refined the presumed dates on which each of the lunar images on the

1 Gingerich 1975 p. 87 identifies the leaf accurately: "a stiff folded piece of drawing paper"; although its stiffness is not because it is "drawing paper", but because it is half of a Royal sheet, the thickness of which is intrinsically greater than that of smaller paper sizes. By some oversight, Gingerich & Van Helden 2003 p. 256 identify the leaf inaccurately as "two single sheets of watercolour paper."

Seven-Moon Bifolium was made, or based.[2] Whatever the differences in opinion in these studies, they share the opinions first that the images of the Seven-Moon Bifolium represent the Moon just as Galileo viewed it through his telescope on specific evenings, that is, that these drawings are implicitly dated records; and second that the drawings are the sources for the etchings in *Sidereus Nuncius*, Moons I–IV. As one of the accounts puts it, "these images suffice to account for the complete basis of the lunar portion of the *Sidereus Nuncius*".[3] Yet there are significant and even striking differences between the *Sidereus Nuncius* etchings and the Seven-Moon Bifolium's drawings which have not gone unnoticed: for instance, that none depicts the "perfectly round crater" as it appears on Moons II and III in *Sidereus Nuncius*.[4]

Recently, Horst Bredekamp has challenged this common ground of belief by bringing into the argument the evidence of the "ML" copy of *Sidereus Nuncius*: a copy signed by Galileo and presented by him to Prince Federico Cesi of Rome, founder of the Accademia dei Lincei. This copy does not contain impressions of the etchings. Instead, in the blank spaces where the etchings should be, there are detailed wash drawings of the Moon, very close in size and detail to the etchings. By Bredekamp's analysis, these drawings are the direct models for the *Sidereus Nuncius* etchings. They retain on their surfaces traces of a material that he believes was used for a partial direct transfer process from drawings to copperplates. According to this analysis, the drawings of the Seven-Moon Bifolium are authentically by Galileo's hand, a question that Bredekamp was the the first scholar to pursue in detail, the earlier writers having assumed it as self-evident. However, the drawings – in Bredekamp's expression the "Florentine Drawings" – are seen as unrelated to the making of *Sidereus Nuncius*, having been made at some time after its publication, perhaps in late 1610; and more precisely, having been made after Galileo moved from Padua to Florence.[5]

Bredekamp's conclusion has been vigorously opposed by Gingerich, who with what he characterizes as "Holmesian persistence" argues that the drawings in the ML copy are the work of a modern forger: "*The M-L image* [that is, the ML drawing related to the Moon IV etching] *is a pastiche derived from elements both in the final etching and from the Florentine sheet.* In his attempt to link the drawings more firmly to Galileo the forger is caught being too clever by half."[6] The idea that the drawings are forgeries has considerable contextual difficulties, for this copy belonged to Galileo, as his title-page inscription makes

2 Righini 1975; Gingerich 1975; Drake 1976; Whitaker 1978; Gingerich & Van Helden 2003; Gingerich 2009. For the identification of the horoscopes as Cosimo II's, see Righini 1976.

3 Gingerich & Van Helden 2003, p. 256.

4 See Gingerich & Van Helden 2003, p. 258: "how can we understand the differences between the seven inkwash drawings and the four engravings ...[?]". To similar effect is Biagioli 2006. On p. 145, he accepts the common opinion without question: "As Galileo's original wash drawings were turned into engravings ...". But a few sentences later, "... the difference between the wash drawings and the engravings is so noticeable that, were Galileo a modern astronomer, he would probably have been charged with scientific misconduct."

5 Bredekamp 2007, p. 146, "Auf der Suche nach Vorlagen für die Stiche des *Sidereus Nuncius* sind die Florentiner Zeichnungen mangels Alternativen durchweg konkurrenzlos ins Spiel gebracht worden. Bei näherer Betrachtung aber kommt nicht eine einzige der Tuschzeichnungen als unmittelbare Vorlage in Frage." And, p. 214, "Mond 4 der Florentiner Zeichnungen ... gehört damit aller Wahrscheinlichkeit nach zu jener erstmals im November 1610 dokumentierten Erkenntnisphase ...".

6 Gingerich 2009, p. 163, the italics his; "Holmesian persistence": p. 141.

clear; and, even more significantly, it is a proof copy, as will be shown in detail in chapter XIII. The hypothesis of a modern forger as the author of the ML drawings entails an assumption that might cause uneasiness: that the forger gained access to the hitherto unrecorded sole proof copy known, one that has been undisturbed in a Rome binding dating from the mid-seventeenth century, and that has a provenance going directly back to Galileo himself. Alternatively, one could, so to speak, climb higher up the ladder of skepticism and argue that the Galileo signature is also forged, the copy itself is forged, the binding is forged, and so on.

We may try to test Bredekamp's opinion, and so by contrast Gingerich's counter-opinion, from a new direction, what may be called text genesis. A question may be put in the following terms: do the Galileian materials that indisputably preceded the printing of *Sidereus Nuncius*, that is the letter of 7 January 1610, the autograph Draft, and the Fair Copy, contain any information that either strengthens Bredekamp's conclusion, or creates problems for it? The letter of 7 January 1610 to an unknown correspondent, mentioned at several points above, becomes especially significant. The now-lost Vatican version of that letter contained pen and wash sketches of the Moon, reproduced in the Edizione Nazionale, that must have been copied from Galileo's own sketches, and these latter must of course have been made in early January. It has long been realized that a major portion of the 7 January letter became, paraphrased and translated into Latin and slightly rearranged, incorporated into Topic III of *Sidereus Nuncius*: essentially, the letter is the skeleton of Topic III. This raises a corollary question: if the words of the 7 January letter influenced the words of *Sidereus Nuncius*, did its Moon sketches similarly influence the Moon illustrations of *Sidereus Nuncius*? The question is to be explored by comparing the 7 January letter not directly with *Sidereus Nuncius* as printed, but rather with Galileo's Draft, which immediately followed the letter.

It is especially revealing to relate the text of the 7 January 1610 letter with the three figures, Delineations 2, 6, and 7, that Galileo originally planned in the Draft as distinct illustrations in *Sidereus Nuncius* but that were instead, as printing proceeded, incorporated into three of the four etchings of that edition. Delineation 2, it will be recalled, was planned to illustrate the "massive shadowy bay" (*ingens sinus tenebrosus*) that intruded into the southern horn of the lighted surface as the first quarter approached, and within which, after a time, a lighted triangular figure (*trigona figura*) arose. In his autograph Draft, fo. 3v, Galileo described this (figure VII.5):

> ... *interim silentio minime inuoluam notabile quoddam à me obseruatum dum Luna ad primam quadraturam properaret, cuius etiam imaginem* ~~2.~~ ^EADEM^ DELINEATIO DEMONSTRAT; *ingens enim sinus tenebrosus in partem luminosam subit positus uersus inferius cornu; quem quidem sinum cum diutius obseruassem, totumque obscurum uidissem, tandem post 2as. ferè horas paulo infra medium cauitatis exurgere cępit uertex quidam luminosus, hic uero paulatim crescens trigonam figuram prę se ferebat* ....

With amplifications of detail, this corresponds directly to paragraph 6 (as printed) of the 7 January letter:

Quando la luna è intorno alla quadratura, si scorge nella parte inferiore, ciò è nella australe, un immenso seno, il quale incava la parte lucida NELLA MANIERA APPRESSO: *nella qual cavità, crescendo la parte lucida, comincia poi a sporgere, in guisa di un promontorio, un'eminenza triangulare …*

pre la parte tenebrosa è verso il sole, et la chiara all'opposto; inditio certo, quella essere una grandissima cavità perfettamente rotonda et da termini eminenti circondata.

Quando la luna è intorno alla quadratura, si scorge nella parte inferiore, ciò è nella australe, un immenso seno, il quale incava la parte lucida nella maniera apresso: nella qual cavità,

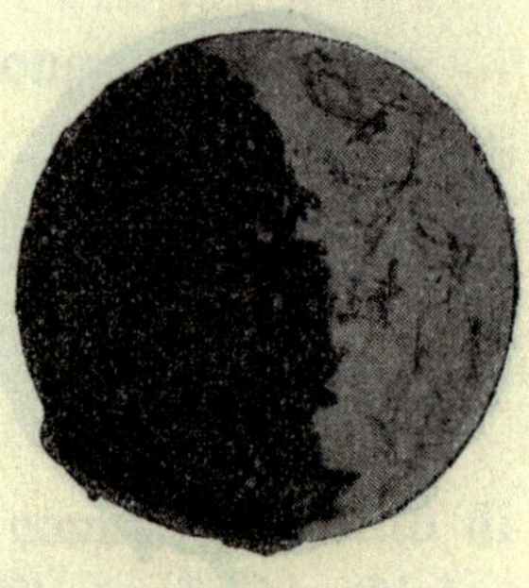

VIII.2: Opere X, p. 275.

Galileo's "nella maniera appresso" refers to his accompanying sketch, which indeed, in the Vatican copy of the letter, shows the Moon approaching but not yet at the first quarter, and with a dark irregular shape, the "immmenso seno", obtruding into the lighted southern horn. Thus, the "secunda delineatio" called for in the earliest state of Galileo's draft would have been an illustration essentially identical to the sketch of the "shadowy bay" included in the 7 January letter. No such illustration is present in the Seven-Moon Bifolium.[7]

As discussed in chapter VII, this separate illustration was cancelled because of the limited space available for the text and pictures that make up the first part of *Sidereus Nuncius*. Instead, Moon I, for which space had been provided on 8r (B4r) was made to do double service. Overall, it featured the jagged, irregular line of the terminator: that was its primary purpose. Then, at the southern horn, was added a second feature, the intrusion of the "shadowy bay" into the horn and, within that dark bay, the sunlit triangular promontory that arose within it after a time; as also the three smaller points of light that emerged somewhat afterward as the Sun continued to rise across the Moon (… *mox circa illum tres aliae cuspides exiguae lucere caeperunt* …). Moon I depicts all of these with seeming precision (see figure VII.8). But there is a problem that this illustration glides past quite unsatisfactorily. Moon I represents, according to Galileo's own words, the fourth or fifth day of the lunar cycle. However, the shadowy bay did not meet Galileo's eyes until the seventh day, just before the first quarter (*dum Luna ad primam quadraturam properaret*). Therefore, the true position of the bay, as observed by Galileo, would have lain significantly west of the position represented on the Moon I etching, for the terminator had moved westward across the Moon's face for two more days. The copy of Galileo's sketch in his 7 January letter (figure VIII.2) clearly shows this later position of the terminator. And so,

7 The sketch in the Vatican copy does not show the lighted triangular "promontory" to which Galileo refers, but this is probably owing to the poor draughting abilities of the copyist.

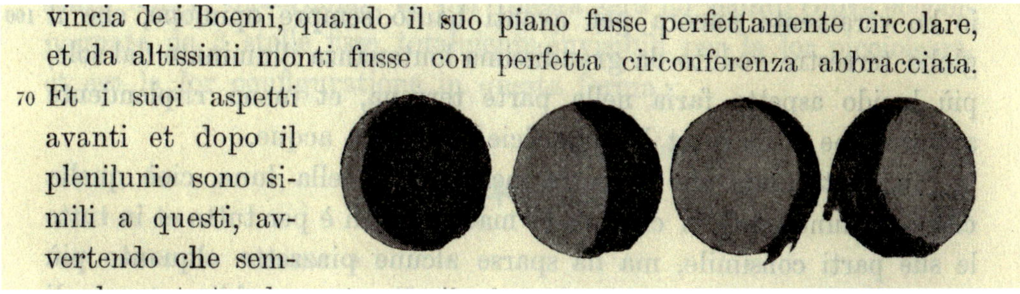
vincia de i Boemi, quando il suo piano fusse perfettamente circolare, et da altissimi monti fusse con perfetta circonferenza abbracciata. Et i suoi aspetti avanti et dopo il plenilunio sono simili a questi, avvertendo che sem-

VIII.3: *Opere* X, p. 275.

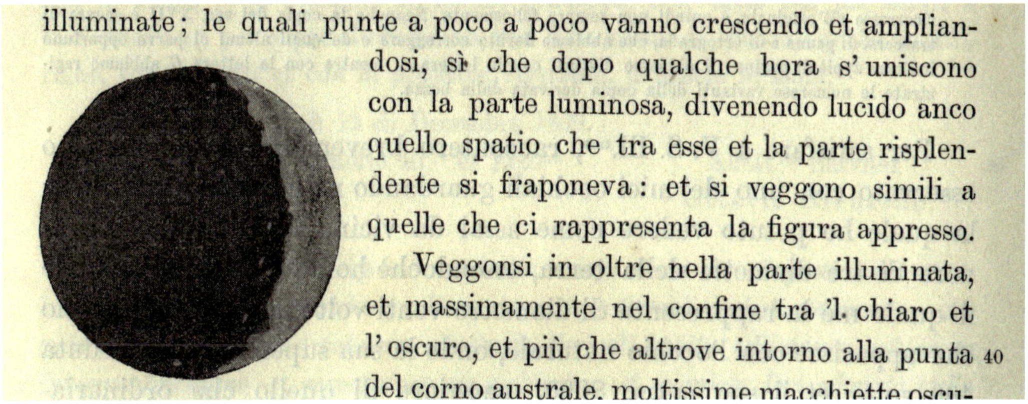
illuminate; le quali punte a poco a poco vanno crescendo et ampliandosi, sì che dopo qualche hora s' uniscono con la parte luminosa, divenendo lucido anco quello spatio che tra esse et la parte risplendente si fraponeva: et si veggono simili a quelle che ci rappresenta la figura appresso.

Veggonsi in oltre nella parte illuminata, et massimamente nel confine tra 'l chiaro et l' oscuro, et più che altrove intorno alla punta del corno australe, moltissime macchiette oscu-

VIII.4: *Opere* X, p. 274.

not only are two different features worked into the same illustration – but also two different times and two different places. Readers who had never seen the Moon through a telescope might not have noticed the direct contradiction between words and image, but Galileo could hardly have been unconscious of how significantly the depictions of Moon I deviated from his observations.

The two other cancelled illustrations, Delineations 6 and 7, designed to show the large, perfectly round crater he had observed, have already been discussed. They too make no appearance on the Seven-Moon Bifolium. As ersatz for the detailed pictures Galileo originally planned, he "pasted" the crater images, enlarged in scale, onto his eventual etchings Moons II and III. The inaccuracy of the result has, unlike that of the "shadowy bay" on Moon I, been noticed by a multitude of readers. Most strikingly, the crater is "pasted" on in two different places in Moons II and III. In Moon II, the center of the crater is located as if at about 40° South of the Moon's notional equator, while in Moon III as if at about 25° South (see figures VII.12, 13).

There remains a point to explore: the relation between the crater discussion on fo. 4v of Galileo's Draft (figure VII.6), and that in the 7 January 1610 letter. The respective passages read:

> *Unum quoque obliuioni non tradam, quod non nisi magna cum admiratione adnotaui; mediam quasi Lunę partem occupatam esse a cauitate quadam reliquis omnibus maiori, ac figura perfectę rotunditatis; hanc prope utramque quadraturam conspeximus, eandemque* IN 6. ET IN 7. FIGURA QUANTUM LICUIT IMITATI SUMUS; *eundem quo*

ad obumbrationem et illuminationem facit aspectum, ac faceret in terris uastissimum amphyteatrum rotundum, vel potius Boemię regio consimilis si montibus altissimus in perfecti circuli periphercam dispositis occluderetur undique ...

And:

Apparendo le sopradette macchiette di diverse figure et molto irregolari, una ve ne ho io, non senza qualche meraviglia, osservata, che è posta quasi nel mezo della luna, la quale apparisce perfettissimamente circolare, et è tra le altre assai grande: nella quale, et quando il sole comincia ad illustrare la sua altezza, lasciando lo spatio di mezo tenebroso, et quando poi, alzandosegli maggiormente, comincia ad illuminare il fondo, et successivamente mutandosi gl'aspetti di esso sole con la luna nel crescere et nel calare di quella, si veggono le medesime apparenze a capello di lume et di ombre, che fa in terra un grandissimo anfiteatro rotondo, o per meglio dire che faria la provincia de i Boemi, quando il suo piano fusse perfettamente circolare, et da altissimi monti fusse con perfetta circonferenza abbracciatta. Et i suoi aspetti avanti et doppo il plenilunio sono SIMILI A QUESTI ...[8]

The words "simili a questi" refer to four small images of the crater, with varying shadow and light, that accompany the text, and that most commentators have misconstrued as full images of the Moon (figure VIII.3):

The reduction of the four crater figures in the 7 January letter to the two called for in Galileo's Draft reflects a conscious reworking of the text in its passage from Italian to Latin. In the 7 January letter, Galileo described the changing light and shadow across this crater as he had observed it over a series of nights bracketing the full Moon (*avanti e doppo il plenilunio*). As Bredekamp vividly puts it, the four images of the 7 January letter present a kind of time-delay or filmstrip sequence.[9] In the Draft, Galileo intentionally reduced this to two (originally planned) images, and changed the text accordingly. The movement of light and shadow across the crater became less dynamic, being "frozen" to

8 Pantin 1992, p. 66 n. 49 has pointed out that Galileo's vivid assimilation of the large crater to Bohemia comes from his Padua colleague Giovanni Antonio Magini's extensively expanded edition of Ptolemy's Geography, with map engravings by Girolamo Porro: *Geographiae universae tum veteris tum novae ... commentariis uberrimis illustratus à Io. Antonio Magino Patavino* (Venice: Heredes Simonis Galignani, 1596, 4°). In discussing Bohemia, Magini wrote (pt. II, 80r) "Hercina autem Sylua [Bohemiam] uniuersam cingit, clauditque circum undique ad amphitheatri faciem: rotundam habet formam, cuius diameter trium dierum itinere expedicto [sic] absoluitur." In the 7 January letter and in the Draft, Galileo used Magini's simile of Bohemia as being like an amphitheatre; in the Fair Copy, he removed the amphitheatre reference. Favaro 1886, no. 194, records that Galileo owned a copy of Magini's Ptolemy, but he mistakenly gave the date as 1595, not 1596. Magini's description is in turn derived from the fifteenth-century *Historia Bohemica* of Aeneas Sylvius Piccolomini (Pius II), who wrote: "Regionis longitudo latitudoque pene parem formam rotundam ferunt, cuius diametrum trium dierum itinere expedito pateat. Sylua uniuersam claudit, quam ueteres Herciniam uocauere ..." (*Aeneae Sylvii ... opera quae extant omnia* [Basel: Officina Henricpetrina, 1551, F°], p. 82).

9 Bredekamp 2007, p. 113: "Die sequentielle Wiedergabe und insbesondere der 'Film' der vier Phasen des Riesenkraters bestechen in der Heraushebung und Motorisierung eines markanten Details." Bredekamp appears to be the first scholar who has understood that the four sketches represent a crater, not the entire Moon.

just the two endpoints, separated in time by half a lunar cycle, first quarter to last quarter. Galileo's observations in the 7 January letter were both more vivid and more accurate than in this reworking. As in the case of the "shadowy bay," Galileo silently adjusted the times of his observations of the crater. The need to center the depiction of the crater on the terminator at first and last quarters gave a kind of false precision to the crater's location, right at notionally (not adjusting for longitudinal libration) 0° longitude. In the 7 January letter, Galileo had written less specifically that the crater was "quasi mel nezo della luna," "more or less" centered. The constraints of time and space imposed by the printing processes – both the letterpress and the later copperplate printing – must have been the primary cause of Galileo's changes at this point in his narrative.

To summarize: When we compare Galileo's several treatments of Topic III, The Moon's Rough Surface, first in the letter of 7 January 1610, next in the Draft, and finally in the printed *Sidereus Nuncius,* we see not only a progression of changes in the illustration schema, but also the reasons for the changes. And when we compare the Seven-Moon Bifolium drawings not directly with the etchings of *Sidereus Nuncius,* but with the planning stages that preceded the finished book, the lack of direct correlation between those drawings and Galileo's illustration plans becomes ever more evident. Differences between the Seven-Moon Bifolium drawings and the *Sidereus Nuncius* etchings had already been seen as puzzling and in need of further explanation in several points, even by the investigators convinced that these drawings must have been the direct source of the etchings. Bredekamp was the first to break with this tradition, and argue that the differences are great because the drawings and the etchings are essentially unrelated to one another. When this hypothesis is tested from an independent angle of approach – the textual study of the progressive changes in Galileo's illustration plans – the disparity between the Seven-Moon Bifolium and Galileo's plans at any stage in the making of *Sidereus Nuncius* becomes all the more pronounced. Regardless of when it was made – a question that perhaps has not been exhausted – the Seven-Moon Bifolium does not figure directly in the creation of *Sidereus Nuncius.* Much closer to the *Sidereus Nuncius* etchings are the sketches, unfortunately not preserved in their originals, that Galileo supplied in his 7 January 1610 letter. Indeed, we understand certain peculiarities of the etchings only when we trace their sources to the 7 January sketches.

A final remark must be made about the changing plans for illustrations. The 7 January letter contains, besides those already discussed, two more Moon illustrations. These exemplify features that Galileo took up and discussed in the Draft text, yet he did not there call for them to be separately illustrated. Thus, the Draft originally specified seven illustrations, but there are two more potential illustrations that Galileo could appropriately have used, yet chose not to.

The first is tied to the text of *Sidereus Nuncius* 8v, lines 1 and after, corresponding to Draft 3v lines 11 and after. Galileo observes here that not only was the terminator on the Moon uneven and sinuous, but also, on the dark side of the line, many bright cusps could be seen, well separated from the lighted side; these gradually expanded over the hours of observation of a given evening, finally merging with the lighted side. In the 7 January letter, this passage reads (figure VIII.4):

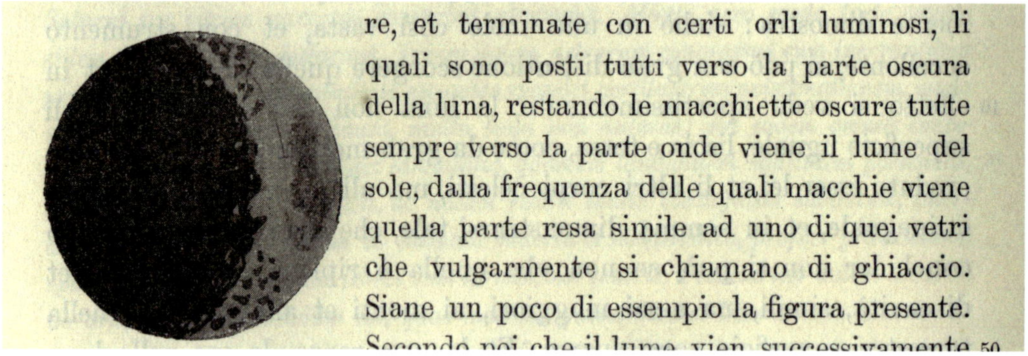
re, et terminate con certi orli luminosi, li quali sono posti tutti verso la parte oscura della luna, restando le macchiette oscure tutte sempre verso la parte onde viene il lume del sole, dalla frequenza delle quali macchie viene quella parte resa simile ad uno di quei vetri che vulgarmente si chiamano di ghiaccio. Siane un poco di essempio la figura presente. Secondo poi che il lume vien successivamente

VIII.5: *Opere* X, p. 274.

> *Di più, non solamente è il predetto confine e termine tra 'l chiaro e'l tenebroso, sinuoso et ineguale, ma scorgonsi vicino ad esso diverse punte luminosissime poste nella parte oscura, et totalmente separate da le corne illuminate: le quali punte a poco a poco vanno crescendo … sì che dopo qualche hora s'uniscono con la parte luminosa, divenendo lucido anco quello spatio che tra esse et la parte risplendente si fraponeva:* ET SI VEGGONO SIMILI A QUELLE CHE CI RAPPRESENTA LA FIGURA APPRESSO.

In the Draft, the equivalent passage reads:

> *Verum enimuero non modo tenebrarum et luminis confinia in Luna sinuosa, et inęqualia cernuntur, sed … permultę apparent luminosę cuspides intra tenebrosam Lunę partem omnino à lucidiori plaga diuisae, et auulsę, ab eaque non per exiguam intercapedinem dissitę, quę paulatim aliqua interiecta mora magnitudine et lumine augentur, post uero secundam horam aut tertiam reliquę parti lucidę et ampliori iam factę iunguntur. interim tamen: alię atque alięhincinde quasi pullulantes intra tenebrosam partem accenduntur, augentur, ac tandem luminosę parti ampliori iam factę copulantur,* HUIUS EXEMPLUM EADEM FIGURA NOBIS EXIBET.

That is, this feature was to be exhibited on Moon I, and indeed that etching depicts five such white "beads" bordering the dark side of the terminator. Thus, the Moon I etching served not double but triple duty, bringing together on its plate three features that Galileo had illustrated separately in his 7 January letter. The copy of this latter sketch preserved in the Vatican version of the letter only obscurely depicts these beads, but that is owing to the mediocre quality of the copy.[10]

The next illustration of the 7 January letter applies to the immediately following paragraph, where Galileo describes the opposite effect, the many dark spots (*moltissime macchiette oscure*) with lighted borders that were visible on the bright side of the terminator, especially near the tip of the southern horn. Their appearance, he wrote, is similar to the patterns in what is called "ice" glass (figure VIII.5):

10 It is possible that the photograph of the copy of the letter that the Vatican Library's prefect, Franz Ehrle, sent to Favaro preserved this detail, and that it was lost in the reproductive process. Apparently, the Vatican photographs can no longer be found, so that the first edition of the *Opere*, vol. X, is our only resource.

> *Veggonsi in oltre nella parte illuminata … et più che altrove intorno alla punto del corno australe, moltissime macchiette oscure, et terminate con certi orli luminosi … dalla frequenza delle quali macchie viene quella parte resa simile ad uno di quei vetri che vulgarmente si chiamano di ghiaccio.* SIANE UN POCO DI ESSEMPIO LA FIGURA PRESENTE.

In the Draft (4r, lines 17 sqq.) this passage was significantly reworked, including a description of how "ice" glass was made, by immersing still-hot glass vessels into cold water, and with a new simile: the dark spots resemble the cerulean eyes of a peacock's fan:

> *… Hęc lunaris superficies, quà maculis, instar Pauonis cauda ceruleis oculis distinguuntur, uitreis illis uasculis consimilis redditur, quę adhuc calentia in frigidam immissa perfractam, undosamque superficiem acquirunt, ex quo à uulgo glaciales ciati nuncupantur.*

However, no illustration was called for or alluded to. In this case Galileo used the striking added simile of the peacock's fan as a verbal substitute for an image.

It should be noted that both these features are carefully and elegantly represented in the Seven-Moon Bifolium. The "bright cusps" on the dark side of the terminator appear all along the crescent-Moon terminator of the drawing labelled as "1", like a rosary of silver beads; and the peacock-fan "eyes" on the bright side of the terminator are rendered on several of the drawings, most strikingly on that labelled as "3". The peacock-fan eyes also are represented on Moon II, printed on 9v (C1v), the page following Galileo's description of the effect. However, he did not refer forward to Moon II at this point, either in the Draft or in finished *Sidereus Nuncius*. The probable reason is that when he was drafting this description, Galileo had not yet settled fully in his mind exactly what details each etching would present. The creation of the etchings belongs to a stage he had not yet reached when writing the draft of part I.

THE "TRIGONA FIGURA"

As noted above, a sequence of studies has attempted to attach specific dates to the drawings of the Seven-Moon Bifolium and to the etchings of *Sidereus Nuncius*. They have taken it as self-evident, requiring no discussion, that each drawing has a proper date: that is, each depicts, over the entire surface of the Moon as pictured, a specific time when Galileo observed the features indicated. The most influential of these studies was made by Ewen Whitaker in 1978, who gave a closely reasoned rationale, which need not be closely examined here, for how he arrived at each date.[11] Whitaker identified the starting date of Galileo's Moon observations as Monday, 30 November 1609. It will be recalled that the same date was arrived at above by a much more approximate and seemingly fuzzier estimation: this would have been the day or evening Galileo must have been referring to in his 7 January 1610 letter to an unknown correspondent, when he described, in greater detail than he

11 Whitaker 1978, see particularly pl. 3 and captions; and compare Gingerich 2009, pp. 154–155.

had done in some previous now-lost letter, what he had observed with his telescope "4 o 5 giorni dopo il novilunio."

Whitaker's results have an almost startling appearance of precision, particularly on this "start date" of 30 November 1609, when by his calculation one of the Moons on the Seven-Moon Bifolium depicts what Galileo saw in Padua at four o'clock in the afternoon, and another at six o'clock in the evening, while the Moon I etching of *Sidereus Nuncius* depicts the Moon at eight o'clock the same evening. Moving north to south along the terminator of Moon I, Whitaker identified seven different features A–G that he correlated with the specific topography of the Moon, the last of these, G, being the lighted "trigona figura" that stood out in the darkness just west of the terminator. By Whitaker's analysis, this figure was the "illuminated centre of Jannsen floor plus Fabricius w[ith] wall" – that is, an area with approximate coordinates of 45° South by 42° East.[12]

Recently, Gingerich has enlarged upon this identification of the triangular feature, which he sees as the "anchor" of Whitaker's arguments for chronology: "By happy coincidence the same illumination angle [as obtained for Padua, 30 November 1609] occurred for the western United States on 9 April 2008, and Whitaker succeeded in photographing the trianglar feature ...". Gingerich illustrated this feature with telescopic photography taken on unspecified dates (his Fig. 4b), and concluded, "Once again a reasonable argument can be made for dating to 30 November 1609 the feature drawn and described by Galileo."[13]

This is an impossible conclusion. As noted above, the Moon I etching incorporates features from two different times and two different places, a contradiction of verbal text and image that readers who had never used a telescope might easily overlook. But even readers with deep experience of observing the Moon telescopically have overlooked the contradiction by neglecting to read Galileo's text. Galileo did not see a lighted triangular feature on 30 November 1609, the "fourth or fifth day after conjunction." He states explicitly that he observed it "as the Moon was hastening toward first quarter," that is, on the evening of 2 December. The Moon I etching depicts the terminator as it was on the evening of 30 November, apparently at roughly longitude 40° East. However, the triangular feature is arbitrarily positioned on the etching. The depiction on the etching was not, and could not possibly be, the location where Galileo actually observed it. When he observed it two nights later, the terminator would have moved westward by roughly 24°, let us say very approximately to about longitude 16° East. Whatever topographic feature created the lighted triangle that Galileo saw, its position would have to be a little further west of that approximate longitude, for it was on the dark side of the terminator.[14] The topography of the Janssen and Fabricius craters had nothing to do with this, for on the evening of 2 December 1609 in Padua they were fully bathed in sunlight. The triangular feature Galileo observed has still not been identified. Its presumed identification within the Janssen crater is illusory.

12 Whitaker 1978, p. 162.

13 Gingerich 2009, p. 154.

14 For the position of the terminator on the evening of 2 December, Padua time, see Gingerich 2009, p. 147: "shortly after dusk in Padua on 2 December 1609 the lunar terminator ... stood at 346 degrees (measured westward from the mean central meridian). Five hours later the Moon set with the terminator not quite at 349 degrees."

IX.
Quire D

In quire D Galileo's most complicated adjustments and rearrangements of the text finally crystallized into types and woodcuts and were locked into formes. To track the changes, we must lay out once more Galileo's successive plans, printing in small capitals the two topics he added to the original plan:

Plan A. Topics I (Introduction) – II (Telescope) – III (Moon's Rough Surface) – VI (Earthshine); then VII (Fixed Stars).
Plan B. Topics I (Introduction) – II (Telescope) – III (Moon's Rough Surface) – VI (Earthshine) –V (HEIGHT OF THE MOON'S MOUNTAINS); then VII (Fixed Stars).
Plan C: Topics I (Introduction) – II (Telescope) – III (Moon's Rough Surface) – IV (SILHOUETTE PROBLEM) – VI (Earthshine) – V (HEIGHT OF THE MOON'S MOUNTAINS); then VII (Fixed Stars).
Plan D (as printed): Topics I (Introduction) – II (Telescope) – III (Moon's Rough Surface) – IV (SILHOUETTE PROBLEM) – V (HEIGHT OF THE MOON'S MOUNTAINS) – VI (Earthshine); then VII (Fixed Stars). That is, as Plan C, but reversing Topics VI and V.

The primary evidence for these changes is the linking symbols that Galileo wrote on various pages of the Draft, indicating which topic should follow which. One of these has already been briefly noted and illustrated: the Mars symbol (♂) with which Galileo marked the continuation, on a separate slip, of his expanded conclusion to Topic III (see figures VII.2–3).

The linking symbols, all or almost all astronomical signs, were, of course, written for Galileo's eyes only. Because this portion of his Draft was both physically and textually complex, divided over five writing units, he used the symbols to guide from one topic in one writing unit to another in another unit, thereby knitting together the text that he would subsequently revise further and transcribe as Fair Copy and setting copy. At the earliest stage, plan A, Galileo did not use linking symbols, presumably because the intended order was entirely clear in his mind, and fully contained in two writing units, of which one could simply be placed in front of the other: a folded single sheet for Topics I and II (Venetian Rampant Lion paper), and a fascicle of two folded sheets for Topics III and VI (Paduan Unicorn paper).

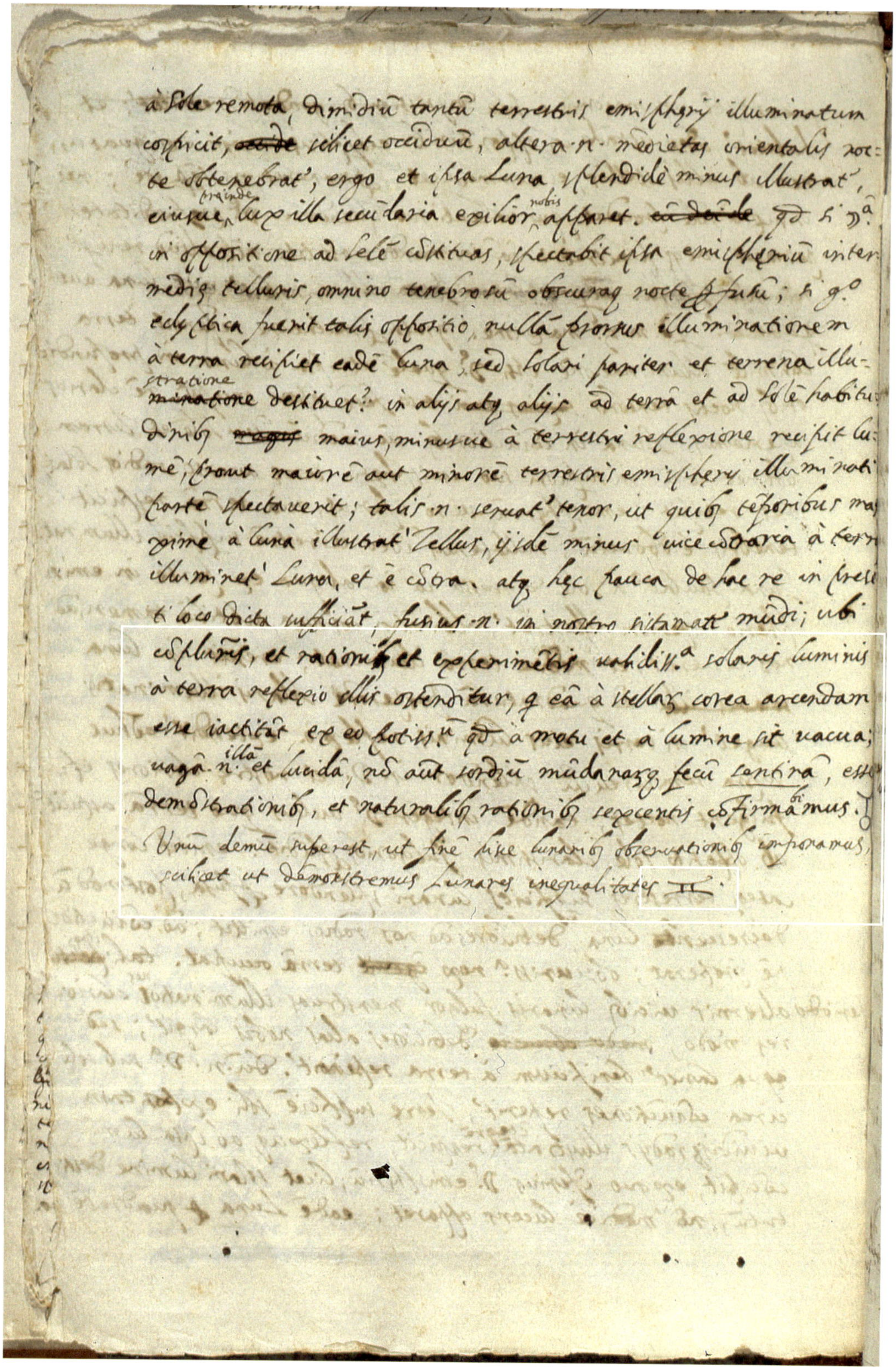
à Sole remota, dimidiu tantu terrestris emisphery illuminatum
conspicit, ~~occide~~ scilicet occiduu, altera n. medietas orientalis noc=
te obtenebrat', ergo et ipsa Luna splendidè minus illustrat',
eiusque proinde lux illa secundaria exilior nobis apparet. ~~eadem de~~ qd si Lunam
in oppositione ad Solè constitueris, spectabit ipsa emisphæriu inter
medij telluris, omnino tenebrosu obscuraq nocte pfusu; si igo
ecliptica fuerit talis oppositio, nullá prorsus illuminationem
à terra recipiet eadè Luna, sed Solari pariter et terrena illu-
~~minatione~~ strationé destituet? in alijs atq alijs ad terrá et ad Solè habitu=
dinibq ~~magis~~ maius, minusue à terrestri reflexione recipit lu=
me, prout maiorè aut minorè terrestris emisphery illuminati
parte spectauerit; talis n. seruat' tenor, ut quibq teporibus ma=
xime à Luna illustrat' Tellus, ijsdè minus uice cotraria à terra
illuminet' Luna, et è contra. atq hec pauca de hac re in prese=
ti loco dicta sufficiat, fusius n. in nostro Systemate mudi; ubi
coplurimis, et rationibq et experimetis ualidiss.a solaris luminis
à terra reflexio illis ostenditur, q eà à stellar corea arcendam
esse iactitat ex eo potiss.m qd à motu et à lumine sit uacua;
uagá n. illam et lucidá, no aut sordiu mudanarq fecu contra, esse
demostrationibq, et naturalibq rationibq sexcentis cofirmabimus.
Unu demu superest, ut finè hinc lunaribq observationibq imponamus,
scilicet ut demostremus Lunares inequalitates II.

IX.1: BNCF, Ms. Gal. 48, 13v.

The linking symbols can be individually allocated to plans B, C, and D. As to the Fair Copy, it should be noted that the quire it was written on ends a little less than halfway through Topic IV, Silhouette Problem, which in the printed edition followed directly Topic III, Moon's Rough Surface. Thus, the Fair Copy belongs to a later stage of Galileo's thinking than plans A or B represent. Because the Fair Copy is only known from this single quire, it seems at first that we cannot be sure whether it belongs to the stage of plan C (where VI would have followed IV) or of D (where V did follow IV). However, as will be shown shortly, we find the answer by a different approach. It may be noted also that, strictly speaking, despite the catchword on the last page of the Fair Copy, we do not know certainly that Galileo continued to write it on a now-lost following quire. It is possible that by or shortly after the time he had finished writing the single quire that survives, he had fully settled the order of topics, and so no longer felt the need for a retained Fair Copy, instead moving directly from the Draft to writing out the setting copy.

The states of Galileo's mind represented by plans B and C would be lost to us if he had not added his linking symbols to the Draft. Plan B is defined by the symbol Gemini (♊). Galileo entered it at the end of Topic VI Earthshine (Draft 6v), and then at the beginning of the new Topic V Height of Moon's Mountains (Draft 9r), written on a single leaf of Paduan Unicorn paper (figures IX.1, 2). Thus, Topic V was in Galileo's mind and drafted before we have any evidence that he had even thought of Topic IV Silhouette Problem; and at first he planned to put it at the end of his discussion of the Moon. In briefest terms, the instruction of the Gemini symbols is: Topic VI > Topic V.

As can be seen, Galileo subsequently added to the end of the Earthshine discussion, writing with a fine-nibbed pen, a bridge passage to introduce the next (and at that stage final) topic regarding the Moon:

> *Unum demum superest ut finem hisce lunaribus observationibus imponamus, scilicet ut demonstremus Lunares inequalitates …*
> "Just one matter remains as we bring our lunar observations to a close, namely that we should demonstrate the lunar inequalities …"

Because the Height of the Moon's Mountains was not, on final consideration, made the last topic of the discussion of the Moon, this sentence never entered *Sidereus Nuncius*. At the plan B stage, Galileo had probably not yet drafted his first two topics, I Introduction and II Telescope, for the sheet with those topics belongs to the period in late January or early February when he moved from Padua to Venice to oversee the printing of his book.

Plan C certainly belongs to the period when Galileo was in Venice, and is defined by the symbol Aquarius (♒). Galileo placed it at the end of Topic IV Silhouette Problem (Draft 8r), which he wrote on a sheet of Venetian Rampant Lion paper, and then at the beginning of the already-written Topic VI Earthshine (Draft 5r); and on 8r he also wrote as a cue the first words of Earthshine: *Lubet hoc loco* (figures IX.3, 4). Thus, at this stage Topic IV > Topic VI. In the original version of the Silhouette Problem, Galileo closed with a bridge passage specifically intended to lead into the discussion of Topic VI Earthshine; in it, he emphasized that he could go much more deeply than he had done into this problem. As with the bridge passage at the end of the Earthshine discussion, it never made its way

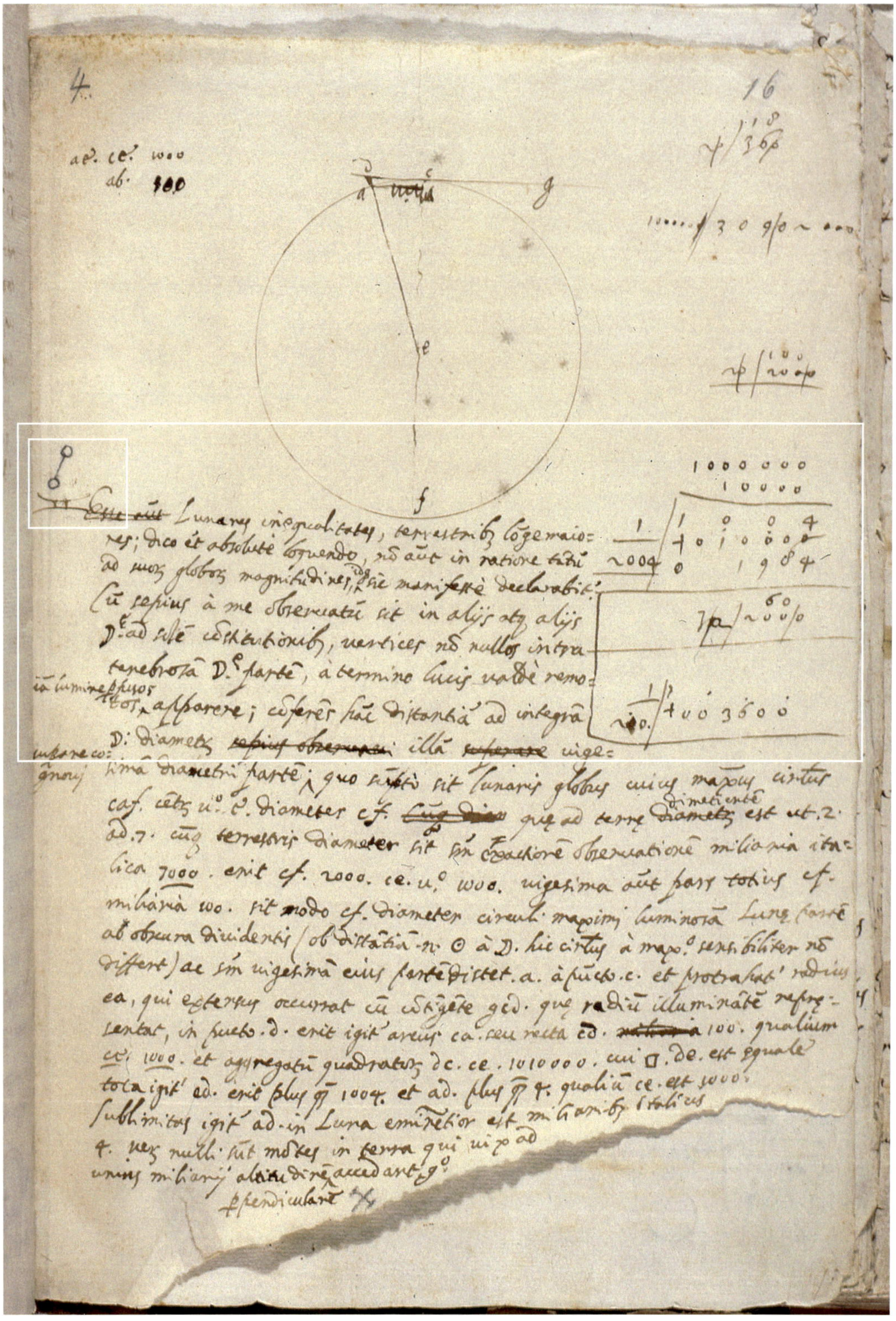

IX.2: BNCF, Ms. Gal. 48, 16r.

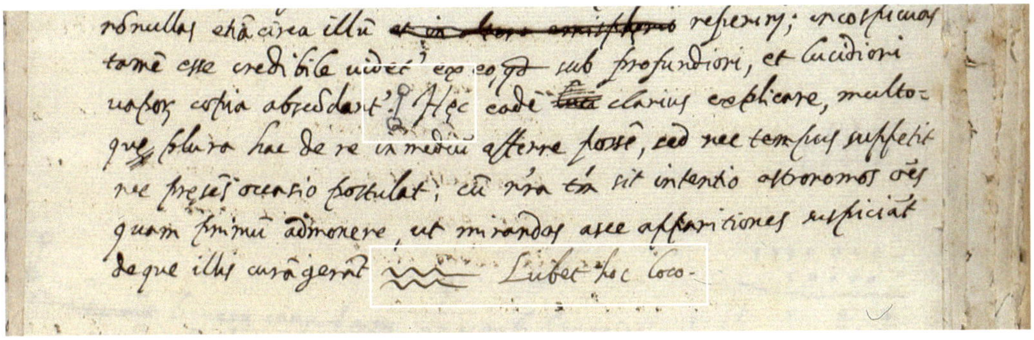

IX.3: BNCF, Ms. Gal. 48, 15r.

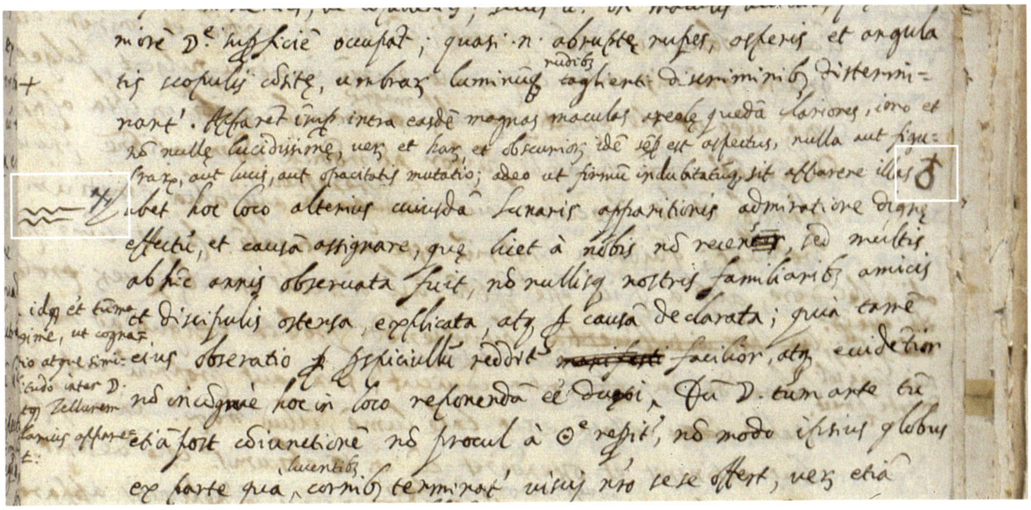

IX.4: BNCF, Ms. Gal. 48, 12r.

into *Sidereus Nuncius*. It is well worth bringing into print for the first time, for it is intensely Galilean:

> *Hęc eadem ~~luci~~ clarius explicare, multoque plura hac de re in medium afferre possem, sed nec tempus suppetit nec pręsens occasio postulat, cum nostra tantum sit intentio astronomos omnes quam primum admonere, ut mirandas esse apparitiones suspiciant deque illis curam gerant.*
> "I could explain this more clearly and bring much more relevant evidence to bear, but neither is there time, nor is this the proper occasion, for our sole purpose is to advise all the astronomers to be aware that there are wondrous apparitions [in the heavens] and to give them due attention."

He might almost have been paraphrasing Hamlet: There are more things in heaven than are dreamt of in your astronomy.

The final arrangement of the texts, plan D, required three pairs of linking symbols, and it is noteworthy that whereas the Gemini and Aquarius symbols were written in ink, these were in pencil. First was the symbol for Opposition of heavenly bodies.[1] Galileo used

1 I am indebted to Prof. Richard Kremer for identifying this sign.

it to link the end of Topic IV Silhouette Problem (Draft 8r) to the beginning of Topic V Height of the Moon's Mountains (Draft 9). In short, Topic IV > Topic V (figures IX.3, 2). Both these places already bore earlier linking symbols reflecting the preceding stages of Galileo's plans. On 8r was the Aquarius symbol from plan C, to cue to the beginning of Topic VI. Galileo did not insert the Opposition symbol immediately before or after Aquarius, but rather four lines up, in front of the bridge passage just quoted, *Hęc eadem clarius explicare*. That is how the passage came to disappear from *Sidereus Nuncius*: when Galileo wrote out the Fair Copy or setting copy of the Draft, on encountering the Opposition symbol he moved immediately to where the Opposition symbol began a new text, skipping over the text that immediately followed the first Opposition symbol. If we were to think of these symbols as the suits of a deck of cards, we would say that Opposition trumps Aquarius. Similarly at the beginning of Topic V Height of Moon's Mountains, the Gemini sign had already been placed, reflecting the plan B stage, when this topic was to follow immediately after VI, Earthshine. Galileo put the Opposition sign right above it: Opposition also trumps Gemini; and, in general, pencil symbols trump ink symbols.

The second symbol-pair of plan D may be astronomical, but I do not recognize it: its shape is a St. Andrew Cross with cross-crosslets on two of the arms. Galileo wrote this mark at the end of Topic V Height of Moon's Mountains (Draft 9r) and at the beginning of Topic VI Earthshine (Draft 5r), that is, Topic V > Topic VI (figures IX.5, 4). The beginning of the Earthshine discussion had earlier been marked with the Aquarius symbol from plan C, when it was to follow the Silhouette Problem. Galileo placed the Cross mark immediately after Aquarius: Cross trumps Aquarius, and pencil trumps ink.

The third symbol-pair is Taurus (♉), which Galileo entered at the end of Topic VI Earthshine (Draft 6v, figure IX.1), and at the beginning of Topic VII (Draft 10r), thus VI > VII (figure IX.6). The end of Earthshine already bore the Gemini mark from plan B, when the next topic was to be V, Height of the Moon's Mountains. Galileo placed Taurus two lines above, thus omitting the draft passage Galileo had added specifically to lead on to Topic V: Taurus trumps Gemini, and for the final time pencil trumps ink.

The "pencil trumps ink" rule shines a little more light on Galileo's conclusion to Topic III, Moon's Rough Surface, where as noted in chapter VII (figures VII.2, 3) he added a long supplementary statement to the end of the topic, squeezing as much of it as he could on Draft 5r, in the space he had left between the end of Topic III and beginning of Topic VI Earthshine, then finishing it on a small loose slip, linked to 5r by the Mars symbol. This Mars symbol is written in pencil in both places; therefore the addendum apparently belongs to the last stage, Plan D. In printed *Sidereus Nuncius* this addendum appears on C3v,

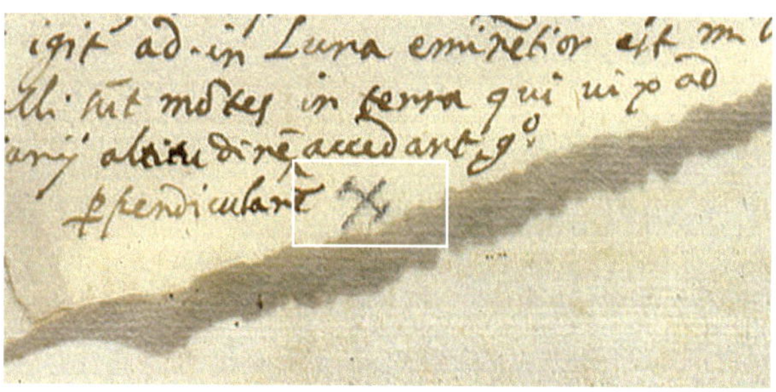

IX.5: BNCF, Ms. Gal. 48, 16r.

and so we may suppose that no later than the time this page was set in type, Galileo had reached his final planning stage. Since the addendum is incorporated with further revisions in the Fair Copy, we may conclude that the Fair Copy also belongs to the time of plan D.

In *Sidereus Nuncius* the setting of Topic IV Silhouette Problem begins on C3v (fo. 11v), line 26, and continues through C4. The foot of C4v (12v) contains the woodcut diagram relevant to this topic, extending deep into the bottom margin and also into the left margin. This diagram illustrates the second of Galileo's two solutions: that an atmosphere surrounds the Moon, making its outline indistinct. The topic ends on D1r (13r) at line 14: … *vaporum copia abscondantur.* A bridge paragraph follows (lines 15–22): *Esse igitur clariorem Lunae superficiem tumoribus … idque sic manifestè declaratur.* Then on lines 23 and after Topic V Height of the Moon's Mountains begins with a new paragraph: *Cum saepius à me obseruatum sit ….* The draft of Topic V presents only a rough sketch of this bridge paragraph, its key words not fully drawn into a sentence, as can be seen by comparing its text with that finally printed (small capitals indicate the words identical in both):

> Draft 9r (figure IX.2): ~~*Esse autem*~~ *Lunares inequalitates, terrestribus longemaiores; dico* ETIAM ABSOLUTE LOQUENDO, NON AUTEM IN RATIONE TANTUM AD SUORUM GLOBORUM MAGNITUDINES, IDQUE SIC MANIFESTE *declarabitur.*
> *Sidereus Nuncius* 13r: *Esse igitur clariorem Lunae superficiem tumoribus, atque lacunis vndiquaque conspersam, ex iam explicatis apparitionibus satis apertum esse reor; superest vt de illorum magnitudinibus dicamus, demonstrantes Terrestres asperitates lunaribus esse longè minores: minores inquam* ETIAM ABSOLUTÈ LOQUENDO, NON AUTEM IN RATIONE TANTUM AD SUORUM GLOBORUM MAGNITUDINES; IDQUE SIC MANIFESTÈ *declaratur.*

Note in particular that in revising, Galileo reversed the direction of his comparison of the Earth's and Moon's surfaces. In the Draft, "the unevennesses of the Moon [are] far greater than those of the Earth"; in *Sidereus Nuncius*, "the roughnesses of the Earth [are] far less than those of the Moon."

The draft of Height of the Moon's Mountains was aparently made rapidly. The recto of the leaf contains not just the verbal text, but also Galileo's relevant arithmetic calculations, written in the margins, and at the top a full-size sketch of the necessary woodcut diagram. The leaf itself was an already-damaged piece of paper, with its lower right corner torn away. On its verso is a sketch of the asterism of Cassiopeia, which lies well to the north of the ecliptic, and it is possible that the leaf was first used to make this sketch.[2] The draft was clearly accommodated to fit within the boundaries set by the calculations, the Lunar diagram, and the already torn-away corner (figure 5 above). In preparing Topic V for the printer, Galileo made further revisions, partly to heighten its style. Thus, "ob distantiam enim Solis a Luna" was changed to "ob maximam enim elongationem Solis à Luna," and "terra" became "Tellus." And, he added a concluding clause: … *manifestum igitur relin-*

2 The verso of the leaf was left out of the reproductions of the *Sidereus Nuncius* Draft in volume III of the *Opere*, but was included in the Supplement, III pt. 2, p. 962.

quitur, Lunares eminentias terrestribus esse sublimiores – "and so it remains manifest that the lunar eminences are loftier than the terrestrial ones."

In *Sidereus Nuncius* as printed, Topic V Height of the Moon's Mountains concludes on D2r (14r) line 4. It took up in all one full page – 34 lines of typesetting – plus just over a half page, the equivalent of 18 lines, for its woodcut diagram. These three pages of quire D, 1r–2r, could not be assembled until the woodcut had been created, presumably by the unknown but accomplished artisan who had made the two preceding woodcuts printed on C4v and B3r.[3]

Topic VI Earthshine, the final topic of the lunar discussions, followed on D2r (13r) at line 5 with a new paragraph, and continued to D4r (16r) line 11: a total of four full pages of typesetting, plus seven lines over. This topic differs from those preceding it, for it does not directly depend on or proceed from the recent telescopic observations. Galileo states that it was an argument he had presented many years ago to friends and pupils (*multis abhinc annis obseruata sit, nonullisque familaribus amicis, & discipulis ostensa*), yet it did not come amiss here, for with the telescope the phenomenon is more easily viewed, and it reinforces the relationship and similarities between Earth and Moon. Galileo had intended to include the Earthshine argument in *Sidereus Nuncius* from his earliest plan, and it is fully set out in the Draft with relatively few changes in the printed edition. Its summation is one of the most striking passages in *Sidereus Nuncius*. Galileo presents a defense of the "dignity" of the Earth as against all the other bodies of the cosmos. The Earth is in movement – a small Copernican hint – and is bright: it is not merely the lowest part, the bilge of the universe. This, he writes, he will demonstrate in his forthcoming *Systema Mundi*, with fully "six hundred" proofs.[4] It will be useful to quote the passage in full, not from *Sidereus Nuncius*, but from the Draft (6v: figure IX.1):

> *Atque hęc pauca de hac re in presenti loco dicta sufficiant, fusius enim in nostro sistamate mundi; ubi complurimis, et rationibus et experimentis ualidissima solaris luminis à terra reflexio illis ostenditur, qui eam a stellarum corea arcendam esse iactitant ex eo potissimum quod à motu et à lumine sit uacua; uagam enim ^ illam^ et lucidam, non autem sordium mundanarumque fecum sentinam, esse demonstrationibus, et naturalibus rationibus sexcentis confirma^bi^mus.*
>
> "And these few words may suffice here, for [we will treat this] at greater length in our System of the World, where, with numerous reasons and experiments, the most powerful reflexion from the Earth of solar light [will be] demonstrated to those who claim that it [the Earth] is to be excluded from the starry dance, on the grounds that it lacks both movement and light. For that it does wander and is bright, and is not merely the

3 The general appearance of the woodcuts is similar to that of the numerous cuts used in Buonaiuto Lorini's *Fortificationi*, published in Venice by Francesco Rampazetto just a few weeks before the completion of *Sidereus Nuncius* (chapter II above).

4 Wootton 2009, pp. 139–140 argues that this passage "may have been the last to be written before the book went to press," one that Galileo dared to insert only after he felt assured that he would be sheltered under the aegis of Cosimo II's patronage (p. 138): "In return for having the moons of Jupiter named after his family, Cosimo was to become a sponsor of Copernicanism." The physical and textual evidence argue strongly against this view.

> bilge for the filth and waste of the universe, we will verify by six hundred demonstrations and natural reasons."

The sentence is both dramatic and awkward. It is as though Galileo's vivid thoughts of his soon-to-come (as he then thought) *Systema Mundi* took his attention away from a careful formulation of his words. The phrase "fusius enim in nostro sistamate mundi" needs a verb in the future tense to complete its sense; at first Galileo wrote "confirmamus" in the present, then realized that it needed the future; but having made that change, he overlooked that for the same reason "ostenditur" needed the future tense.[5] In his revision before printing, Galileo changed only a few words, most notably when he added an emphasis that not only is our Earth bright, it excels the Moon in brightness (*ac Lunam splendore superantem*). Regarding Earthshine, that is, the secondary light on the Moon (in Galileo's phrase, *secunda Lunae claritas*), Noel Swerdlow has tentatively suggested that nonuniformities of shading in the etching Moon I could have been intended to indicate this light. Independently, Bredekamp has also given attention to this possibility.[6] However, against this possibility, it is probably significant that Galileo's discussion of Earthshine makes no reference to this or any illustration. As has been amply discussed, the Moon etchings are very clear, and even exaggerated, as to the features they were meant to exemplify, which could not be said of the dark-Moon shading lines of Moon I. Secondary light is, however, beautifully depicted on several of the Moons on the Seven-Moon Bifolium.

The fixed stars (D4r–6v)

Topic VII, The Fixed Stars, occupies the remainder of quire D: five full pages plus two thirds of a sixth, these being occupied by 119 printed lines, and four woodcuts of asterisms. Galileo's Draft contains this topic on a separate sheet of (Paduan) Unicorn paper: that is, it belongs to the earliest parts of the Draft. As noted above, it is prefixed with the Taurus sign, linking it to the end of Topic VI Earthshine (figure IX.6). Like the leaf used for the Height of Moon's Mountains draft, the sheet of the Fixed Stars draft was already partly damaged when Galileo took it up: portions of the upper and lower corners of the second leaf of the sheet had been torn away. Again, Galileo accommodated his writing to these flaws, proving that they were present before he wrote.

Topic VII makes conspicuous the page-space problems that became urgent after Galileo added two new topics to his discussion of the Moon. The Fixed Stars topic did not even begin until line 12 of D4r, so that not even the most stringent editing could bring the topic to a close by the end of D4v. There was room remaining in sheet D for only 57 lines of typesetting, less than half the quantity of text the topic required, and besides this illu-

5 The omission of a verb from "fusius enim in nostra sistamate mundi" is partly to be explained because Galileo was, surely without realizing it, repeating himself. Twenty or so lines earlier in his discussion of Earthshine (Draft fo. 5v), he had already written more completely, "qua de re fusius in libris de sistemate mundi pertractabimus." I am grateful to Isabelle Pantin for her thoughts on this passage.

6 Swerdlow 1998, p. 249; Bredekamp 2007, pp. 138–140, 143. The history of Galileo's thoughts on secondary light is discussed very thoroughly by Reeves 1997, passim, in the context of other contemporary opinions.

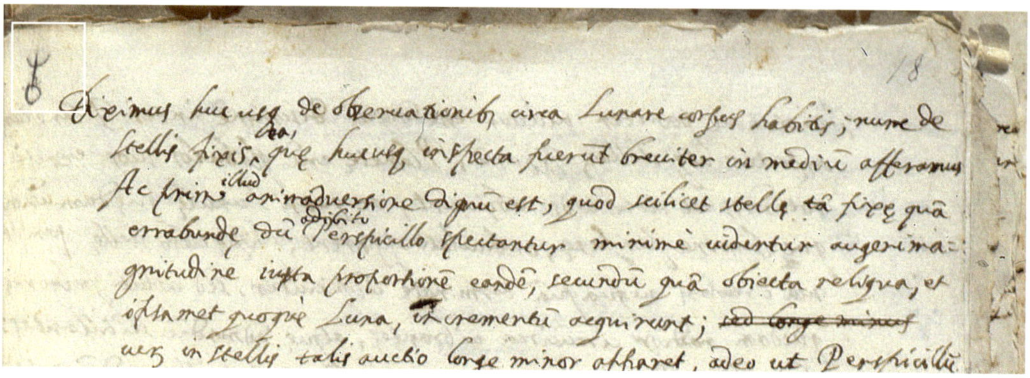

IX.6: BNCF, Ms. Gal. 48, 18r.

strations were essential. If a single leaf were added to quire D, that would be sufficient to include the full written text of the Fixed Stars topic, but there still would be no space available for illustrations. The next possibility would be to add two leaves, that is a bifolium or half-sheet, with the second leaf providing the space for the asterism illustrations. Although the asterism woodcuts are laid out over three pages, not two, the space calculation may have been in terms of their combined (planned) dimensions, the exact equivalent of two pages.

This additional bifolium could have been laid out as a two-leaf quire following D4 and preceding E1, to be signed Dd or some other variant. But there was less chance of it going astray when the sheets were gathered and copies were bound if, as was done, it was sandwiched within the full sheet of quire D, and signed as D3. Thus, the full sheet of D contained outer and inner formes comprised of pages 1r, 2v, 5r, 6v; and 1v, 2r, 5v, 6r respectively. The forme structure of the half-sheet is a much more difficult problem, and will be discussed further in chapters XI and XIII.

The Fixed Star draft was written on three pages of its sheet (Draft fos. 10r–v, 11r), the last page left blank. The draft explicitly describes the four illustrations that were eventually supplied in *Sidereus Nuncius* by woodblocks: the asterisms of the belt and sword of Orion; of the Pleiades within Taurus; of the nebulous star in the head of Orion; and of the nebulous star Praesaepe (Manger) within Cancer. As noted by Whitaker and others, Galileo's original observations for the first two illustrations can be quite closely dated from sketches that he entered in his Jupiter Observation notes (the notes themselves are discussed in the next chapter, on quires E–G). On Saturday January 23, Galileo drew a partial sketch of stars to the east of the belt of Orion, locating them within a grid marked by units of longitude and latitude (figure IX.7). On Sunday January 31, now in Venice, he made a sketch of the Pleiades (figure IX.8). Finally, a week later, Sunday 7 February, he made a fuller sketch of the belt and sword of Orion (figure IX.9).[7]

Besides these sketches within the Jupiter Observation sheets, three other leaves in the Galileo collections of the Biblioteca Nazionale Centrale of Florence, described briefly in chapter V, contain autograph drawings related to the illustrations of quire D. One leaf

7 Whitaker 1978 pp. 165–166; Pantin 1992 adds important details, see especially p. XXIII n. 34, p. 79 n. 96, p. 81 n. 100.

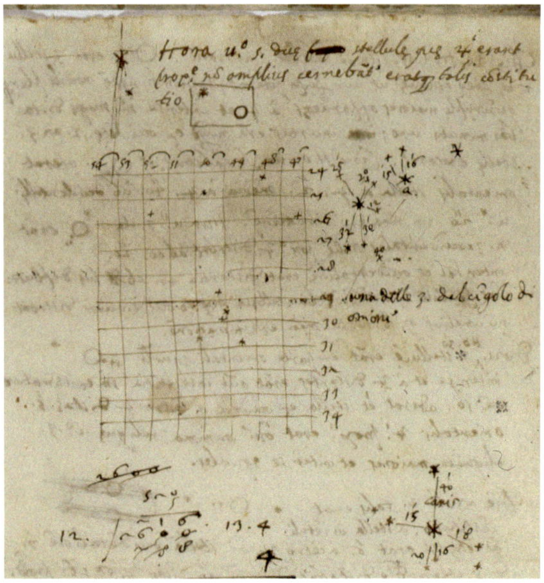

IX.7: BNCF, Ms. Gal. 48, 31v.

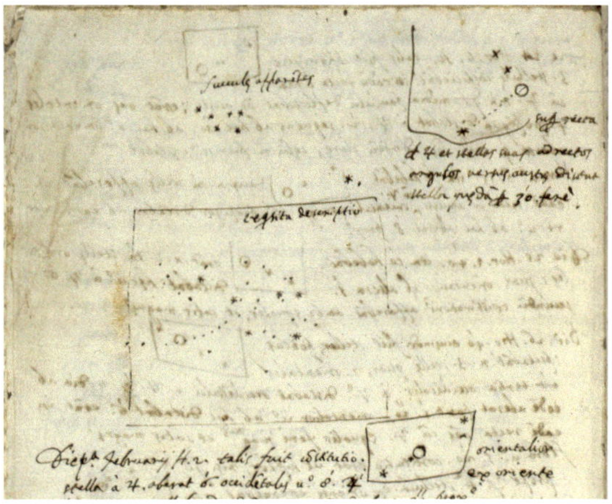

IX.8: BNCF, Ms. Gal. 48, 32v.

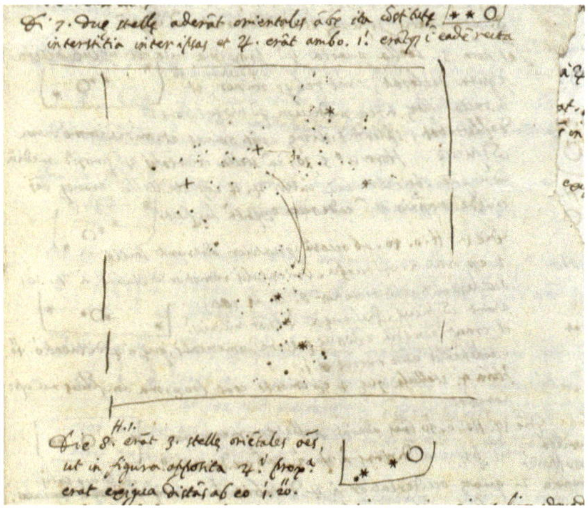

IX.9: BNCF, Ms. Gal. 48, 33r.

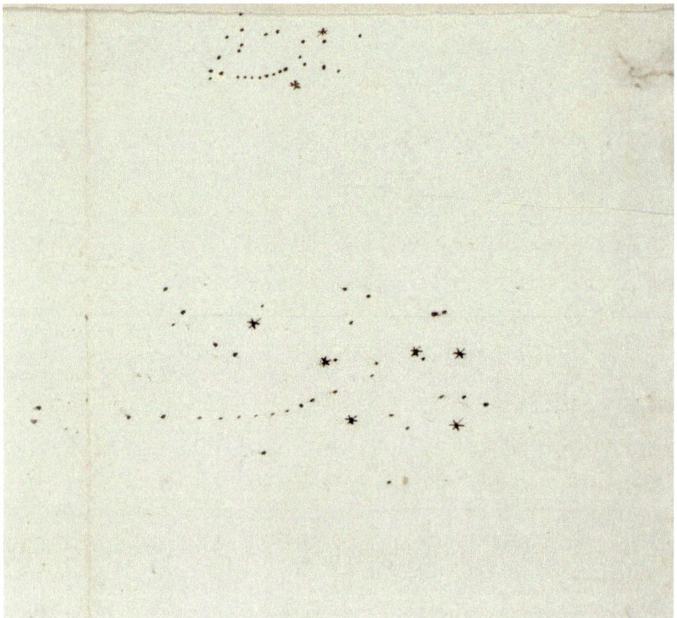

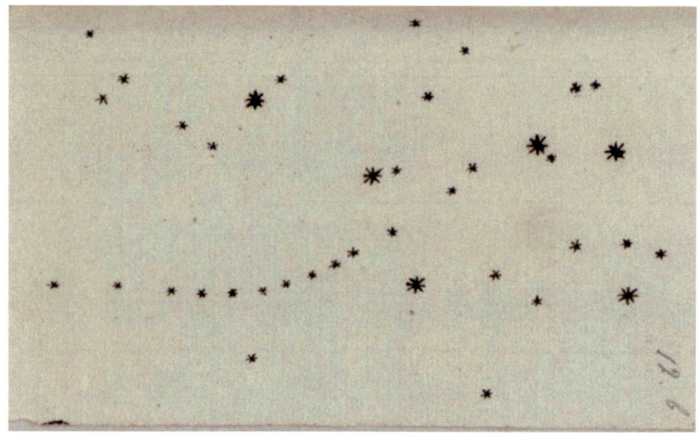

Figure IX.10: BNCF, Ms. Gal. 70, 3r.

Figure IX.11: BNCF, Ms. Gal. 70, 12Br.

(Ms. Gal 70, fo. 3) contains two additional sketches of the Pleiades (figure IX.10).[8] A second leaf (Ms. Gal. 70, fo. 12b) bears a finished sketch that appears to be the precise model for the Pleiades woodblock, for its dimensions and star placements are essentially identical (figure IX.11). A third leaf (Ms. Gal. 70, fo. 18) bears the finished sketch that was the model for the Orion woodblock (figure IX.12). These three leaves, overlooked when Ms. Gal. 48 was assembled, are nonetheless part of Galileo's autograph dossier for *Sidereus Nuncius*. We have the reverse situation from that for the Seven-Moon Bifolium, which was mistakenly brought into Ms. Gal. 48 because of its seeming close connection with *Sidereus Nuncius*. Observation notes or sketches for the last two illustrations of the Fixed Star draft, the asterisms of Praesaepe and of the nebulous star in the head of Orion, seem not to be preserved.

In the Fixed Stars draft the desired placements for all four illustrations are carefully marked. It is noteworthy that Galileo originally planned that the Pleiades illustration should precede that of the belt and sword of Orion.[9] This order is indicated not only by the caption in the Draft (figure IX.13), *Loci constellationum Pleiadum et Orionis*, but also by the Draft text, where Galileo wrote,

8 It should be noted that although Galileo's Pleiades sketch in his Jupiter Observation sheet was made on 31 January, when he was in Venice, this double sketch was made on (Paduan) Unicorn paper.

9 As noted by Pantin 1992, p. 70 n. 95. Van Helden 1989, p. 59 n. 65, drew incorrect conclusions because of believing that fos. 5–6 of quire D are "very late" insertions, whereas they form part of the full sheet of D. All of quire D was planned as a unit. In fact, on the evidence of the Draft, there is nothing "late" about any part of the Fixed Stars discussion: from the beginning it was planned to supply the conclusion to part I of *Sidereus Nuncius*.

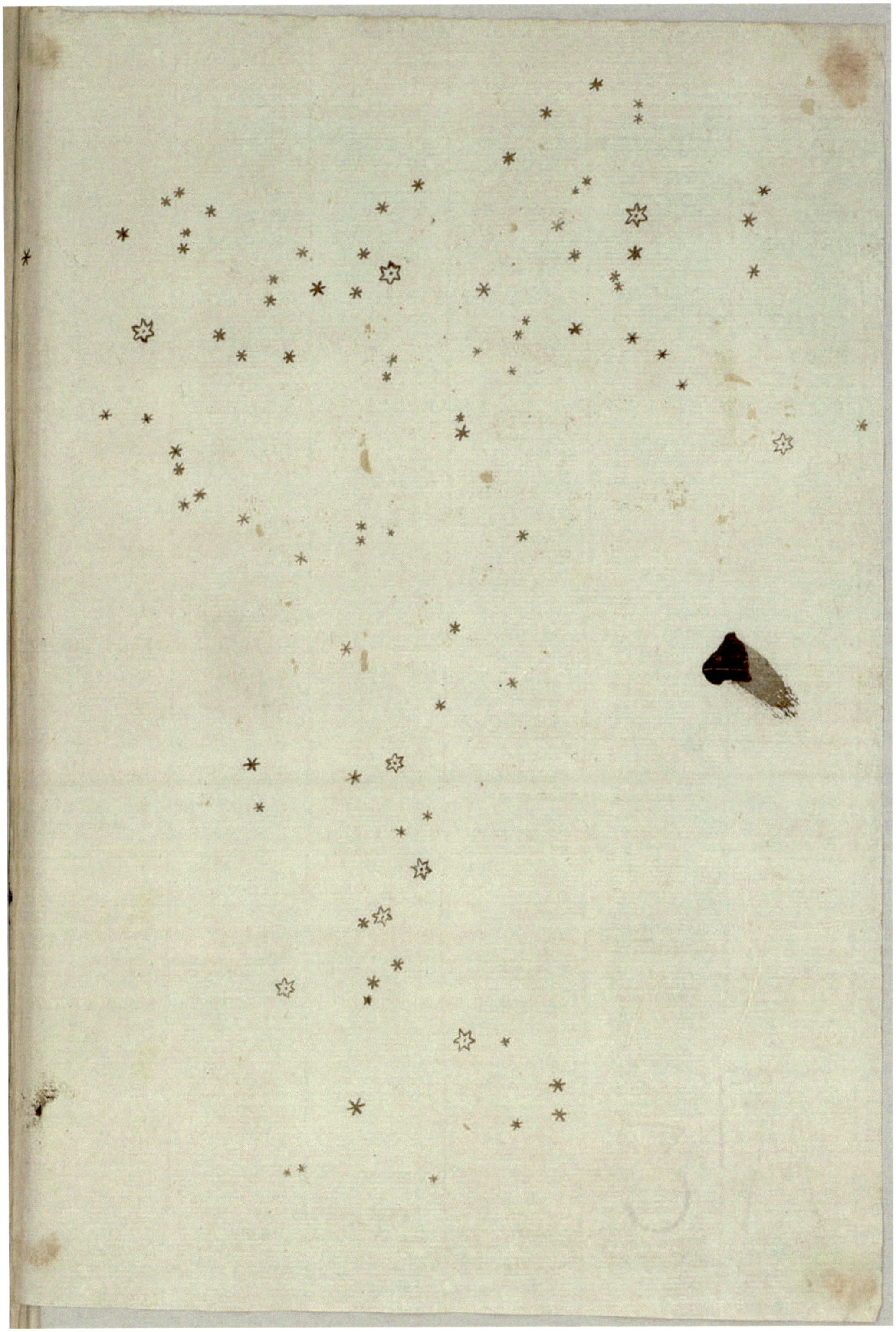

IX.12: BNCF, Ms. Gal. 70, 18r.

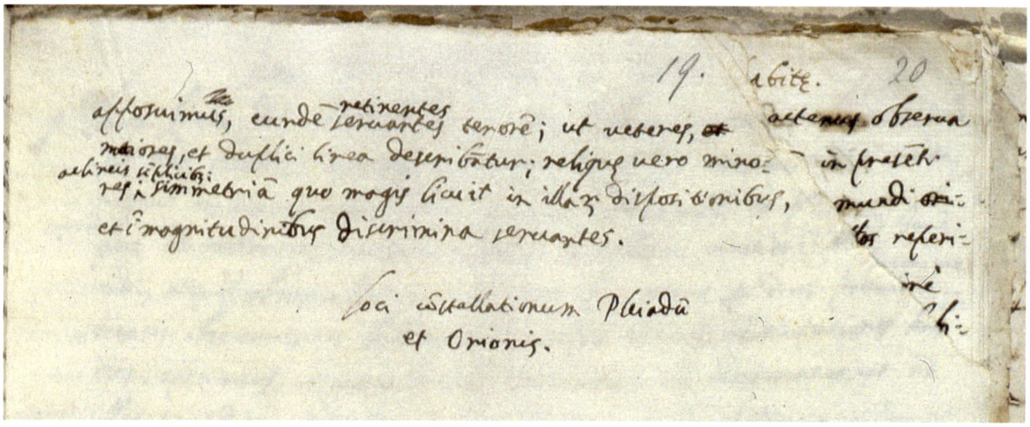

IX.13: BNCF, Ms. Gal. 48, 19r.

> *... constellationes, seu asterismos duos subscribere placuit ... In primo descripsimus 6. stellas Tauri Pleiadas dictas ... in altero exemplo integram constellationem Orionis depingere decreueram ....*
>
> In printed *Sidereus Nuncius* the order was reversed (D5r lines 9–24): *Asterismos duos subscribere placuit ... In primo integram Orionis Constellationem pingere decreueram ... In altero exemplo sex Stellas Tauri, PLEIADAS dictas depinximus ...*

The repositioning of the two illustrations, and the textual changes necessary to the reversal of order, must belong to the late period of printing production, when the asterism woodblocks had been made, and the layout of quire D was being arranged page by page. The full-page cut of Orion was the most inflexible of the woodblocks. Its minimum size, the notional rectangle containing all the stars depicted, was about 205 mm in height by 175 mm in width. In reality the physical block would have been somewhat larger in both dimensions, for we may presume that the outermost stars at each of the four borders were not placed absolutely on the edges, like small castles on vertiginous cliffs. The block dimensions are strikingly greater than those of the standard type-page of *Sidereus Nuncius* which, including the headline and signature-catchword line, measured about 170 mm in height by 95 mm in width (figure IX.14). The Orion illustration is the one page of *Sidereus Nuncius* that took full advantage of the "Mezana Grande" paper that Galileo chose for his publication. A quarto printed on common Chancery paper, in original binding and not overly trimmed, would have had leaf dimensions of about 20–21 × 15 cm. The Orion woodblock could not have been completely printed on a leaf with those dimensions. And in fact, the majority of surviving copies of *Sidereus Nuncius*, because of excessive trimming by both original binders and rebinders, have suffered the loss of one or more stars of this asterism. The Orion asterism is displayed properly, with all the stars and with ample margins, only in the four uncut copies and in a handful of other copies – such as that at Wellesley College (leaf size 25 × 17.8 cm) – that were only slightly trimmed when put into temporary boards or wrappers.

Even when printed on Mezana Grande paper, the Orion woodblock was so large that it could not conveniently be placed under the standard headline. For this forme – the inner forme of full sheet D – the verso headline "OBSERVAT. SIDEREAE" was not set. In its

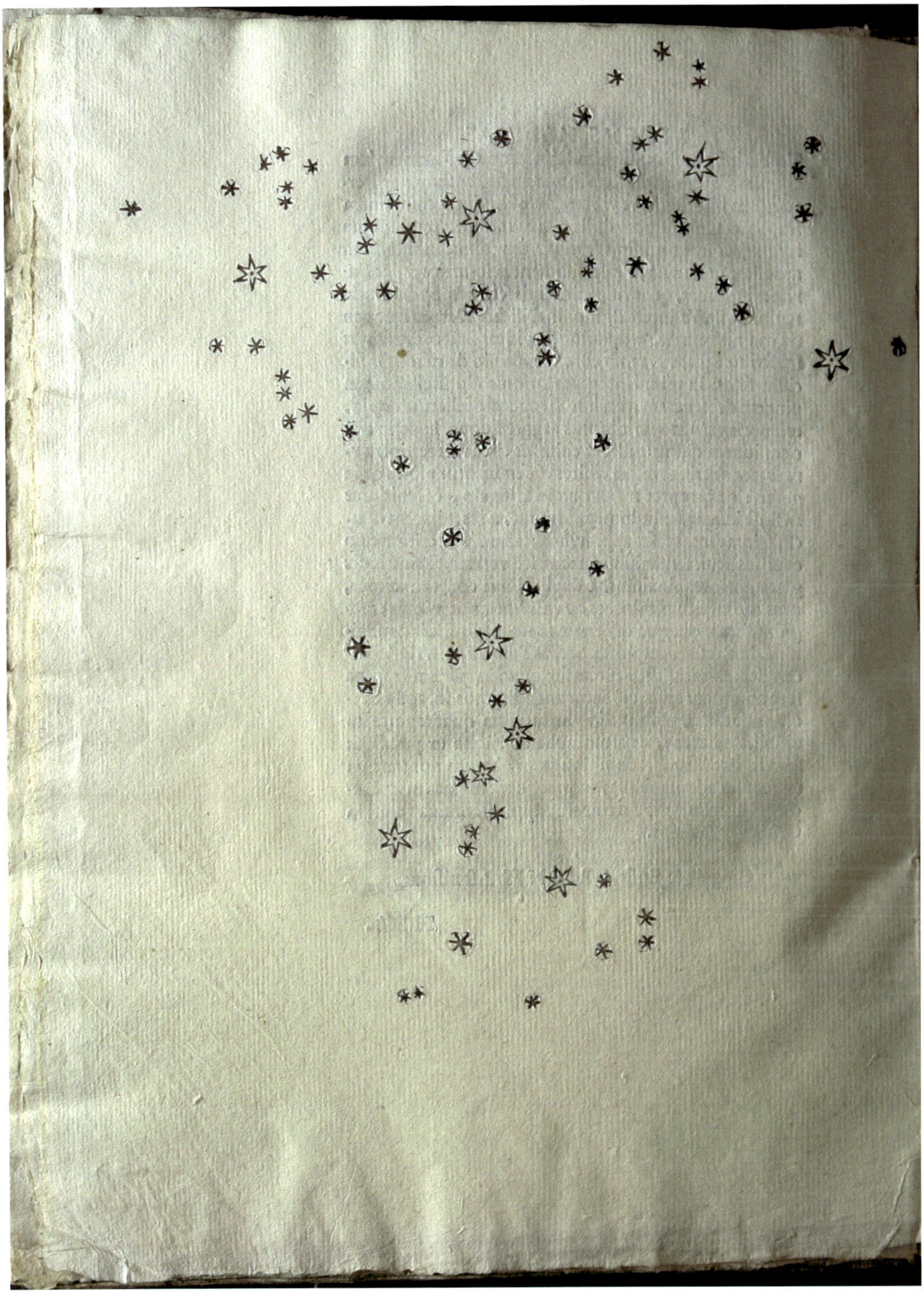

IX.14: Sidereus Nuncius, D5v, Library of Congress.

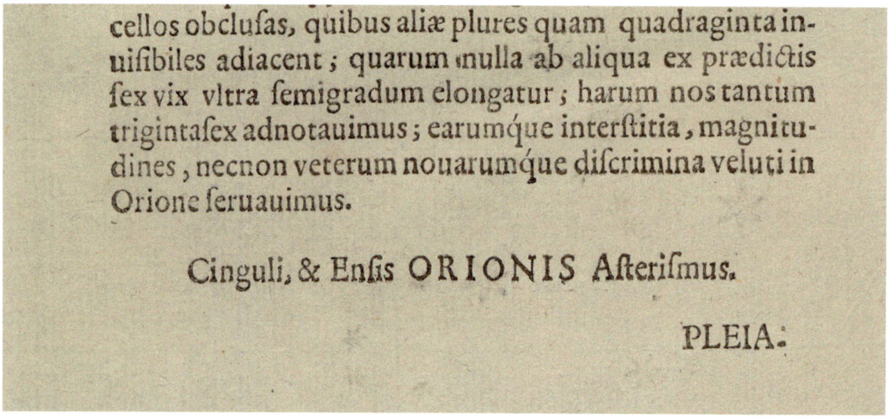
cellos obcluſas, quibus aliæ plures quam quadraginta in-
uiſibiles adiacent; quarum nulla ab aliqua ex prædictis
ſex vix vltra ſemigradum elongatur; harum nos tantum
trigintaſex adnotauimus; earumque interſtitia, magnitu-
dines, necnon veterum nouarumque diſcrimina veluti in
Orione ſeruauimus.

Cinguli, & Enſis ORIONIS Aſteriſmus.

PLEIA.

IX.15: Sidereus Nuncius, D5r, Institute for Advanced Study.

place, quite awkwardly, a caption for the cut was set as a widowed line at the foot of the preceding page, D5r: "Cinguli, & Ensis ORIONIS Asterismus." (figure IX.15) This can be seen as a makeshift solution introduced in the printing shop when the inner forme of the full sheet of D was being imposed. The caption for the facing Pleiades woodcut "PLEIADVM CONSTELLATIO" was handled more elegantly, by setting it in place of the standard recto headline "RECENS HABITAE." (figure IX.16) Curiously, the catchword of D5r, "PLEIA.", points to this headline and not, as would have been typical practice, to the first text word on D6r, "Quod." Strictly, of course, D5r did not need a catchword, as there is no text on D5v; the D5r catchword skips that page, and points forward to the second page on.

There is a hint in Galileo's Draft that the Orion block was made, or at least fully designed, in advance of the other three blocks of this section. Galileo wrote in the Draft that besides the three conspicuous stars in the belt of Orion, and the six in its sword, he had depicted another eighty new adjacent stars (*nouas alias adiacentes 80.*) The printed text keeps that number, and the block does indeed depict just eighty smaller stars around the conspicuous ones.

The Pleiades illustration presents a problem, for there is a disparity between the number of stars depicted, and the number called for in the text. In the Draft Galileo originally wrote that besides the six stars visible to the naked eye, "more than forty other invisible stars are in its vicinity … of which we have taken note [in the illustration] of only thirty-three, although more can be discerned …" (*aliȩ plures quam 40. inuisibiles adiacent … harum nos tantum 33. adnotauimus, licet plures cernantur …*). The printed text enlarges the stated number to thirty-six (… *harum nos tantum trigintasex adnotauimus*). But this does not correspond to the block, which depicts not thirty-six "invisible" stars, but rather the six naked-eye stars plus thirty "invisible" stars, for a total of thirty-six in all. The disparity is due to copying errors by the cutter of the woodblock. Galileo's Maquette (figure IX.11 above) does indeed depict thirty-three small stars arranged among the six conspicuous ones. However, the blockcutter overlooked three of the small stars, one at the upper left corner of the block as printed, and two among the seven stars that form a long tail in the lower left quadrant of the block as printed. We may conjecture that the number stated in *Sidereus Nuncius*, thirty-six, was a late modification. When the block was finished, Galileo counted its stars, saw that they came to thirty-six, and so added in that number to substitute for the "thirty-three" of the draft. He failed to notice that his sentence as written was referring only to the hitherto invisible stars, and so the six conspicuous stars should not have been included in the count. The correct word at this place would have been *triginta*.

Finally came the two quarter-page blocks, placed side by side but separated by a type rule at the foot of the last page of the quire, D6v (figure IX.17). For the nebulous star in the head of Orion, Galileo wrote in the Draft that of more than thirty observed stars, he had depicted twenty-one (*In nebulosa capitis Orionis sunt stellę plures quam 30. nos 21. descripsimus*), the same number as in the printed text and as represented on the block. However, regarding the nebulous star Praesaepe, he left a blank space in the Draft to be filled in with the exact number of stars depicted, out of more than fifty that he had discerned (*aggregatio stellarum est, in angustissimo illo spaciolo constipatarum, quę numero plures quam 50. sunt, nos tantum . descripsimus*). In the printed edition, this became reduced to "more than forty" stars, and the block depicts as he said thirty-six of them (*nos praeter Asellos trigintasex notauimus*), that being the number of stars on the block besides the two conspicuous Aselli or Ass-colts, the stars that were notionally feeding at the nebulous Manger. It will further be noted that at the end of this Fixed Stars discussion, Galileo drew a Sagittarius reference symbol, in pencil, to link it to the beginning of the next section of his Draft, the Jupiter observations (figure IX.18).

In the impressions of all four woodblocks, there are visible traces of the way the blocks were constructed. Each star must have been marked out on the surface, and then the intervening spaces were chiseled away from around them, leaving a scattering of raised, approximately round pegs or "mesas" standing up. These were then worked more carefully to create asterisk shapes on their surfaces, except for the larger stars visible to the naked eye, which were given full outside borders, the space within the borders being cut away except for a central dot. The work must have been rapidly done, with little attempt to make the asterisks, of varying sizes according to Galileo's model, uniform in shape. In various places, the raised pegs were not fully cut around, so that portions of their borders take additional ink. In unpressed copies of *Sidereus Nuncius*, the blind impressions of the pegs push strongly into the paper.

One aspect of quire D seems to have been slightly misunderstood or misperceived by several investigators. It is true that quire D, which brought to a close the first part of *Sidereus Nuncius*, had to be expanded from its original calculation, to fit in all of Galileo's planned topics. It is true also that at various places both in *Sidereus Nuncius* itself and in letters he wrote during this period, Galileo referred to the press of time in getting his book completed. However, it is not true that Galileo "rushed" the conclusion of quire D, bringing in the two final woodcut asterisms as an afterthought to fill in an unplanned-for space at the end of the quire.[10] The treatment of these two nebulous stars, as examples, was integral to Galileo's treatment of the Milky Way, and he had already announced the two

10 Gingerich & Van Helden 2003, p. 263: There was "a gap at the end of the second unnumbered leaf [D6], and Galileo had to figure out how to fill that empty space. This he solved by adding wood blocks of Praesaepe and the nebulous head of Orion ... Apparently he rushed out one evening, mapped the stars in these nebulae, and sent the diagrams straight to the block cutter ...".

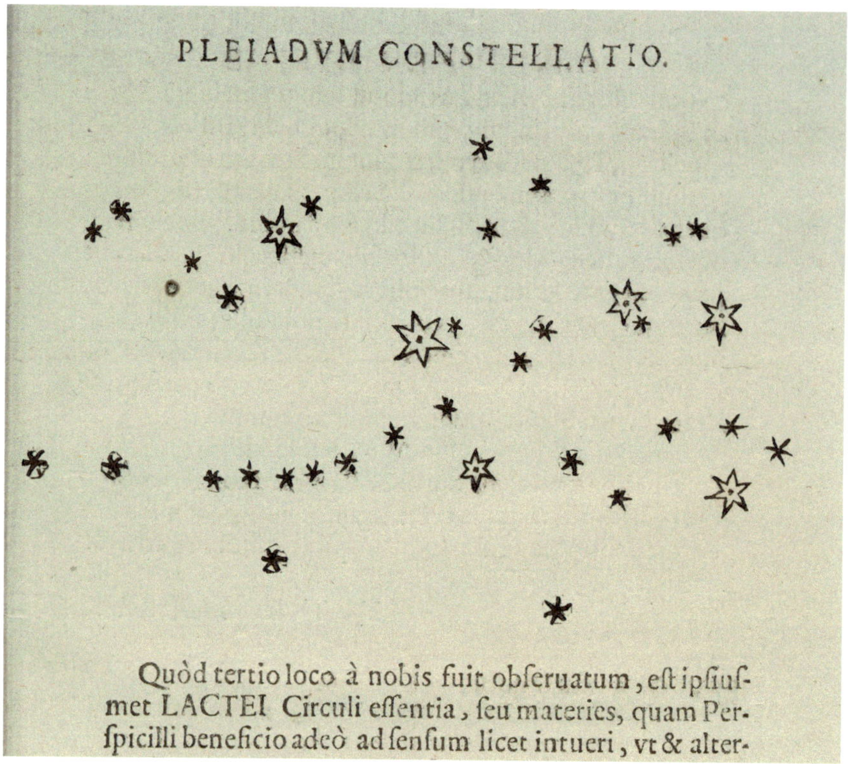

IX.16: *Sidereus Nuncius, D6r, Institute for Advanced Study.*

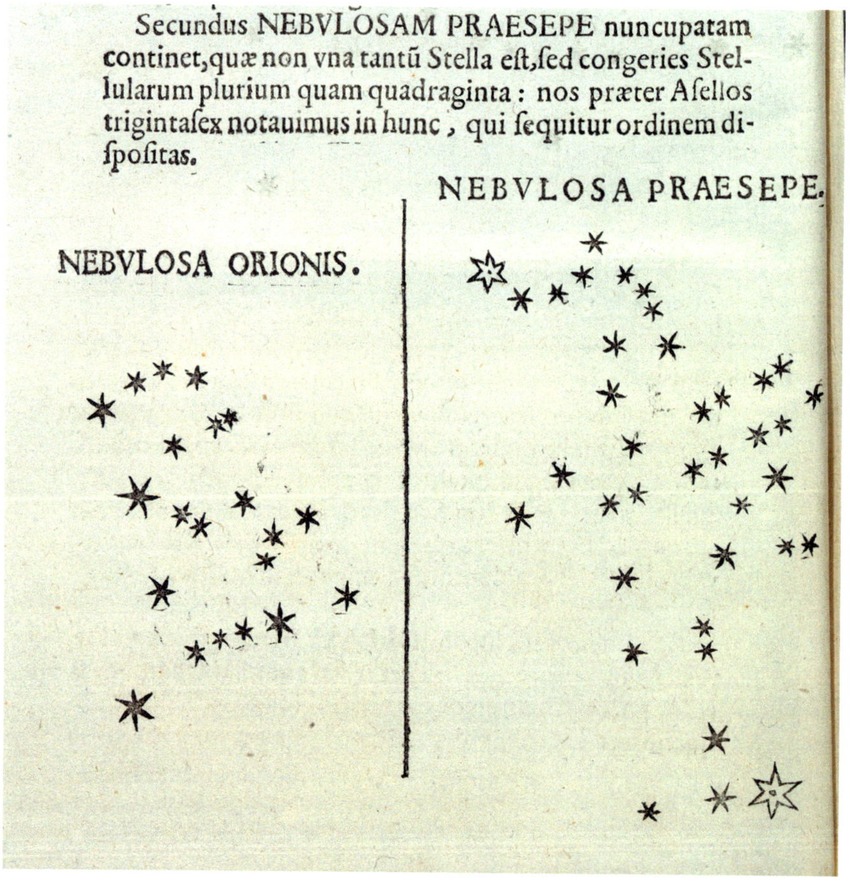

Figure IX.17: Sidereus Nuncius, D6r, Institute for Advanced Study.

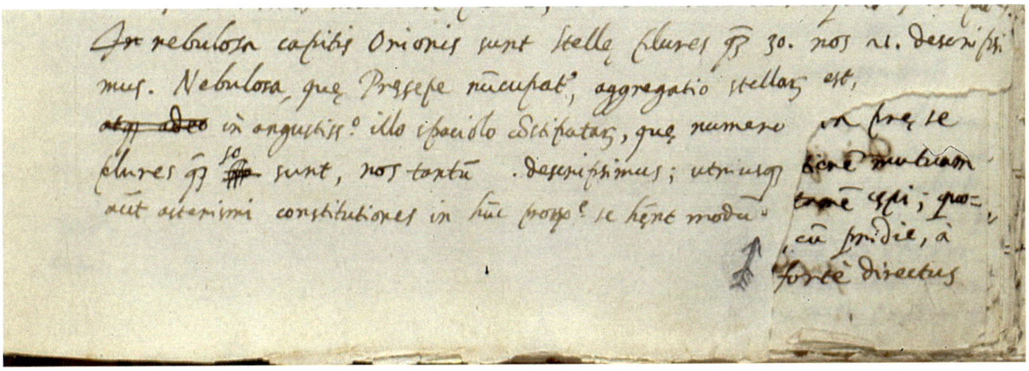

IX.18: BNCF, Ms. Gal. 48, 19r.

subjects in just this order in the drop-title of the book (5r: B1r), in the first sheet of the edition to be printed: "Observationes ... in ... Lacteo circulo STELLISQUE NEBULOSIS ...". And, as noted in chapter VI, Galileo's intention from the beginning to give attention to the Milky Way and nebulous stars was emphasized further on the second page of the text (fo. 5v: B1v) where he interpolated a clause specifically highlighting these topics: "... likewise to demonstrate that the substance of the stars that astronomers have called nebulous is wholly different from what has hitherto been believed will be pleasant and indeed most elegant." Finally, Galileo's Draft shows that the two subjects, Milky Way and nebulous stars, flowed directly from one to the other, and that the treatment of the nebulous stars was not an interpolation.

Galileo's treatment of the Milky Way, though extremely brief – considerably fewer than a hundred words – contains one of his boldest assertions. He has, he writes (D6r), resolved a question over which the philosophers had been torturing themselves for centuries with tedious disputations from which the world was now relieved (*altercationes omnes, quae per tot saecula Philosophos excruciarunt ab oculata certitudine dirimantur, nosque à verbosis disputationibus liberemur*): the Milky Way is in fact no more than a dense collection of small stars. Everywhere one trains the telescope upon it, these countless stars swim into view.

Even some modern scholars appear to be almost irritated by the definitiveness with which Galileo dismissed as irrelevant the opinions of so many unnamed authorities, ancient and modern: in the words of one eminent historian, his tone was "exaggeratedly melodramatic."[11] Here, opinions will differ. It is equally possible to admire a man who could be so unequivocal in the face of long and powerful authority. In any case, despite its

11 Pantin 1992 p. 81 sqq., notes 101, 102 ("un vocabulaire éxagérement mélodramatique"), and 103, gives a detailed survey of the various opinions on the Milky Way of Aristotle, Averroes, Albertus Magnus, Tycho Brahe, Christopher Clavius, and even of the pre-Aristotelian Democritus, whose opinion was transmitted primarily through Macrobius's *In Somnium Scipionis*. For another modernist "correction" of Galileo's statement, see Chaberlot 2003, p. 55: In the spring of 1611 the best Jesuit mathematicians including Clavius were asked to comment to Cardinal Bellarmine on telescopic observations of the skies. With regard to the Milky Way their opinion was that although it could not be denied that there were a multitude of small stars, some parts seemed to show a "more continuous density." This statement, "plus prudente et circonspect" than Galileo's, showed the latter to have been "présomptueux." The opinion of the Jesuit mathematicians is printed in *Opere* XI, p. 92 (Ep. 520: 24 April 1611).

brevity Galileo's argument was not so bald as at first may appear, for he immediately extended it to other parts of the heavens. The Milky Way was not a unique phenomenon, he emphasized. The same resolution of lighted cloudy areas into a multitude of small stars is widespread, particularly in the "nebulous stars." The two examples he discussed and illustrated, Praesaepe and the head of Orion, thus served as "manageable" and displayable examples of the phenomenon of the Milky Way. As such, they formed from the beginning part of Galileo's statement on the Milky Way. Nothing about them is an afterthought.

The adjacent woodblocks for Praesaepe and the head of Orion occupy almost exactly the lower half of D6v; with a combined width of about 115 mm, they extend beyond the type-page area into both the outer and the inner margins. In the last stage of getting the two pages D6r-v in type, Galileo expanded the accompanying explanatory text beyond his original draft, adding just enough words to make D6v a completely full page. On fo. 11r of the Draft, Galileo's discussion of these two nebulous stars totals about ninety words; as printed in *Sidereus Nuncius*, the discussion is half again as long, entirely rewritten, and even repetitious in parts. Without those expansions, the letterpress on D6v would have ended not at line 16, but approximately at line 9 or 10, and below the woodcuts there would have been an empty space.

The apparent delay associated with quire D may have been owing, at least in part, to extra time needed for getting the woodblocks of the asterisms made. The verbal text of the quire could be set, but could not be arranged into formes until the woodblocks were present. This presumed delay is partly corroborated by the paper stocks of the half-sheet D3.4, which, purely typographical, seems to have been printed after the full sheet (see further chapter XI). As discussed in chapter II, quires B–G of *Sidereus Nuncius* were printed on a single stock of paper (stock II, LA monogram with Tp cornermark), but the half-sheet D3.4 is a partial exception. The sheets used to print D3.4 were partly the standard stock of quires B–G, but in a number of copies a second stock (stock IV, GZ monogram, no cornermark) was used. The plausible implication is that the chief stock was used up during the print run of D3.4, and the remaining copies were printed on this remnant stock that was available in the printing shop.[12] In one other way the half-sheet D3.4 distinguishes itself from the full-sheet printing of *Sidereus Nuncius*: its type impressions are in all copies noticeably blacker and more deeply impressed than those of the full sheets. The contrast is, of course, most evident at the two openings where the half-sheet of D directly faces the full sheet: D3r against 2v, and D4v against 5r (figures IX.19–20). The implications of this are discussed in chapter XIII.

12 The GZ monogram watermark of stock IV is often difficult to see; the clearest example I have found is in the Florence, Biblioteca Nazionale Centrale's ordinary-paper copy. In certain large copies, it can be determined that the D3.4 half-sheet is the GZ Monogram stock, despite the absence of a watermark, because in those copies, if the half-sheet were on stock II, it would bear either the LA monogram main mark or its Tp monogram cornermark.

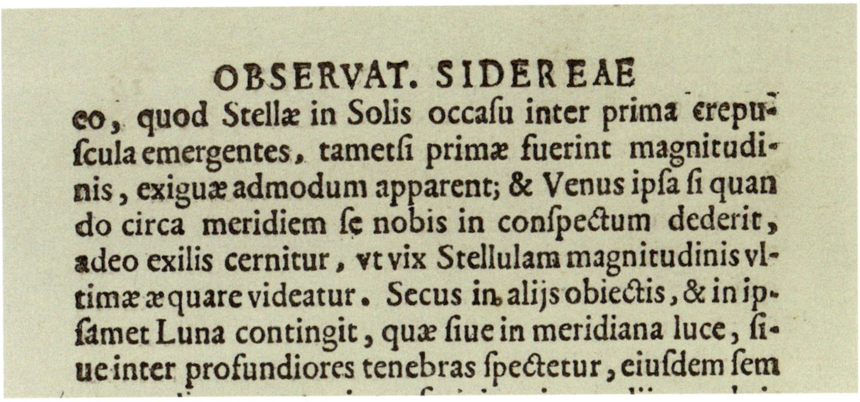
OBSERVAT. SIDEREAE
eo, quod Stellæ in Solis occaſu inter prima crepu-
ſcula emergentes, tametſi primæ fuerint magnitudi-
nis, exiguæ admodum apparent; & Venus ipſa ſi quan
do circa meridiem ſe nobis in conſpectum dederit,
adeo exilis cernitur, vt vix Stellulam magnitudinis vl-
timæ æquare videatur. Secus in alijs obiectis, & in ip-
ſamet Luna contingit, quæ ſiue in meridiana luce, ſi-
ue inter profundiores tenebras ſpectetur, eiuſdem ſem

XI.19: *Sidereus Nuncius, D4v, Institute for Advanced Study (half-sheet, heave type impression).*

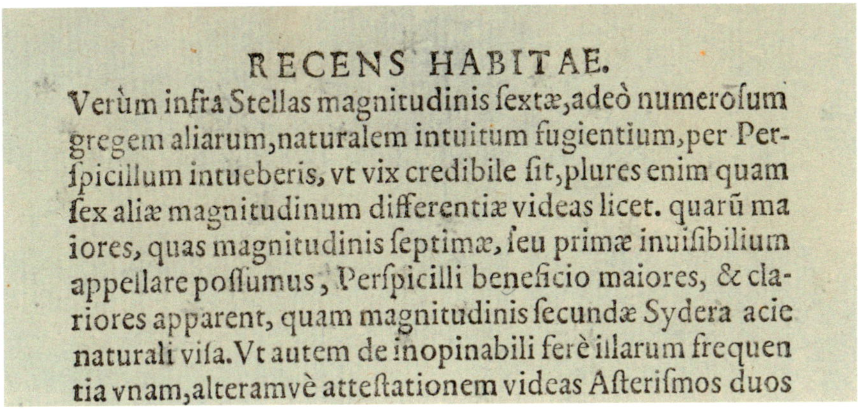
RECENS HABITAE.
Verùm infra Stellas magnitudinis ſextæ, adeò numeroſum
gregem aliarum, naturalem intuitum fugientium, per Per-
ſpicillum intueberis, vt vix credibile ſit, plures enim quam
ſex aliæ magnitudinum differentiæ videas licet. quarũ ma
iores, quas magnitudinis ſeptimæ, ſeu primæ inuiſibilium
appellare poſſumus, Perſpicilli beneficio maiores, & cla-
riores apparent, quam magnitudinis ſecundæ Sydera acie
naturali viſa. Vt autem de inopinabili ferè illarum frequen
tia vnam, alteramvè atteſtationem videas Aſteriſmos duos

XI.20: *Sidereus Nuncius, D5r, Institute for Advanced Study (full sheet, ordinary type impression).*

CALENDAR COVERING THE FIRST FOURTEEN WEEKS OF THE YEAR 1610

SUN	MON	TUE	WED	THU	FRI	SAT
					JAN 1	2
3	4	5	6	7 ♃	8 ♃	9
10 ♃	11 ♃	12 ♃	13 ♃	14 ♃	15 ♃	16 ♃
17 ♃	18 ♃	19 ♃	20 ♃	21 ♃	22 ♃	23 ♃
24 ♃	25 ♃	26 ♃	27 ♃	28	29	30 ♃
31 ♃	FEB 1 ♃	2 ♃	3 ♃	4 ♃	5	6 ♃
7 ♃	8 ♃	9 ♃	10 ♃	11 ♃	12 ♃	13 ♃
14	15 ♃	16 ♃	17 ♃	18 ♃	19 ♃	20
21 ♃	22	23	ASH WED 24	25 ♃	26 ♃	27 ♃
28 ♃	MAR 1 ♃	2 ♃	3	4	5	6
7	8	9 ♃	10 ♃	11	12 ♃	PRINTING FINISHED 13
14	15	16	17	18 ♃	19	20 ♃
21	22	23	24	25 ♃	26	27 ♃
28 ♃	29 ♃	30	31	APR 1	2 ♃	3 ♃
PALM 4	5 ♃	6 ♃	7	8	9 ♃	10
EASTER 11 ♃						

♃ Jupiter observations — cloudy night — dated documents and letters — Galileo's 46th birthday

X.
Quires E – F – G

Part II of *Sidereus Nuncius*, the final three gatherings, contains Galileo's Jupiter observations and analysis, followed by a summation of 64 lines. The whole ends precisely at line 34 of G4v (28v), extended to a 35th line containing just the centered word F I N I S. The layout and forme construction of quires E–G are considerably simpler than of B–D, for there were no woodcut or copperplate illustrations to be calculated and allowed for. Galileo reported his observations in chronological sequence, exemplified with typographic diagrams of the Jupiter-moon dispositions marked off from the text by white spaces, all of which fit evenly within the standard type-page area of 34-line height and 95-mm width.

However, few books printed before this, if any, could have shared the unusual textual circumstances of this part of *Sidereus Nuncius*. The foundations of the text on these pages were laid night by night, at the same time that the pages themselves were being drafted and set in type. We know, in fact, that the author was more or less haunting the printing shop throughout production – if not, he would have had no need even to go to Venice. He himself stated "I sent it out in the form of a newsletter, the greater part written while printing of the earlier parts was underway" (*l'ho mandato fuori in forma di avviso, scritto la maggior parte mentre si stampavano le cose precedenti …*).[1] In principle, the type-pages of sheets E to G could have been set gradually, so to speak, one after the other in sequence, as Galileo gathered and redacted his Jupiter observations, the pages kept standing until the last page necessary to complete the sheet was ready. In a well equipped printing shop it would not have drawn unduly on the typecases to keep eight quarto pages standing in type – about 5,500 ems – before printing the sheet. And, it may be noted that because of the numerous spaced-out Jupiter-moon diagrams, discussed in greater detail presently, the quantity of type required for these pages was distinctly less than if the pages had been set solid to 34 lines each, that is 252 lines per eight-page sheet. Sheet E contains 192 lines of text setting, less than six solid pages, and sheet F 167 lines, less than five solid pages. In particular, by this procedure all of sheet E, whose observations cover 7 through 22 January, could have been set more or less immediately after Galileo reached Venice at the end of that month, and on Monday 1 February in Venice his observations would have extended still further, through the first three pages of sheet F. And yet, as will be shown in detail in the following chapter, the running-title evidence of the edition argues

1 *Opere* X p. 300 (Ep. 277), Galileo to Belisario Vinta, 19 March 1610.

decisively that this was not the work procedure: sheet E did not go to press at the beginning of the production process.

As summarized in chapter V, there is a substantial body of autograph material relating to the formation of the second part of *Sidereus Nuncius*. First, there is Galileo's Draft, hereafter referred to as Draft II to distinguish it from the first part of the Draft, almost all of which was written earlier. As with Draft I, Draft II can be divided into several separate units, distinguishable by their writing and layout, by their paper stocks, and by linking symbols that tie the end of one unit to the beginning of the next. Besides this we have the sheets containing Galileo's original Jupiter observations (hereafter, Observation Sheets). Third, as briefly noted in chapters I and V, an autograph leaf at the University of Michigan contains very early notes – by Stillman Drake's argument, the earliest extant notes – on Galileo's Jupiter observations of 7 to 15 January 1610 (figure I.4). Drake may have been wrong in his interpretation, but the leaf is nonetheless important for understanding the first days of the Jupiter observations.[2] Finally, there are three leaves of a Maquette for laying out the typographic diagrams of Jupiter and its moons (Ms. Gal. 50, fos. 70–72).

To recapitulate chapter V, Draft II consists of a folio fascicle of two sheets or four leaves, containing the draft of the Jupiter observations from 7 January to 16 February; and a folio sheet, that is, two leaves (the second of which blank) containing the draft of the Conclusion. These three sheets are all (Venetian) Rampant Lion paper, the stock used in Draft I for the Introduction and Telescope topics (one sheet) and for the Silhouette Problem topic (one sheet); for the Jupiter-moon Maquette; and for Galileo's 30 January 1610 letter to Vinta, written in Venice. Between the two-sheet fascicle of observations and the single-sheet conclusion were eventually fitted in two single leaves. The first of these is a leaf torn from the third Observation Sheet, containing the observations of 17 February to 1 March; the second a smaller quarto leaf, that is with horizontal chainlines, containing the observation of 2 March, and two drafts of a transition that formed a bridge to the Conclusion. No watermark is preserved on the quarto leaf, but its chainline spacing suggests, without certainty, that it may be the Rampant Lion stock (see figure V.7). The Jupiter Observations were written on three successive folio sheets. The first two of these, containing the original observations from 7 to 23 January, and then from 24 January to 16 February 1610, are (Paduan) Unicorn paper, the stock used in Draft I for the topics of The Moon's Rough Surface and Earthshine (a two-sheet fascicle), for the Height of the Moon's Mountain (a single leaf), and for the Fixed Stars (a single sheet); and for three letters that Galileo wrote in Padua between 30 October and 7 December 1609. The third Observation Sheet, with the observations of 17 February to 1 March 1610, and then, after a hiatus, of 9 March to

2 Drake 1976. Gingerich & Van Helden 2003 argue against Drake that the Observation Sheets contain the primary record of Galileo's Jupiter observations. However, as they do not mention the Michigan Leaf at all, it is uncertain whether they overlooked it, or simply considered Drake's argument to be prima facie wrong. Pantin 1992, p. 87 (note 127), not cited by Gingerich & Ven Helden, had already presented an argument that the Michigan Leaf ("brouillon Drake") "était une copie (peu soigneuse) du journal, et non l'inverse." But as Drake sufficiently showed, even if it derives from the Observation notes, this leaf contains certain calculations concerning the movements of the moons that are unique to it, and so it remains a significant, but still unresolved, part of the history of *Sidereus Nuncius*.

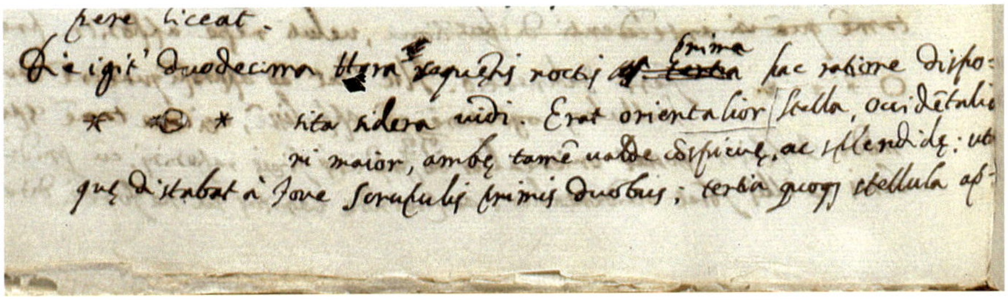

X.1: BNCF, Ms. Gal. 48, 20v.

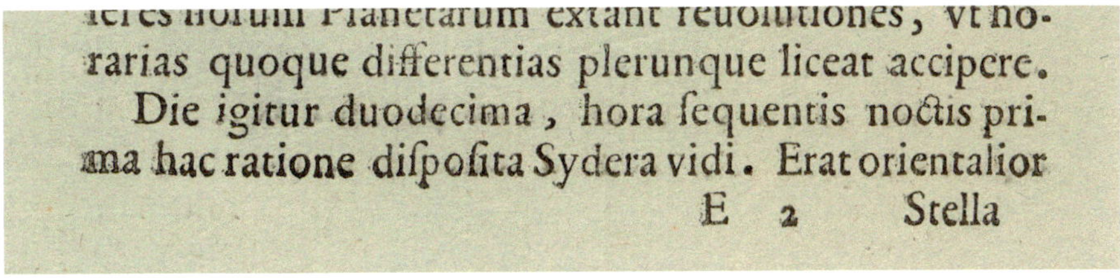

ieres horum Planetarum extant reuolutiones, vt ho-
rarias quoque differentias plerunque liceat accipere.
Die igitur duodecima, hora ſequentis noctis pri-
ma hac ratione diſpoſita Sydera vidi. Erat orientalior
E 2 Stella

X.2: *Sidereus Nuncius*, E2r, Institute for Advanced Study.

18 April 1610, is (Venetian) Crossbow in Circle paper, used also by Galileo for four letters he wrote in Venice from 13 to 19 March 1610 (again, see figure V.7).

As with part I, we do not have the setting copy for part II. It may be presumed that it was delivered to the printing shop and eventually became waste paper there. Galileo's own working archive, what we have called his autograph dossier, was so substantially – it would hardly be wrong to say, so religiously – preserved by himself, his son, and his disciple Viviani, that if the setting copy had been returned, it would have had unusually favorable odds for survival. There is a strong hint in Draft II that while he had it in hand, Galileo also saw proofs of some kind of the first few pages of quire E. At the foot of fo. 1v of Draft II, Galileo (we presume, by this explanation) added in ink a bracket to demarcate the end of E2r and beginning of E2v (figures X.1, 2). This treatment corresponds precisely to that of fo. 8r of the Fair Copy of Part I, as discussed and reproduced above (figures VII.16–18), with a scratched bracket demarcating the end of C3v and beginning of C4r, that is, within the inner forme of sheet C.

It has been suggested or stated by most students of *Sidereus Nuncius* that the first (quires B–D) and second (quires E–G) parts were worked on in the printing shop concurrently, or perhaps, rather, in some alternating pattern whose exact details of concurrency cannot be precisely determined. The following chapter will examine that problem more closely, with special attention to the evidence of the book's running titles. The "expansion problem" of quire D, with the last two of its six leaves lacking printed foliation, clearly indicates that quire E, with foliation beginning at 17, had been set, printed off and distributed before D. Because the Jupiter observations are dated, they provide a series of termini before which each sheet could not have been printed, or rather before which Galileo could not have prepared the full text fitted within each sheet. For sheet E, the terminus is Friday 22 January; for sheet F, Saturday 13 February; and for sheet G, Tuesday 2 March. More realistically, the terminus in each case must have been at least the following day, for these limiting dates are the nights when the respective observations were made. There would have been two copying stages – the Draft, and the lost setting copy – between the observations and the printing shop. Moreover, the handwriting of the Draft, taking into account

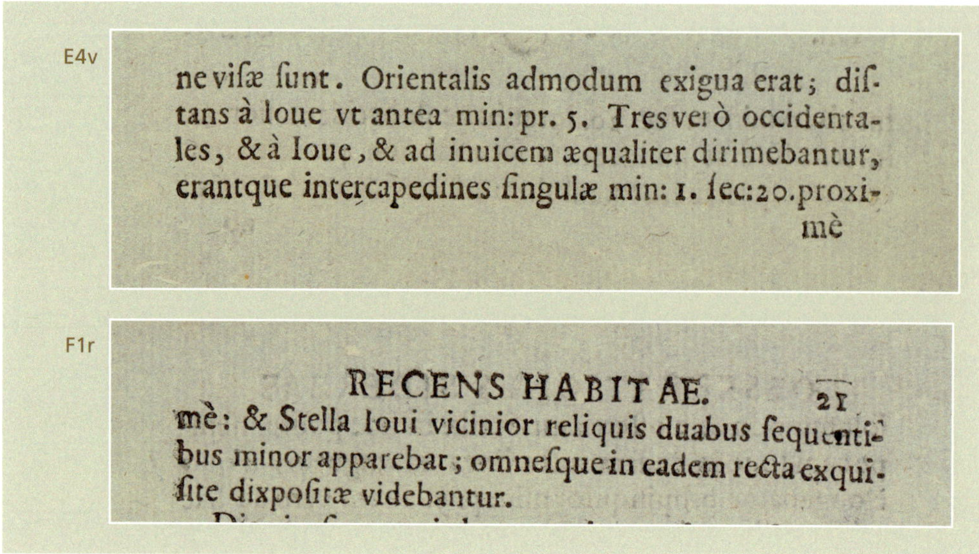
E4v

ne viſæ ſunt. Orientalis admodum exigua erat; diſ-
tans à Ioue vt antea min: pr. 5. Tres verò occidenta-
les, & à Ioue, & ad inuicem æqualiter dirimebantur,
erantque intercapedines ſingulæ min: 1. ſec: 20. proxi-
mè

F1r

RECENS HABITAE. 21
mè: & Stella Ioui vicinior reliquis duabus ſequenti-
bus minor apparebat; omneſque in eadem recta exqui-
ſite dixpoſitæ videbantur.

X.3: Sidereus Nuncius, Institute for Advanced Study.

also breadth of the quill's nib and ink tints, seems to indicate that in many places, a single writing stint would include the observations of several evenings. That is, the Draft itself was not for the most part written day by day.[3]

Sheets E–G are not independent of each other, each containing a self-standing set of observations. Sheet E ends in the midst of Galileo's second observation of 22 January, and indeed in the midst of a word, which is completed on F1 (figure X.3). Sheet F likewise ends in the midst of a word within the observation of 13 February, the word being completed on G1r. In terms of printing of formes, it is to be noted that the latest observations of sheets E and F are of course contained on their last page (4v) and therefore are in their outer formes, which were printed after the inner formes.[4] We do not know whether this inner-forme-first procedure simply reflected a common shop practice of the printer, but in any case, it meant in principle that the inner formes of sheets E and F could have been made ready for printing before the latest observations incorporated in those sheets had been set in the outer formes.

The first page of Draft II reveals three noteworthy features (figure X.4). First, it was marked by Galileo with the Sagittarius symbol, linking its beginning to the end of the text of part I (fo. 11r of the draft: see figure IX.18). Second, in paler brown ink at the top is written, including a partial deletion: "~~Sidere~~ Obseruationes Sidereę – Recens habitę"; that is, the running title of the printed edition. Third, the introductory paragraph of part II contains a "prospective" statement that is in a sense the reverse of Galileo's earlier "retrospective" statement (*Sidereus Nuncius*, 6r) about first hearing of the telescope, where "eight months ago" in the Draft, correct for its time of writing, late January 1610, became in print "ten months ago", correct for the publication of *Sidereus Nuncius* in March 1610.

3 My own observations of the original of the Draft have not been detailed enough to allow more than general remarks.

4 The final Jupiter observation, of 2 March 1610, is printed on the inner forme of sheet G (3v).

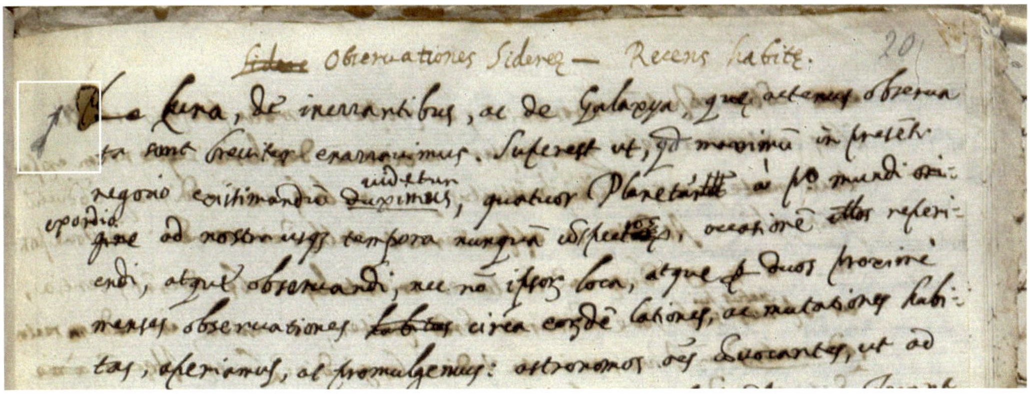
Observationes Sidereæ — Recens habitæ.

X.4: BNCF, Ms. Gal. 48, 20r.

In the introduction to part II, Galileo states that his observations of the four planets, that is Jupiter's moons, had been gathered over nearly two months (*per duos proximè menses obseruationes … habitas*). Thus, when he first drafted this section, Galileo was already planning to include many more weeks of observation. Whatever the exact date of the last Jupiter observation that Galileo hoped to incorporate in his book, he could never have intended to publish *Sidereus Nuncius* earlier than March. The point is worth keeping in mind, for the common accounts of this book have emphasized almost exclusively Galileo's haste to get it out. Yet the Aldine motto *Festina lente*, make haste slowly, applies precisely to *Sidereus Nuncius*. There was haste in its making, but this must be counterpoised against an intentional delaying element: the book was not to be a February publication, for it was always intended to bring in "almost two months" of Jupiter observations. In the event, the Jupiter observations extended to just one day less than eight weeks.

The layout of part II, with diagrams of Jupiter and its moons – sixty-five in all – incorporated within the text itself, shows a major change from Galileo's original plan. It will be recalled (see figure I.3) that on the night of 15 January, Galileo envisioned that his diagrams, corresponding to the sketches in the Observation Sheets, would be cut in woodblock, with white stars on a black ground. A series would be executed on a single block, then cut into individual strips to be placed where needed. This would, indeed, have been an elegant representation if the blocks were cut by a competent craftsman.

However, this plan was abandoned before printing began, leaving no traces within Galileo's surviving papers. Instead, the Jupiter-moon diagrams were laid out typographically. Jupiter is represented by an upper-case O, set on its side rather than upright, from one of the printing shop's titling fonts – the font, in fact, with which the words "GALILEO GALILEO" were set on the title-page. The moons are represented by asterisks from several fonts, all positioned according to Galileo's instructions. The reason for the change must have been the pressures of time. For instance, the woodcut diagrams necessary for the inner forme of sheet E, ten in all, could not have been finished until Galileo handed over to the blockcutter his model for his last observation within that forme, 20 January. And, if the diagrams were cut on a larger block, to be eventually sawn into strips, none of the pages needing the blocks could be fully prepared until that date. Typographic setting, less elegant than the woodcut plan, removed the need for this secondary production stage which would have added an element of uncertain control.

Several other details of the Jupiter-moon diagrams will be better understood after the relation of the Observation Sheets (figure X.5) to Draft II has been reviewed. Fundamentally, Draft II recasts the text of the Observation Sheets in smoother and more formal language, with interpolations in which Galileo explains at varying length his reasonings and even doubts, creating a compelling intellectual narrative out of the daily observations. A simple example is that of 15 February, where the Observation record's "there were three eastern stars, none western" becomes in the Draft "there were assuredly three eastern stars, but no western was, indeed, to be seen": *tres erant orientales stellę, nulla occidentalis* > *tres* NEMPE *erant orientales stellę, nulla* UERO CERNEBATUR *occidentalis*.

At other places Galileo added emotion to the terse and cool language of the Observation Sheets. For the 8th of January, he wrote in Draft II, "I do not know what fate led me" (*nescio quo fato ductus*) to look again at the three small but bright stars that he had seen in a line bracketing Jupiter the night before. In the Observation record, of course, there were no musings on fate. As to the next evening: in the Observation record, Galileo wrote simply, "the 9th was cloudy": *Adi .9. fù nugolo.* In Draft II this became, "with highest longing I awaited the following night, but truly I was frustrated in my hopes, for the sky was covered over with clouds": *quapropter maximo cum desiderio sequentem expectaui noctem, uerum à spe frustratus fui, nubibus enim undiquaque obductum fuit cęlum.*

There are several indicative disparities between the Observation Sheets and Draft II within the early days of observation that give clues to when the latter began to be drawn up. The first Observation Sheet records no specific times of evening until the second observation of 15 January, that is, at precisely the point where Galileo changed the language of his notes from Italian to Latin. His first words in Latin on this sheet were, "The former observation was at the third hour of the night, but at the seventh hour only three stars were present …": *Fuit pręcedens constitutio hora noctis 3. sed hora .7. tres tantum aderant stellulę* …. Thus, Galileo's decision to publish his findings, and consequently to change his notes to Latin, correlates precisely with the decision to keep a more accurate record of the observation times. In Draft II Galileo "pre-dated" this decision, taking it back to the 11th of January. At the end of his narrative for that day he stated his intention to note thereafter both the times of observations, and the distances separating the "new planets" at those times, for their swift movements produce discernible changes hour by hour: *interstitia quoque inter nouos planetas per Perspicillum, superius explicata ratione, dimetitus sum: horas quoque oberuationum apposui, presertim cum plures in eadem nocte habitę fuerunt adeo enim celeres horum Planetarum extant reuolutiones, vt horarias quoque differentias plerunque liceat accipere.*[5]

Thus, in the Draft Galileo placed under the eleventh of January an idea that he had not formed until the fifteenth of January. He moved it earlier to prepare for his narrative of 12 January, the first day on which he recorded two observations, at the first hour of the night (that is, after sunset) and at the third hour. The times he gave for those observations must have been retrieved from memory, for they are not recorded in the Observation Sheet, which refers only obliquely to the second observation. Galileo wrote that two moons (as they will be called hereafter, for simplicity), one to the east and one to the west, were visible,

5 In *Sidereus Nuncius* 18r (E2r) the passage was slightly reworked.

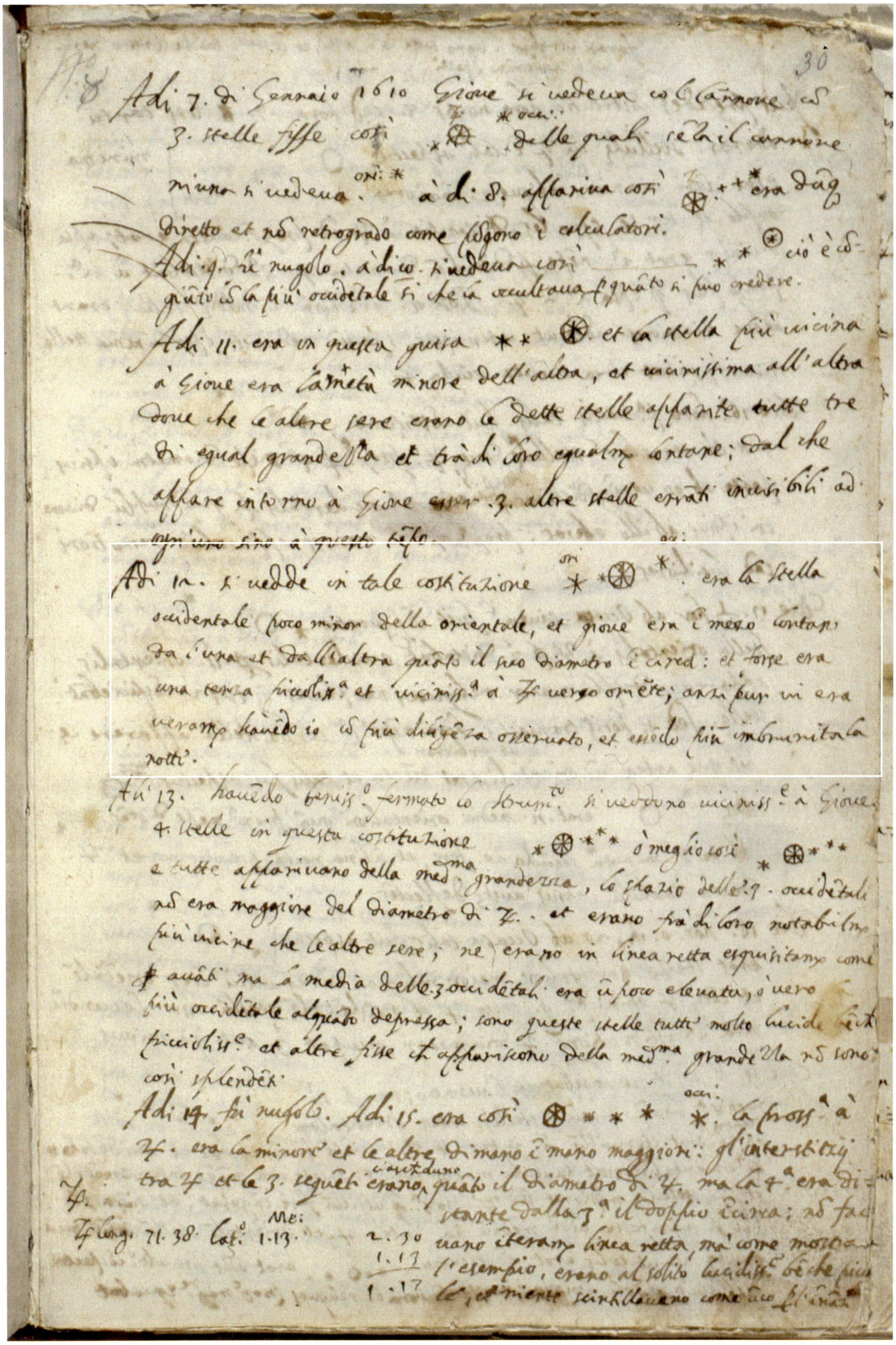
Adi 7. di Gennaio 1610 Giove si vedeva col Cannone con 3. stelle fisse così * * ⊛ * delle quali senza il cannone niuna si vedeva. à dì 8. appariva così ⊛ * * * era dunq diretto et non retrogrado come vogliono i calculatori.
Adi 9. si nugolò. à dì 10. si vedevan così * * ⊛ cioè congiunto con la più occidentale sì che la occultava per quanto si può credere.
Adi 11. era in questa guisa * * ⊛ et la stella più vicina à Giove era la metà minore dell'altra, et vicinissima all'altra dove che le altre sere erano le dette stelle apparite tutte tre di equal grandezza et trà di loro egualmente lontane; dal che appare intorno à Giove esser 3. altre stelle erranti invisibili ad ogn'uno sino à questo tempo.
Adi 12. si vedde in tale costituzione * ⊛ *. era la stella occidentale poco minor della orientale, et Giove era in mezo lontano da l'una et dall'altra quanto il suo diametro in circa: et forse era una terza piccolissima et vicinissima à Giove verso oriente; anzi pur vi era veramente havendo io con più diligenza osservato, et essendo più imbrunita la notte.
Adi 13. havendo benissimo fermato lo strumento si veddono vicinissime à Giove 4. stelle in questa costituzione * ⊛ * * * ò meglio così * ⊛ * * * e tutte apparivano della medesima grandezza, lo spazio delle 3. occidentali non era maggiore del diametro di Giove. et erano frà di loro notabilmente più vicine che le altre sere; ne erano in linea retta esquisitamente come per avanti ma la media delle 3 occidentali era un poco elevata, overo la più occidentale alquanto depressa; sono queste stelle tutte molto lucide bench piccolissime et altre fisse che appariscono della medesima grandezza non sono così splendenti.
Adi 14. fu nugolo. Adi 15. era così ⊛ * * * *. la prossima à Giove era la minore et le altre di mano in mano maggiori: gl'interstizii tra Giove et le 3. seguenti erano ciascheduno quanto il diametro di Giove. ma la 4.a era distante dalla 3.a il doppio in circa; ne facevano esattamente linea retta, ma come mostra l'esempio, erano al solito lucidissime benchè piccole, et niente scintillavano come anco per l'innanzi.
Giove long. 71.38 Lat. 1.13
2.30
1.13
1.17

X.5: BNCF, Ms. Gal. 48, 30r.

"and perhaps there was a third, minute and very close to Jupiter on the east; or rather indeed it was definitely there, my having observed with greater care and the night being darker": *et forse era una terza piccolissima et uicinissima a Gioue uerso oriente; anzi pur ui era ueramente hauendo io con più diligenza osseruato, et essendo più imbrunita la notte* (figure X.5). The darker night implies, of course, a later time for the second observation. In Draft II there is a strong hint that when he wrote about the twelfth of January, say on the sixteenth or so, Galileo indeed had to consult his memory to provide the times of the observations. He recorded the first observation as having been made at the third hour of the night then changed this to the first hour (figure X.1); and the second observation as made at the fourth hour of the night, then changed to the third hour (figure X.6).

There is a similar disparity between the Observation Sheets and Draft II with regard to the separations between the moons and Jupiter, further emphasizing that in its narrative of the earliest observations, Draft II is, so to speak, retrospective. In the first Observation Sheet, distances between Jupiter and its moons were not recorded for 7, 8, 10, and 11 January. Galileo refers to this omission explicitly in Draft II, at the end of the 7th January narration. Believing at the time that he was seeing fixed stars, "I was not concerned with the distances between them and Jupiter": *de distantia inter ipsas et Iouem minime sollicicus* [sic] *fui*. However, as promised at the end of the 11 January narration, from 12 January onward he regularly expressed those distances in terms of minutes and seconds of arc, although his first such statement used the alternative "scrupulum", found in many astronomical tables, in place of "minutum", and in several other places he used the fuller terms "minutum primum" and "minutum secundum" for minute and second respectively.

The statements of distances between the moons in Draft II (and hence *Sidereus Nuncius*) for 12, 13, 15, and 16 January are all recalculations into minutes and seconds of what in the first Observation Sheet was expressed in terms of the diameter of Jupiter. For instance, in the Observation record for 12 January, "Jupiter was in the middle, distant from the one [moon] and the other by roughly its own diameter": *gioue era in mezo lontano da l'una et dall'altra quanto il suo diametro in circa*. In Draft II and *Sidereus Nuncius* this was stated as 2 minutes each. In a careful analysis Isabelle Pantin has pointed out that the recalculations from Jupiter-diameters to arc-minutes in Draft II contain a small and still unresolved mystery. Although Jupiter's angular diameter was implicitly equated in Draft II with two minutes, in the Observation record under 18 January Galileo estimated the diameter at three minutes "or something less": *Secundum meam existimationem Iovis diameter occupat 3'. aut quid minus*. It appears that very soon after, perhaps even by the next day 19 January, Galileo had revised this estimate downward.[6] The estimates of arc-minutes recorded in Draft II for 12 to 16 January reflect, therefore, a post-18 January opinion of the diameter of Jupiter.

After 18 January the Observation Sheet regularly recorded distances in minutes and seconds, not according to Jupiter's diameter. It is uncertain whether Galileo developed at this point a superior system for recording his observations, or simply continued to esti-

6 Pantin 1992, notes 125 (p. 86) and 133 (p. 88). Drake 1983, p. 218, hypothesized that Galileo at this time estimated Jupiter's diameter as one minute, but this raises conflicts with the evidence of the Observation Sheets, the Maquette, and *Sidereus Nuncius* itself that he did not fully encounter.

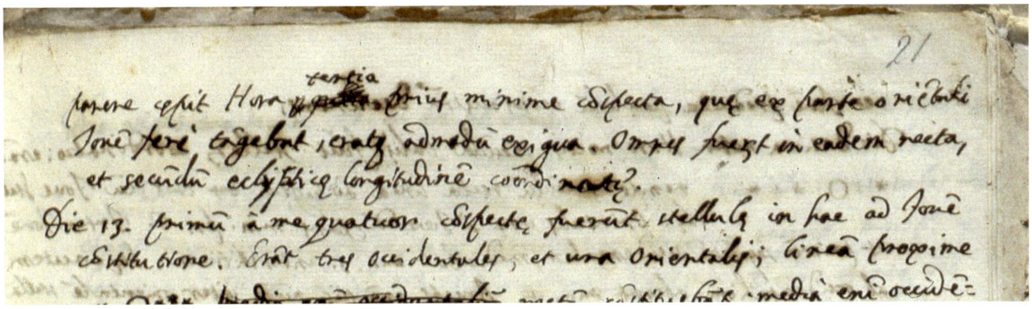

X.6: BNCF, Ms. Gal. 48, 21r.

mate on the basis of a revised angular diameter of Jupiter of 2 minutes. This figure was likewise considerably too great. As Drake has noted, by mid-1612 Galileo had established a better system of measurement and a much more accurate estimate of about 40 minutes.[7] It is probably relevant to this question, even if a precise explanation cannot be given, that Draft II contains no observation at all for 18 January, proceeding directly from 17 to 19 January. In *Sidereus Nuncius*, the 18 January observation is present (E3v) but very brief, with only slight paraphrase of the Observation record, and omitting the latter's estimate of the diameter of Jupiter.[8]

Any adjustments Galileo made with regard to the diameter of Jupiter were invisible to readers of *Sidereus Nuncius*, which did not allude to the question. They had before them only the typographic Jupiter-moon diagrams, with Jupiter represented by a sideways uppercase O whose printed width is about 6 mm, so that notionally an interval of one arcminute would correspond to about 3 mm. Thus, a moon that Galileo recorded as 6' from Jupiter should be represented by an asterisk positioned about 18 mm from one side of the Jupiter-O. Overall, the diagrams roughly follow this scale, but day by day and moon by moon, there are a considerable number of "out-of-scale" placements. For example, on 26r (G2r), in the two consecutive diagrams for 18 and 19 February, the first represents a Jupiter-to-moon separation of 10' by a spacing of 36 mm, somewhat too much; and the second represents a separation of 13' by the same 36-mm spacing, somewhat too little (figure X.7).

Similarly, on E3r, the diagram for 16 January represents the western moon, 8 minutes from Jupiter, at a distance of 25 mm, which is according to the notional scale. But the following diagram for 17 January represents the western moon, 11 minutes from Jupiter, at a distance of only 23 mm rather 33 mm: a disparity large enough that we may perhaps speak of the diagram as containing a typographical error.

The placements of asterisks very close to the Jupiter-Os or to other asterisks tend to be out of scale. All such placements were, of course, created by means of setting blank quads, standing less than type high, to various widths, a procedure that only allowed for discrete, "quantized" distances. These were far from ideal for representing distances that Galileo sometimes expressed in increments of 10 seconds, or sixths of a minute. On the rough average in *Sidereus Nuncius* of 3 mm for one minute, a 10-second unit would be about

7 Drake 1976, p. 166. For Galileo's later calculations of the apparent diameter of Jupiter see also Shea 1990.

8 Pantin 1992, note 133 (p. 88): she points out that the Observation record states the distances of the two moons as 9 and 11 minutes respectively from the center of Jupiter, whereas in *Sidereus Nuncius* these are given as 8 and 10 minutes, that is, from the edge of Jupiter with its radius calculated as 1 minute.

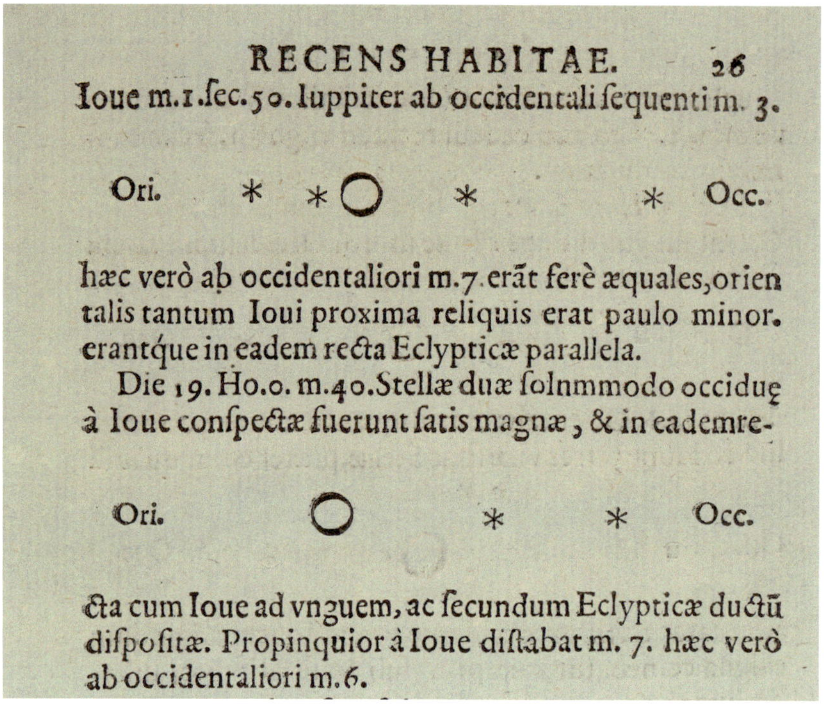
RECENS HABITAE. 26

Ioue m.1.ſec.50. Iuppiter ab occidentali ſequenti m. 3.

Ori. * * ○ * * Occ.

hæc verò ab occidentaliori m.7.erãt ferè æquales,orien
talis tantum Ioui proxima reliquis erat paulo minor.
erantque in eadem recta Eclypticæ parallela.

Die 19. Ho.o. m.40.Stellæ duæ ſolnmmodo occiduę
à Ioue conſpectæ fuerunt ſatis magnæ, & in eademre-

Ori. ○ * * Occ.

cta cum Ioue ad vnguem, ac ſecundum Eclypticæ ductũ
diſpoſitæ. Propinquior à Ioue diſtabat m. 7. hæc verò
ab occidentaliori m.6.

X.7: Sidereus Nuncius, G2r, Institute for Advanced Study.

0.5 mm. No quad was that narrow. Moreover the surface area of the asterisk types themselves would not allow for accuracy in close placements. Through the telescope Galileo saw the Jupiter moons as points of light, practically speaking dimensionless, but these had to be represented on the page by clumsily large types. When the stated distances between two of the moons was less than about a minute, the breadth of the asterisk types would not allow them to be placed close enough to each other to keep to the notional scale of 1 mm = 20 seconds. A good example is the observation of 1 March, when Galileo saw three moons not widely separated. The first was about 2 minutes east of Jupiter, the second 3 minutes east, and the third, brighter than the other two, 3 minutes 20 seconds east: … *Ioui proximi aberat ab eo m. 2. sequens ab hac m. 1. tertia m. 0 sec. 20. eratque reliquis clarior.* This was how he saw three points of light near Jupiter's disk, but on the printed page the physical distance between the centers of the second and third asterisks, representing 20 seconds, is very little less, if at all, than between the first and second asterisks, 60 seconds (figure X.8). Frequently, but with variations from copy to copy, some of the type quads used to space out the asterisks worked up in their formes and caught a small amounts of ink on their shoulders. These printed as vertical strokes or brackets, and, amounting to extraneous noise, must be mentally removed from their diagrams (figure X.9).

As already noted, Galileo prepared a Maquette to aid in the typographic construction of the Jupiter-moon diagrams. On three pages in double columns, it lays out the Jupiter-moon positions, night by night, with more precision than in the sketches of the Observation Sheets and Draft II, based on a scale of minutes written at the top of the first page. The figures of Jupiter were neatly drawn with a compass, with a diameter indeed corre-

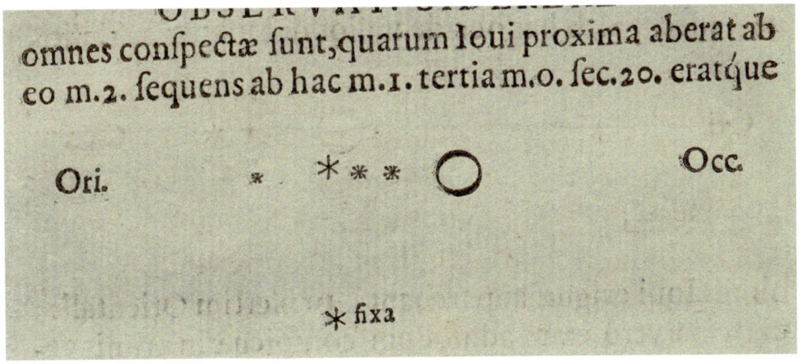
omnes conſpectæ ſunt, quarum Ioui proxima aberat ab eo m.2. ſequens ab hac m.1. tertia m.o. ſec.20. eratque

Ori. Occ.

fixa

X.8: *Sidereus Nuncius*, G3v, Institute for Advanced Study.

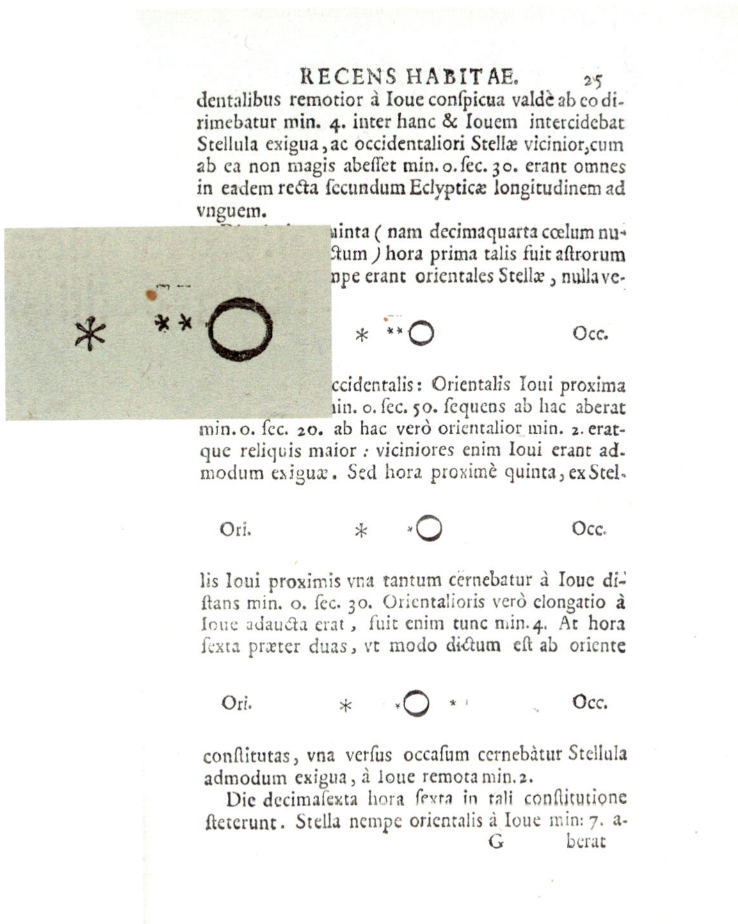
RECENS HABITAE. 25

dentalibus remotior à Ioue conſpicua valdè ab eo dirimebatur min. 4. inter hanc & Iouem intercidebat Stellula exigua, ac occidentaliori Stellæ vicinior, cum ab ea non magis abeſſet min. o. ſec. 30. erant omnes in eadem recta ſecundum Eclypticæ longitudinem ad vnguem.

uinta (nam decimaquarta cœlum nu-
ctum) hora prima talis fuit aſtrorum
npe erant orientales Stellæ, nulla ve-

Occ.

ccidentalis: Orientalis Ioui proxima
in. o. ſec. 50. ſequens ab hac aberat
min. o. ſec. 20. ab hac verò orientalior min. 2. eratque reliquis maior: viciniores enim Ioui erant admodum exiguæ. Sed hora proximè quinta, ex Stel-

Ori. Occ.

lis Ioui proximis vna tantum cernebatur à Ioue diſtans min. o. ſec. 30. Orientalioris verò elongatio à Ioue adaucta erat, fuit enim tunc min. 4. At hora ſexta præter duas, vt modo dictum eſt ab oriente

Ori. Occ.

conſtitutas, vna verſus occaſum cernebàtur Stellula admodum exigua, à Ioue remota min. 2.

Die decimaſexta hora ſexta in tali conſtitutione ſteterunt. Stella nempe orientalis à Ioue min. 7. a-

G berat

X.9: *Sidereus Nuncius*, G1r, Institute for Advanced Study.

sponding to about 2 minutes as represented by the scale (figure X.10).[9] With a few small exceptions, the layouts of the Maquette correspond very closely to the diagrams in *Sidereus Nuncius*, and the indicated spacing of the moons is very close to that of the typesetting. A close scrutiny of the original ink of the Maquette might reveal stages in its writing.

9 See Drake 1983, pp. 217–219; in his translation of the Jupiter portion of *Sidereus Nuncius* in this edition, Drake reproduced the Jupiter-moon diagrams from this Maquette rather than from the typography of *Sidereus Nuncius*.

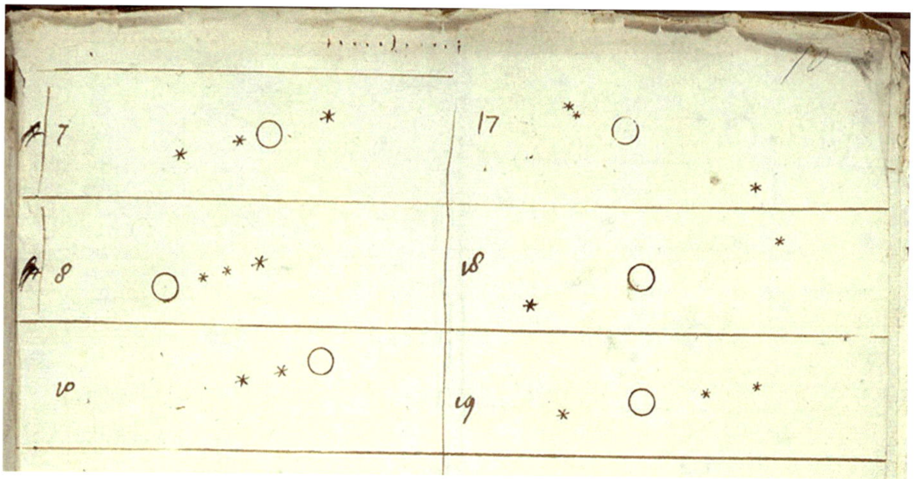

X.10: BNCF, Ms. Gal. 50, fo. 70.

The Maquette typically gives a somewhat more precise representation of the respective magnitudes of the moons than do either the Observation Sheets or Draft II, although even in it there are ambiguities. In *Sidereus Nuncius* there cannot in principle be any ambiguities, since different type-sorts were used for each of the magnitudes. Galileo recorded eventually four magnitudes of the moons, each represented by a designated sort. In descending magnitude they are: (1) an asterisk (that is, six-rayed character) with diameter of about 1.5 mm, perhaps from a larger font than the text font; (2) an asterisk with diameter of about 1.2 mm, which apparently is the asterisk belonging to the text font; (3) a type whose face bears eight rays in the shape of a rectangle, broader than it is tall; (4) a small five-rayed star. The latter two cannot strictly be called asterisks, yet it is convenient to refer to the group as Asterisks 1, 2, 3, 4. Asterisk 3 in particular does not look like a conventional character for any font, and seems to have been crudely prepared ad hoc. It rarely prints cleanly: conceivably, it is a collection of blank sorts whose top surfaces were turned into eight-rayed shapes by individual hand filing, rather than by multiple castings from a matrix. This asterisk was sometimes set on its side, making a taller, narrower character. For the observation of 13 January, the first night that Galileo saw all four moons, four examples of Asterisk 3 were set together, two broad and two tall. These moons, Galileo wrote, "were all of the same magnitude, and although tiny, they shone most brightly, much more splendidly than stars of the same magnitude": *omnes eandem prae se ferebant magnitudinem; ac licet exiguam, lucidissimae tamen erant, ac fixis eiusdem magnitudinis longe splendidiores.*[10] In the observation of 15 January all four grades of asterisk were used: Galileo recorded that the four moons were all to the west of Jupiter, each brighter than the one before, moving outward: *propinquior Ioui erat omnium minima, reliquae consequenter maiores apparebant* (figure X.11).

Asterisk 1 fell out of use well before the end of the Jupiter observations. It first appears in the observation of 11 January (18r: E2r) where two moons were seen, "that further to

10 See Pantin notes 115 (p. 84) and 126 (p. 87) at end: Galileo seems to merge two different senses of magnitude, size and brightness.

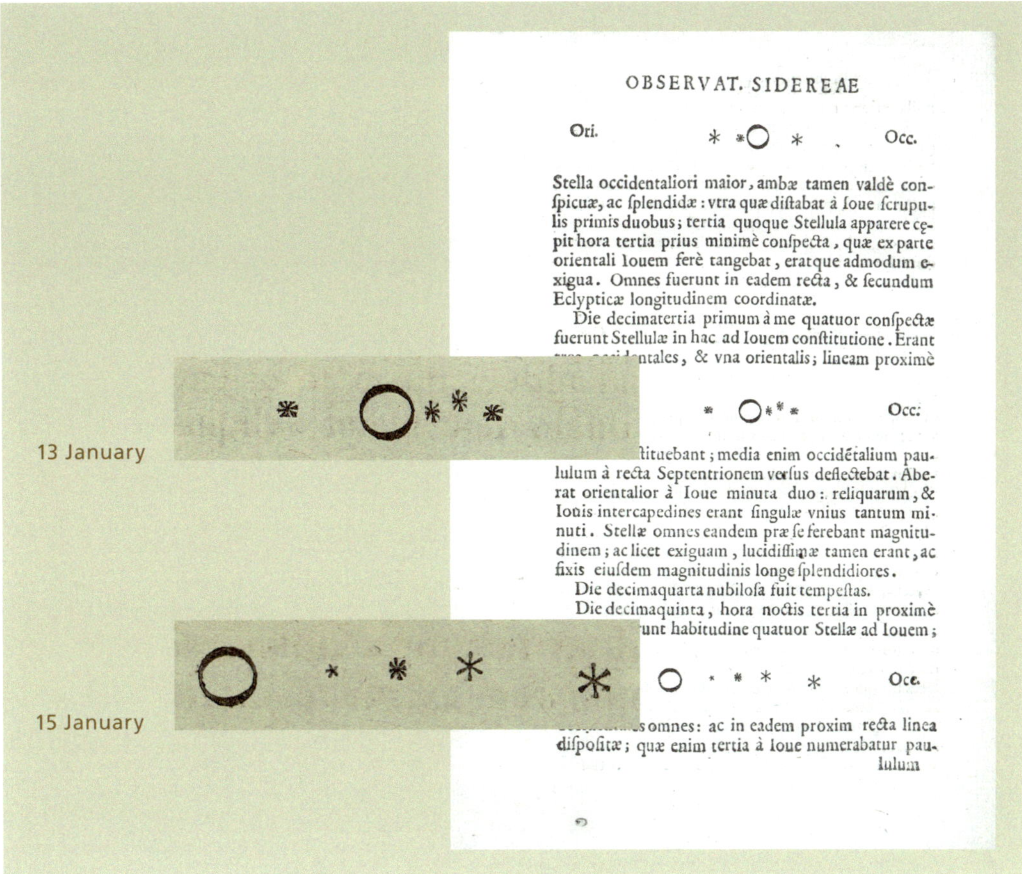
OBSERVAT. SIDEREAE

Ori. Occ.

Stella occidentaliori maior, ambæ tamen valdè conſpicuæ, ac ſplendidæ: vtra quæ diſtabat à Ioue ſcrupulis primis duobus; tertia quoque Stellula apparere cępit hora tertia prius minimè conſpecta, quæ ex parte orientali Iouem ferè tangebat, eratque admodum exigua. Omnes fuerunt in eadem recta, & ſecundum Eclypticæ longitudinem coordinatæ.

Die decimatertia primum à me quatuor conſpectæ fuerunt Stellulæ in hac ad Iouem conſtitutione. Erant ... ntales, & vna orientalis; lineam proximè

Occ.

... ituebant; media enim occidẽtalium paululum à recta Septentrionem verſus deflectebat. Aberat orientalior à Ioue minuta duo: reliquarum, & Iouis intercapedines erant ſingulæ vnius tantum minuti. Stellæ omnes eandem præ ſe ferebant magnitudinem; ac licet exiguam, lucidiſſimæ tamen erant, ac fixis eiuſdem magnitudinis longe ſplendidiores.

Die decimaquarta nubiloſa fuit tempeſtas.

Die decimaquinta, hora noctis tertia in proximè ... unt habitudine quatuor Stellæ ad Iouem;

Occ.

... s omnes: ac in eadem proxim recta linea diſpoſitæ; quæ enim tertia à Ioue numerabatur paululum

X.11: Sidereus Nuncius, E2v, Institute for Advanced Study.

the east almost twice as large as the other": *eratque orientalior duplo ferè maior reliqua.* This difference was depicted by Asterisks 1 and 2. For the second observation of 15 January, at the seventh hour (19r: E3r), Asterisks 1 and 2 were set side by side, but strictly speaking this was a mistake. The text of *Sidereus Nuncius* says nothing about a difference of magnitude, while the original Observation record explicitly states that the two moons were equal (*inter se ęquales*). For the observation of 18 January (19v: E3v) Asterisks 1 and 2 were set in accordance with the text, "the eastern star was larger than the western." Asterisk 1 was used for the last time, twice, for the observation of 25 January (21v: F1v), in accordance with the text's statement that the two moons were "quite large", *satis magnae.* This was the twenty-fifth diagram. The remaining forty diagrams do not draw on Asterisk 1, not even that of 19 February (26r: G2r), where the two moons were again defined as *satis magnae.*

In this last instance we know, in fact, that by Galileo's own standard, Asterisk 1 ought to have been set. From 15 February onward, Galileo entered the magnitudes of the moons on his Maquette, numbering them 1 to 4, and the two moons of 19 February are both marked as magnitude 1. Thus, the setting of Asterisk 2 in this diagram qualifies, in the tiniest way, as a compositorial error. Perhaps the likeliest explanation for the disappearance of Asterisk 1 is banal: the compositor forgot that it was available to him. Because of

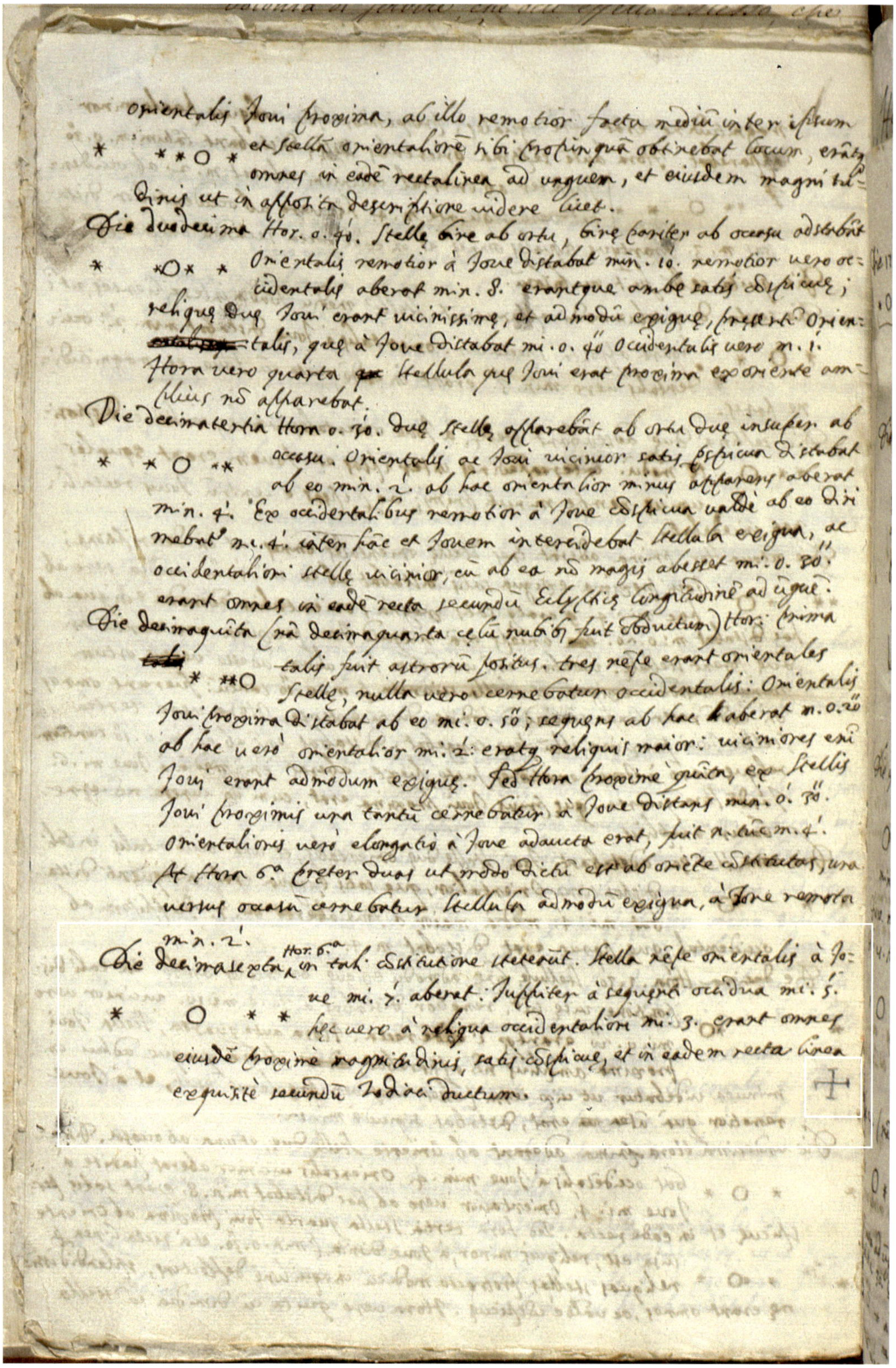

orientalis Jovi proxima, ab illo remotior facta mediu inter ipsum et stella orientaliore sibi propinqua obtinebat locum, eratq omnes in eadē rectalinea ad unguem, et eiusdem magnitudinis ut in apposita descriptione videre licet.

* **O *

Die duodecima Hor. 0. 40. stelle binae ab ortu, binae pariter ab occasu adstabant Orientalis remotior à Jove distabat min. 10. remotior vero occidentalis aberat min. 8. erantque ambe satis conspicue; reliquae duae Jovi erant vicinissime, et admodu exigue, presertim Orientalis, quae à Jove distabat mi. 0. 40 Occidentalis vero m. 1. Hora vero quarta stellula quae Jovi erat proxima ex oriente amplius nō apparebat.

* *O* *

Die decimatertia Hora 0. 30. duae stellae apparebāt ab ortu duae insuper ab occasu. Orientalis ac Jovi vicinior satis conspicua distabat ab eo min. 2. ab hac orientalior minus apparens aberat min. 4. Ex occidentalibus remotior à Jove conspicua valde ab eo dirimebat m. 4. inter hanc et Jovem intercidebat stellula exigua, ac occidentaliori stellae vicinior, cu ab ea nō magis abesset mi. 0. 30. erant omnes in eadē recta secundu Eclyptis longitudinē ad unguē.

* *O **

Die decimaquinta (nā decimaquarta celu nubibus fuit obductum) Hor. prima talis fuit astroru positus. tres nēpe erant orientales stelle, nulla vero cernebatur occidentalis: Orientalis Jovi proxima distabat ab eo mi. 0. 50; sequens ab hac aberat m. 0. 20 ab hac vero orientalior mi. 2. eratq reliquis maior: viciniores enī Jovi erant admodum exigue. Sed Hora proxime quinta, ex stellis Jovi proximis una tantu cernebatur à Jove distans mi. 0. 30. orientaliores vero elongatio à Jove adaucta erat, fuit n. tūc m. 4. At Hora 6a preter duas ut modo dictu est ab oriente constitutas, una versus occasu cernebatur stellula admodu exigua, à Jove remota min. 2.

* **O

Die decimasexta Hor. 6a in tali constitutione steterūt. stella nēpe orientalis à Jove mi. 7. aberat. Juppiter à sequenti occidua mi. 5. haec vero à reliqua occidentaliore mi. 3. erant omnes eiusdē proxime magnitudinis, satis conspicuae, et in eadem recta linea exquisite secundu Zodiaci ductum.

* O * *

X.12: BNCF, Ms. Gal. 48, 23v.

these magnitude numbers, it should be noted, the Maquette became, from 15 February on, part of the original observation record for Jupiter; the Observation Sheets record no magnitudes under these dates. Galileo did not begin entering magnitudes in the third Observation Sheet until Tuesday 9 March, a week after the 2 March observation incorporated in *Sidereus Nuncius*, and the first observation made since that date.

Various records over a week or so around the middle of February bring detail and life to our picture of Galileo's activities as *Sidereus Nuncius* was in process of being made. Day by day, the events are:

FRIDAY 12 FEBRUARY: Galileo travels from Venice to Padua, where he finds Belisario Vinta's letter of 6 February, assuring him of the Grand Duke's eager interest to see Galileo's publication. The same delivery would have brought him Enea Piccolomini's letter of 6 February (with apologies for lateness: Enea has been out hunting almost every day), likewise attesting the Grand Duke's more than benevolence. That evening, Galileo makes a Jupiter observation not long after sunset.

SATURDAY 13 FEBRUARY: Galileo writes to Vinta, asking for a quick and confidential reply to be sent directly to Venice, on the question of whether it would be better to name the new planets (moons) Cosmici, after Cosimo, or Medicei, after Cosimo and his three brothers. The answer will affect the title-page and dedication (that is, sheet A), which are being held in abeyance until the answer is given. Galileo observes Jupiter not long after sunset

SUNDAY 14 FEBRUARY: Galileo returns to Venice; the sky that night is cloudy.

MONDAY 15 FEBRUARY: On Galileo's 46th birthday he makes three Jupiter observations that are eventually printed on the first page of the last quire, G. For the first time he quantifies numerically the magnitudes of the moons, entering the numbers not in the Observation records but in the Maquette. The observations Galileo made in Padua on the 12th and 13th February are eventually printed on the last page of sheet F, that of the 13th spilling over to lines 1–6 of G1r.

TUESDAY 16 FEBRUARY: Galileo makes a Jupiter observation quite late in the evening (*Hora 6.a*), entering it in the second Observation Sheet, whose last page leaves ample space for an entry to follow this one. But the space was not filled: this was the last observation that Galileo transferred to Draft II (4v), and he made only minimal changes from the wording of the Observation record (figure X.12).

WEDNESDAY 17 FEBRUARY: Galileo makes a Jupiter observation an hour after sunset. This begins the third Observation Sheet, on Venetian paper, whose first page contains records extending through Monday 1 March (figure X.13).

THURSDAY 18 FEBRUARY AND AFTER: Galileo continues to enter his observations in the third Observation Sheet. Draft II has been set aside and no longer figures in the transmission history of the text. Looking forward, it should be noted that because of cloudy evenings, Galileo was only able to make one observation over the five days Saturday 20th to (Ash) Wednesday 24th of February. On Saturday 20 February Vinta replied to Galileo's request of the 13th: on Vinta's advice, Cosimo wished the new planets to be called "Medicei," for "Cosmici", of the same formation as the word "cosmos," might

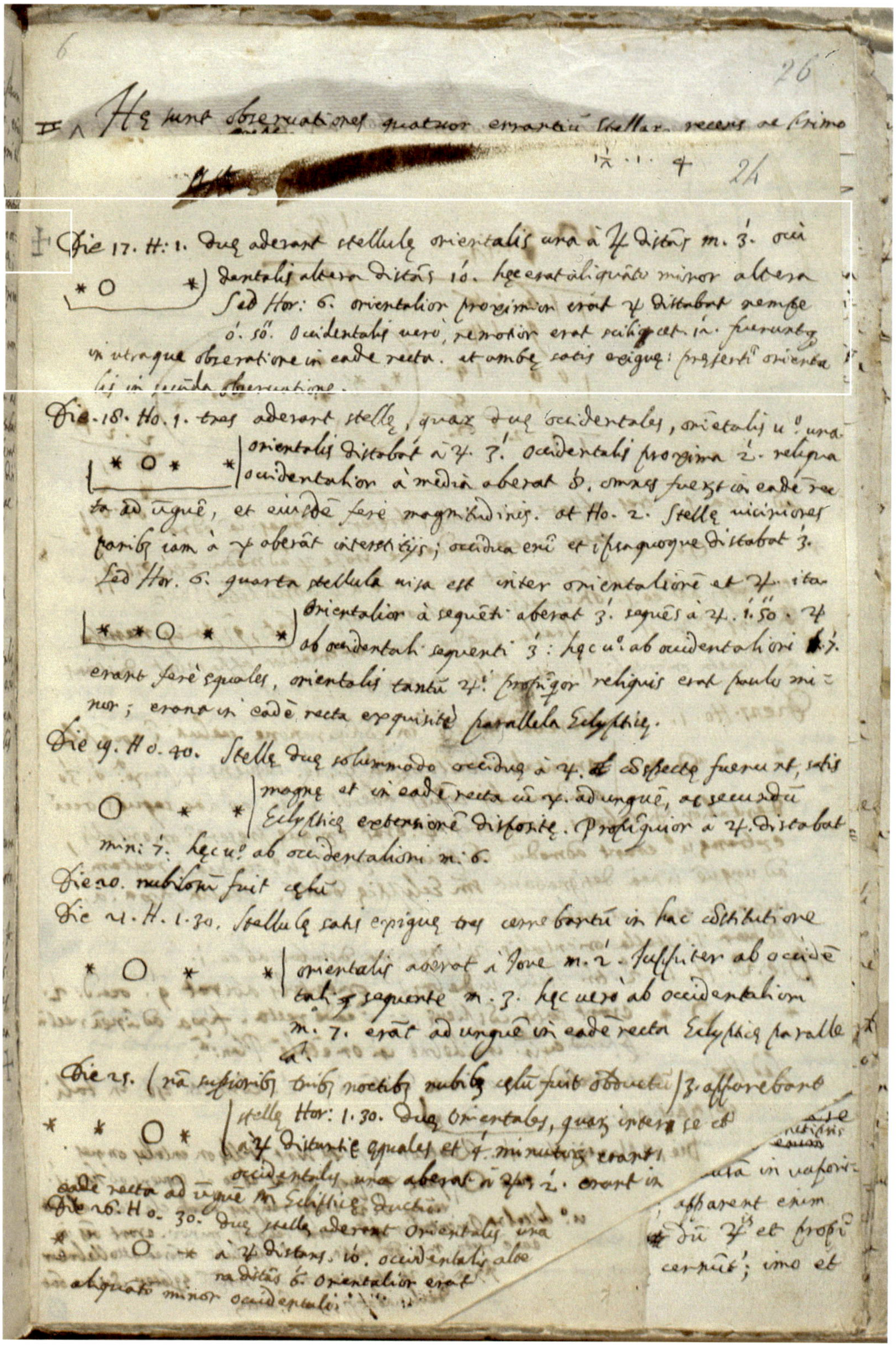

X.13: BNCF, Ms. Gal. 48, 24r.

lose its point with readers. Galileo probably would have received this reply on about Friday 26 February.

It is apparent that during the days just reviewed, Galileo had the completion of *Sidereus Nuncius* well in mind and sight. The question of whether to call the new planets "Cosmici" or "Medicei" has been interpreted in several ways, none with real proof. By the interpretation of this essay, Galileo's first thought had been to call them Cosmici in direct compliment to Cosimo II, and he had proceeded on that basis, having used the expression "Cosmica Sydera" on the drop-title of sheet B, the first sheet to be printed off. Not until 12 February, or shortly before, did the alternative "Medicei" occur to him. It should be noted that in the Draft of the conclusion to the Jupiter observations Galileo similarly refers to the "Cosmica sidera" and "Planetę Cosmici". Thus, the Draft II conclusion (which, it will be recalled, was written on an independent sheet) was probably finished by about 25 February or earlier, that is, before he received Vinta's reply that the "Medici" compliment was the preferable one, and a week or more before he made the last Jupiter observation incorporated into *Sidereus Nuncius*. In the printed edition, the Draft's references to "Cosmica" / "Cosmici" were changed to "Medicea" – in the second instance, creating a careless grammatical error that Galileo did not notice until well after sheet G had been printed off (see chapter XIV).

The decision after 16 February to stop transferring observations into Draft II, first recognized by Isabelle Pantin, is fundamental to understanding the last two weeks of Jupiter observations, and Galileo's wish at that time to streamline the progress from nightly observation to preparing the setting text for the printing shop. The Observation leaf containing the observations of 17 February through 1 March is, in fact, a kind of hybrid between Observation Sheets and Draft: their distinguishing features merge on the leaf. As Pantin has noted, the first three diagrams on the leaf, of 17 and 18 February, were bracketed in ink. This was the usual way the diagrams were set off in the Observation leaf, but not in Draft II. However, with the observation of 19 February the bracketing ceased, and the Observation leaf became in this sense more draft-like. There are other signals of hybridization. The diagrams of this leaf were all written at the left margin of the entries, this being the layout formula of Draft II, whereas the preceding two Observation Sheets had placed them consistently at the right margin. Conversely, in Draft II, the date-numbers were always written out – "decimatertia", "vigesimaprima", and so on. This practice was evidently carried over into the setting copy, and so, regularly, into *Sidereus Nuncius*. On the "merged" leaf of Observations, the first date was written in arabic numerals (as had been all preceding dates in the Observation records) as "Die 17.", while in *Sidereus Nuncius* the date was spelled out: "Die decimaseptima". That is, Galileo must have made the usual change when he wrote the setting copy for this observation. But the remaining dates in *Sidereus Nuncius* are all set as numerals, suggesting that from this point on, Galileo, in preparing the setting copy directly from the Observation leaf and not from Draft II, either forgot or intentionally abandoned his earlier date-styling.

At some indeterminate time after Galileo entered his 17 February observation on the third Observation Sheet, he removed the leaf and placed it after the just-finished leaf of Draft II. As he had done when he assembled the separate physical sections of Draft I, he placed a distinctive mark, a potenced cross, at the end of the Draft II leaf, and again at the beginning of the separated Observation leaf, to show their linkage (figures X.12, 13 above). Galileo wrote a heading on this Observation leaf as a further identifier. Although heavily inked out, it seems that the heading was "Sidereę Obseruationes". This is related to the heading that he originally wrote at the top of the first page of Draft II (figure X.4). There too he at first began to write "Sidereę Obseruationes", and then by deletion of the first word changed to "Obseruationes Sidereę – Recens Habitę", identical to the headline of *Sidereus Nuncius* itself.

A little more attention may be given to the sixty-five Jupiter-moon diagrams of sheets E–G. Their setting would have been closely supervised by Galileo himself, according to his Maquette and his measurements. The preservation of the Maquette in Galileo's archive suggests that he kept it with him, and did not leave it in the printing shop to be used in his absence. The magnitude numbers that began to be written in the Maquette on 15 February may be seen as entered there for practical convenience. By this point in production, the first page of sheet G, Galileo knew well each of the asterisk sorts that were used, and so the numbers told him directly which sort was to be selected for each moon. Sheet E has seventeen settings of the awkward and unsymmetrical Asterisk 3, seven of which appear on E4v. If as suggested above this asterisk was created by a process of filing rather than casting, and if both formes of sheet E were standing in type concurrently, as is almost certainly the case, then there must have existed at least seventeen individual types for Asterisk 3, fourteen of which were set in E's outer forme. As for Asterisk 2, the asterisk character of the text font, in sheet F it was set sixty times.

There is a difficult error in one of the Jupiter-moon diagrams. At hora 6 of 24 January Galileo observed two moons east of Jupiter. Or more precisely, two points of light, one of which, as Galileo stated, must have been the coalescence of two moons that earlier in the evening had been slightly separated. In *Sidereus Nuncius*, most copies accurately depict this on the corresponding diagram at the foot of 21r (F1r), but in a substantial number of copies – thirty-three of the eighty-four currently recorded with respect to this feature – the easternmost moon asterisk does not appear (figure X.14). The cause of this error is discussed further, if inconclusively, in chapter XIII.

As already indicated, the Jupiter of these diagrams was represented by the upper-case O of one of the printing shop's titling fonts. Setting the type on its side seems to have rendered it subliminally less obvious as being a pure letter form. Because this O comes from a titling font, not used for setting lengthy text, its typecase would have been relatively small, with many fewer sorts of each character than a case of text type would contain. This smaller supply of sorts reveals itself in the printing of *Sidereus Nuncius*. Sheet E contains twenty settings of the Jupiter-O, sheet F twenty-six, and sheet G eighteen. In the course of composing sheet F the O-box apparently was emptied out entirely, so that one of the O's – that of the last diagram on the sheet – was set with a larger, awkwardly shaped ersatz character that is apparently not the sort of a font. This same ersatz character was distributed after sheet F was printed off, then reset in sheet G (figure X.15). Besides the ersatz O, four of the O-types of this titling font have distinctive earmarks, and are seen to repeat on these sheets. Two were set once each on all three sheets, two were set on two of the three sheets (figure X.16). Naturally, if the box of upper-case O of this font had only twenty-five char-

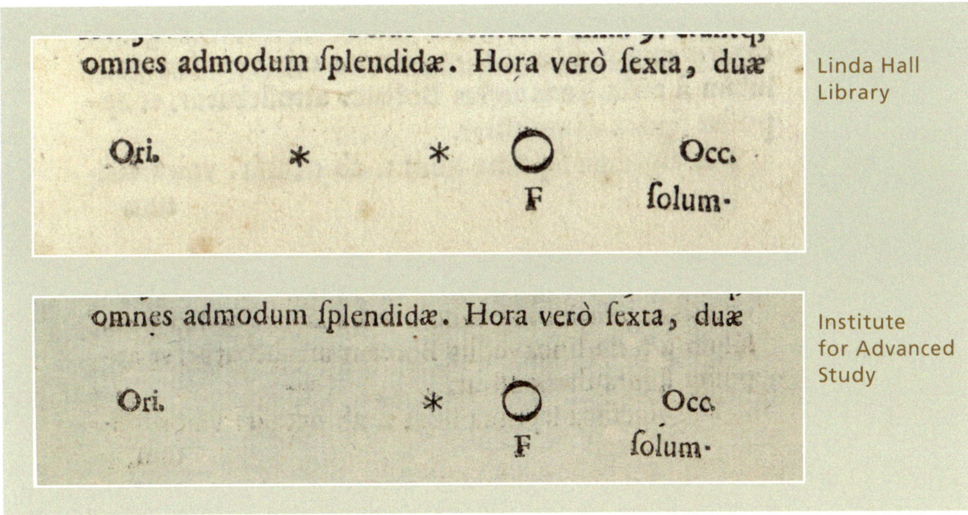

X.14: *Sidereus Nuncius*, F1r.

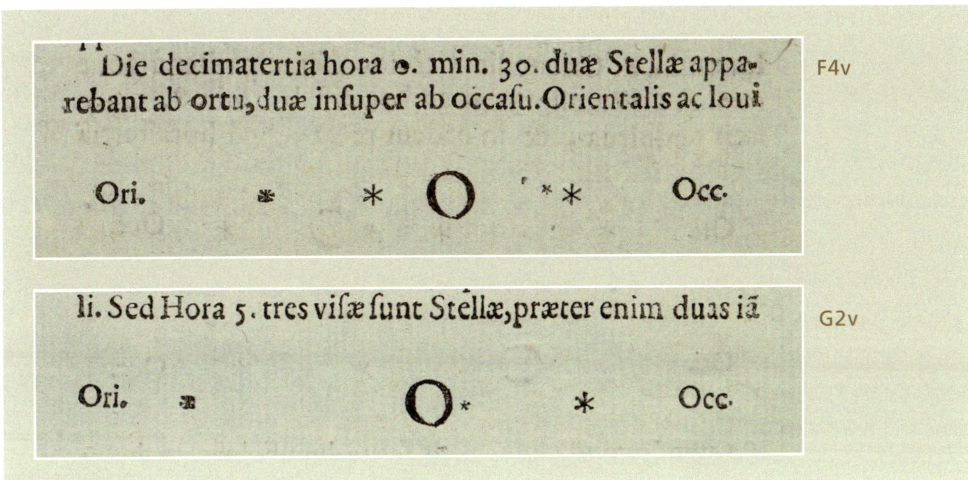

X.15: *Sidereus Nuncius*, Institute for Advanced Study.

acters, or even a few more than that, the majority of those used in sheet F must have been drawn also when E and G were set. It is a happy chance that four sorts from this small number print distinctively because of small flaws on their printing surfaces.

Some time was devoted in the preceding chapter to the changes and recalculations Galileo made to ensure that quire D, extended by an added bifolium, would contain all the final text and images necessary to part I of *Sidereus Nuncius*. Similar calculations and adjustments were necessary to fit the final text and diagrams of part II onto quire G, although here no additional leaf or leaves were needed. As with part I, an examination of the Draft throws light on the process. One clear signal of Galileo's need to compress his material to keep within the bounds of the eight pages of sheet G has been noted by several investigators: the increased abbreviation rate of the final two pages, G4r–v.[11] These typographic abbreviations, to be surveyed shortly, may be seen as the final refinements of larger textual adjustments that emerge when comparing the Draft with the printed text.

11 Pantin 1992, XXX; Gingerich & Van Helden 2003, p. 263.

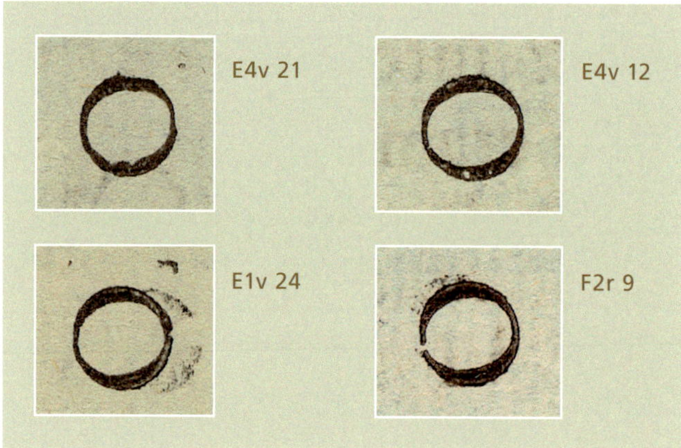

X.16: Sidereus Nuncius, Institute for Advanced Study.

Galileo's autographs reveal several discrete composition stages within the last two weeks of Jupiter observations, 16 February to 2 March. The first, already noted, was the abandonment of Draft II proper as an intermediate text between the records of nightly observations and the preparation of the lost setting copy. Beginning with Wednesday 17 February, the Observation leaf itself became the immediate source for the setting copy, and as already stated its layout was changed to conform it more closely to that copy, making it a hybrid: written on Observation Sheet paper, but textually part-Observation record, part-Draft. The recto of the leaf contains Galileo's observations from 17 February to the first observation of 26 February, made a half hour after sunset. The verso contains the observations from 26 February, five hours after sunset, through Monday 1 March. At the late-evening observation of 26 February Galileo marked for the first time the presence of a nearby star, somewhat south of the ecliptic in Taurus, which served as a fixed point for measuring the eastward drift of Jupiter and its moon system along the ecliptic. At the top of the verso of this leaf is a diagram showing the relative position of this fixed star against Jupiter and its moons over the three days 26–27–28 February, the first two of which positions are marked as "a" and "b" (figure X.17).

Because they are spaced out above and below, the standard Jupiter-moon diagrams in *Sidereus Nuncius* each take up the equivalent of slightly more than four lines of solid typesetting. On pages 26v, 27r and 27v (G2v–3v), the five diagrams that include the fixed star each take up slightly over six lines of solid typesetting. That is, the addition of the fixed-star adds about thirty lines, virtually a full page. Yet this expansion did not, it seems, cause Galileo to worry about being able to complete *Sidereus Nuncius* within the four leaves of quire G. In fact, on 26v, Galileo created duplicate diagrams for the observation of 26 February hora 5: the first without the fixed star, and the second with it. The identical information could have been conveyed with a single diagram, saving about four lines. The fixed star was captioned with "fixa" in a considerably smaller font than that of the text: this is the font used on the title-page to set lines 15–18 (*Circa IOVIS Stellam … primus; atque*). It was a small oversight that this diagram omitted the word "Occ." at the right (figure X.18).

The necessity of compressing text to stay within the bounds of sheet G arose from Galileo's decision to extend his Jupiter observations by one day, from 1 March to 2 March.

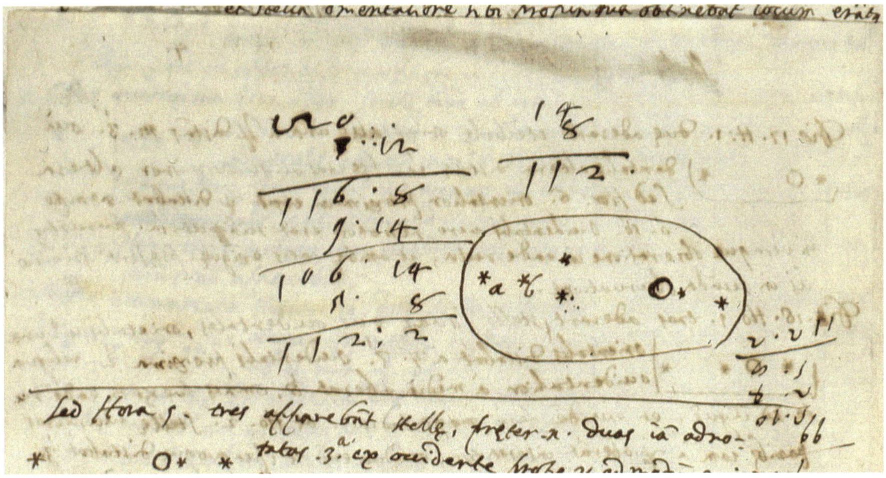

X.17: BNCF, Ms. Gal. 48, 24v.

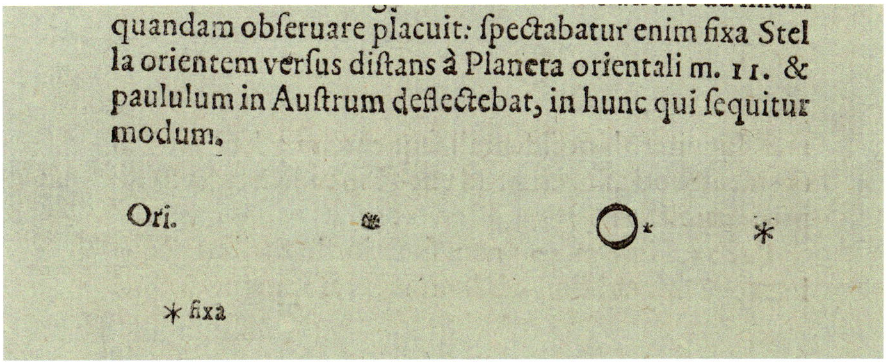

quandam obſeruare placuit: ſpectabatur enim fixa Stel la orientem verſus diſtans à Planeta orientali m. 11. & paululum in Auſtrum deflectebat, in hunc qui ſequitur modum.

X.18: Sidereus Nuncius, G2v, Institute for Advanced Study.

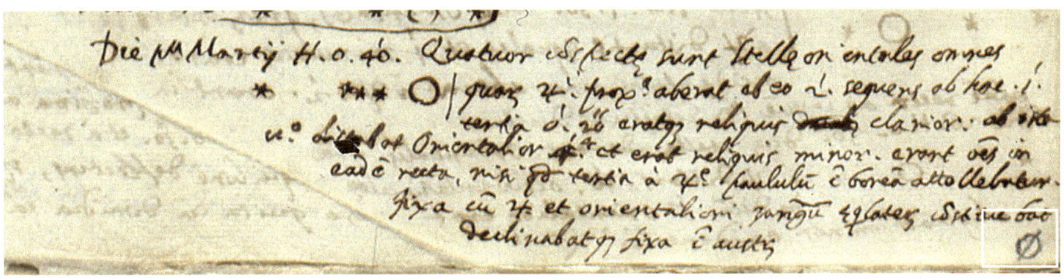

X.19: BNCF, Ms. Gal. 48, 24v.

The 1 March observation was the last entry on the verso of the "hybrid" Observation leaf. Galileo wrote smaller at this point, so that he could complete the observation on the leaf – the available space being diminished because the outside corner of the leaf had been torn away. At the end of the leaf he wrote the Greek letter Φ as cue to the text that was to follow (figure X.19). This Φ, written in pencil, forms a link with a loose leaf, smaller in size and in quarto orientation, on the recto of which he entered the 2 March observation (figure X.20). This loose leaf contains three distinct texts, which were not however written in order from top to bottom. At the top is the observation of 2 March; in the middle of the page is a transition paragraph of sixty-five words intended to join the end of the Jupiter observations to the already- drafted conclusion; at the bottom of the page is a briefer transition paragraph of thirty-five words, serving the same purpose. This briefer paragraph was the first text to be written on the leaf. It is prefaced by Galileo's Φ to mark its intended con-

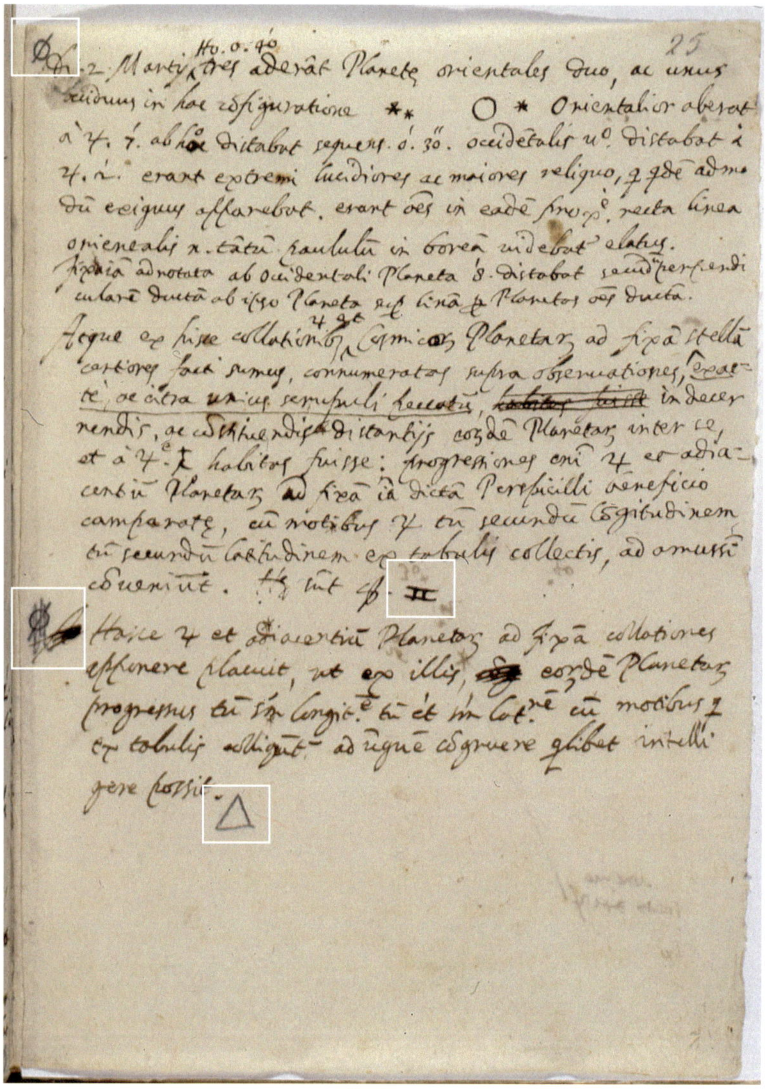

X.20: BNCF, Ms. Gal. 48, 25r.

tinuation from the observation of 1 March. That is, the Φ was written before Galileo knew that he would be able to include a 2 March observation. When Galileo decided to add the 2 March observation into *Sidereus Nuncius*, he struck through the Φ, and wrote it instead at the beginning of the 2 March observation itself, again in pencil, at the top of the leaf.

The sequence of actions must have been essentially the following. Galileo first wrote the briefer transition paragraph at the foot of the page, the Φ before it indicating that it would follow directly the 1 March observation. This transition text was probably written on the evening of 1 March or morning of 2 March, in any case at a time when Galileo was not sure if the sky would be clear after sunset on the second. If it was cloudy that evening, the observation of 1 March would be the last to be included in *Sidereus Nuncius*, and so the transition paragraph could lead directly to the conclusion (beginning *Hę sunt obseruationes quatuor errantium stellarum*), which had already been drafted but not finalized on the first leaf of its separate sheet (figure X.21). By writing the transition paragraph at the

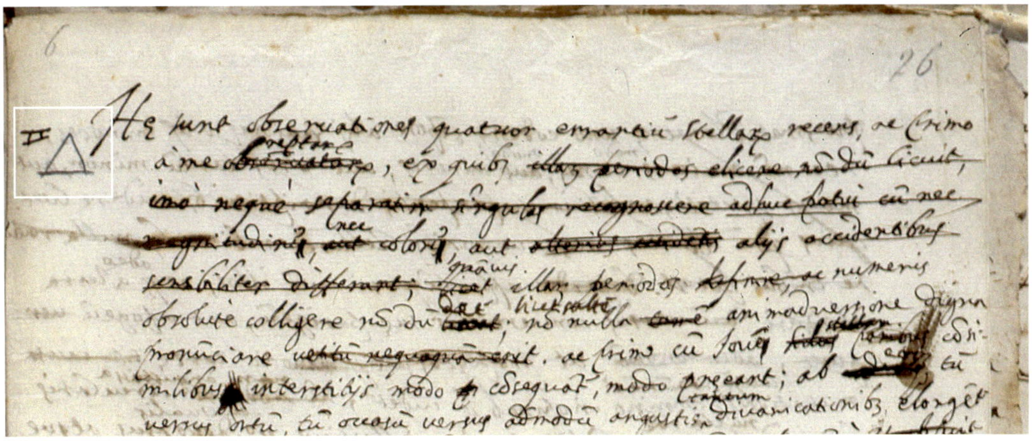

X.21: BNCF, Ms. Gal. 48, 26r.

foot of the page, he ensured that there would be room above for the 2 March observation record, if the heavens were favorable.

The heavens were favorable, and the observation of 2 March was duly drafted at the top of the leaf. Immediately thereafter, Galileo drafted also an expanded transition paragraph of nearly twice the length, including a rather redundant reference to his use of the telescope, and an attestation that the measures of the distances between Jupiter, its moons, and the fixed star had been made carefully, with an error of less than a minute: *exactè, ac citra unius scrupuli peccatum*. Galileo apparently intended this expanded transition to replace the briefer one he had already drafted below it, and so at its end he wrote the incipit of the draft conclusion (*Hę sunt etc.*) and a linking symbol, the sign of Gemini (♊), which is duly written in ink both here and at the beginning of the draft conclusion (figure X.21).

A small peculiarity of this expanded transition must be noted. Its position on this extra leaf for the 2 March observation, and the consistency of its ink with that of the observation proper, strongly suggest that Galileo drafted the paragraph either on the evening of 2 March, or perhaps the morning of the third. Yet in this paragraph Galileo still referred to the "Cosmici" rather than "Medicei" planets: *ex hisce collationibus Iouis et Cosmicorum Planetarum*. Vinta's reply to Galileo's urgent query of 13 February, that Cosimo wanted the Jupiter moons to be called Medicean and not "Cosmic" was dated from Florence on 20 February, and the ordinary course of delivery between Florence and Venice should have brought the reply to Galileo well before 3 March. We cannot, finally, be sure whether on the second or third of March Galileo had not yet received this reply due to a delay in the post, perhaps from bad weather; or, when he drafted this paragraph, he forgot about the revised formulation that he was to use.

The incorporation into *Sidereus Nuncius* of the 2 March observation put in jeopardy the plan to complete the work within sheet G. Without that observation, the text would easily have ended on G4v, probably even with space remaining for an ornamental tailpiece to make a visually more elegant conclusion. But the 2 March observation, set on G3v, added the equivalent, including the blank spaces of the Jupiter-moon-fixed star diagram, of twenty lines of solid setting, pushing the text remaining – transition paragraph, and conclusion – beyond the remaining limits of of the sheet. To pull back the text to within that limit, Galileo reverted to the briefer, thirty-five word transition paragraph, which occupies five lines in *Sidereus Nuncius* (G3v last line to G4r line 4), compared to the nine or ten lines that the longer transition would have needed. Within his drafts, Galileo marked this by placing a Δ in pencil at the end of the shorter transition, and wrote a second pencilled Δ at the beginning of the draft conclusion (figure X.21): thus Δ trumps Ⅱ, and again, as in Draft I, pencil trumps ink.

This saved four or five lines, but was not nearly sufficient to solve the problem of an excess of text for the less than two full pages remaining. The second step was to abridge the text of the conclusion. In *Sidereus Nuncius* as printed, the conclusion takes up the remainder of G4r, lines 5 and after, and all of G4v, a total of 64 lines and of about 485 words. The Draft of the conclusion shows substantial authorial deletions, essentially all of which reflect the steady paring away of words deemed extraneous. In global terms, the conclusion as drafted contains about 730 words, and of these Galileo deleted about 240, roughly one word in three. The typesetting of *Sidereus Nuncius* in its sections of continuous prose averages a little less than seven words per line. If the conclusion had not been substantially abridged, it would have taken up about 34 more lines, a full extra page.

A comprehensive collation of Galileo's Draft with the printed *Sidereus Nuncius* remains a significant desideratum, for reasons that go far beyond word-counting. For present purposes, two examples of Galileo's abridgements in his conclusion may be looked at. The first sentence of the conclusion was cut in half by Galileo's score-throughs:

> Draft (the deleted words in strike through): *Hę sunt obseruationes quatuor errantium stellarum recens ac primo à me repertarum, ex quibus ~~illarum periodos elicere non dum licuit, imò nequè separatim singulas recognoscere ad huc potui cum nec magnitudine, nec colore, aut alijs accidentibus sensibiliter differant; licet~~ quamuis illarum periodos ~~definire, ac~~ numeris ~~absolutè~~ colligere non dum detur; licet saltem ~~non nulla tamen~~ animaduersione digna pronunciare ~~netitum nequaquam~~ erit.* [60 words]
> *Sidereus Nuncius* 28r: *Hae sunt obseruationes quatuor Mediceorum Planetarum recens, ac primò a me repertorum, ex quibus quamuis illorum periodos numeris colligere nondum detur; licet saltem quaedam animaduersione digna pronunciare.* [27 words]

Note that as printed in *Sidereus Nuncius,* besides the change from the four "wandering stars" to "Medicean planets," Galileo's statement that he had not yet determined the periods of these moons is strangely abrupt. In the original draft, he explained some of the reasons: at the moment, he could not differentiate these points of light by size, color, or other conspicuous features. The draft reads more smoothly and engagingly, but space for the fuller statement was no longer available.

Particularly noteworthy are the deletions where, in the conclusion, Galileo refers to the Copernican system. In *Sidereus Nuncius* as printed he referred to those who had no problem with the general idea that the planets revolve around the Sun, yet were troubled by the contingent notion that the Moon revolves around the Earth, so that it is the two together that make an annual revolution: *qui in Sistemate Copernicano conuersionem*

Planetarum circa Solem aequo animo ferentes, adeò perturbantur ab vnius Lunae circa terram latione, interea dum ambo annuum orbem circa Solem absoluunt. He pointed out that Jupiter, with its four moons moving together as a system, demonstrates that the seeming paradox is not in reality paradoxical. But in the Draft, Galileo originally had a notably stronger remark. After referring to the Copernican system, he added, and then struck through, *quod apprimè ueritati consonum existimo* – the Copernican system, "which I judge to be strikingly in accord with the truth." Thus, in parallel passages toward the end of the two parts of *Sidereus Nuncius*, in the Earthshine discussion of part I (see preceding chapter), and in the general conclusion, Galileo expressed strong pro-Copernican hints. But originally, his last statement would have been more than a hint, would in fact have been the crowning conclusion. It is uncertain whether the deletion here arose from cautious second thoughts, or simply from the overall necessity to prune words.

Even after this rigorous pruning the conclusion was too lengthy for the 64 lines available on G4 recto and verso. If set to the conventional pattern of *Sidereus Nuncius*, with infrequent abbreviation, the text would have required something over 70 lines of setting. The compositor resorted to two tricks to fit the remaining text to the space available. First, the line width of G4 recto and verso was increased, from the 95-mm standard to about 99 mm. This added to the two pages roughly the equivalent of 4 × 60 (lines), or 240 mm of setting: that is, the equivalent of about two and a half lines.[12] Second, the compositor could take advantage of the opportunities for abbreviating words, transmitted from the medieval scribal tradition for Latin, that was embodied in hundreds of fifteenth-century printing fonts, and still left traces in the fonts of the early seventeenth century. The simplest of these was provided by types with horizontal strokes over the letter forms to indicate the omission of a following *m* or *n*, and sometimes of more than just those letters. The text font of *Sidereus Nuncius* contained, besides ordinary types for the vowels, versions with top strokes: ā, ē, ī, ō, ū. On G4r–v these were used to set dozens of words, each thus shortened by at least the width of an m: *Solê* (Solem), *Planetâ* (Planetam), *maximû* (maximum), and so on; but also *rône* (ratione), *ôes* (omnes), *oîno* (omnino), *câm* (causam), *êt* (etiam). The greatest compression was afforded by an r with a top stroke (r̄) which was used twice on G4v to compress "videtur" to *vr̄*. This abbreviation might have caused some readers to pause slightly, as *vr̄* was the standard abbreviation for "vester." (figure X.22) The setting "12. annorum" on the first line of G4v is also to be seen as an abbreviation, for the rule of continuous prose in *Sidereus Nuncius* called for spelling out numbers: *duodecim annorum.*

By such means the compositor drew in his copy by some six or seven lines, and there came a point where it was clear to him that he could, so to speak, relax and set full words. The last abbreviation, *êt* for *etiam*, appears on line 25 on G4v, and the drawing-in had been so successful that the final four lines of text could be artfully shaped as an inverted triangle, leading to the spaced-out "F I N I S." that concludes the work (figure X.22).

12 I am indebted to David Vander Meulen, who first noticed the increased line width of these pages.

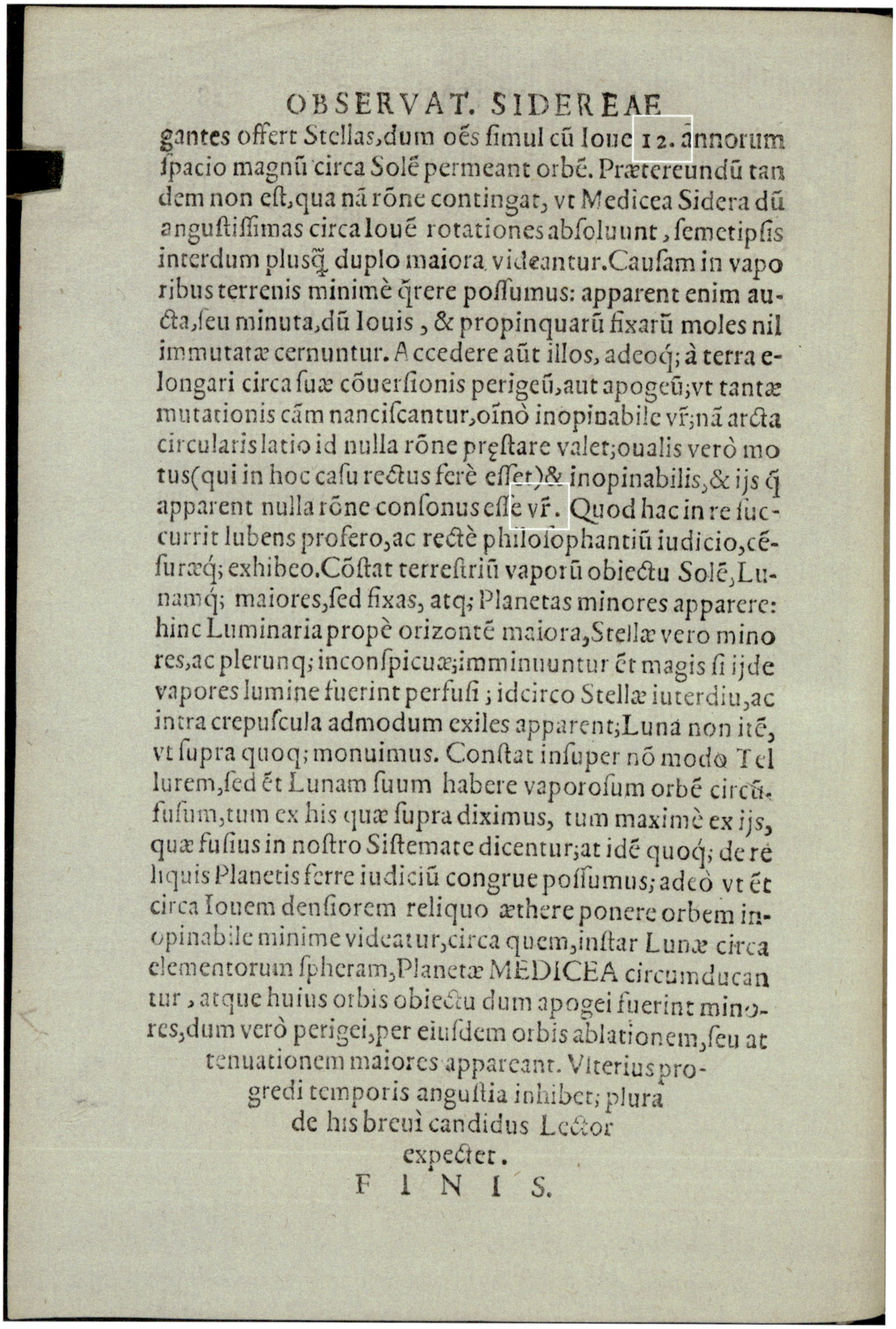

OBSERVAT. SIDEREAE

gantes offert Stellas, dum oẽs simul cũ Ioue 12. annorum spacio magnũ circa Solẽ permeant orbẽ. Prætereundũ tandem non est, qua nã rõne contingat, vt Medicea Sidera dũ angustissimas circa Iouẽ rotationes absoluunt, semetipsis interdum plusq̃ duplo maiora videantur. Causam in vaporibus terrenis minimè q̃rere possumus: apparent enim aucta, seu minuta, dũ Iouis, & propinquarũ fixarũ moles nil immutatæ cernuntur. Accedere aũt illos, adeoq; à terra elongari circa suæ cõuersionis perigeũ, aut apogeũ; vt tantæ mutationis cãm nanciscantur, oĩnò inopinabile vr̃; nã arcta circularis latio id nulla rõne prȩstare valet; oualis verò motus (qui in hoc casu rectus ferè esset) & inopinabilis, & ijs q̃ apparent nulla rõne consonus esse vr̃. Quod hac in re succurrit lubens profero, ac rectè philosophantiũ iudicio, cẽsuræq; exhibeo. Cõstat terrestriũ vaporũ obiectu Solẽ, Lunamq; maiores, sed fixas, atq; Planetas minores apparere: hinc Luminaria propè orizontẽ maiora, Stellæ vero minores, ac plerunq; inconspicuæ; imminuuntur ẽt magis si ijdem vapores lumine fuerint perfusi; idcirco Stellæ interdiu, ac intra crepuscula admodum exiles apparent; Luna non itẽ, vt supra quoq; monuimus. Constat insuper nõ modo Tellurem, sed ẽt Lunam suum habere vaporosum orbẽ circũfusum, tum ex his quæ supra diximus, tum maximè ex ijs, quæ fusius in nostro Sistemate dicentur; at idẽ quoq; de reliquis Planetis ferre iudiciũ congrue possumus; adeò vt ẽt circa Iouem densiorem reliquo æthere ponere orbem inopinabile minime videatur, circa quem, instar Lunæ circa elementorum spheram, Planetæ MEDICEA circumducantur, atque huius orbis obiectu dum apogei fuerint minores, dum verò perigei, per eiusdem orbis ablationem, seu attenuationem maiores appareant. Vlterius progredi temporis angustia inhibet; plura de his breuì candidus Lector expectet.

F I N I S.

X.22: *Sidereus Nuncius,* G4v, Institute for Advanced Study.

Ghost errors in the Jupiter-moon diagrams

Besides the genuine error of an omitted asterisk in some copies of *Sidereus Nuncius* in the diagram of 24 January hora 6 (figure X.14 above), false reports of other diagram errors have burrowed their way into the modern literature and need to be eradicated. Three of the reports are very firmly rooted, two somewhat less so, and a third is quietly waiting its turn, a trap not yet sprung. The well-rooted false reports are for the diagrams of 1 February (F2v), 13 February (F4v) and 27 February (G3r). According to publications of 1989 and after, all or some copies of *Sidereus Nuncius* lack a required moon asterisk in each of these diagrams. Three works (Van Helden 1989, Shea 2009, Bascelli 2009) have made good the supposed deficiencies by adding the missing asterisks within brackets in their illustrations.

The source of the false reports is the anonymous photo-retoucher employed by Fotolito Emiliana in Bologna, when that firm printed the 1964 Domus Galilaeana facsimile of *Sidereus Nuncius*, a facsimile produced in commemoration of the four-hundredth anniversary of Galileo's birth. The facsimile states that it was based on the Palatina copy of the first edition in the Biblioteca Nazionale Centrale, Florence.[13] However, in that copy, now shelfmarked 22.B.5.55, these diagrams are correctly printed. The situation is further complicated by the bizarre circumstance that despite the publisher's statement, the Milan Osservatorio copy was the basis of the facsimile (see Appendix). The significant point, however, is that the Jupiter-moon diagrams in question are correct not only in the Florence Palatina copy, but in all copies of the original edition. The asterisks went missing not in 1610, but in 1964.

After the Milan copy was photographed for the Domus Galilaeana facsimile, the prints or negatives were extensively retouched to produce "clean" black-on-white plates for printing. That extensive retouching was applied is obvious at dozens of places. For example, as noted above (figure X.9), erratically in all copies of the original, shoulders of some of the quads used to space out the asterisks picked up ink and left marks on the pages. In the Domus Galilaeana facsimile, such marks are entirely absent. At several places the retoucher tried to improve words that were unclear in the photographs, because of light inking in some places, creases of the leaves in others. In doing so, the retoucher created such nonsensical pseudo-variants as (10r line 2) "compicitur" (conspicitur) and (14r li. 11) "posuconiunctione[m]" (post coniunctione[m]). At the diagrams in question, the retoucher removed in each case a setting of Asterisk IV, the small five-rayed star. Each time, the thought in his mind would have been: "I will remove that unsightly smudge."

But another step was necessary to bring the 1964 retoucher's hand into the scholarly apparatus of Galileo studies. This step was supplied by the University of Chicago Press in 1989 in its edition of Albert Van Helden's translation of *Sidereus Nuncius*, which has been frequently reprinted. The preface (p. xii) states that its "original illustrations … are repro-

13 Domus Galilaeana 1964, p. [5]: "La copia qui riprodotta anastaticamente, è conservata presso la Biblioteca Nazionale Centrale di Firenze sotto la segnatura Pal 1200/23."

duced from the copy at Wellesley College." It further states, "The first edition of the *Sidereus* omitted one of the 'Medicean stars' from four figures ... even though these stars were shown in Galileo's manuscript. I have added these missing stars in square brackets." The assertion is that not just the Wellesley but all copies of the first edition have these deficiencies. One of the four figures is the 24 January hora 6 diagram, that is, a diagram that in a number of copies (including that in Milan) truly is deficient – only not in the Wellesley copy, where the diagram is correct. The other three are the "ghost lacunae" of the Domus Galilaeana facsimile. The explanation is that the Chicago edition did not take its Jupiter-moon diagrams from the Wellesley copy of the original edition, but, silently, from the retouched Domus Galilaeana facsimile.[14] A Galileo researcher giving credence to the prefatory statements of Domus Galilaeana 1964 and Van Helden 1989 would believe that the entire edition had four deficient diagrams, and have apparent evidence of at least two copies documenting this: the Wellesley and the Florence Palatina copies (figure X.23).

The misinformation was partly corrected by Pantin 1992, who observed (p. 85, note 116) that there are some copies in which all the diagrams are correct, yet accepted that at least the Wellesley and Florence Palatina copies must have deficient diagrams at these four places (cf. Pantin notes 141, 145, 151, 155). The translations of Shea and Bascelli, 2009, accept the remarks of Van Helden 1989, and again artificially supply the supposedly missing stars in brackets.

The affair of the lost-and-found moons demonstrates the power of the press, and also of the conventions of scholarly practice, to broadcast and multiply error. Consider: the Domus Galilaeana facsimile states that it was based on photographs of the Florence Palatina copy of *Sidereus Nuncius*, but was not. It pictures four Jupiter-moon diagrams, each apparently lacking a necessary asterisk. One of these (24 January, hora 6) is an authentic representation not of the Florence Palatina copy, whose diagram is complete, but rather of the Milan copy. The other three deficient diagrams were created in 1964 by a printing house craftsman. The Chicago edition similarly misstated its copy source. By supplying the missing asterisks within brackets, to mark them out, Van Helden 1989 followed good scholarly practice: the editorial interventions were made explicit. But that practice in this case solidified the error by making it conspicuous: it seemed that a precise visual documentation of the Wellesley copy's supposed errors was provided. Pantin's notes need to be corrected as to the Wellesley and Florence Palatina copies, but the unambiguous yet false statements on copy sources in Domus Galilaeana 1964 and Van Helden 1989 could not possibly have invited larger doubt and awakened a need to verify.

Two more diagram errors have worked their way into the notes of Pantin 1992 by means of the 1967 facsimile of *Sidereus Nuncius* published by Culture et Civilisation, Brussels. In this facsimile, the Jupiter-moon diagram of 21 January at the top of fo. 20v (E4v) lacks the small asterisk to the left of the Jupiter O (cited Pantin note 136), and that of 27 January at the top of 22r (F2r) lacks its only asterisk, that is, the diagram represents Jupiter but without moons (cited Pantin note 144). The Culture et Civilisation facsimile does not identify its source copy, but it is virtually certain that it was the copy at the Bib-

14 I am grateful to Prof. Van Helden for consulting his memory on this.

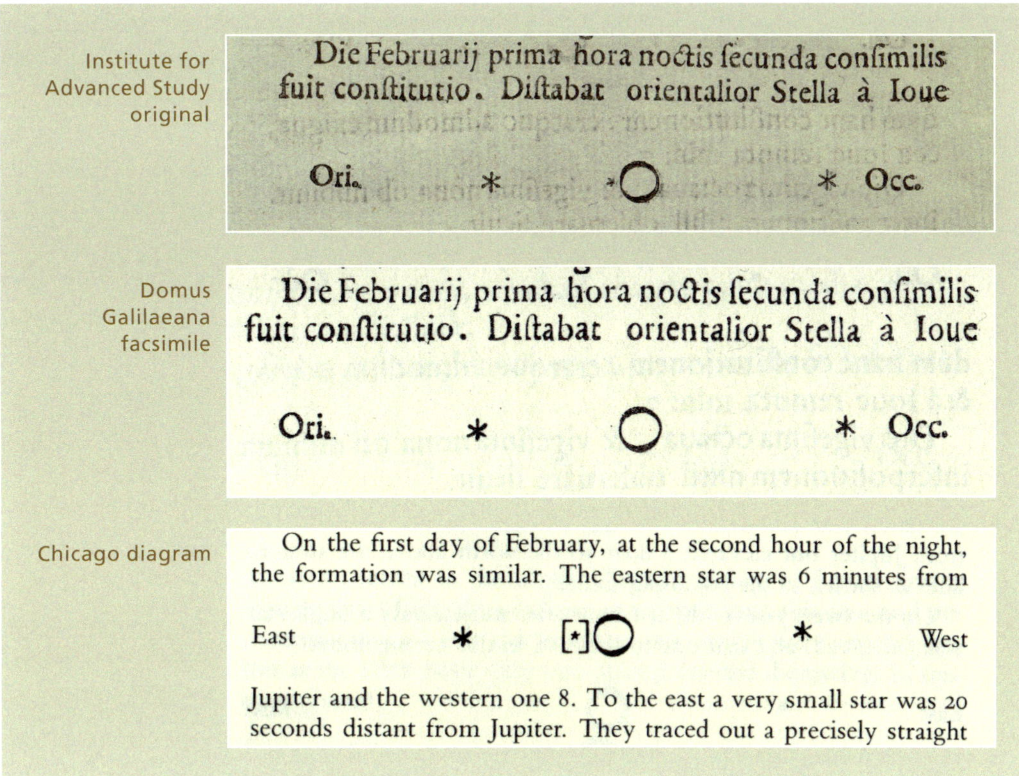

X.23: *Sidereus Nuncius*, F2v, Institute for Advanced Study.

liothèque Universitaire, Liège.[15] In that copy, both diagrams are correct: the missing asterisks are the handicraft of another, presumably Belgian photo retoucher. That retoucher in fact erroneously removed a third asterisk, that closest to the word "Ori." in the diagram for 1 March at the top of 27v (G3v), present in the Liège original. Fortunately, this faulty reproduction has not until now been cited.[16]

There is at the moment no explanation of why or how Domus Galilaeana misidentified the source copy of its facsimile. Conceivably, someone who worked either at the printing shop in Bologna or the publishing house in Pisa in 1964 still retains a memory that would shed light on this. As for the photo retouchers in Belgium and Italy, they did only what they had been trained to do. Intellectual responsibility for the accuracy of the facsimiles resided with the publishers. As regards the University of Chicago Press and its disguised use of the 1964 Domus Galilaeana facsimile despite a stated credit of the Wellesley College copy, there is an obvious explanation: $$. Van Helden 1989 contains eleven images truly

15 I am grateful to Prof. Carmélia Opsomer for examining this copy.

16 The story of incorrect Jupiter-moon diagrams continues. In Bascelli 2009, p. 151, the diagram for 1 February puts in brackets the small asterisk just to the left of the Jupiter O; as noted above, all copies of *Sidereus Nuncius* have this asterisk, but the retoucher of the Domus Galilaeana removed it. However, in Bascelli's diagram, the asterisk just to the left of the word "Occ." is also placed within brackets. Again, all copies of the original edition have it, and so does the Domus Galilaeana facsimile. The source of the brackets is therefore a mystery. Equally mysterious is that in Shea 2009 (p. 83), which in principle uses the same illustrations as Bascelli, this latter asterisk is missing.

based on the Wellesley copy, thus on eleven photographs acquired from Wellesley. To use that copy also for the Jupiter-moon diagrams would have required buying twenty-two more photographs. The cost of silently using instead the Domus Galilaeana facsimile was $0.

XI.
Parallel Printing and Running Titles

(with David Vander Meulen)

As noted in the preceding chapter, it has been agreed by essentially all who have thought about the printing of *Sidereus Nuncius* that the two major parts, B–D for Moon and stars, E–G for Jupiter, were printed more or less concurrently, the two unnumbered leaves of quire D being the conspicuous signal that E had been printed off before D. "Concurrently," however, may have several interpretations, and it is worth the effort to look for evidence that may improve our knowledge. One investigator has stated that *Sidereus Nuncius* was "printed on two presses simultaneously," and argued that no printing began until after Wednesday 17 February 1610. This may be called the dilatory view.[1] Another investigator has suggested that the first sheet of *Sidereus Nuncius* to be printed was quire E and not B (the last of the Jupiter observations included in sheet E was made Friday, the 22nd of January). By this latter interpretation quire B, the first of part I, was not set and printed until after the weekend of 13–14 February, for that is when Galileo received Vinta's letter bringing more or less official word that Cosimo II approved of the book project. In this view, it would not have been until then that Galileo would have dared to refer on the drop-title at the beginning of quire B to the "Cosmica Sydera."[2]

There are some difficulties with both viewpoints. If quire B were not printed until, say, Monday the 15th of February at the earliest, then quires C and D must also belong to the later days of February. Quire F, the last of whose observations was made in Padua on the 13th of February, could likewise not have been printed before the week of the 15th February. Quire G could not have been printed before Wednesday the 3rd of March, its last observation having been made the evening before. And, as will be discussed more closely in chapter XII, sheet A with title-page and dedication was not printed until Saturday, the 13th of March. Thus, if quire E had been printed first, say at the end of January or

1 Wootton 2009, p. 132: "… the second half of the manuscript was given to a copyist on 17 February"; p. 133: "We can thus be fairly confident that Galileo had a complete manuscript on 17 February, which was copied for … the printer …"; p. 134: "Once the printer had decided that the book would be printed on two presses simultaneously …". This implies that printing began even later than 17 February, allowing time for the hypothetical copy to be made and brought to the printing shop. It is not clear why this "copyist" is invoked; the Fair Copy shows that Galileo himself was able to write an extremely legible setting copy.

2 Pantin 1992, p. XXVIII: "La composition du livre commença probablement par le cahier E … le titre de départ du feuillet B1 comporte déjà l'expression 'cosmica sydera', or, si sûr de lui que fût Galilée, il ne pouvait s'engager ainsi avant d'avoir reçu de Florence le moindre signe d'approbation: la composition de la forme extérieur de ce cahier B n'a donc pas dû s'achever avant le 14 février …"

the early days of February, there must have been a considerable hiatus before any other printing of *Sidereus Nuncius* was carried out.

As for the dilatory view that no printing whatever took place until later February, this invites the question of what Galileo was doing in Venice in the first half of the month. On 30 January, he had written to Vinta, "I find myself currently in Venice, bringing into print certain observations which I have made among the celestial bodies by means of my telescope". The words imply activity that called for his presence in Venice, but by this view, printing did not begin for some twenty days, in which interval he returned to Padua. The implied scenario is almost paradoxical.

Neither of the viewpoints just reviewed sits entirely comfortably with what Galileo wrote to Vinta from Padua on the 13th of February. He asked for a reply, soonest, to the question of whether to call Jupiter's moons "Cosmic" for Cosimo personally, or "Medicean" for Cosimo and his brothers. The matter required both urgency and confidentiality: "For this matter alone I am holding up the printing, as the wording of the title and the dedicatory letter depend" on the answer: *perchè per tal rispetto solo fo trattener le stampe, restandomi da determinar questo punto nel titolo, et nella dedicatoria.* That is, the text for sheet A could not be written until Galileo knew which nomenclature Cosimo favored. This seems to suggest that in Galileo's mind the printing was nearing its final stages, rather than that it was, in the week to come, just about to begin, possibly apart from sheet E. We may consider an alternative view of Galileo's state of mind. He had been confident of Cosimo's encouraging interest, which Vinta's letter just received confirmed to him, knowing that as early as September 1609 Cosimo had been favorably informed of Galileo's development of a superior telescope. But, after first thinking of naming the moons "Cosmic", as he had done on the drop-title of B1r, he had second thoughts: the "Medicean" alternative came to him, and he wanted to feel certain. It was in itself a compliment to bring the question before the Tuscan court, and to defer to its wishes in the matter.

The fairest conclusion is that from these letters of 30 January and 13 February, despite the vividness and directness of their language, we cannot really draw any firm picture of the course of printing. To someone convinced that Galileo would not dare to begin printing until he had explicitly received Tuscan approbation, his words of 13 February "fo trattener le stampe" could be taken to mean that he had held up all printing, and not just his preparations for the final sheet, despite the qualifying phrase "per tal rispetto solo." We need less malleable evidence than Galileo's words.

Concerning the idea that quire E went to press before quire B, two features of *Sidereus Nuncius* have already been noted in chapter VI that seem to argue for the opposite order, B before E. The first is that B1v was set to 33 lines, one line shorter than what on B2r and after became the standard of 34 lines. The second is the setting at the top of B2r of a two-line M, followed by full capitals to complete the word, that began the paragraph introducing the second topic, the Telescope: "(M^2)ENSIBUS abhinc decem ferè ..." (figure VI.6). By the argument of the present analysis, these two features, coming close together, reflect the earliest plan of the book's layout, before the challenge arose of incorporating the several stages of text expansion in Part I, which imposed on Galileo and the printing shop the need to conserve space. In this early plan, it is here argued, the major sections of the book were all to be so marked out, almost as chapters though without chapter numbering. This

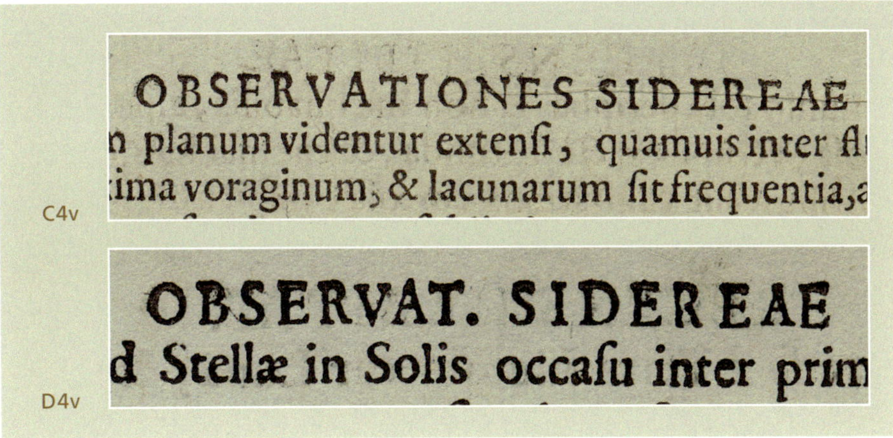

XI.1: *Sidereus Nuncius*, Institute for Advanced Study.

useful *ordinatio* of the text then went by the wayside, so that even at the beginning of quire E, where the topic of Jupiter and its moons began, there was no change in layout to mark the transition to a major topic. Its opening sentence "De Luna, de inerrantibus stellis …" was treated simply as an indented paragraph. If printing had begun with quire E, it seems unlikely that experiments envisioning a more spacious layout should have arisen not there but later, in quire B.[3]

But these considerations are, indeed, only hints. The practice of the printing shop, in its treatment of the headlines, helps to test the hints. The running title set in the headline of *Sidereus Nuncius* regularly reads, across each opening, "OBSERVAT[IONES] SIDEREAE | RECENS HABITAE." As already noted in chapter III, the specific setting of a running title may be distinctive because of its use of one or more distinctively damaged types. In quires B–G there are twenty-four verso headlines reading OBSERVAT. (or OBSERVATIONES) SIDEREAE, and twenty-four recto headlines reading RECENS HABITAE. In principle, the book's compositor or compositors might have set these headings forty-eight times, one for each page requiring one or the other text. But even in the fifteenth century it became common practice to retain running-title settings, and re-use them from forme to forme when there was no need for their wording to change.[4] The patterns of recurrence, and accidents in the transfer of running titles, can throw light on the order in which pages or formes were printed.

From close examination of the types and of their spacings, seven settings can be specified of the recto running title RECENS HABITAE. Two of these appear once only, the others were each used in two or more formes. Of the verso headine OBSERVAT[IONES] SIDEREAE six settings can be specified, all used in more than one forme each. One of the latter (designated V5), placed in three formes, is especially conspicuous, for it spells the

3 Textual transmission is often reflected in layout, and it may be noted that the Frankfurt 1610 and Bologna 1655 editions of *Sidereus Nuncius* likewise set a two-line M at this place, while the two 1653 London reprints set the word in full caps: MENSIBUS.

4 The pioneer study of headlines is Bowers 1941. For fifteenth-century examples see also Bühler 1949 and Needham 2004, and for a wide conspectus of studies using headline evidence Tanselle 2002, pp. 268, 293–297.

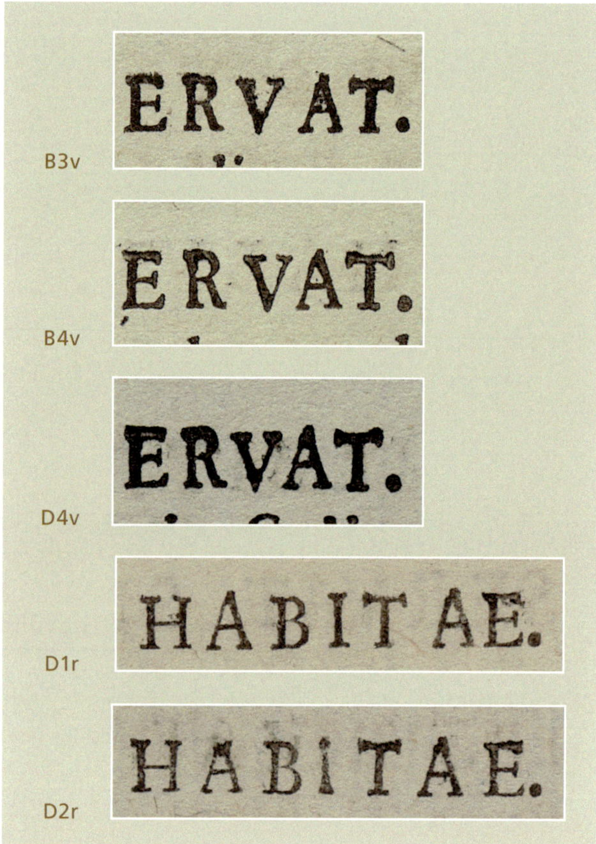

XI.2: Sidereus Nuncius, Institute for Advanced Study.

word OBSERVATIONES in full (figure XI.1). Another verso running title, V6, is easily recognized by its final A, which belongs to a smaller font than the other letters (figure XI.1). Variations in spacing can also be distinctive. The headlines were set with spaces between each letter, to give them an inscriptional appearance. The spacing quads used were not, however, necessarily equivalent in each setting, so that different settings may by that feature alone be distinguishable. Note, for instance, the variations in spacing of the three letters RVA in the running title settings V2, V3, and V6 (figure XI.2).

Most settings of the recto running title did not space out the final AE of HABITAE. The idea was that without extra spacing the AE would appear to be more or less ligatured, corresponding in capital letters to the lowercase ligatured type æ. But one setting, R7, did distinctively set a space between the two final letters (figure XI.2).

In identifying damaged types, a chief difficulty is distinguishing between differences in appearance due to variables of inking and impression, and differences that reflect intrinsic damage to the face of a character. The fibers of handmade paper, which spread ink by capillary action; knots in the paper; the presence or absence of a small amount of persistent dirt on a letter: all these may create variations in impression that resemble type damages but are not. However, if the same apparent damage appears in multiple copies, the probability that it reflects a flaw on the printing surface of the type increases. One common form of damage or flaw arises in the original casting of a piece of type. A small bubble of air caught under the molten typemetal as it is poured into a mould may prevent the rapidly hardening metal from pouring smoothly over the entire surface of the matrix at the foot of the mould, leaving a portion of the face uncast. A piece of dirt within the matrix may have a similar effect. The resulting "pit" on the face of the type will regularly leave a small lacuna in the same place every time it is inked, although an especially viscous ink will sometimes cover it up. Several apparent examples of such miscasting were illustrated in the preceding chapter in the O's of the titling font used to represent Jupiter in quires E to G (figure X.16). In the running-title settings, one of the most conspicuous type damages is that of a letter A in setting **V1**, where the foot of the left arm of the A seems not to have cast properly. Two examples of this setting were reproduced as figure III.6.

The running-title settings of *Sidereus Nuncius* can be displayed in several ways, each arrangement highlighting a different feature: by the pages on which the different settings appear; by the exact running-title arrangements of each forme; and by the running titles that were used in each sheet (or, for D3.4, half sheet). Tables I and III exclude the D5r setting PLEIADVM CONSTELLATIO, abbreviated as Pl. Cons. in Table II:

I. Running-Title Settings

Recto 1	B4r	C2r		D4r	E2r	F2r	G1r
Recto 2	B2r	C1r			E3r	F3r	
Recto 3	B3r		D1r	D3r	E4r	F1r	
Recto 4	B3r		D5r		E1r	F4r	G3r
Recto 5		C3r					
Recto 6							G4r
Recto 7			D2r				G2r
Verso 1	B1v	C3v	D6v		E3v	F2v	G3v
Verso 2	B3v	C1v	D2v	D3v	E2v	F3v	
Verso 3	B4v		D1v				G2v
Verso 4	B2v				E4v	F4v	G1v
Verso 5		C4v			E1v	F1v	
Verso 6				D4v			G4v

II. Running Titles, Forme by Forme

Table II places the two parts of the book roughly in parallel. For each sheet, the inner forme is shown first, as it was printed first. Notionally, each box of four pages represents a mirror image of the forme, that is, as the pages would be arranged on an unfolded sheet printed from that forme, with leaves 1 and 4 upright at the bottom and leaves 2 and 3 upside-down above them. The first printed page (1r) would be at the bottom right of the outside of the sheet, and the last printed page (4v) at the bottom left of the the sheet, and so forth.

Forme	Page	Page	Forme	Page	Page
B inner	2^r: R2	3^v: V2	**E inner**	2^r: R1	3^v: V1
	1^v: V1	4^r: R1		1^v: V5	4^r: R3
B outer	3^r: R3	2^v: V4	**E outer**	3^r: R2	2^v: V2
	4^v: V3	1^r: -o-		4^v: V4	1^r: R4
C inner	2^r: R1	3^v: V1	**F inner**	2^r: R1	3^v: V2
	1^v: V2	4^r: R4		1^v: V5	4^r: R4
C outer	3^r: R5	2^v: -o-	**F outer**	3^r: R2	2^v: V1
	4^v: V5	1^r: R2		4^v: V4	1^r: R3
Forme	**Page**	**Page**	**Forme**	**Page**	**Page**
D inner	2^r: R7	5^v: -o-	**G inner**	2^r: R7	3^v: V1
	1^v: V3	6^r: Pl. Cons.		1^v: V4	4^r: R6
D outer	5^r: R4	2^v: V2	**G outer**	3^r: R4	2^v: V3
	6^v: V1	1^r: R3		4^v: V6	1^r: R1
D ½-sh. inner	4^r: R1	3^v: V2			
D ½-sh. outer	4^v: V6	3^r: R3			

III. Running Titles, Sheet by Sheet

Table III lists all the running titles that were used in each sheet, regardless of forme and of their arrangement within the formes.

B	R1	R2	R3					V1	V2	V3	V4		
C	R1	R2		R4	R5			V1	V2			V5	
D			R3	R4			R7	V1	V2	V3			
D(½)	R1		R3						V2				V6
E	R1	R2	R3	R4				V1	V2		V4	V5	
F	R1	R2	R3	R4				V1	V2		V4	V5	
G	R1			R4		R6	R7	V1		V3	V4		V6

Several features of the running title distribution are noteworthy. No two formes (see Table II) share a single layout of running-title settings, nor do any two formes contain the exact same four running titles. Therefore, when the formes were taken apart and the types distributed, the running titles were also removed and set aside for re-use: they did not remain in their original positions in the chase or chases that held the type-pages. When new formes were imposed, the running-titles at the top of each page could in principle be supplied by any of several existing settings so long as the recto pages got recto settings, and verso pages verso settings. Moreover (Table II), no running title appears in both the inner and the outer forme of any given sheet. This agrees with the suggestion made in chapter X that all eight pages of a given sheet were typically standing in type at once. It does not directly prove that the inner and outer formes of the sheets existed concurrently; but if any running-title setting did appear in both inner and outer formes, that would disprove the hypothesis.

Although no two formes contain the same four running titles, there is a striking pattern in sheets E and F, that is, the first two sheets of the Jupiter observations, that brings them close together (see Table II). Picture one side of an unfolded quarto sheet as it comes from the press. It has four printed pages, two side-by-side in one half of the sheet, and two side-by-side in the other. Across the ream fold, the tops of the pages abut, and are upside-down with respect to each other. Thus, for the outer forme printing, the verso headline of fo. 2 is aligned with but upside down with respect to the recto headline of fo. 1; the same is true of fo. 3r as against 4v. We may think of these as the abutting, or head-to-head, headlines, there being two pairs in each forme: 1r + 2v and 3r + 4v for the outer forme, 1v + 2r and 4r + 3v for the inner forme. In sheets E and F, in each half of the forme, these abutting headlines stayed together consistently. On one side of the formes **R1** + **V5** abut in E (inner) and F (inner), while **R2** + **V4** abut in E (outer) and F (outer). On the other side of the formes, **V1** + **R3** abut in E (inner) and F (outer), while **V2** + **R4** abut in E (outer) and F (inner). Realistically, this pattern could only have arisen if these particular headlines had been kept together in abutting pairs, which in turn implies that the two sheets were printed in close succession.[5]

This gives a strong clue to the positive chronology of printing, one which lends support to what has been called the dilatory view; or rather to a modification of it. The terminal dates of sheets E and F, based on the last observations in each, are 22 days apart: E ends within the Jupiter observation of Friday 22 January, F the observation of Saturday 13 February. That is the minimum separation, but since Galileo was in Padua on 13 February, we must add a few more days to allow for his observation to be drafted and given to the compositor in Venice for setting the last page of sheet F. Indeed, by the chronology of its contents, sheet E could in principle have been printed at the end of January or early in February, shortly after Galileo's arrival in Venice to oversee production of his book. But the tight connection between its running titles and those of sheet F makes this highly unlikely. If we hypothesized that all these headline pairs were kept together, in the same order, over more than three works, so that they could be used in the formes of both E and F, then how could it be that the same settings, though in different orders, were used in printing sheets B, C, D, and the half-sheet of D? One would have to suppose that the printing of all these sheets was postponed until after sheet F had been printed and its formes distributed, at which time the running-title settings would be "free" for other use.

On the other hand, the running-title distribution of all the sheets provides a strong negative to the corollary of the dilatory view quoted in the first note of this chapter: that the two parts of the *Sidereus Nuncius* text, B–D and E–G, were "printed on two presses simultaneously." If composition and imposition had been strictly segregated, with each compositor serving a different press, we would expect that sheets B–D would have one set of running titles, and E–G another. But as the listing above shows (Table II), all the running-title settings appear in the formes of both parts except for the two, **R5** and **R6**, that were used only once.

Table III may therefore be slightly rearranged to show what seems to be the likeliest order of printing the sheets of *Sidereus Nuncius*:

B	R1	R2	R3					V1	V2	V3	V4		
E	R1	R2	R3	R4				V1	V2		V4	V5	
F	R1	R2	R3	R4				V1	V2		V4	V5	
C	R1	R2		R4	R5			V1	V2			V5	
D			R3	R4			R7	V1	V2	V3			
G	R1			R4		R6	R7	V1		V3	V4		V6
D(½)	R1		R3						V2				V6

5 In quires B–D two abutting pairs repeat within the six four-page formes: the pair R1 + V2 in the inner forme of B and inner forme of C; and the pair R4 + V1 in the inner forme of C and outer forme of D.

By this hypothesis, sheet B was the first to be printed, and it is likely to have been finished during the week of 8 to 12 February 1610. On Friday of that week, as Galileo explicitly states, he travelled back to Padua, where he found a letter from Vinta awaiting him. We may suppose that he brought with him a copy of sheet B, including the drop-title with its reference to "Cosmica Sydera," and that this awakened as an afterthought the question that he put to Vinta in his reply the following day: would "Medicea" be the preferable complimentary adjective? Galileo's reply made clear that he was soon returning to Venice: Vinta was to send the answer about Cosmica versus Medicea directly there, not to Padua. Galileo's Jupiter observation in Padua on Saturday night, the thirteenth, provided the final text necessary for filling sheet F, and the two sheets E and F were printed in succession. On the evidence available, it is equally possible that sheet E was printed in the week of 8 to 12 February, and F in the following week 15 to 19 February, or that both were printed in the latter week. In either case the formes of sheet E, except for the retained headline settings, were distributed before the pages of sheet F were set, for as already noted, four distinctive "O" types appear in the Jupiter diagrams of both sheets (figure X.16).

This proposed sequence, with sheets E and F printed after B but before C, accords well with the reasonings outlined in chapters VI through VIII about the complicated additions and changes in order made to the text part I. It will be recalled (p. 86) that topic IV of part I, The Silhouette Problem, was the latest addition to the Draft and a kind of afterthought, written on (Venetian) Rampant Lion paper. In *Sidereus Nuncius* as printed, the Silhouette Problem text begins at the end of C3v and continues on throughout C4r–v, coming to an end on the twenty-second line of D1r. The Silhouette Problem was the single major reason why the original calculation of fitting the text of part I onto three quarto sheets (B–D), to be foliated 5 through 16, turned out to be impossible. Suppose sheet E had not been taken up until after sheet C was printed. By that point, with the lengthy Silhouette Problem already in process of setting, it would have been obvious that quire D would inescapably need extra leaves, and it would have made no sense to number E1 as leaf 17. Thus, we may say with some confidence that sheet E was printed before sheet C.

The running-title patterns highlight the parallel production of the two parts of *Sidereus Nuncius* also in the final quires of those parts, D and G. Three new running-title settings, not used in quires B, C, E, or F, appear in quires D and/or G: the already mentioned **R6**, used once only in the inner forme of G; **R7**, used in the inner formes of both D and G; and **V6**, used in the outer forme of G and in the inner half-sheet forme of D. We know already that sheet G, containing a Jupiter observation of Tuesday 2 March in its outer forme, could not have been completed earlier than Wednesday 3 March, and possibly a day or so after that. It is plausible that the two full-sheet formes of quire D (1v–2r–5v–6r and 1r–2v–5r–6v) were printed before quire G. For reasons already stated, quires G and D could not have been concurrent in the strictest sense, that is printed on two presses at the same time: their respective formes contain six of the same running titles (**R1, R4, R7, V1, V3, V6**). The headlines also contribute to our knowledge of the half-sheet of D, leaves 3 and 4. It is likely that the two formes of this half-sheet were printed even after G, and thus with some distinct separation from the printing of D's full sheet. One strong hint to this effect has already been reviewed in chapter IX: part of the print run of the half sheet was on the main stock of *Sidereus nunceus*, the monogram LA paper used also for D's full sheet and for sheet G. The remainder of the run was on monogram GZ paper that otherwise makes no appearance in the edition. The obvious explanation is that the supply of LA paper originally

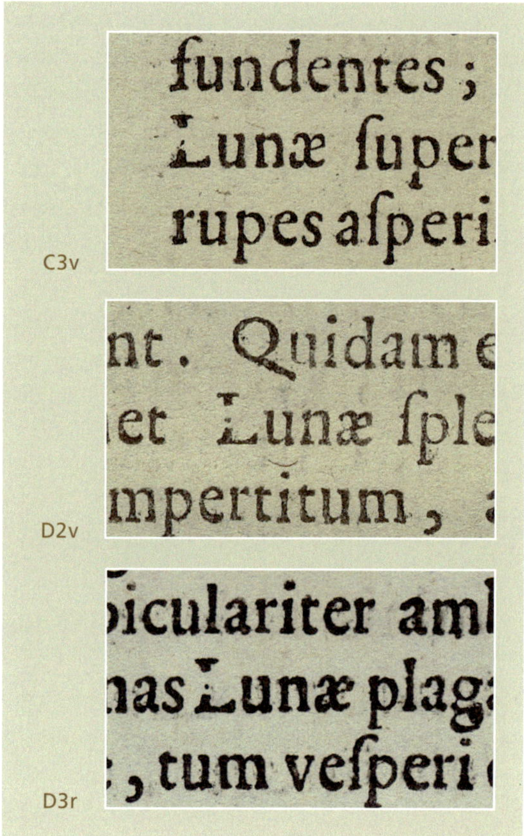

XI.3: Sidereus Nuncius, Institute for Advanced Study.

purchased for the edition (at a time when the half-sheet of D was not yet planned) ran short, and had to be eked out at the very end of printing.

But the headlines tell us something more that we would not know without their testimony: that there must have been some hiatus between the printing of D's full sheet and of its half sheet. On that half-sheet the two-page formes contain two headline settings from the outer forme of G (**R1** and **V6**), but also one headline setting from the outer forme of D's full sheet (**V2**). Thus, all twelve pages of quire D could not have been locked up in formes at the same time: the outer forme of D's full sheet – printed after its inner forme, it will be recalled – must have been taken apart before the forme of D3v–4r was assembled. This raises a question about the typesetting of quire D: were all its pages set serially, as seems to have been the case for the remainder of the edition? That is, when the full sheet of quire D (leaves 1, 2, 5, 6) was printed, could the pages of the half-sheet (leaves 3, 4) also have been standing in type, although not yet imposed in formes? There is strong evidence against this possibility. If all twelve pages of D (one of which, D5v, contained no type, only a woodcut) existed at once, then all the types creating those pages must be different physical objects. However, it is clear that one conspicuously damaged type, a capital L, was set both on D2v, the inner forme of the full sheet, and on D3r, the outer forme of the half-sheet (this type had also been been set once earlier, in the inner forme of sheet C) (figure XI.3).[6]

6 I am indebted to Wolfgang Undorf, who first noticed this damaged form.

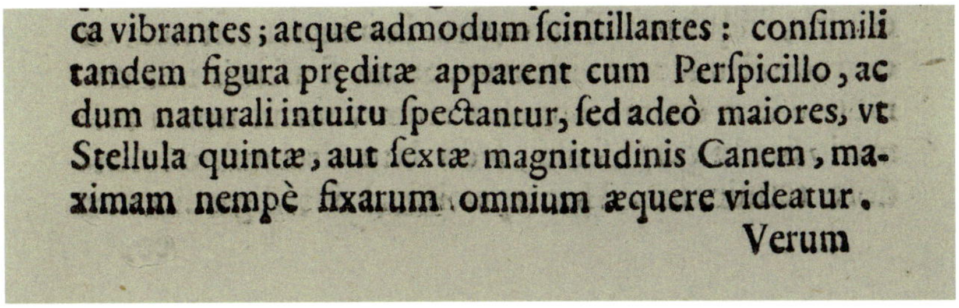
ca vibrantes; atque admodum ſcintillantes: conſimili
tandem figura prȩditæ apparent cum Perſpicillo, ac
dum naturali intuitu ſpectantur, ſed adeò maiores, vt
Stellula quintæ, aut ſextæ magnitudinis Canem, ma-
ximam nempè fixarum omnium æquere videatur.
Verum

XI.4: Sidereus Nuncius, D4v, Institute for Advanced Study.

Therefore, the types of the full sheet of D must have been distributed before the half-sheet was set. This entails, in turn, that D5r was set before the four pages D3r–4v, and that a calculation was made to determine how much text to skip over in the setting copy between the end of D2v and the first word of D5r: the intervening text was to fill evenly four type-pages or 136 lines. We can, to a degree, look over the compositor's shoulder as he set these four pages. On D3r–v he increased the line width from 95 to 100 mm. This may be construed as a literal "safety measure", in case the text calculated for these pages would not fit quite comfortably into 136 lines of typesetting. The greater line width allowed the equivalent of another three and a half lines of ordinary width to be fit onto the first two of the four allotted pages. In setting D4r–v the compositor fell back to the standard 95-mm width, knowing that he would not now have to deal with a surplus of text. In fact, he found he had a very slight deficiency of text for filling smoothly the 68 lines of D4r–v. On close examination a minute trace can be seen of an adjustment that the compositor made, as he came to the end of setting D4v, to make the last words he set join up neatly with the beginning of the already-set, -printed and -distributed D5r. D5r begins with a new sentence, "Verum infra Stellas magnitudinis sextae …", which is itself a hint, no more, that it could have been set before D4v. As the compositor came to the end of D4v, he noticed that his remaining text was not quite long enough. We can see that at several places in the last five lines he added extra space between the words, so that line 34 would reach all the way to the right margin and not stop short. By scattering these added spaces over several lines, the compositor made the adjustment less conspicuous. Without them the final line would have ended roughly 10 mm short of the margin (figure XI.4).

XII.
Quire A; The Completed Book

The running-title evidence of gatherings B through G has produced a number of results, some of which may fairly be called necessary or compelled, some probable but not rigorously proved, and some plausible but with probabilities difficult to assess. It is striking, withal, that so inconspicuous an element of *Sidereus Nuncius*, when added to the complex mixture of other textual, typographical, and paper-stock evidence, allows us to form a considerably clearer general picture and chronology than we have until now had of the production tempo of *Sidereus Nuncius*. The summary that follows, culminating in the production of sheet A, can hardly be definite, but it does attempt to maintain steady contact with the evidence laid out in the preceding chapters.

On the late evening of Friday 15 January 1610, feeling confident that he had identified four satellites revolving around the planet Jupiter, Galileo decided that he must publish a report in Latin, addressed to the world of learning, of his recent telescopic discoveries. His first thought was of how to depict these moving satellites: by means of white-on-black woodcuts, each one representing a distinct observation. Over the next two weeks, while in Padua, he worked on his draft, which progressed eventually in two channels: the first a report on the Moon and on the fixed stars, the second a report on the satellites of Jupiter. As to the latter, at the time he decided to publish his discoveries, Galileo had made only seven Jupiter observations. By the end of January he had made fourteen more, and had refined his observations, taking more careful note of the observation times, and of the apparent distances between Jupiter and its satellites. As for the Moon and fixed stars, his original plan was to treat of the Moon under two headings: the evidence that its surface is rough, irregular and earthlike, including craters, mountains and plains; and the evidence that its "secondary light" must be a reflection from the Earth. For the "rough surface" discussion, Galileo reworked and expanded the reasonings he had laid out in his letter of 7 January 1610, and the small Moon sketches included in that letter guided his earliest plans for the Moon illustrations of *Sidereus Nuncius*. For the Earthlight discussion, he planned no illustration.

While still in Padua, Galileo also drafted, or began to draft, his discussion of the fixed stars. In his first Jupiter Observation sheet, on 23 January, he included an asterism sketch based on Orion, which would exemplify one part of the discussion. He also, before leaving Padua, added a new topic to the Moon discussion: a calculation of the heights of the Moon's mountains. He would place this after the Earthlight topic, then proceed to the fixed stars. Finally, though less certainly, there is a reasonable probability that Galileo began to write his Fair Copy while still in Padua, yet the state of the planning stage represented in its last pages shows that they must have been written in February, while in Venice. The Fair Copy was probably written to provide for himself a clean reading copy, corresponding closely with the clean setting copy that he delivered, apparently in stages, to the printing shop.

At the end of January Galileo traveled to Venice, taking his notes with him. He met with the publisher Tommaso Baglioni, who a few years earlier had published his *Difesa*, and made arrangements for production of the book. We know very little of these arrangements, but it is likely that Galileo specified the format, quarto, and the paper size, Super-Median; and almost certain that he took on the costs of printing a limited issue on finer paper. Baglioni would have had a large say in the edition size apart from the fine-paper copies. We do not know if Galileo shared in the costs of printing.

From the time of the original publication plan, there was an internal battle of time in *Sidereus Nuncius*. Galileo wanted to continue and incorporate his Jupiter observations as long as possible – he was hoping that with sufficient records, he could find some clue to the periods of the moons. In the end, he had to be satisfied, in his closing words, with an abridged apology for being unable, so far, to do so. Yet both Baglioni and Galileo would have needed, for good sales and good publicity, the book to be completed in time to allow shipment of sheets to the spring book fairs of Frankfurt and Leipzig. Purely textually, *Sidereus Nuncius* was not a substantial printing job. If Galileo had brought with him to Venice a finished text of some ten thousand words, ready for the compositor, its seven (as originally planned) quarto sheets could easily have been set and printed off in two weeks or less. Generally speaking, the considerably longer time it took, six weeks from Galileo's presumed arrival in Venice, was owing to the "bottlenecks" created by Galileo himself, who was still writing, adding to and rearranging portions of the Moon and Fixed Stars text; adding to and analyzing his Jupiter observations; planning seven woodcut illustrations; and probably himself making the four Moon etchings, a reduction from his original plan of seven.

As to the woodcut illustrations, it may be noted that Galileo had direct experience with the process, and with the Venetian block-cutters: the many woodcut diagrams of his Padua 1606 *Operazioni del Compasso*, published in the summer of that year, had been made in Venice. He noted explicitly, in a letter of 11 November 1605 to Christina of Lorraine, Grand-duchess of Tuscany, that he was having the blocks cut there, and as soon as they were ready, "I will quickly print the book."[1] This letter hints at the desirability of having the woodblocks in hand before typographic composition began, a procedure not available for *Sidereus Nuncius*, where his maquette for the large Orion block could not have been made before Monday 8 February, and where two diagrams belong to his late additions to the Moon discussion: Height of Moon's Mountains, and The Silhouette Problem.

It is especially noteworthy that Galileo did not begin drafting part II of the text, the Jupiter observations, until after he arrived in Venice, for it is written on the Venetian Rampant Lion paper. Thus, although the observations contained in sheet E end with 22 Janu-

1 *Opere* X, p. 149 (Ep. 127): "In Vinezia ho fatto dar principio ad intagliare le figure che vanno nel discorso circa l'uso di esso mio instrumento; et intagliate che siano, farò subito stampar l'opera."

ary, a week or more before he left Padua, the paper evidence argues that when he got to Venice, he had not even begun to draft the text for that sheet. This fits with the headline evidence, which strongly suggests that B was printed before E, the beginning of the Jupiter observations; that E and F were printed in close sequence; and that C followed F. It seems moreover that when sheet B (fos. 5–8) was printed, the space calculations for the topics of the Moon and Fixed Stars still seemed to be under control, calculated to fit on two more sheets, fos. 9–12 and 13–16 respectively, so that it made sense to number E1 as fo. 17. Since (see chapter IX) it was the Silhouette Problem that definitively required more space than the original calculation allowed, we may suppose that the Silhouette Problem arose in Galileo's mind after quire B had already been printed.

The close sequence of sheets E and F means that the terminal date of the observations printed on E, 22 January, was well in the past by the time the sheet was printed. The terminal observation of sheet F being Saturday 13 February, that sheet could not have been printed before Monday the 15th or later. On present evidence, it is uncertain whether E was printed before the weekend of 13–14 February and F in the week after; or both E and F were printed in the week after. Either way, this hints at another hidden cause for delay in the production of *Sidereus Nuncius*: a delay in printing the Jupiter-moon observations. When Galileo first resolved to print his report, the treatment of this part was clearest of all in his mind: elegant white-on-black woodcut rectangular blocks, fitted at their proper places into the type-pages, would depict the changing dispositions of the satellites of Jupiter. The relatively late printing of sheet E may indicate that some time was lost in trying to carry out this procedure, which then culminated in the realization that the time needed for making the many dozens of blocks was slowing down production and would have to be abandoned. The typographic diagrams of the finished book can be seen as a makeshift, whose chief advantage over against their inelegance was the directness and rapidity with which they could be set.

Finally, the headline evidence suggests that there was some hiatus between the printing of quires E and C, and of quires G and D. Quire G, it will be recalled, could not have been printed before Wednesday 3 March or after, its final Jupiter observation having been made on 2 March. This could have been a week or more after E and C were printed. The signal of the hiatus is the apparent loss or misplacement of existing running-title settings, and their replacement with new settings: three settings used in quires B, E, F and C disappeared (**R2**, **R5**, **V5**), and new settings took their place (**R6**, **R7**, **V6**). As for the half-sheet of D which apparently was printed with some hiatus between it and the full sheet, it may be noted that two of its running titles are from the outer (second) forme of sheet G (**R1**, **V6**), and two from the outer (second) forme of sheet D (**R3**, **V2**).

Quire A, containing title-page (1r, verso blank), dedication (2r–4r), and licence (4v), was the last to be printed, and as already noted is on a different paper stock from the text. The clean division of the paper stocks, separating quire A entirely from the rest of *Sidereus Nuncius*, could reflect the hiatus, probably of a week or so, between its printing and that of quire G. Or it could be that this stock, possibly of slightly higher quality than the main stock of the ordinary-paper issue, was specifically reserved for the first gathering. The Inquisition's license to print, and the record of registration of the license, are dated on successive Mondays, the first and the eighth of March, but the process began during the last week of February. The details of the licensing process, which required successive certifications by the Office of the Inquisition, the "Riformatori" of the University of Padua, and Venice's Council of Ten, have been closely studied by Antonio Favaro. Favaro uncovered two records in the state archives of Venice relating to the license. In the files of the Riformatori is a certificate dated Friday 26 February, in which Galileo's treatise was referred to under the title "Astronomica denuntiatio ad astrologos etc." In the original record of the Council of Ten, dated 1 March, the work bore the same inaccurate title. Thus, the record as printed on A4v, with title as SYDEREVS NVNCIVS, is technically speaking inauthentic, for it deviates from the original document.[2] The title Galileo apparently supplied to the licensers is actively misleading. It may be suggested (only) that his design was to maintain as long as possible the shadow of secrecy around his discoveries, until the book was truly finished and made public under his name.[3] It should be noted that we do not know what text, what form of text – manuscript or printed – or how much text, Galileo supplied to the Riformatori. By the end of February sheets B, E and F had almost certainly been printed, and possibly also sheet C.

As noted in chapter VI, the title of *Sidereus Nuncius*, of some seventy-three words disposed over twenty lines, is, so to speak, in dialogue with the drop-title on B1r, of some thirty words over nine lines. The latter's *Astronomicus nuncius* now became *Sidereus Nuncius* – a rhetorical heightening through use of the more poetic term. The "baptism" of Jupiter's moons went from *Cosmica sydera* to *Medicea sidera* in accordance with Vinta's wishes – really, a command – and it was made explicit that these new planets revolve around Jupiter, whereas in the drop-title they had been unlocated, with nothing said of their orbits. And, the title-page states dramatically that the "spectacle" of the report – treating, that is, the entire sweep of the cosmos as a theater – was "presented to all, but especially, indeed, to the philosophers and astronomers": *Spectacula ... proponens vnicuique, præsertim vero philosophis, atque astronomis*. The statement, or a variation on it, was long planned. On 13 February Galileo had written to Vinta, "As for my new observations, truly I send them as an *avviso* to all the philosophers and mathematicians, but not without the auspices of our most Serene Master ...".[4] Visually, the title was also more

2 Favaro 1889. At pp. 101–102 Favaro misread the date of the certificate of the Riformatori as 26 February 1607, which led him to some convoluted hypotheses that need not be reviewed. In the *Opere*, XIX p. 227, he silently corrected the date to "26 Febbraio 1609" Venetian style, that is, 26 February 1610. Poppi 1998 has performed a useful service in making the correction explicit, and confirming the correct reading from the original document. For a broader account of the licensing process in Venice and Padua see Grendler 1977, pp. 147–161.

3 Wootton 2009, p. 126 suggests that the drop-title of B1r, beginning "Astronomicus nuncius", "was sufficiently unusual to confuse the authorities, who rendered it *Astronomica denuntiatio ad Astrologos*." But it is difficult to conceive that a university official could look at the very familiar word "nuncius" and subconsciously transmute it to "denuntiatio", then supply a word not present at all in the drop-title, "astrologos." In any case, it should be noted that in the original record of the Council of Ten, printed Favaro 1889, p. 101, Galileo's "Astronomica denuntiatio" was one of three works to be licensed. The stated titles of the other two, Giovanni Bellarino's "Doctrina [...] catechismi Romani", and Antonio da Gama's "Decisionum supremi Senatus regni Lusitaniae" [... centuria quatuor], are free of error and correspond closely to the printed titles of the respective works.

4 *Opere* X, p. 283 (Ep. 265): "Quanto alle mie nuove osservazioni, le mando bene come per avviso a tutti i filosofi et matematici, ma non senza gl'auspicii del nostro Serenissimo Signore ...".

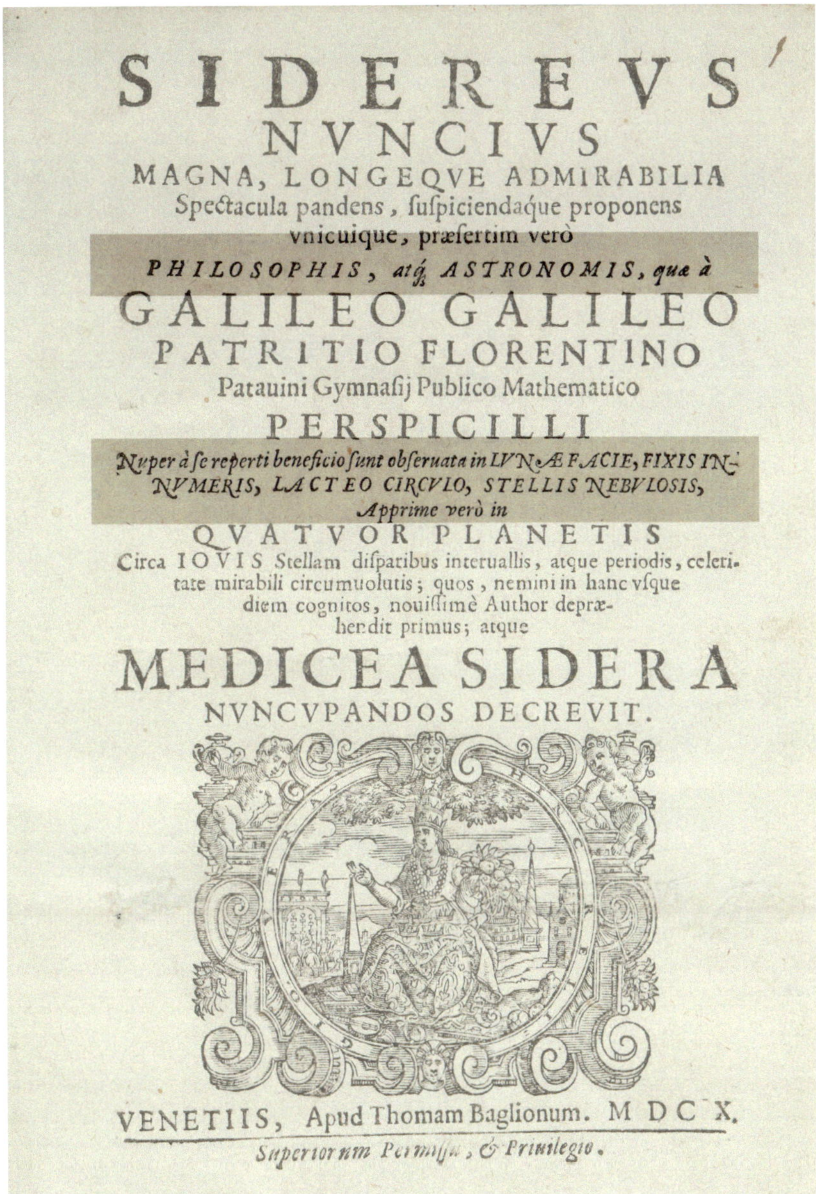

SIDEREVS
NVNCIVS
MAGNA, LONGEQVE ADMIRABILIA
Spectacula pandens, suspiciendaque proponens
vnicuique, præsertim verò
PHILOSOPHIS, atq; ASTRONOMIS, quæ à
GALILEO GALILEO
PATRITIO FLORENTINO
Patauini Gymnasij Publico Mathematico
PERSPICILLI
Nuper à se reperti beneficio sunt obseruata in LVNÆ FACIE, FIXIS INNVMERIS, LACTEO CIRCVLO, STELLIS NEBVLOSIS,
Apprime verò in
QVATVOR PLANETIS
Circa IOVIS Stellam disparibus interuallis, atque periodis, celeritate mirabili circumuolutis; quos, nemini in hanc vsque diem cognitos, nouissimè Author depræhendit primus; atque
MEDICEA SIDERA
NVNCVPANDOS DECREVIT.

VENETIIS, Apud Thomam Baglionum. M D C X.
Superiorum Permissu, & Priuilegio.

XII.1: Sidereus Nuncius, A1r, Institute for Advanced Study.

complex as against the drop-title: the four typographic modes or grades of the drop-title were expanded here to about a dozen modes, taking into account the possibilities of variation provided by fonts of seven different body sizes; setting words in full caps versus upper- and lowercase; and by use of italic versus roman characters. Whether intentionally or not, there even appear to be two different italic fonts, similar in size but one with a number of distinctive swash capitals among other differentia (figure XII.1).

The dedication to the young Cosimo II, whose nineteenth birthday was approaching on 12 May, was dated from Padua on the 4th Ides of March, that is, Friday the 12th. No draft of this text survives. There is reason to suppose that the dedication was completed at least a few days earlier, and awaited for its date to be coordinated with the day when sheet

A was actually printed. When, on 13 February, Galileo had raised with Vinta the question of whether to name the Jupiter moons "Cosmic" or "Medicean," he suggested that the second alternative, which had not been his preference, would associate the moons, "exactly four in number," with Cosimo II and his three brothers, as the "fraterna" or heirs of the late Cosimo I (*mi resta un poco di ambiguità, se io deva consecrargli tutti quattro al G[ran] D[uca] solo, denominandogli Cosmici dal nome suo, o pure, già che sono a punto quattro in numero, dedicarli alla fraterna con nome di Medicea Sydera*).[5] However, in the dedication as printed, this was not his rhetorical approach. The planets are indeed referred to as the "Medicea sydera", but Cosimo's brothers make no appearance. "Medicea" is related to Cosimo's princely house, but no reader would find a hint of brotherhood in the name. The "fraterna" motif did not, however, entirely disappear. Among Galileo's papers is a copy, in his autograph, of a commendatory sonnet that Piero de' Bardi composed for the planned Italian edition of *Sidereus Nuncius*, which explicitly associates Cosimo and his brothers Ferdinando, Giovanni and Francesco with the four moons, "al ciel volando, | Nel sesto giro …".[6] It is to be noted that the placement of Jupiter at the "sesto giro" or sixth sphere is that of the Ptolemaic system.

It is thus possible that by mid-February Galileo had already drafted the dedication, with "Cosmica" nomenclature, before he received an answer to the Cosmica-Medicea question. When he received the reply, he would have realized that he needed to change only two or three words. At the central place, "Cosmica sydera" became "Medicea sydera". And, in the sentence preceding, Galileo could slightly reword his intention to dedicate the moons: not to "thy most august name", but to "the name of thy most august house" (*eas Augustissimo Prosapiae tuae nomine insignire decreui*). In essence, the change was only from Cosimo's baptismal to his family name.

By Galileo's own statement to Vinta, the printing, that is of sheet A, was completed on Saturday 13 March: "Not until today, quite late, was it possible to get a copy of my Avviso Astronomico …" (*Non prima che oggi, et ben tardi, si è potuto havere alcuna copia del mio Avviso Astronomico …*).[7] As will be reviewed in the next chapter, this was a few days later than he had originally expected. Also on Saturday, as was noted in the Introduction, Sir Henry Wotton, British ambassador to Venice, procured a copy "come abroad this very day," and sent it off to London, with promise of a telescope soon to follow. Wotton's dispatch to James I's chief minister, Lord Salisbury, included a lucid summary of the discoveries, as filtered through the mind of the common, though more than commonly intelligent, reader, one in whom the picture of the Aristotelian-Ptolemaic cosmos was deep-dyed. One of his immediate reactions was that "we must have a new sphere to save the appearances." Wotton further recorded that the appearance of the *Sidereus Nuncius* had attracted immediate attention, and not just from him: it was news "whereof here all corners are full."

5 Ibid.
6 *Opere*, X, p. 399 n. 1.
7 *Opere* X, pp. 288–289 (Ep. 271).

XIII.
The Martayan Lan Proof Copy

During the weeks in which *Sidereus Nuncius* was being produced in its ordinary-paper and fine-paper copies, another copy separate from these was brought into existence by a slightly different printing process: the Martayan Lan copy. At the beginning of this study it was asserted that this copy (hereafter SNML), which contains Galileo's pen and wash drawings of the Moon used as models for his Moon etchings, is a proof copy: that is, it was assembled from a series of proof printings. The evidence lies in three interlocking physical features: the paper stock(s) of the copy, its typographic settings, and the manner in which its settings were printed on its paper supply. A proof leaf, or sheet, or collection of leaves of a given edition may be defined as any such item that preceded the ordinary print run of the edition.[1] Such leaves may be hand-marked for changes to be made in the setting, or not; but in either case, they belong to an earlier stratum of work than that of the print run proper. Printing proofs are relatively uncommon, being intrinsically ephemeral. They have two chief survival channels. They may become part of the printed waste of a shop that ended up being re-used by bookbinders as pastedowns, flyleaves, sewing guards, or other structural elements of bookbindings. Or, rather than being thrown away, they may get put into the stacks of the ordinary printing, and so get bound in some finished copy or copies. For example, three copies at the Folger Shakespeare Library of the 1623 collected plays of Shakespeare each contain a sheet marked for correction; and a small number of other copies contain the same sheets in their first, uncorrected state, but unmarked.[2]

The fact that SNML is printed throughout on a different paper stock from either the ordinary-paper or the fine-paper copies of the main print run is of itself a strong hint that it is a collection of proofs in the sense just defined. An active printing shop will naturally accrue various sheets of paper left over from previous printing jobs, and it made sense to use up these remnants, already paid for, in ephemeral work such as proofing a subsequent job. For similar reasons, in more recent times galley proofs were very commonly pulled on

1 McKerrow 1935, p. 347: "... what are usually known as proofs, namely, prints taken from the type before the working of the particular sheet begins." This definition is adequate for the case of *Sidereus Nuncius*, but for completeness another important category of proof must be added: sheets or leaves from the print run itself that were marked for correction, the printing of the forme in question being temporarily suspended until the correction (and possibly, some new error) was introduced by loosening the forme and moving some type or types. An example – apparently the earliest known marked proof – is preserved in the Cambridge University Library copy of the Gutenberg Bible, see Needham 1986 (b), pp. 59–60.

2 Hinman 1963, I, pp. 233–234. A loose single leaf at the Folger also contains marked corrections.

newsprint or other cheap paper. The use of a cheaper material for proof-printing goes back to the earliest days of European typography. In the case of the Vulgate Bible printed in Strasbourg by Johann Mentelin, not after 1460, there is in Cambridge University Library a copy of volume I that was printed not as a folio on Royal-size paper, but rather as a sequence of broadsheets printed, leaf by leaf, on Chancery paper, half the size of, and less than half the cost of, Royal paper. Mentelin, or someone in his shop such as a corrector, apparently realized that these proof sheets, gathered together, would make up a marketable copy.[3] In short, they survived because they had commercial value. The edition of Durandus, *Rationale divinorum officiorum*, printed in Mainz by Johann Fust and Peter Schöffer, 6 October 1459, presents a closely analogous example. The entire edition was printed on vellum except for the copy at Munich State Library, which is printed on paper, eked out by two and a half vellum sheets. This copy is likewise a collection of proofs, some with marked corrections but many not, pulled on paper. The paper used, in fact, was very probably a group of remnants left over from the paper stocks used four to five years earlier to print the Gutenberg Bible.[4] These proofs survived for the same reason as those of Mentelin's printing of the Vulgate Bible.

As already noted in chapter II, the paper stock of SNML is closely related to the chief stock of the ordinary print run, and must have been manufactured at the same paper mill. The two stocks display the same primary watermark, an LA monogram, and cornermark, a Tp monogram, but they were made on different pairs of moulds. Their watermarks, though identical in type, were all separate hand-shaped wire constructions. The word "same" with regard to paper and watermarks has been so imprecisely employed by scholars, even by many with experience in the typographical aspects of printing, that these distinctions of mould-pairs and wiremarks must be specially emphasized. The paper stock of SNML (with a variation in one of its bifolia to be mentioned shortly) was made on a pair of moulds of Mezzana Grande (Super-Median, to give it an English name) size. To the wire-screen surface of each of those moulds was sewn a main mark, LA, hand-shaped from wire, and likewise a cornermark Tp monogram. The two moulds of the stock can be distinguished from each other by small differences in their wiremark and chainline positionings. Most conspicuously, the wire shapes of the main marks and of the cornermarks were sewn to their respective moulds in mirror images of each other, and on opposite sides of the moulds. In the sheets made by one of the moulds, the main mark, looked on from the mould side, reads as LA, in the sheets from the other mould, as a backwards LA. In one, the cornermark looks like a Tp monogram, in the other like a Tq monogram. The corresponding paper stock of the ordinary-paper copies was made on a different pair of moulds, each with its own differently constructed wire shapes for cornermark and main mark, and those two moulds have their own distinctive features which make them different from each other, and from the twin moulds of the SNML stock. In what is now a large group of copies of *Sidereus Nuncius* that have been examined closely as to paper, bifolium by bifolium, the stock of SNML appears only in that copy; the second stock with LA watermark appears

3 Needham 1986 (a). In this Cambridge copy, there is one full sheet supplied from the ordinary run, and three pages – the verso of one printed leaf, and another interpolated leaf – were supplied in contemporary hand, copied from the printed edition.

4 Hellinga 1986; further analysis of the paper stocks by Needham 1986 (a).

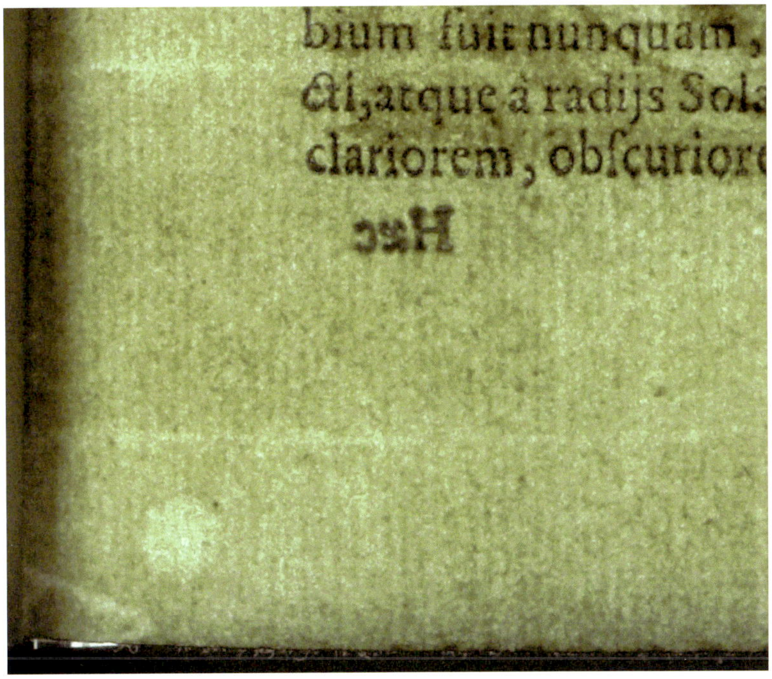

XIII.1: Sidereus Nuncius, C1r, SNML.

only in quires B–G of the ordinary-paper copies; and as already noted in chapter II, an Anchor in circle stock appears only in quire A of the ordinary-paper copies.

The likely situation is that SNML's LA-marked stock was made in some unidentified northern Italian mill some months earlier than, from the same mill, the LA stock of the ordinary-paper copies. The printing shop had a remnant of the first LA paper, left over from some preceding printing job, which was put to use for the proofing of *Sidereus Nuncius*. One half-sheet of SNML, C1.4 (fos. 9.12), falls outside this pattern and is indeed quite mysterious. It is notably thicker and stiffer than the other bifolia, and its watermark, seemingly the letter A (and so very probably part of an LA monogram), is placed, partly shaved by the binder, at the lower inner margin of C4 (figure XIII.1).[5] This placement requires that the mark was placed either near one of the short edges of its mould or, just possibly, in the center of the mould.[6]

Because proof printing belongs to a stage preceding the print run, the typesetting of SNML must in principle present the earlier state of any variant settings that may exist within the edition as a whole. This is, in fact, the primary criterion for concluding that, say, a printed fragment of some known edition is a proof, and not simply the remains of an ordinary copy. SNML passes this chronological test. There is relatively little typographical variation among the many copies of *Sidereus Nuncius* including SNML. The relative pau-

5 This unusual paper was discovered by Irene Brückle and Manfred Mayer, who made transmitted-light photographs of all the leaves of SNML.

6 An analogy of the latter possibility, but almost certainly unrelated, is found in Geneva papers. Briquet, p. 14, cites a 1562 regulation of the council of Geneva that papers made in that region should have watermarks centered in the sheet so as not to interfere with the printing ("affin que la marque n'empesche point l'ouvrage de l'impression"), and papers so marked are indeed found in Geneva sixteenth-century imprints.

city of variation between the proofs of SNML and the main-run copies is very probably because there were earlier, now-lost proof stages before the SNML printings were made as final proofs: what in English printing shops of the time were called revises. The majority of the variations that have been noticed separate SNML from all other copies. Most of these can be explained as the replacement of damaged types in the SNML settings, made before the main run of each forme was printed off. For example, at B3r line 8, "CD. bracteas, aliàs maioribus", in SNML the comma is damaged, and was replaced before the main print run began with an intact comma:

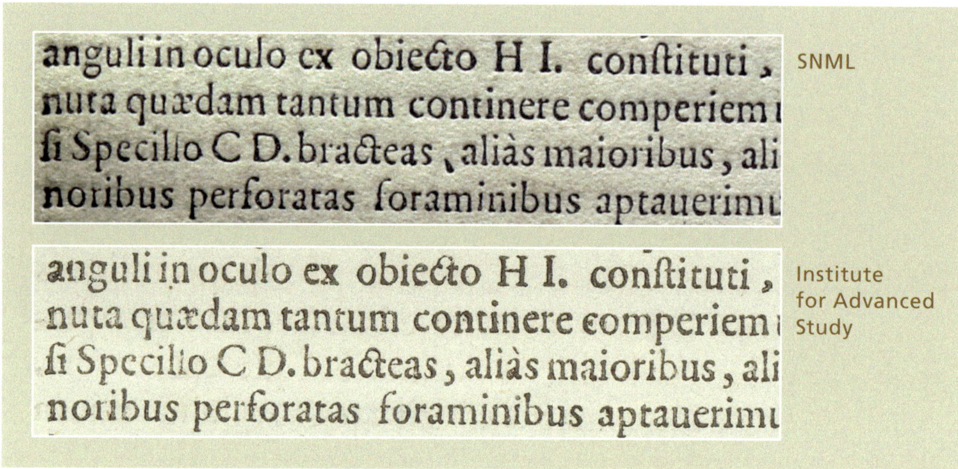

XIII.2: Sidereus Nuncius, B3r.

Likewise at C3r line 12 "tenebrosae": in SNML, the *æ* digraph is damaged, in all copies from the main print run it is clear:

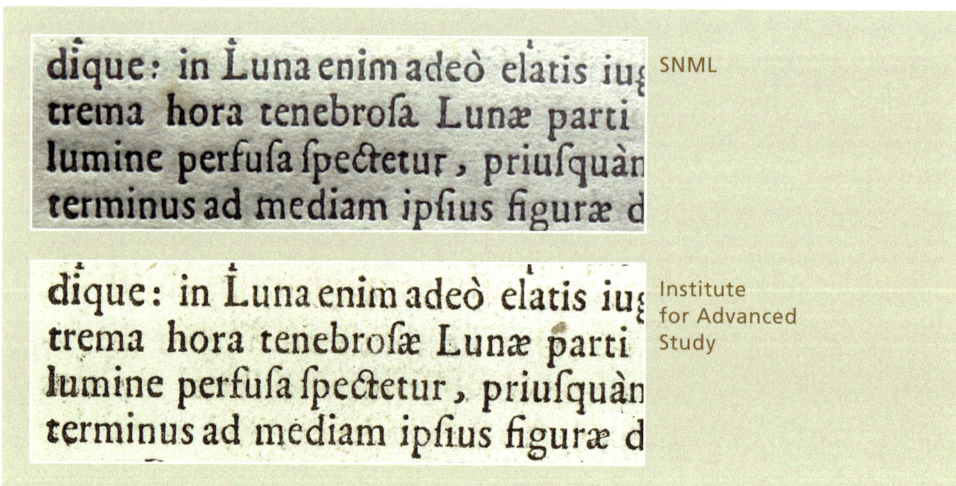

XIII.3: Sidereus Nuncius, C3r.

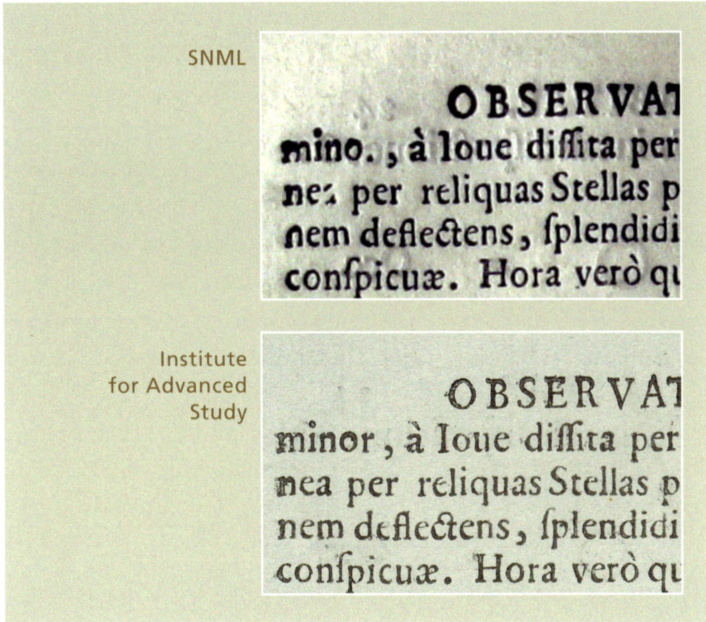

XIII.4: Sidereus Nuncius, F4v.

On F4v line 1, the first word "minor," contains what appears to be a period instead of the "r" (in fact it is probably a heavily damaged "r"). This was corrected before the main print run with a perfect "r" (figure VIII.4). Similarly, in the first Jupiter-moon diagram on G3r, the leftmost asterisk (Asterisk 2, see chapter X) is damaged in SNML, undamaged in all copies of the main print run (figure XIII.5).

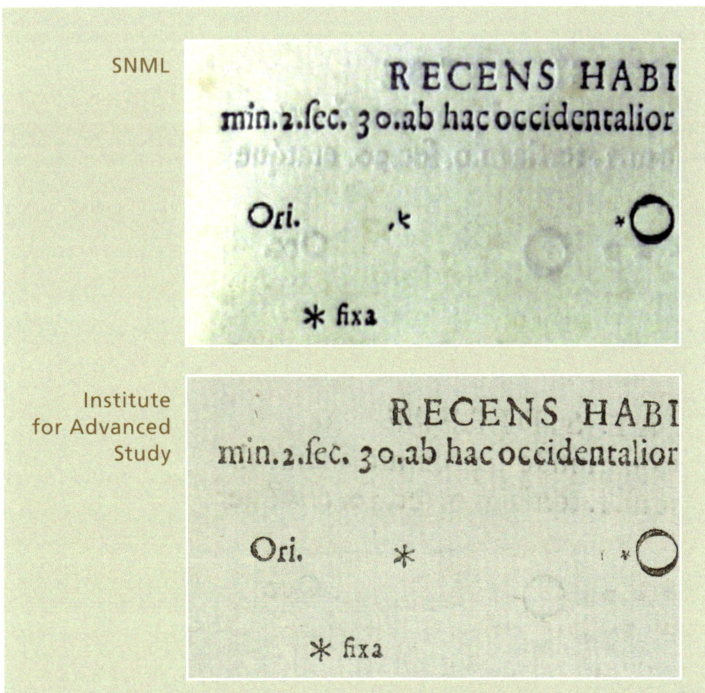

XIII.5: Sidereus Nuncius, G3r.

Just one change corrected a textual error. On G3v the second diagram of the Jupiter-moon configuration, 2 March, is faulty in SNML. Galileo observed two moons to the east of Jupiter, a larger one (set with Asterisk 2) and a very small one quite close to the larger. In the state of setting represented by SNML, the compositor had neglected to set this smaller moon. In the main print run, the necessary smallest asterisk (Asterisk 4) was added to the forme. This corresponds exactly to the Maquette, which indicates the larger moon as magnitude 2, and the smaller as magnitude 4 (figure XIII.6).

One typographic change between SNML and the main print run highlights a difference between Galileo's orthography and that of the printing shop; it may be seen as a borderline correction. D6r's catchword ought to have been "coetum", the first word on D6v. In

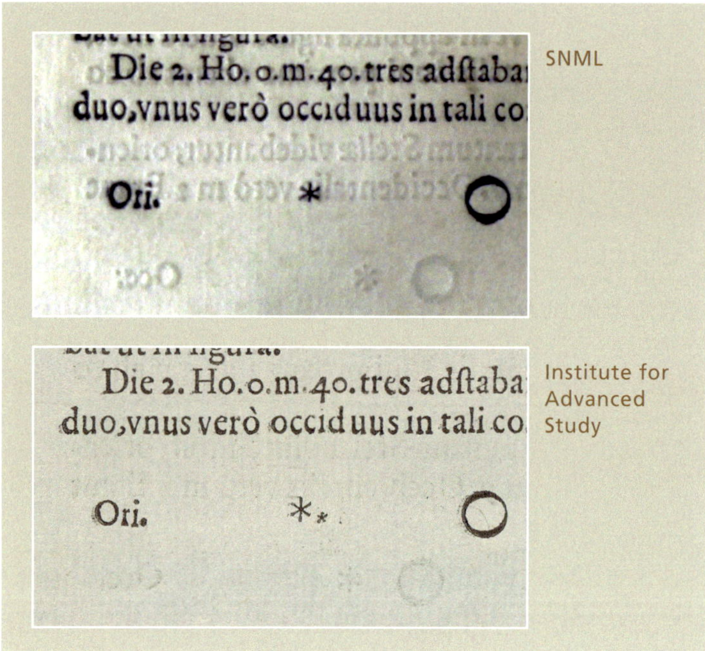

XIII.6: Sidereus Nuncius, G3v.

SNML the word is set as "cetum", in the main print run as "cętum", with the caudate e substituting for the "oe" diphthong (figure XIII.7). As discussed in chapter VI, Galileo typically wrote the "ae" diphthong as a caudate e, a letter form with a long history. The compositor or compositors of *Sidereus Nuncius* generally changed all these to an *æ* digraph, that is, a type on whose face the "a" and the "e" are coalesced. However, again as reviewed in chapter VI, at a number of places the compositor forgot to regularize to *æ*, typically setting caudate e instead – for the font also had that character available – under the subliminal influence of Galileo's handwriting. Galileo used caudate e also when writing the words "caelum" and "caelestis". *Sidereus Nuncius*'s compositor spelled them consistently as "cœlum" and "cœlestis", using the font's *œ* digraph. By that standard, the catchword on D6r should have been set as "cœtum". The gain in correctness in changing "cetum" to "cętum" was minimal, but it caught Galileo's or, much more likely, the printer's eye nonetheless.

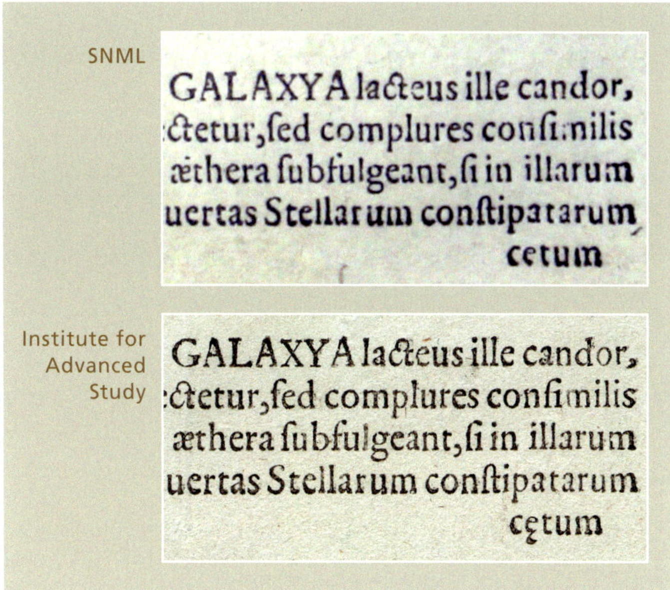

XIII.7: *Sidereus Nuncius*, D6r.

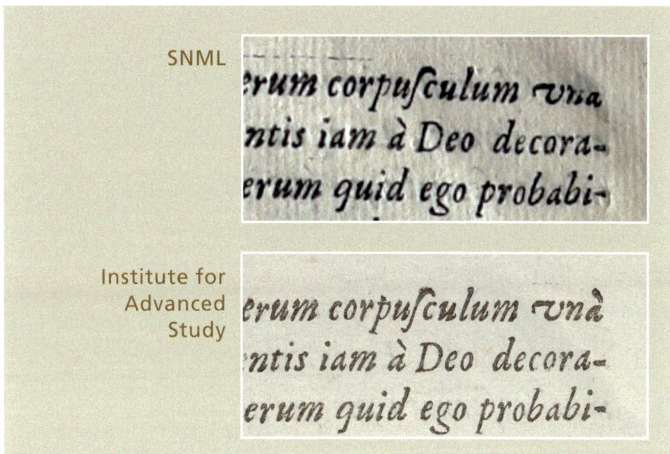

XIII.8: *Sidereus Nuncius*, A3v.

Various misspellings and other typographical errors in the formes of *Sidereus Nuncius* never got corrected over the course of the printing cycle. Examples include the misspellings "nuncunpantur" (*nuncupantur*) at C1r line 23, "quaeda" (*quaedam*) at C1v line 23, "sollicicus" (*sollicitus*) at E1v line 4, and "consequantia" (*consequentia*) at E3r line 3. The upside-down "t" in "nationibus", C3v line 21, was not corrected, nor "essenr" (*essent*) at D4r line 32, which probably resulted because at an earlier stage of the printing cycle, an r-type got distributed to the t-box.

Two places have been noticed where faulty settings in the SNML pages were corrected not before, but during the course of the main print run, that is by what is called stop-press or in-press correction. Both are on the same page, A3v (inner forme), and so must belong to the last day or two of printing. SNML shows two errors on this page. The last word of line 1, "vna", ought by the typesetting conventions of *Sidereus Nuncius* to have been set with an *à* character, long final vowels (as elsewhere *sanè*, *longè*, and *adeò*) being typically set with grave-accented characters. In SNML the "a" of "vna" lacks an accent (figure XIII.9). Or, more precisely, the compositor reached correctly into the "à" box, but the type he drew out was damaged, its accent stroke did not print. Similarly the *n* appears to be a damaged type. In almost all copies belonging to the main print run, "vnà" was set with a clean *n* and a clean *à* (figure XIII.8), but at least four ordinary-paper copies from the main

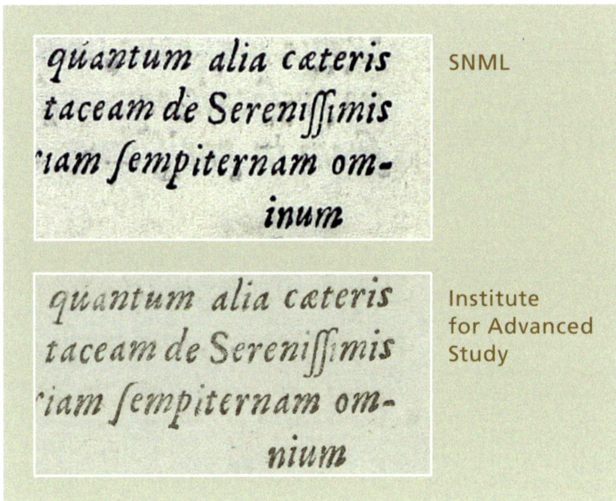

XIII.9: Sidereus Nuncius, A3v.

run have "vna" in the same setting as SNML. Therefore, the main print run must have begun without correction, but quite early on the flaw was noted, printing was paused, and "vna" was corrected within the forme; after which printing continued.

The second error on A3v is the catchword, set as "inum" in SNML instead of "nium" – the divided word between A3v and 4r being "om- | nium."(figure XIII.9) This error was not noticed and corrected until later in the printing of the inner forme of sheet A. The four ordinary-paper copies just listed as having the poorly set "vna" of line 1 also have, as they must, the incorrect catchword "inum". But at least nineteen other ordinary-paper copies have "vnà" corrected, but "inum" uncorrected. Finally, at least fifty-nine copies, including all the identified fine-paper copies, have both words corrected, "vnà" and "nium". In terms of the printing cycle, this means that when the inner forme of sheet A was printed, it was interrupted at two different times for correction, and so copies variously represent three correction states according to when their sheets were printed over this forme:

vna and *inum*	SNML (proof); 4 ordinary paper copies
vnà and *inum*	19 ordinary-paper copies
vnà and *nium*	50 ordinary-paper copies; 9 fine-paper copies

Because the fine-paper copies are in the last state, there is good evidence that generally the fine-paper sheets went to press, forme by forme, after the ordinary-paper sheets had already been printed off. This corresponds to the practice found as early as the Gutenberg Bible, whose vellum sheets were printed after its paper sheets.

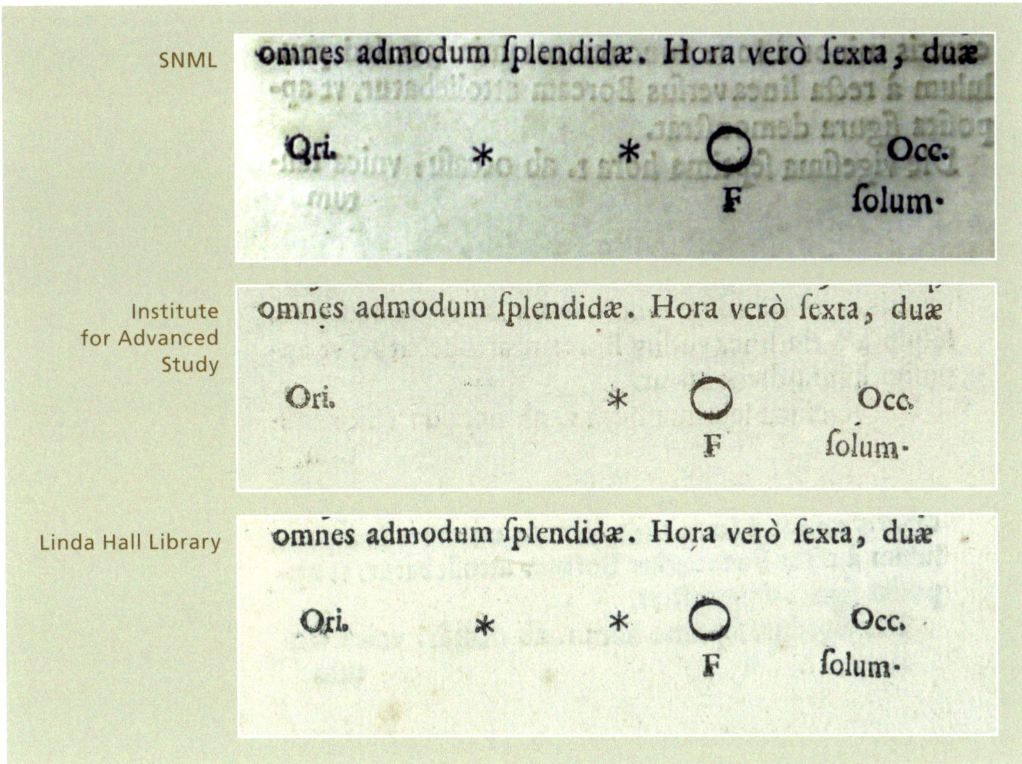

XIII.10: Sidereus Nuncius, F1v.

If the fine-paper sheets were printed last, this suggests that a larger group of copies can be identified as having late impressions of the inner forme of the full sheet of quire D: that is, all the copies showing frisket slippage on D6r, where the frisket partially occluded the printing of two lines (see figure III.8). The 9 fine-paper copies all show this "frisket bite," and so do at least 30 other ordinary-paper copies, with variations in the degree of occluding.[7] SNML and at least 32 ordinary-paper copies show no frisket bite.

The outer forme of sheet F presents a problem that cannot easily be explained, for the testimony of copies seems to indicate that its setting was correct at the beginning, then acquired an error, then got corrected again. The Jupiter-moons diagram at the foot of F1r (21r) shows the configuration Galileo observed on 24 January, at the sixth hour of the night. Earlier that evening three moons had been visible to the east of Jupiter; at the sixth hour, only two were to be seen (*duae solummodo sese offerebant Stellae*), one 3 minutes and the other 11 minutes from Jupiter. In SNML, the diagram is correct, with two asterisks to the left of the Jupiter O, more or less in scale with Galileo's stated separations (figure XIII.10). However, in at least 33 ordinary-paper copies, the asterisk for the moon 11 minutes east of Jupiter does not appear, or did not print. Then in at least 41 ordinary-paper copies and all 9 fine-paper copies, that same asterisk is present once again. If all the sheets

7 In the Beltrame (Vicenza) copy of *Sidereus Nuncius*, frisket bite or frisket slip has occluded almost all of the bottom line of D4v, and blocked out entirely the catchword. This accident must have been noticed quickly as the forme was being printed, for the accident has not been noticed in a large number of other copies.

lacking the asterisk were printed after all the sheets with the asterisk, there would be no particular problem, even if we cannot say how the asterisk got "lost". If the fine-paper sheets were printed last of all, as is almost certainly the case, there is a perplexing sequence of lost and found. There appear to be no other signs of damage or disarrangement on this page, for instance of growing gradual damage to the asterisk character. In the copies lacking the asterisk, there are no traces of it, such as a blind impression, at all. Theoretically, it is possible that the frisket of this page slipped a little, as it did on F6r, covering the asterisk for a certain number of impressions, then got noticed and shifted back into correct position. Yet this is not satisfying. The frisket, whatever its shape, would have covered the entire page. If it shifted at one place sufficiently to cover up the asterisk, then at some other place it ought also to have intruded on the inked types. At the moment, we have no convincing explanation, only the facts of the matter.

The third feature of SNML that shows it to be a collection of proofs is the manner in which its pages were printed, a feature independent of both the typesetting and the paper stock itself. The ordinary copies of *Sidereus Nuncius* were printed on full sheets of paper laid over four-page formes, whereas SNML was printed on half-sheets of paper over two-page formes. Thus A1.4 is a half-sheet, A2.3 is another half-sheet, and so on for the remaining gatherings. In no gathering of SNML do the bifolia 1.4 and 2.3 belong to the same sheet of paper. This exceptional feature of SNML was discovered by Irene Brückle and Manfred Mayer, who made transmitted-light digital photographs of all its leaves, and assembled them in groups of four to create notional representations of the (putative) original sheet of each quire. They found that under transmitted light the paper of the first bifolium of a quire (1.4) often looked different in weight from the paper of the second bifolium (2.3).[8] This difference or disparity is most striking for quires B and C. Consider the transmitted-light reconstruction of quire B (figure XIII.11) As is apparent, the bifolium B2.3 (above) is considerably darker than B1.4 (below), because it is a denser, less transparent piece of paper. Indeed, different sheets of paper from the same stock may vary in thickness, and thickness may vary, sometimes even considerably, from one place to another on the same sheet. But paper made on a single mould cannot divide evenly at the center into two halves, one consistently thicker, the other consistently thinner. For the present writer, who had already examined (not closely enough, obviously) the SNML paper, the Brückle-Mayer result was a considerable surprise! It was necessary to start over again with examining the paper of SNML.

On re-examination, the difference between B1.4 and B2.3 in SNML was evident to the touch: B2.3 is unmistakably a thicker piece of paper. The difference between C1.4 and C2.3, again as brought out by the transmitted-light images, is even more striking, C1.4 being much stiffer than C2.3. Furthermore, SNML's D1.6 bifolium is quite browned, while D2.5, the putative other half of the sheet, is not. Two more related tests were made of the paper in SNML. First: if the four leaves of a quarto quire are quarters of a twice-folded single sheet, their mould sides need to match up consistently. For example, if E1r is the mould side, so too must be E2v, E3r, and E4v. Or if E1v is the mould side, then E2r, E3v and Er must all be the mould sides. In quires A, B, E, F, and G of SNML, the bifolia failed this

8 See *Galileo's O*, vol. I, *The Paper*, pp. 127–142 for full details.

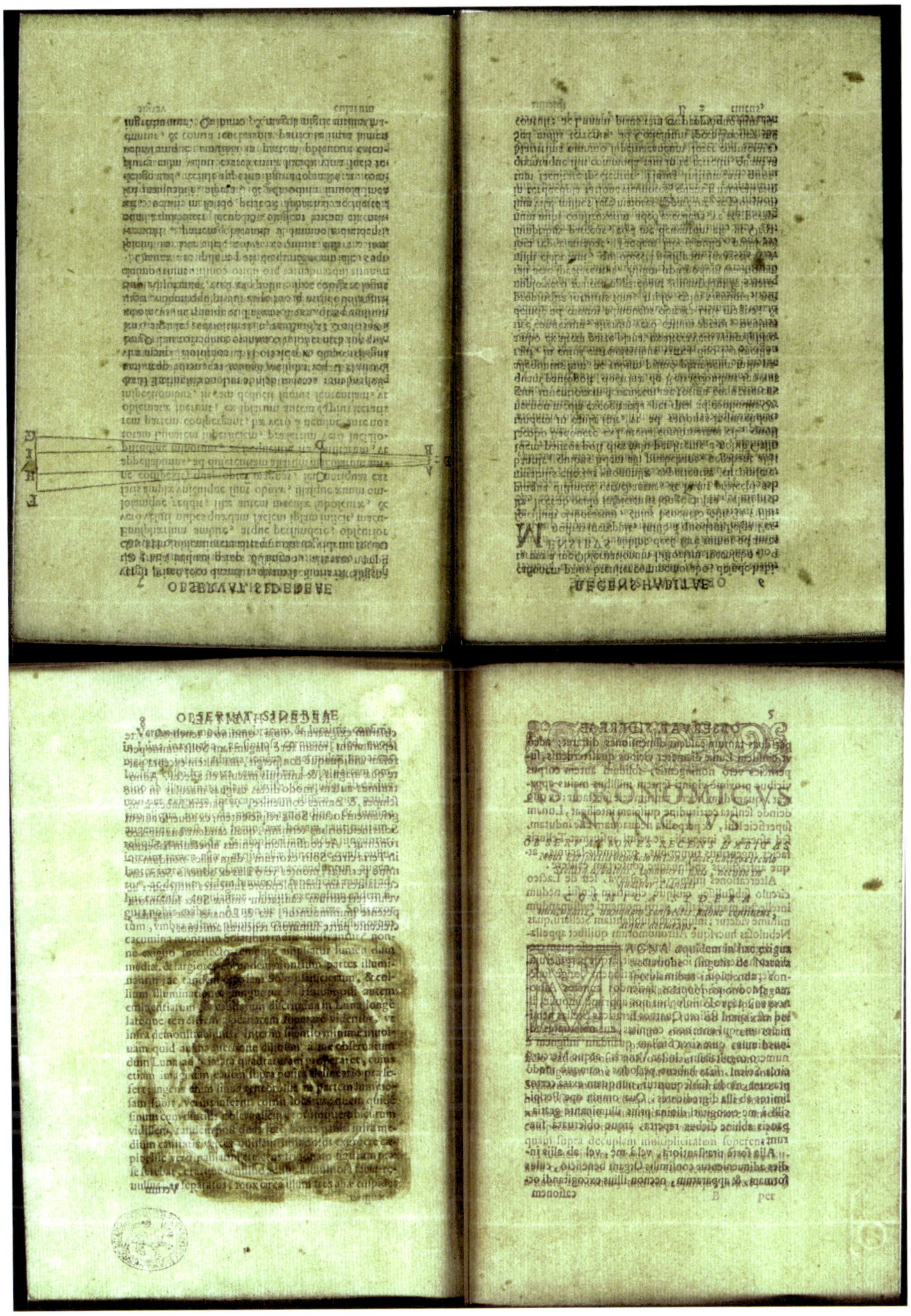

XIII.11: Sidereus Nuncius, B1–4, SNML.

test, and so these quires, by this test alone, could not have been constructed from twice-folded single sheets. As for quire C, which passes the test, it too cannot be a single sheet of paper, for as just noted, C1.4 is distinctly thicker and stiffer than C2.3. In summary, after several rounds of what may be called format-testing, the only pair of bifolia that remain as possibly forming two halves of a single sheet are D1.6 and 2.5: D1r and D2v are both mould sides, as they must be. However, as will be indicated shortly below, a further test proves that D1.6 / 2.5 are likewise not two halves of the same sheet.

The result of the test of mould sides is definitive and agrees fully with the Meyer-Brückle test of visual comparison of the weight or density of the papers forming the bifolia of the quires: SNML was printed on half-sheets of paper, not on full sheets. There is still another test, independent of both of these: that of the sequence of watermarks and countermarks. Consider a quarto quire printed or written on a single sheet, whose paper has a main watermark centered in one half and a cornermark in the other half. Evidently, if the bifolium 1.4 contains the main mark, 2.3 must contain the cornermark, or vice versa. If the two bifolia both contain the main mark, or both the cornermark, obviously they cannot be the same sheet of paper. Curiously, quires A, B, C, E, F, G and the bifolia D1.6 / 2.5 of SNML all meet this test: in each quire, if bifolium 1.4 has the main mark, 2.3 has the cornermark, or vice versa. For each quire, there is an equal probability that this will or will not happen. Thus the probability that in seven quires no main mark-cornermark conflict will occur is 1 in (2^7), or 1 in 128. We cannot be certain whether this is coincidence, or whether something in the handling of paper as these separate bifolia were printed compelled or strongly encouraged this result. The probability of coincidence is small enough that we must prefer the latter alternative, even if we have no idea what that paper-handling process might have been.

The final test is that of the order of printing the two sides of each bifolium, as determined by the sequence of type indentations. This test was discussed in chapter III with regard to the question of whether the inner or the outer formes of the ordinary edition were printed first. In the case of SNML, the test shows that in every gathering, including D, the outer side of the bifolium (1r or 2r as the case may be) was the first to be printed, and the inner side second. The necessary conclusion is that in every gathering the proofs were printed on two-page "interim" formes, put together before the full four-page formes were made up. It should be explicitly noted that the headlines of the pages in these interim formes are in every case the same as those in the full formes of the main print run. Since each bifolium of SNML contains pages from both the eventual inner and outer formes, we have in them a further indication that all eight pages of a given sheet stood in type at the same time, as was suggested already in chapters X and XI.

The interim two-page formes of SNML reflect also on the printing of the inserted half-sheet D3.4, whose analysis was postponed in chapter IX. It was noted there that in all copies of the main print run, both ordinary- and fine-paper, the type impressions of D3.4 are blacker and more deeply impressed than those of the full sheet within which it is sandwiched. The probable explanation is that, like the interim formes of SNML, it was printed on a smaller press, that is, a press with a smaller platen, accommodated to the size of a single quarto page; and that this press printed the two pages of the formes by two pulls. The depth and strength of the type impressions, which is also a conspicuous feature of every page of SNML, must reflect the higher pressure exerted by the screw action of the press upon the smaller area of its platen. So little of detail has been published on the printing of proofs in Italian shops of this period that we cannot know whether such interim formes reflected a common practice. Survival of proofs that might give the necessary evidence for books other than *Sidereus Nuncius* must in the nature of things be exceedingly uncommon. But the consistency across all copies of deeper type impressions on the *Siderius* half-sheet gives its own evidence.

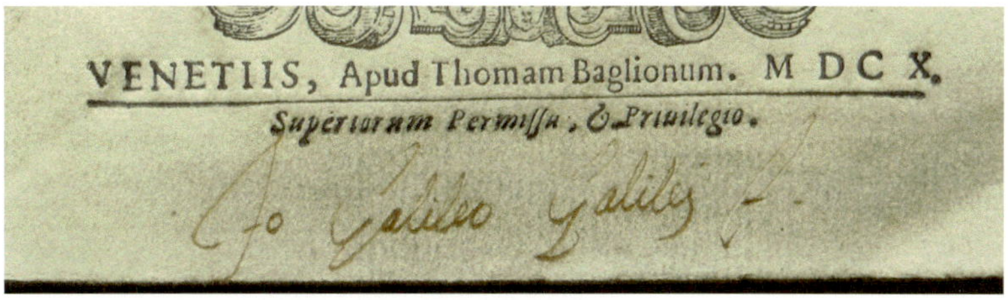

XIII.12: Sidereus Nuncius, A1r, SNML.

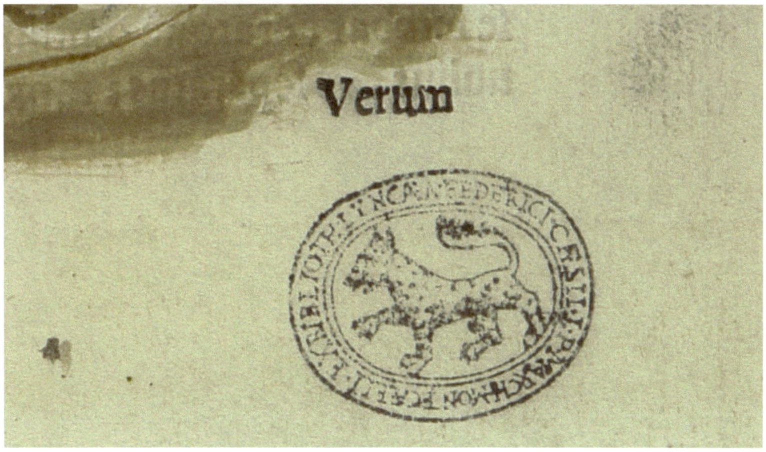

XIII.13: Sidereus Nuncius, B4r, SNML.

Somewhat indirectly, the SNML proofs do relate to a question that several scholars of English printing in this period have discussed: did proofing take place before formes were fully made up, as well as – as many scattered survivals of marked proofs show – in-forme? The anonymous English author of the licentious collection of verses *Ex ungue leonem* (London, 1656) seemed to refer to such a practice when he wrote at the head of his errata list (E7r), *"I possibly may be aspersed, for not revising the* Proofs, *before they were finally locked up in the* Forms."[9] The detailed account in Joseph Moxon's *Mechanick Exercises on the Whole Art of Printing* (London, 1683–1684) states that proofing was done before final formes as well as after, though Moxon did not describe interim formes such as we have in SNML.[10] As elsewhere in Moxon's descriptions of printing practice, there are uncertainties of whether he was recording typical practices, or was laying down the ideal procedures of a utopian printing shop. Equally uncertain is whether practices he wrote about in the 1680s were in force in the late sixteenth and early seventeenth centuries; or at least, whether we may assume that they were. Several examples of English marked proofs (1594, 1639–1640) printed before the formes were finalized on the press bed have been well analyzed by David Foxon and Donald F. McKenzie, but again these do not represent interim formes in the manner of SNML. But as Foxon wrote, "procedures may vary and … almost any book may be a special case in one way or another."[11]

9 Quoted by Simpson 1935, p. 15.
10 Moxon 1962, pp. 238–239 etc.
11 Foxon 1970, p. 154; see also McKenzie 1973.

In the case of *Sidereus Nuncius*, it was Galileo himself who preserved the proofs which, as printing progressed from late January to early March 1610, allowed him to retain handily a tangible record of the growth of the edition, which he could consult as needed. In particular, in the final proofs for sheets B and C, he had in his hands the exact reserved spaces into which the eventual Moon etchings would have to be fitted. It may be suggested that the drawings he made in these spaces elevated the proofs, in his mind, to the status of a unique document of his discoveries worthy of preservation. His inscription on the title-page, "Io Galileo Galilej f[eci]" (figure XIII.12) is close in spirit to the proud words he drafted when he intended to send to Cosimo II the telescope that he had used for his observations: assuredly neither artifact was magnificent, but they were primary. Because the SNML copy contains the ownership stamp of Prince Federico Cesi, inaugurator of the Accademia dei Lincei, it may be presumed that Galileo took this copy to Rome in the spring of 1611, where he met Cesi, was inducted into the academy, and presented the copy to his new friend (figure XIII.13).

We do not know what binding, if any, the proofs then had; nor whether Cesi himself had a binding put on them. At Cesi's death in 1632, the larger part of his library was purchased from his widow by Cassiano Dal Pozzo, who died in 1658.[12] The SNML copy received its present gilt vellum binding in Rome in 1655 or after: it is bound with four of the constituent components of volume I of the quarto *Opere Di Galileo Galilei*, published in Bologna. The binding includes tools originating in the shop of the Soresini family, papal binders from the 1590s to about 1630 (figure XIII.14).[13]

Possibly and plausibly, this was the first of a set of three or four volumes in Cassiano's library, identically bound, containing all the parts of the Bologna edition, but substituting for that edition's *Sidereus Nuncius* Galileo's personal copy of the original edition, brought by him to Rome more than four decades earlier. We know that Cassiano had a lively interest in the Bologna edition of Galileo's *Opera*, was eager to see it, and was supplied with a copy by the Florentine scholar Carlo Roberto Dati, librarian to the Grand Duke of Tuscany.[14]

12 Schettini Piazza 2005.

13 Vianini Tolomei 1991, see especially Tav. I row 11 no 2 (cornerpiece with two Tritons) and row 10 no. 6 (centerpiece pendant: Winged Caryatid).

14 Mirto 2001, p. 38, Dati (Florence) to Cassiano, 12 October 1652: "In Bologna si sono cominciate a ristampare tutte le opere del famosissimo Galileo …"; p. 92, Dati to Cassiano, 29 January 1656: "S'aspetto ogni giorno la pubblicazione delle opere del signor Galileo intagliandosi il frontespizio, subito si possino avere, le farò pervenire una copia."; pp. 95–96, Dati to Cassiano, 7 March 1656: "Sabato passato consegnai a Mario Cangioli un involto diretto a V.S. Ill.ma entrovi l'opere del Galileo ristampate a Bologna, sendo la stima ch'ella fa di quel grande intelletto."

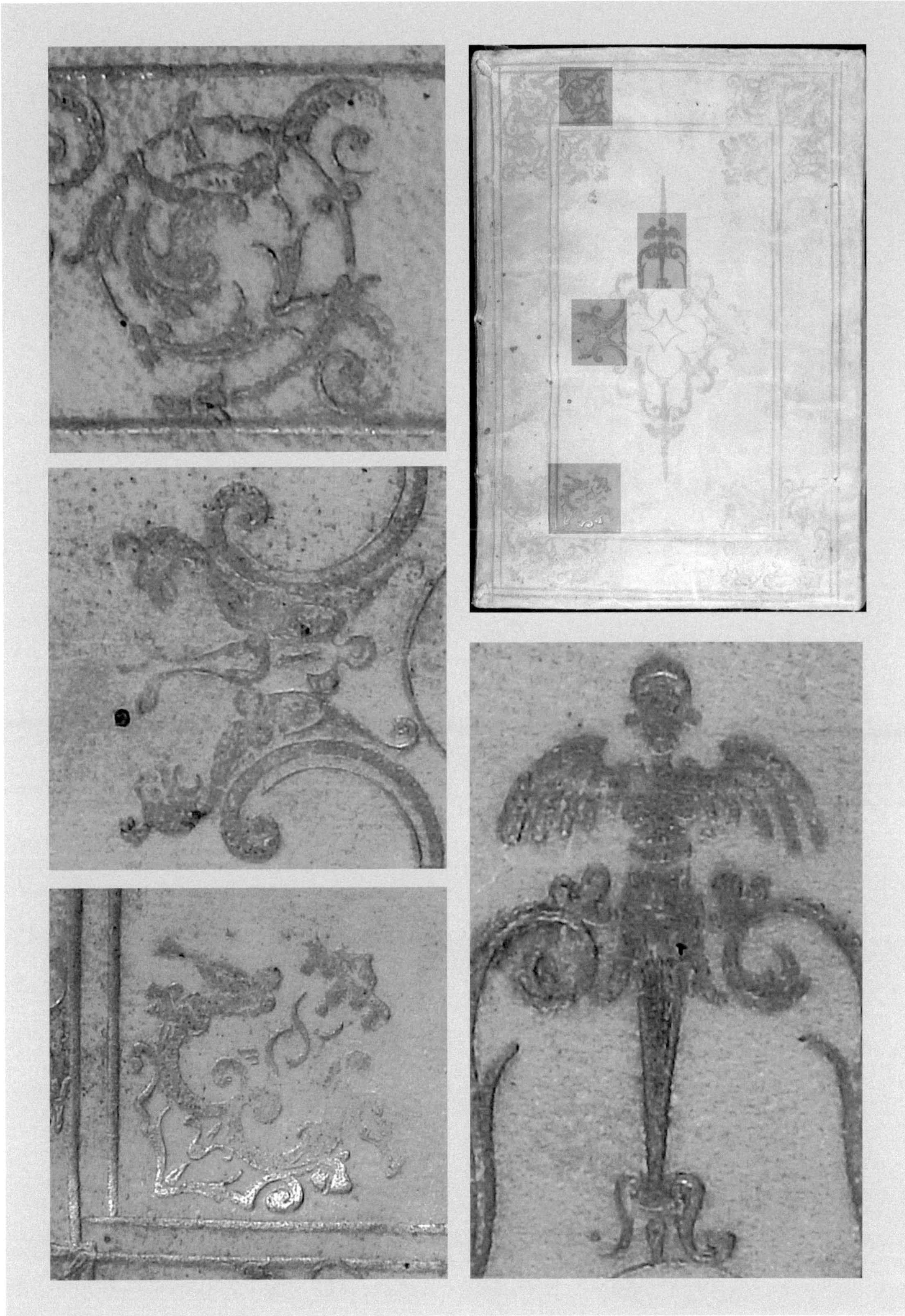

XIII.14: Sidereus Nuncius, binding, SNML.

XIV.
Publication and Distribution; Galileo's Presentations

Galileo's letter to Vinta of Saturday 13 March 1610 documents the pivotal moment of *Sidereus Nuncius,* its completion in the printing shop and the first release of copies to readers. Writing to Vinta that evening, Galileo's overriding concern was that the presentation to Cosimo should go smoothly, and he felt anxious. Cosimo's official wish, transmitted by Vinta on 6 February 1610, had been to see as soon as possible both the book and the new, improved telescope that Galileo had used: [the duke and duchess] *sono entrate in eccessivo desiderio di veder quanto prima dette osservazioni et l'altro occhiale più eccellente.* Within Galileo's letter the chief cause of anxiety comes out. There had been delays in the last days, and this had complicated his plans for getting a suitably bound copy of the book, together with the telescope, into Cosimo's hands in good time. Galileo wrote:[1]

> Not until today, and quite late, has it been possible to get a copy of my Avviso Astronomico. Thus, there has not been time to get one bound for His Majesty. I am forced to postpone this until next week. Besides, to be able to send the telescope together with the book, I must return to Padua; because, hoping to have finished four days ago, with time to return to Padua and send everything from there, I have let myself fall behind schedule. But in any event, and with all good respect, I have not wanted to neglect to send you a copy, although unbound and still damp.
>
> *Non prima che oggi, et ben tardi, si è potuto havere alcuna copia del mio Avviso Astronomico, tal che non ci è tempo da poterne far legare uno per S.A.S., ma sarà forza che io indugi alla settimana ventura; oltre che mi bisogna tornare a Padova per poter inviar l'occhiale insieme con l'opera; perchè, sperando di essere spedito sin quattro giorni sono et di haver tempo di tornare a Padova et inviare il tutto di là, mi son lasciato traportare avanti. Tutta via non ho voluto mancare di inviarne una copia a V.S. Ill'ma., così sciolta et ancora bagnata, per ogni buon rispetto.*

That is: Galileo had expected the printing to be finished four days earlier, say on Tuesday 9 March, the day after the book's license was registered. By that expectation he could have

1 For help with this passage I am grateful to the informal committee of Drs. Cristina Dondi, Daniele Danesi, and Antonio Ricci, who met in the British Library's cafeteria on 9 June 2009 to discuss it; I remain responsible for any inadequacies in the translation.

had a copy, undoubtedly one of the fine paper issue, quickly bound in handsome manner, taken it to Padua to be joined with the required telescope and, by the end of the week, have forwarded the ensemble to Florence. Because the printing was not completed until late on the Saturday, these arrangements were necessarily pushed into the following week. And so it happened that on this Saturday Galileo was in Venice, while the presentation telescope was in Padua. Parenthetically, the third Observation Sheet has records, during this week before printing was finished, of three Jupiter observations, after a week's hiatus, on the evenings of Tuesday the 9th, Wednesday the 10th, and Friday the 12th. We cannot know whether he was in Padua or Venice or a combination of places on those dates; they were followed by another hiatus of six days.

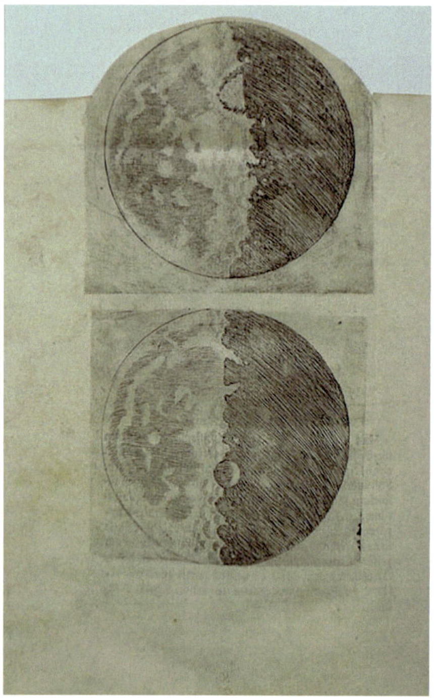

XIV.1: Sidereus Nuncius, C2v, Göttingen.

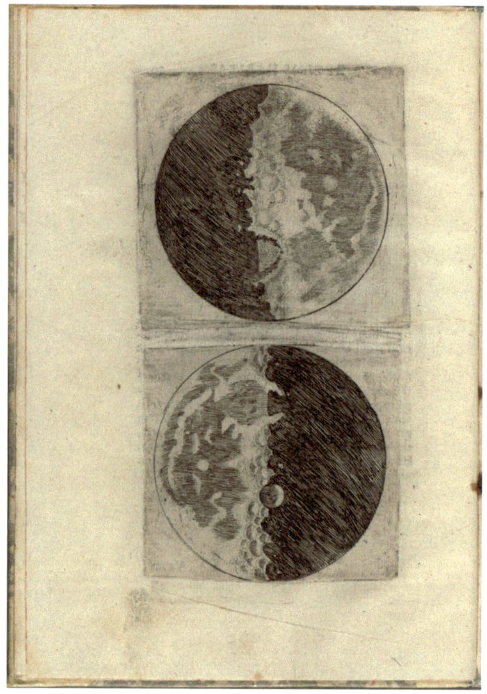

XIV.2: Sidereus Nuncius, C2v, Harvard copy 2.

Thus, we have a lively picture of Galileo's original plan for finishing *Sidereus Nuncius*, and of how it was overtaken by events. His final Jupiter observation was made on Tuesday 2 March, and was interpolated into the text and fitted into the bounds of sheet G by means of a rigorous abridgement of the concluding remarks. Galileo expected the last printing, preliminary sheet A, to be finished a week later, on (probably) 9 March. We cannot know the specific cause of the delay, but it will be recalled that besides the typographic printing of sheet A, sheets B and C needed to be printed again, repeatedly, to insert the five Moon etchings. For each copy, this pair of sheets had to go a total of five times under a rolling press, almost certainly to be found in some printmaker's shop rather than in the typographic shop. This was a considerably slower process than that of the typographic printing press. In one of the copies at the Observatoire de Paris, Moon II on fo. 9v (C1v) was printed upside-down, the large crater shown as if in the northern hemisphere. Fo. 10v (C2v), it will be recalled, contains two etchings, and the constant small variations in register from copy to copy show clearly that the two plates were separately positioned and printed. In the Göttingen copy, the etchings were misprinted on this page. Because there was no headline at the top of the leaf to mark an upper boundary, the upper etching was carelessly positioned against the headline of C1r, on the other side of the sheet's top fold (figure XIV.1). The blank page C2v may have been particularly confusing for the workers in the printseller's shop. In Harvard's second copy the upper etching, Moon IV, was imposed upside-down (figure XIV.2); and in Columbia's copy both etchings on this page are upside-down.

At this time, when the etchings were first printed onto the *Sidereus Nuncius* sheets – perhaps to begin with in only a small number of copies – it would have been discovered that the Moon III plate was too large for its space; or rather, that its allotted space on C2r was too small. Weeks earlier the compositor had created a space of only 17 blank lines, where 20 lines would have been preferable (see chapter VI). At the printseller's, a somewhat brutal fix was introduced: the plate itself was cut away at top and bottom with shears, removing all the plate margin and small portions of the north and south poles of the Moon. This allowed the plate to be squeezed, barely, into the space available, without too much overlap with the already-printed types. The bow-shape of the cuts is apparent at both the top and bottom of the plate (figures XIV.3, 4).

Galileo wrote Vinta that the copy he sent was "still damp". That dampness could refer to the printing that day of sheet A; to the secondary rolling-press printing of sheets B and C (for the sheets would have been redampened for this purpose); or to both. Whatever the cause of the delay about which Galileo complained, if he originally expected printing to be finished on or by Tuesday the ninth of March, he would have completed drafting his dedication before that day; its printed date could be adjusted according to the actual day of printing. One possibility is that sheet A was worked on the day of its dedication date, Friday 12 March, and that the etchings were printed on B and C of a group of copies on the following day, Saturday the thirteenth.

The 13 March letter to Vinta reveals a second concern of Galileo's that should, for context, be followed through. Galileo was obliged to send Cosimo both a copy of the book and the improved telescope. He immediately worried that when the telescope arrived in Florence, himself absent, Cosimo and his courtiers would have no success in espying the moons of Jupiter. The effective use of the telescope required considerable skill in which they would have no experience, for instance in steadying it properly. Galileo voiced his worries to Vinta, and made a suggestion. By original plan he was not to have visited Florence until June, when Padua's academic term ended. He now suggested that, if it were acceptable to the Grand Duke, he could come during the Easter holidays, and laid out a plan whereby he would be able to depart Padua on Friday the second of April and be in Bologna on Monday the fifth, if a mule litter could be arranged for him there to take him over the Apennines.

On Friday 19 March, Galileo had his bound copy of *Sidereus Nuncius* ready for Cosimo, and also the telescope, and he forwarded them and a letter to Vinta, subscribed from Padua, of which both a draft and the finished version are preserved. In the preceding days, still unsure whether he would be asked to come to Florence, he had written out detailed instructions on using the telescope, which he addressed to Enea Piccolomini. Now that *Sidereus Nuncius* was complete and could no longer be changed, he was conscious of its

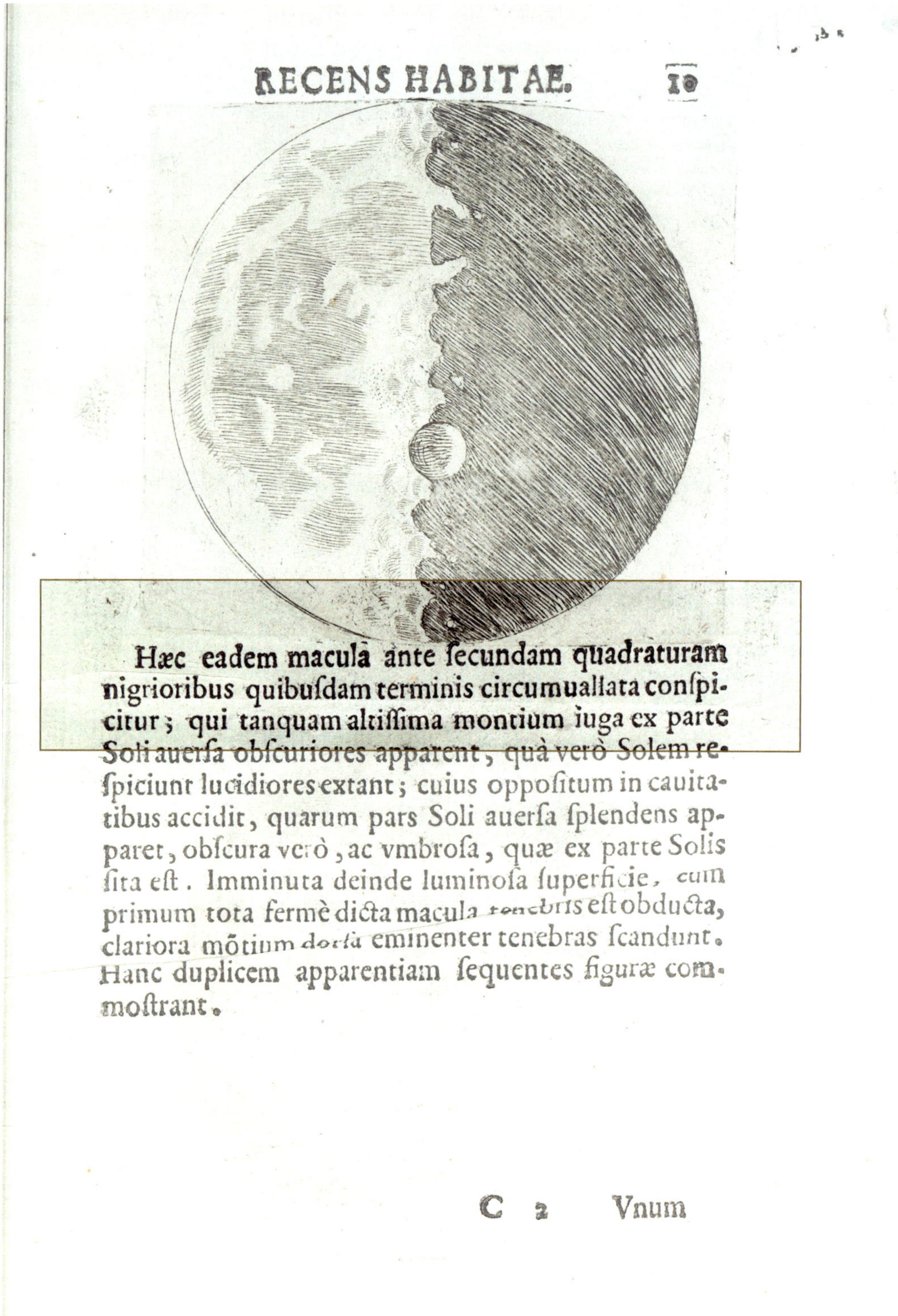

RECENS HABITAE. 10

Hæc eadem macula ante ſecundam quadraturam nigrioribus quibuſdam terminis circumuallata conſpicitur; qui tanquam altiſſima montium iuga ex parte Soli auerſa obſcuriores apparent, quà verò Solem reſpiciunt lucidiores extant; cuius oppoſitum in cauitatibus accidit, quarum pars Soli auerſa ſplendens apparet, obſcura verò, ac vmbroſa, quæ ex parte Solis ſita eſt. Imminuta deinde luminoſa ſuperficie, cum primum tota fermè dicta macula tenebris eſt obducta, clariora mõtium dorſa eminenter tenebras ſcandunt. Hanc duplicem apparentiam ſequentes figuræ commoſtrant.

C 2 Vnum

XIV.3: Sidereus Nuncius, C2r, Institute for Advanced Study.

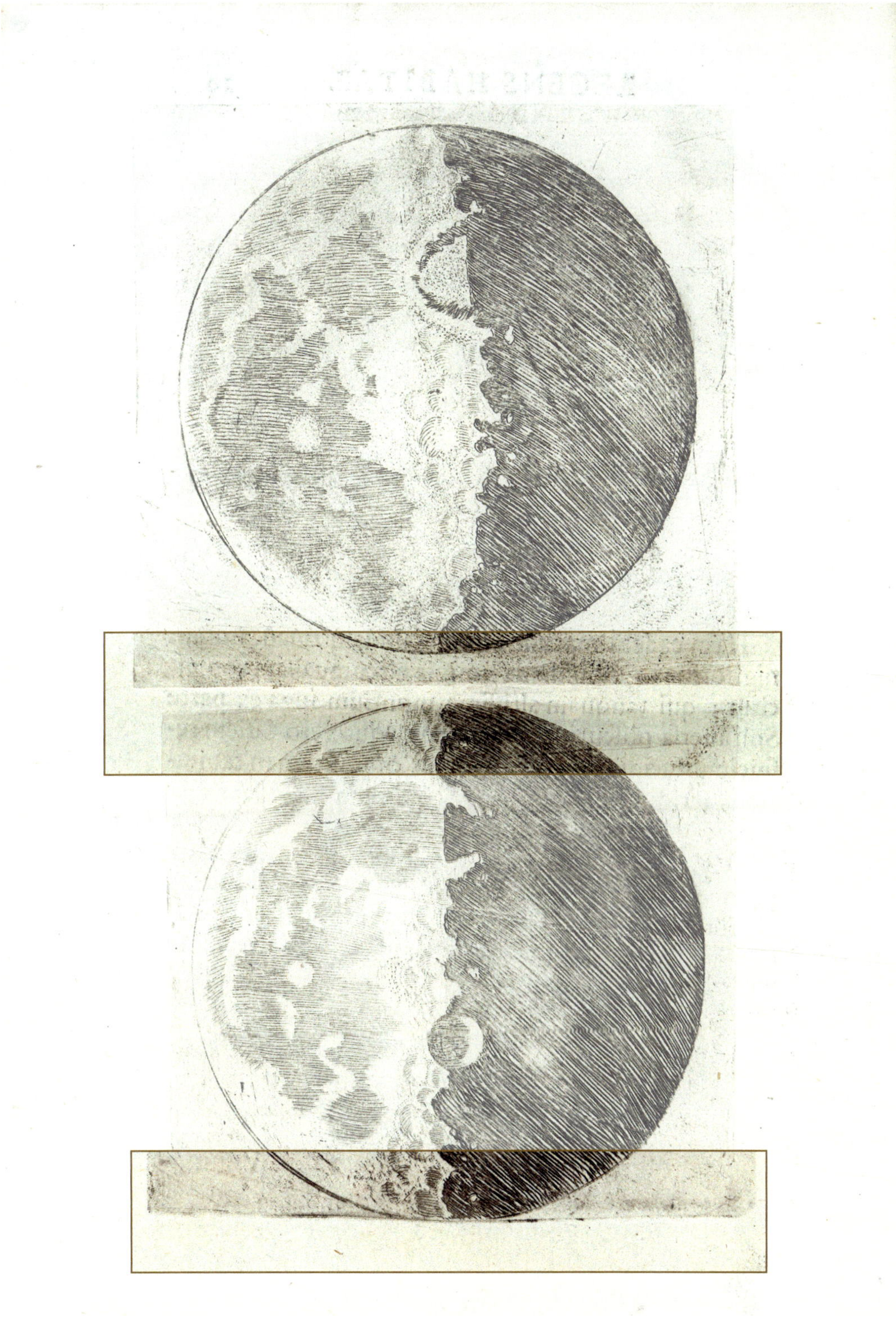

XIV.4: Sidereus Nuncius, C2v, Institute for Advanced Study.

physical flaws: Vinta must apologize on his behalf that due to the press of time, the book might lack all the magnificence and decorum that its great subject matter deserved (*Sarà necessario ... faccia mie scuse ... se l'opera non vien fuori stampata con quella magnificenza et decoro che alla grandezza del suggetto saria stato necessario, perchè l'angustia del tempo non l'ha permesso*).[2]

In this letter Galileo sketched out plans for a second edition in Italian, which would be grander and contain professionally engraved plates, including a precise depiction of a full lunar cycle. This second edition, which never eventuated, would, he wrote, be all the more necessary, as the 550 copies of the present edition had already been distributed in all directions (*il che è anco necessario farsi, perche 550, che ne hanno stampati, sono già andati via tutti*). Some writers have interpreted this as meaning that the edition was "sold out" by 19 March, but that was not Galileo's statement. He indicates rather that the copies had been sent out by the publisher along the several trade routes that would bring them into the hands of booksellers in many cities. The edition was not sold out until those booksellers no longer had sheets in stock, as apparently was the case (discussed further below) by the end of the summer 1610.

Of thirty copies due to him, he had received by 19 March only six; the remainder were still with the publisher in Venice and had not yet had their plates printed; and he doubted on that day whether he would ever see them (*anzi di 30, che ne dovevo havere, non ne hauti altre che 6, nè veggo verso di potere havere il resto, havendogli lasciati in Venezia in mano del libraio, perchè vi mancavano a stampar le figure in rame*). These figures give us more information than is typically recorded about printing projects of this period; documentation of edition sizes is rare in early printing. The thirty copies owing to Galileo would presumably have been or have included the fine paper issue, for which he would have paid at least the paper costs. In terms of its distribution, the *Sidereus Nuncius* was partly a commercial and partly a private publication. It is uncertain whether Galileo counted the fine paper copies among the 550 copies which had been printed and distributed; if not – his wording hints that he saw them as a separate category – the full edition run could have been 580 copies.

We have no specific documentation relating to Baglioni's distribution of copies, but it is useful to keep in mind that for *Sidereus Nuncius*, as for various other Venice editions in Latin as well as a smaller number in Italian, there were two main distribution channels, transalpine and cisalpine. Some substantial group was sent more or less immediately north of the Alps for sale at the soon-approaching Frankfurt Lenten Fair, as also at the Leipzig Spring Fair, which opened shortly after the Frankfurt Fair recessed. The *Sidereus Nuncius* was listed in the printed catalogues of books offered at both:[3]

2 *Opere* X, p. 300 (Ep. 277). Galileo probably sent a now-lost letter to Piccolomini on the same day.

3 *Catalogus universalis pro nundinis Francofurtensibus vernalibus, de anno M. DC. X ... in der Fastenmeß* (Frankfurt: Typis Sigismundi Latomi, 4°), C3v, classed among the "Libri philosophici & aliarum artium". The entry is repeated in the Leipzig catalogue: *Catalogus ... nundinis Francofurtensibus vernalibus de anno 1610 ... Mit vermehrung anderer Bücher, so in der Leipziger Oster Meß außgehen ...* (Leipzig: Abraham Lamberg, 4°), D1v.

> *Sidereus Nuncius, magna longeque admirabilia spectacula pandens suspiciendaq; proponens maximis philosophis atq; Astronomis, quae à Galileo Galileo, patritio Florentino, Mathematico Patuino, perspicilli nuper à se reperti beneficio sunt obseruata in Lunae facie fixis innumeris, &c. apprime vero in quatuor Planetis circa Iouis stellam disparibus interuallis atque periodis celeritate mirabili circumuolutis, quos nemini huc usque cognitos autor primus deprehendit. ib.* [i.e., Apud Societatem Venetam] *in 4.*

The session dates of the Frankfurt Fair are not known to certainty, but it has been said that at this period the usual session was from the fifth Sunday in Lent to the Tuesday after Easter, which for 1610 would have been from 28 March to 13 April.[4] The shipment through the Brenner Pass must have been made more or less immediately after printing was completed, and it is likely that Galileo's frustration at the delays in the final days of printing was in part a concern that the *Sidereus Nuncius* could not be completed in time for inclusion in the printed catalogues of the 1610 Lenten and Spring Fairs, for this was the chief opportunity for a timely delivery of copies for sale north of the Alps.

The Frankfurt Book Fair has such fame as a hub of European book distribution in the early modern period that there is some risk of believing that we understand it better than is the case. Many details of its operations are undocumented, and in particular, no substantial study has been made of the Italian books sent to the fairs.[5] As for the Leipzig Fair, it is even less completely known, especially as concerns its book supplies from outside the German states. The following paragraphs are only a preliminary attempt to give context to *Sidereus Nuncius*'s appearance for sale at the spring 1610 fairs of Frankfurt and Leipzig.

In the years from 1607 to 1610, the number of titles listed annually in the combined spring and fall catalogues of the two fairs varied between about 1,350 and slightly over 1,500 titles. In principle, a title would only be listed once in the catalogues, in fall or spring, but in fact there are not infrequent instances of relistings. Again, in principle, reprintings would not be listed at all, unless the new edition was substantially revised. These totals are, of course, a small fraction of all the books published in Europe in those years, for a large proportion of publishing, especially in vernacular languages, was aimed at national or regional markets, and would not benefit from exposure to the booksellers and agents who attended and traded at the fairs. In the printed fair catalogues of these years, the chief categories were Latin books in theology, law, medicine, history, philosophy, poetry and music; and, separately listed, German books in the same categories. The Leipzig catalogues for its Easter and Michaelmas fairs reprinted the Frankfurt catalogues, followed by a supplement of titles, especially from Leipzig and Wittenberg presses, that had not been sent to Frankfurt.

4 On the sessions of the Frankfurt fairs, which varied considerably at different times, see Dietz 1910, pp. 37–38.

5 For a broad overview with extensive references see Flood 2007. Schwetschke 1850 gives a statistical conspectus of the books offered annually at the combined Frankfurt and Leipzig Fairs. The authors and titles of the books are not given, but their subject categories, languages, cities and, where stated in the fair catalogues, publishers are noted. Because of a change of his sources at 1600, Schwetschke's data before that year cannot be reliably compared with those of 1600 onward.

We are very ignorant of the processes by which the publishers of the fair catalogues, at this time Sigismund Latomus (Maurer) in Frankfurt and Abraham Lemberg in Leipzig, collected and organized the titles of the books offered. It is obvious, however, that the titles as listed were typically close paraphrases of the title-pages of the editions, as is the case for *Sidereus Nuncius*. The treatment of the Venice books entered in the book catalogues for 1609 and 1610 provides at least a hint of the process by which many publishers in many cities got their book titles to Latomus in Frankfurt, who arranged them for typesetting. Under each subject category – *Libri theologici, Libri utriusque iuris, Libri medici,* and so on – the Venice titles are grouped together and listed consecutively, without interruption by titles from other towns. This suggests, at least, that Latomus received a full list from Venice, perhaps in advance of the sheets of the books themselves, already grouped into the standard categories. Full analyses of the ordering of all the titles in the fair catalogues, season by season, would undoubtedly reveal more, but it is immediately obvious that groups of titles from particular cities tended to be listed consecutively in the fair catalogues. One hint of the speed with which the catalogues were assembled and printed for each fair is that Latomus kept his title-page standing in type, at least for the four catalogues of 1609 and 1610, swapping in and out as necessary only the words that needed changing, *autumnalibus* and *vernalibus*, and the dates.

In the years 1607 through 1610 about four-fifths of the books listed in the Frankfurt and Leipzig catalogues were from German towns, and about one-fifth from outside the German states. In these four years Venice (198, or about fifty per year) stood behind Antwerp (246) and just behind Paris (202) in the number of titles sent, and was itself followed by Geneva (156). The other significant towns in these years were Lyons (78), Amsterdam (72), and Leyden (51). It is notable that Venice's totals for the last years of the decade are strikingly short of the years 1600 through 1605, when its shipments to the fairs averaged more than 110 titles yearly, well over double that of the 1607 through 1610 average. This is probably to be connected with Pope Paul V's interdict of Venice, 1606–1607, consequent to the Republic's refusal to acknowledge the superior authority of the ecclesiastical courts in matters touching on heresy and the importation of forbidden books. Concurrently, the prominent publisher Roberto Meietti, who was probably the chief figure in smuggling banned books to Venice, was excommunicated.[6] These events suggest the mechanism by which Venice's export book trade would be constrained, even if none of its exports were of questionable titles. Much of the book commerce at the fairs involved not monetary payments but barter, that is, the balanced exchanges of printed sheets between one merchant and another. If Venice's imports of northern books were increasingly restricted after 1606 because of greater or more effective inquisitorial scrutiny, exports of its own books would be correspondingly impeded. In the year 1611 only sixteen Venetian titles were sent to the fairs, after which the numbers gradually increased again.

A distinctive feature of the Venetian books sent to the Frankfurt and Leipzig fairs from 1607 through 1610, again correlating closely with the papal interdict, is that with one exception, all were entered in the fair catalogues as "Societas Veneta" titles, rather than, as was common in preceding years, by named publishers. The Societas Veneta must have

6 Grendler 1978; Grendler 1977, pp. 280–285.

been a consortium of publishers who in ways invisible to us pooled their resources. In 1610, participants in the Societas Veneta included publishers from Bologna, Pavia, Treviso and Vicenza, eight of the forty-eight titles being from those towns and not Venice. As for Baglioni, he sent besides *Sidereus Nuncius* four other titles to the spring fairs: the folio treatise *De morbis occultis* of Eustachio Rudio, professor of medicine at Padua (1610), and three 1609-dated Italian works that were entered in the fair catalogues in the category of Books in Foreign Languages ("Libri peregrini idiomatis") including the second Italian edition of Pedro de Medina's *Arte del navigare*.[7]

We cannot know how many copies of *Sidereus Nuncius* Baglioni sent to the Frankfurt and Leipzig fairs, but there is some indication that the copies sold out rapidly enough. On 8 July 1610, Bartholomaeus Schröter, mathematician at the gymnasium in Zerbst, Saxony, wrote Galileo that he had seen the notice of the *Sidereus Nuncius* in the Frankfurt Fair catalogue, yet the local booksellers told him that they had never had a copy. As it happened, the Prince of Anhalt possessed a copy which he loaned to Schröter.[8] As for sales in Venice itself, Paolo Sarpi wrote from there on Tuesday 16 March to the Paris jurist Jacques Leschassier, forwarding to him a copy of the *Sidereus Nuncius*; presumably, on that date copies were actually available for sale in Venice bookshops.[9]

Although our only documentation of the northward sales of *Sidereus Nuncius* is the entries in the Frankfurt and Leipzig catalogues, we can literally examine, in surviving copies of early northern provenances, sheets that were in the shipment, and these provide further information. Two variations reflect, one very neatly, the other only slightly less so, a distinction between copies sold in Italy and those sold north of the Alps: the presence or absence of the Moon etchings, and the correction or non-correction of "COSMICA SYDERA" on the drop-title, fo. 5r (B1r). As already noted in earlier chapters the drop-title of sheet B preserves the earlier text state, when Galileo intended to name the Jupiter moons "Cosmic" in personal compliment to Cosimo II. By later February, after Galileo learned from Vinta that the "Medicean" nomenclature was preferable, sheet B had already been printed off and its types distributed. It must have been at some point after the entire edition was completed, thus roughly mid-March or after, that Galileo noticed the conflict between the early "Cosmica" of sheet B and the revised "Medicea" of sheets G and A. The solution was to print a small slip "MEDICEA" to be pasted over "COSMICA", as was done,

7 Rudio's *De morbis occultis* was entered again in the fall 1610 catalogues, and among Baglioni's vernacular titles Giovan Vincezo Piccino's prayerbook *Ghirlanda di dodici rosari* had already been entered in the fall 1609 catalogues.

8 *Opere* X, p. 393 (Ep. 355**): "Perlegenti mihi ... nundinarum proximarum vernalium Francofurtensium catalogum librarium, inter alios libros philosophicos et mathematicos,... sese obtulit titulus libelli cuiusdam, sane novus, mirus et insolitus ... mox vero mihi ex bibliopolis nostris inquirenti, anne libelli istius exemplum attulerint, responsum fit, nullum sibi visum."

9 *Opere* X, p. 290 (Ep. 272*). Sarpi's letter was enclosed with the copy of the book. Leschassier loaned both book and letter to the canon lawyer Christophe Justel, who loaned it in turn to the diarist Pierre de l'Estoile. In his journal, under date 18 April 1610, L'Estoile wrote in detail about the copy, stating that the Venetian ambassador had brought it to Paris and given it to Leschassier. He noted explicitly that "Au commencement de ce livre, y avoit une lettre de Fra Paolo a M. Lescale [viz. Leschassier] ecrite a la main ...," and he copied out the portion of Sarpi's letter relevant to Galileo's book: Sarpi 1961, p. 266 n. 72; Pantin 1993, pp. XIV–XV. Sarpi commented further on Galileo's telescope and discoveries in a letter of 10 May 1610 to the Huguenot Jerome Groslot de L'Isle: Sarpi 1931, I, p. 122.

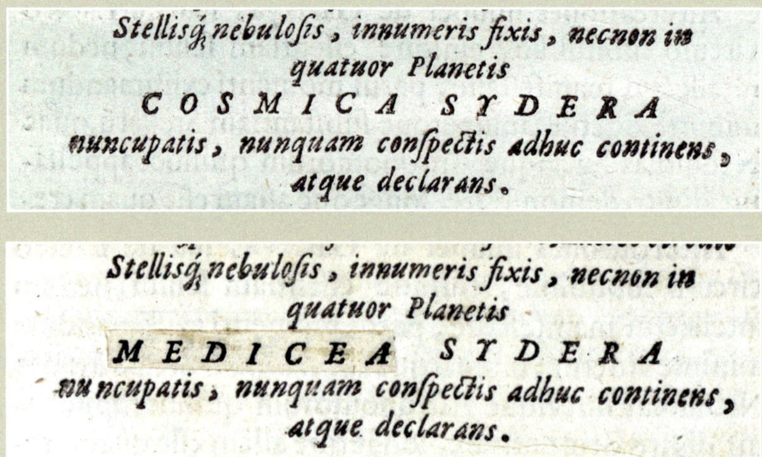
Stellisq; nebulosis, innumeris fixis, necnon in
quatuor Planetis
COSMICA SYDERA
nuncupatis, nunquam conspectis adhuc continens,
atque declarans.

Stellisq; nebulosis, innumeris fixis, necnon in
quatuor Planetis
MEDICEA SYDERA
nuncupatis, nunquam conspectis adhuc continens,
atque declarans.

Institute for Advanced Study

Linda Hall Library

XIV.5: Sidereus Nuncius, B1r.

But not all copies were so corrected (figure XIV.5). One writer has stated that this correction was made in "virtually all copies," but of 83 copies recorded, 51 lack the cancel.[10] Of the 32 that have or once had it, 7 are fine-paper copies that were not part of the commercial distribution and should be treated separately. A detailed census of all surviving copies, greatly to be desired, will eventually give complete results, but the large sample is sufficient to show that most copies never received the cancel. This indicates that Galileo did not notice the error of B1r until a substantial number of copies had left Baglioni's shop, both to Italian booksellers and to the German fairs. We cannot know whether Galileo made the correction in the earliest copies he sent out personally, to Vinta and to Cosimo II. Most copies of *Sidereus Nuncius* with early Italian provenances bear the cancel slip, but there are at least five that lack it, for instance a copy in the Biblioteca Nazionale Centrale, Rome, formerly owned by the Carthusians of Santa Croce in Gerusalemme. Of the copies with early northern provenances, the great majority lack the cancel slip. It is further notable that Zacharias Palthenius's octavo reprint of *Sidereus Nuncius*, published in Frankfurt in the autumn of 1610, copied the drop-title in uncancelled form: *Cosmica sydera*.[11] There is reason to suppose, therefore, that the copies sent north to the spring fairs left Venice before Galileo noticed that the drop-title needed correction. It may be that the few "northern" copies bearing the cancel slip came to their owners not directly from the Frankfurt and Leipzig fairs, but from correspondents in or visits to Italy: for example the copy with cancel in the Bibliothèque de l'Arsenal, Paris, originally owned by Jacques-Auguste de Thou.

The distinction between "northern" and "southern" is even more sharply drawn in the case of copies of *Sidereus Nuncius* lacking the Moon etchings. When Galileo complained

10 Battistini 1993, p. 184 note 44: "quasi tutte le copie." Battistini may have been more familiar with copies with original Italian homes than with those that had been more broadly disseminated. Parenthetically, as is well known, a printed cancel slip supplying an omitted passage is similarly found in most copies of Galileo's *Dialogo*, Florence, 1632, on p. 92. For a broad discussion of cancel slips see Barbieri 2005.

11 "Cosmica sidera" appears also in the Bologna quarto *Opera* of Galileo, 1655–1656; and in the London octavo editions of James Flesher, 1653 (Wing G167, G167A).

to Vinta that he had at hand only six of the thirty copies he was owed because the remainder, still with Baglioni, had not yet received the plate impressions, he highlighted a fundamental difference between the printing of typographic composition and the printing of "prints" – woodcuts, engravings, etchings, and similar media. In typographic printing, the formes that went on the printing press typically only lived as unities until all the planned copies of the relevant page or group of pages had been printed off. That accomplished, the formes were broken up and their types distributed back to the cases for re-use.

Evidence for the rapid breaking up and distribution of formes appears as early as the first large-scale printed book, the Gutenberg Bible (Mainz, 1455). When it was decided, fairly early in the Bible's production, to increase the edition size, all the pages thus far printed off had to be reset later and printed to the number of the edition increase, because the original type pages had all been lost to distribution. In one section of the book (it was concurrently composed in four sections) this meant, even, that at the time the verso of a leaf was printed to the larger edition number, the recto setting of that leaf, probably printed the day before, had already been distributed.[12] Woodblocks and metalplates, by contrast, are complete units and short of being cut up (as happened) or entirely destroyed they remain so, and can be printed from at any time. Various woodblocks of Dürer, for instance, survive today. These could in principle still be printed from, producing authentic, if little-valued, Dürer prints on 21st-century photocopy paper or for that matter on T-shirts.

Consider then the sheets of *Sidereus Nuncius* on 13 March 1610, when both Galileo and Sir Henry Wotton acquired copies. There were presumably about 550 or perhaps 580 copies of each of the seven sheets and of the half-sheet D3.4. But not all those copies would have had the etchings separately printed onto sheets B and C. Probably only a small number of fully printed copies with etchings were available that day, and so it was that Galileo complained six days later that most of his allotted copies were still awaiting their plate printing. Baglioni could well have had the plate printing done in batches, as he gathered distinct groups of sheets for distribution to different towns, which is what Galileo's report suggests.

This gives us a significant clue to the twelve copies of *Sidereus Nuncius* known so far (not counting, for the moment, the SNML proof copy) that never had the plates printed in the provided spaces of sheets B and C. One copy is in American private ownership, the others in libraries of Berne, Cologne, Copenhagen, Cracow, Durham, Liège, Oxford, Paris (Bibliothèque Nationale de France, and one of the three copies at the Obervatoire), St Andrews, and Zurich. All are strong candidates for having come into the book trade at the Frankfurt and Leipzig fairs. The Zurich and Cologne copies bear seventeenth-century Cologne ownerships; the private American copy formerly belonged to the well known book collector Marquard Gude of Schleswig; the St Andrews copy has a 1646 ownership inscription of Caspar Marschalk, parish priest of Leinach; the Copenhagen copy belonged in the seventeenth century to a Danish scholar Petrus Scavenius; the Oxford copy belonged to the antiquary Brian Twyne, who bequeathed his library to Corpus Christi College at his death in 1644; the copy at the Bibliothèque Nationale de France was in Cardinal Mazarin's library in the seventeenth century; and the Berne copy was acquired by the

12 Needham 1985 pp. 320–21, 333–34.

French Huguenot diplomat Jacques Bongars. It is especially indicative that none of these twelve copies has the cancel-slip correction on B1r. We may suppose that Baglioni, whether with knowledge or by oversight, quickly sent to the German fairs a substantial number of sheets before they had visited the printseller's shop, and with reasonable confidence we may surmise that the sheets left Venice during the week of 15 March.

The distribution of the fine-paper copies forms a complicated problem, for the evidence is scattered and imperfect. Nonetheless, it can be argued to high confidence that these must have been Galileo's own copies and were never part of the commercial sale. It is useful to review what we know from early sources about copies of *Sidereus Nuncius* in whose distribution Galileo had a direct hand. Let us posit that the dedication copy to Cosimo, whether now lost or only lying unrecognized, must have been on fine paper. The "token" copy Galileo sent to Vinta on 13 March 1610 would probably have been on ordinary paper. Galileo's letter to Vinta of 19 March gives more information. It is heavily concerned with plans for sending examples of both his book and his telescopes to high dignitaries and courts: to Ernest of Bavaria, elector of Cologne and Kepler's patron; to Cardinal dal Monte, closely allied to the Medici court; to the courts of France, Spain, Poland, Mantua, Modena and Urbino, and perhaps to others according to Cosimo's suggestions. If these, like the dedication copy to Cosimo, were not prime candidates for fine-paper copies, it is hard to imagine why there would have been a fine-paper issue at all. On the other hand, we cannot be sure that all these plans were carried through.

In short, once the printing was completed, Galileo was focused on the need to get his book, and a telescope, appropriately presented to Cosimo and to other major European courts. The sale of the ordinary copies would have been Baglioni's commercial concern, and we have no reason to think that Galileo was involved in this. Indeed, he may not have wanted copies to be widely available to Italian stationers until after Cosimo had received the dedication copy. On Saturday 27 March Galileo's friend in Florence Alessandro Sertini wrote his thanks for the copy of *Sidereus Nuncius* that Galileo had sent him in a parcel from Venice. Sertini recounts how he had collected the parcel the day before from Filippo Mannelli in the Mercato Nuovo, and how the delivery turned into a public occasion. There were other gentlemen around, consumed with vocal curiosity: could the parcel contain a telescope? When it turned out that it was the *Sidereus Nuncius*, curiosity did not cease: all the men of letters wanted to know its contents, and, Sertini wrote, Don Antonio de' Medici would be busy explaining the discoveries (*... io non mi poteva difendere dalle persone, che volevan sapere che cosa era, pensando che fosse un occhiale; e quando si è saputo ch'egl'era il libro, non è cessata la curiosità, massime negli'huomini di lettere, e credo che'l Sigr. D. Antonio harà che fare a mostrarlo*).[13]

Galileo would have posted this copy to Sertini at about the time he sent off the presentation parcel to the Grand Duke. Sertini's words seem to imply that copies of *Sidereus Nuncius* had not yet reached Florence through ordinary book trade channels. This single copy, personally sent by Galileo, aroused general amazement, as being the first that the gentlemen in the Mercato Nuovo had seen. Thus: some fully printed copies of *Sidereus Nuncius*, with the etchings supplied, were ready on 13 March, but the broader publication

13 *Opere* X, p. 305 (Ep. 282). On Cosimo II's illegitimate cousin Antonio de' Medici see also chapter I note 5.

of the work outside Venice may have been delayed for a week or so at Galileo's request, so that the presentation copy to Cosimo would not be stale by the time it reached his hands. As it happened, that presentation copy and the accompanying telescope did get delayed. Enea Piccolomini wrote to Galileo from Pisa, where the Tuscan court was in residence for the Easter holy days, on Saturday 27 March (*Opere* X, p. 304, Ep. 281*). Enea was on this day still awaiting the arrival of the telescopic instructions that Galileo had directed to him, and he expressed his great desire also to have a copy of the book: *Io poi sto con molto desiderio di havere uno de' suoi libri.* The following Tuesday, 30 March, again from Pisa, Vinta wrote Galileo (the letter directed to Florence, to await his arrival) that although Galileo's letter of the 19th had arrived, neither the dedication copy, nor the telescope, nor the detailed instructions to Enea Piccolomini had as yet appeared.[14] Galileo was not, Vinta wrote, to worry overmuch about this – but at this most critical moment of his life, how could he not have worried?

Besides the early copies Galileo sent to Vinta, to Cosimo II, to Sertini, and surely also to Enea Piccolomini, we have record of various other copies that he must have posted to friends and supporters before he left Padua for Florence on Friday, 26 March. Ottavio Brenzoni in Verona received a copy from Galileo on or by Saturday, 3 April, as did Benedetto Castelli in Brescia.[15] Cardinal Conti, in Ancona, received his copy on or by Easter Sunday, 11 April.[16] At least some of the court presentations that Galileo discussed in his 19 March letter to Vinta were also surely put in train before he left Padua at the end of the month. In Munich, Galileo's brother Michelangelo, a court musician, was his agent. Michelangelo reported that he had received presentation copies on Good Friday 9 April, and had had them bound immediately, although the presentations to Maximilian Duke of Bavaria, and to his uncle Ernst, Electoral Archbishop of Cologne, had to be postponed because of the Easter celebrations until early in the following week. Michelangelo's very circumstantial letter, in which he requested additional copies for presentations, adds one particularly interesting detail: the Elector related that he had already acquired a copy, but lacking the plates (*Il Sererenissimo Elettore poi, come principe umanissimo, si misse a parlar meco, et mi disse haver già hauto un de' vostri libri, ma senza figure*).[17] From so curious a channel – a proud account of the gracious condescension of an ecclesiastical prince to the author's brother – do we receive an additional hint that the copies without plates belong to Baglioni's northern shipment.

In his *Dissertatio cum nuncio sidereo*, Kepler provided exact, vivid details of when and under what circumstances he first learned of, then saw and read Galileo's book. Around the Ides of March (the 15th), an imperial courier had arrived with news from Venice that Galileo had discovered "four new planets" – news that must have originated even before

14 *Opere* X, pp. 307–308 (Ep. 284).

15 Brenzoni: *Opere* X, p. 309 (Ep. 286*), "Finalmente ho ricevuto le sue, tanto disate da me, Osservacioni celesti, quali, già alcuni giorni sono, lei mi havea inviate ...". Castelli: ibid. p. 310 (Ep. 287*), "Quanto mi sia stato caro l'Aviso Astronomico mandatomi da V.S. ... havendo di già letta e riletta più di dieci volte ...". For a detailed survey of early readers and their responses see Pantin 1993, pp. XI–XXVII.

16 *Opere* X, p. 311 (Ep. 289*): "Questi giorni santi appena mi permettono di accusare a V.S. la ricevuta del libro che si è compiaciuta mandarmi ...".

17 *Opere* X, pp. 312–314 (Ep. 290).

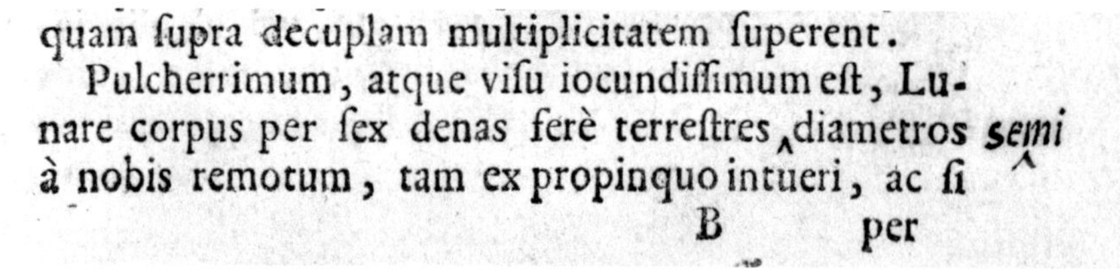

quam ſupra decuplam multiplicitatem ſuperent.
Pulcherrimum, atque viſu iocundiſſimum eſt, Lu-
nare corpus per ſex denas ferè terreſtres ‸diametros semi
à nobis remotum, tam ex propinquo intueri, ac ſi
B per

XIV.6: Sidereus Nuncius, B1r, Linda Hall Library.

the *Sidereus Nuncius* appeared, and such as might have leaked out from someone who had seen or heard about only the drop-title of sheet B, not knowing yet of Jupiter's connection with the planets.[18] The imperial councillor Wacker von Wackenfels drove to Kepler's lodgings to give him this spectacular report. Both men were so bowled over that they could make no sense of it; they roared with excited laughter (*Wackherius ... de curru mihi ante habitationem meam nunciaret, tanta utrumque incessit admiratione, absurdissimi acroamitis consideratione, ut cacchinis et gaudio confusi, neque ille narrando, neque ego audiendo, sufficeremus).*[19] Only later did Kepler deduce that this must mean not a new planet, which would challenge his own idea that each of the planets was connected with one of Plato's five regular solids, but four "lunae minutissimae" revolving about one of the known planets. Early in April the presentation copy of *Sidereus Nuncius* for Rudolph II reached Prague, and Rudolph afforded Kepler an opportunity to skim it quickly. Finally, on Thursday 8 April the Tuscan ambassador to the Emperor, Giuliano de'Medici, delivered a copy to Kepler according to Galileo's direct wish.[20] Other copies had reached Prague by the third of April: Martin Hasdale reported to Galileo that on that day the Augsburg banker-antiquary Marcus Welser had brought the *Sidereus Nuncius* to the Spanish ambassador.[21] This is probably too early for the copy to have come through ordinary book trade channels, Prague being more than four hundred kilometers east of Frankfurt.

Some of these early presentation copies may well survive, though none so far has been clearly identified as such. One good candidate is at Christ's College Cambridge: a copy in contemporary Italian vellum binding and on ordinary paper, with an unidentified early owner's note, "Authoris dono". Of the nine fine-paper copies currently identified, three are candidates for having originally been directed to royalty and high dignitaries in the cities where they remain: those at the Bayerische Staatsbibliothek in Munich, the Österreichische Nationalbibliothek in Vienna, and one of the copies at the Biblioteca Nazionale Centrale of Florence, shelfmarked Postillato 110. The Munich copy was almost certainly one of the presentation copies distributed by Michelangelo Galilei. It is bound in contem-

18 This is the same state of knowledge as was possessed by Marcus Welser in Augsburg on 12 March 1610, as cited in the Introduction of the present study; and likewise of Raffaello Gualterotti in Florence on 6 March 1610 when he wrote to Galileo, "I would hope, before I die, to see that great star with its four planets" (*Desidererei, havanti che io morissi, di vedere quella grande stella co i quattro pianeti che V.S. osservati ...*): *Opere* X, p. 287 (Ep. 268*).

19 *Opere* X, p. 320 (Ep. 297); Kepler revised this somewhat in the printed edition of his letter, *Dissertatio de nuncio sidereo* (*Opere* III, p. 105). See Pantin 1993, pp. 49–51 for an extensive discussion.

20 *Opere* III, p. 101.

21 Opere X, p. 314 (Ep. 291): "Io mi trovai, XII giorni fa, a desinare dal Sig.re Ambasciatore di Spagna, dove il Sig.re Velsero portò al detto Ambasciatore uno di questi libbri ...".

porary vellum and contains two early engraved bookplates of the Munich court library, the Bibliotheca electoralis, one over the other; the visible bookplate is one of those designed ca. 1629–30 by Raphael Sadeler, the bookplate below it one of those made in and dated 1618.[22] The presumption is less strong for the Vienna and Florence copies. The Vienna copy is in nineteenth-century marbled boards, considerably cut down with no indication of early ownership. As for Florence, it is uncertain whether Postillato 110 was in Florence from the time of its first owner, or was acquired at some later time from elsewhere. Its owner, though favored with a fine-paper copy, was no strong supporter of Galileo. He entered extensive marginal notes on the Moon discussions, a number of which are aggressively skeptical: "ridiculous, ridiculous, ridiculous: for what is lacking in proof is rightly called ridiculous" (*nugę, nugę, nugę, quę enim probatione carent, nuge merito appellari possunt*).

Of the six other fine-paper copies, that at Stanford was owned in the eighteenth century by the Genoa bibliophile the Marchese Durazzo, with his engraved bookplate.[23] The Linda Hall and Strasbourg copies contain no early ownership marks, nor do those at Hiroshima and Naples. The Münster copy has a most curious provenance. In the later seventeenth century it belonged to a Frater Sebastian Menge, Premonstratensian of Wedinghausen in Westfalia, and confessor to a small house of Premonstratensian nuns nearby. There is no obvious channel that would have brought into his hands a fine-paper copy of *Sidereus Nuncius*, and yet it happened, and certain corrections of the Latin in this copy show that it had an attentive early reader.

Whatever their individual copy histories, these nine copies, seven with the MEDICEA cancel slip and all with the Moon etchings, must descend directly from Galileo. In contrast with the ordinary-paper copies (with three exceptions noted below), they all contain one or more of three text corrections that may be defined as Galilean. The first is at the beginning of the text, fo. 5r (B1r) lines 16–18. Galileo wrote concerning the resolving power of his most recently improved telescope that with its aid the lunar body, distant from us by sixty earth-diameters, appeared to be distant by only two (… *Lunare corpus per sex denas ferè terrestres diametros à nobis remotum* …). In all the fine-paper "diametros" was corrected to "semidiametros" by means of an insertion mark and a marginal "semi" in Galileo's hand (figure XIV.6).

That is, the distance from Earth to Moon is restated from "almost 60 [earth] diameters" to "almost 60 half-diameters". This passage has been frequently discussed, most thoroughly by Edward Rosen.[24] Galileo's printed statement of the distance, sixty times the Earth's diameter, is plainly wrong, and twice too great. It seems that when writing this paragraph, he mistakenly conflated his estimates of the resolving power of his telescope with a careless mis-statement of the Moon-Earth distance. By Copernicus's calculations,

22 Dressler 1972, "Typ A" and "Typ B". I am grateful to Dr Bettina Wagner, who arranged infrared photography that revealed the earlier, covered bookplate. Dr. Wagner has also found that in the earliest manuscript catalogue of the Bibliotheca electoralis, dating to the first half of the seventeenth century, there is an entry, "Galilei Nuncius Sydereus 4a.95."

23 On Durazzo as book collector see Petrucciani 1988, who notes, p. 28, that in Durazzo's book inventory of 1772, Galileo was among the authors listed.

24 Rosen 1954; see further Pantin 1992, p. 56 note 5.

improving on those of Ptolemy, the Moon's average distance was, with good approximation, about 60.3 Earth radii: quite close, indeed, to modern calculations.[25] The Earth's radius or semidiameter, not the diameter, had for centuries been the common unit for calculations of cosmological distances.

The erroneous phrasing did not arise in the printing shop. Galileo's Draft and Fair Copy make the same erroneous statement and as Rosen pointed out, the error is also in the letter of 7 January 1610 that provided so much of the framework for Galileo's discussion of the Moon. The seventeenth-century reprints of *Sidereus Nuncius* all continued the error, but the 1718 Florence quarto *Opere* silently changed to "semidiametros" as did the Florence 1843 *Opere* (III, p. 59), and the Edizione Nazionale (*Opere*, III, p. 59). In the last, Favaro (Ibid. p. 10, note 2) saw Galileo's misstatement as a slip of the pen, but Rosen's observation that the error is in the 7 January letter shows that it was more than casual: it persisted from the letter to Draft to Fair Copy to setting copy to printed edition. Recent editions and translations – Van Helden 1989, Pantin 1992, Hallyn 1992, Battistini 1993, Shea 2009, Bascelli 2009 – have kept the word "diametros", adding notes of correction or explanation. Rosen further observed that in the 1632 *Dialogo*, Galileo approved a calculation of the distance as 56 earth-radii, adding (p. 348), "There is nothing to indicate, so far as I know, when the blunder [in *Sidereus Nuncius*] came to his attention." From the fine-paper copies of *Sidereus Nuncius* we have the answer: Galileo noticed and corrected his mistake soon after *Sidereus Nuncius* was published, thus almost certainly before the end of March 1610, by which time he had sent off various presentation and gift copies.

The second correction is on fo. 17v (E1v) at line 14, where Galileo discusses his observation of 8 January, when he realized that either the three small stars around Jupiter seen the preceding night had moved; or Jupiter itself had moved. On 8 January the three stars were west of Jupiter, whereas the preceding night, two had been on the eastern side. The passage reads in part: "*... exitare tamen cæpit, quonam pacto Iuppiter ab omnibus prædictis fixis posset orientalior reperiri, cum à binis ex illis pridie occidentalis fuisset ...*". Given that *exitare* is not a Latin word, what could this phrase mean? In the *Opere* (III, p. 80) Favaro emended to "haesitare tamen coepi": "I began to hesitate." The same emendation, if we are to consider it as such, had appeared in the Bologna 1655 edition, whereas the Frankfurt 1610 edition had carried over with slight change of spelling the corrupted wording of the Venice edition. The London 1653 editions made a very clumsy emendation to "excitare tamen caepit", which can be painfully construed, if we further emend "excitare" to "excitari," as roughly "it began to arouse."[26] Pantin 1992 (p. 28) first brought some clarity to the question by indicating in her apparatus that the Draft reading is "esitare tamen cepi", that is, orthographic differences aside, "haesitare tamen coepi." The source of the error must be the compositor's misreading of Galileo's handwriting. We can say this with some confidence, although the setting copy is lost, for both in the Draft and in the more

25 Van Helden 1985, pp. 44–45: Copernicus's "model predicted that the Moon's greatest and least distances at the quadratures were 68 1/3 e.r. [earth radii] and 52 17/60 e.r., respectively. The corresponding values at the syzygies were 65 1/2 e.r. and 55 8/60 e.r."

26 The "excitare" of London 1653 seems to have left a residue in Van Helden 1989, p. 65, "yet I was aroused by the question of how Jupiter could be to the east ...", which implies an emended reading "excitari tamen coepi."

formally written Fair Copy, there are various places where Galileo's internal "s" is formed as an "x" (figure XIV.7). At least two more setting errors arose from this same peculiarity of Galileo's handwriting: at lines 3 and 14 of 20v (E4v) appear *dixtantes* and *dixtabant* for *distantes* and *distabant*.

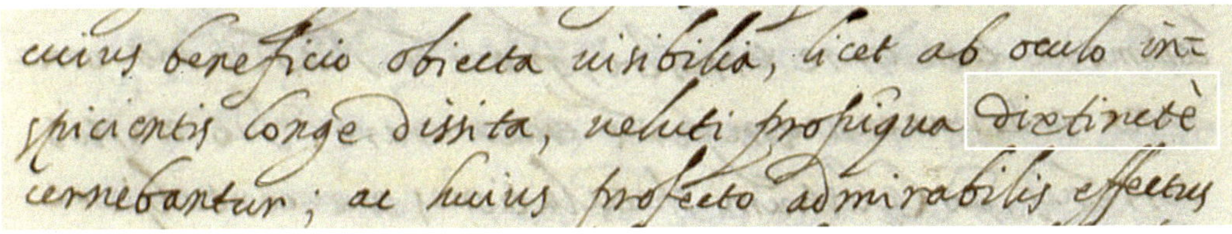

XIV.7: BNCF, Ms. Gal. 48, 39r.

As just suggested, the corrupt setting "exitare tamen caepit" must have been noticed and corrected by Galileo at some point when he was distributing his fine-paper copies. The Strasbourg fine-paper copy is fully corrected in ink by striking through the final "t" of *caepit* and the "ex" of *exitare*, the latter replaced by a marginal "hęs" in Galileo's hand, to produce *hęsitare*. The Münster and Munich fine-paper copies are partially corrected, with the final "t" of *caepit* similarly struck through. Parenthetically we note that in the Arsenal ordinary-paper copy, de Thou emended the passage correctly in the course of his reading, as did an unknown early reader of the ordinary-paper copy now at the Yale Medical History Library.

The third Galilean correction comes at the end of the text, 28v (G4v) line 28, where *Sidereus nunceus* has "Planetae MEDICEA". This is a grammatical error: "planeta" being a masculine noun, the adjective in nominative plural must be "Medicei." The error probably arises in some way from the pressure of time in the last days of copy preparation and printing. As was noted in chapter XI, as space ran out in sheet G, Galileo had to abridge the draft of his conclusion severely, and also had to work in the new nomenclature of "Medicean" for the Jupiter moons. At the corresponding sentence of the Draft he had written correctly: "... *Planetę Cosmici circumducantur.*" When he revised, he may have unconsciously associated "Medicea" not with "planetae", but with the neuter plural "sidera", as used on the title-page: MEDICEA SIDERA. This error was corrected even in the Frankfurt 1610 reprint, and in all the subsequent printings. Galileo must have noticed the error shortly after publication. He corrected it in the same way in all the recorded fine-paper

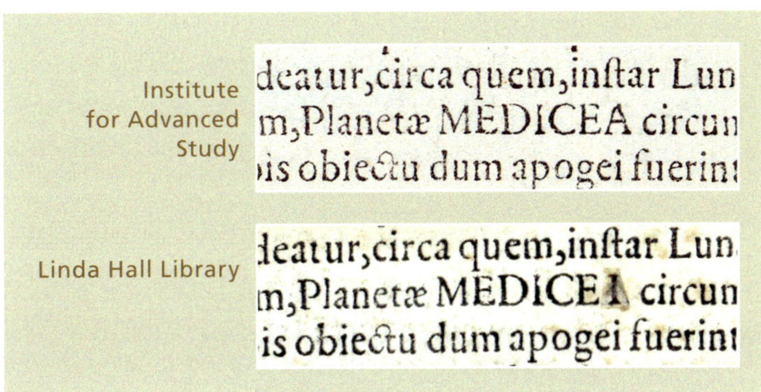

XIV.8: Sidereus Nuncius, G4v.

copies: he wrote a capital "I" on a slip of paper, trimmed the slip to a trapezoidal shape so that it would neatly cover the final "A", and pasted it on (figure XIV.8).

We have, in short, direct evidence that Galileo took a literally personal hand in improving the fine-paper copies. There is another proof that these three corrections are indeed Galilean: they were also entered in Galileo's annotated "desk" copy of *Sidereus Nuncius*, now at Brown University. This copy, on ordinary paper, was preserved by his "last disciple" Vincenzo Viviani, whose signature is on the title-page. At his death in 1702 Viviani left his library to the hospital of Santa Maria Nuova in Florence, and much of it then passed into the Biblioteca Magliabechiana. The Brown copy may be the same as that listed by Viviani in an undated manuscript inventory of his books now in the Fondo Palatino (Ms. 1195) of the Biblioteca Nazionale Centrale in Florence.[27] In modern times, it was sold in London at Sotheby's on 30 May 1960, lot 110, and came to Brown University as part of the fine history of science collection bequeathed by Albert E. Lownes (d. 1978).

One other hand correction in the Galileo-Viviani copy is directly relevant to the present inquiry. Rosen pointed out that *Sidereus Nuncius* repeats the fo. 5r error of "diameter" for "semidiameter" on fo. 6v (B2v) line 1, where Galileo wrote that with his 30-power telescope he had observed the Moon "scarcely two earth-diameters distant": "*... per duas Telluris diametros abesset.*" The correct word would be *semidiametros*. The 1718 Florence edition corrected this later place to "semidiametros", whereas in the Edizione Nazionale (*Opere* III, p. 61) Favaro kept the word "diametros" without comment, apparently overlooking or at least silently accepting the inconsistency. Once more, Pantin (1992, p. 56 note 5) added clarity by observing that Viviani's incomplete and for long unpublished Italian translation, which dates to 1656, corrected both passages. At the first place Viviani translated: "*remoto per quasi 60 semidiametri della terra*", and at the second: "*e prima veddi ... così da vicino la Luna como se appena ci fusse distante per due soli semidiametri della Terra.*"[28]

Galileo did not correct the "diametros" of fo. 6v in any of the fine-paper copies, but he did in his desk copy. With the added information of this copy, we have a well-documented picture of Galileo's post-publication corrections to the first edition of *Sidereus Nuncius*, and of the order in which the errors came to his attention. First, he noticed the incorrect wording of the drop-title, COSMICA SYDERA, and had it repaired by means of a printed cancel slip MEDICEA, to be pasted over COSMICA. By the time Galileo saw the error, however, a substantial number of the ordinary-paper copies – the commercial issue – had been distributed: that is, most if not all of the copies that were shipped immediately to the Frankfurt and Leipzig fairs, and a number of other copies that stayed in Italy but had left Baglioni's shop. Galileo also applied the cancel slip to his fine-paper copies.[29] Next, he noticed that on the last page, "Planetae MEDICEA" should be "... MEDICEI." The commer-

27 Favaro lists a copy of *Sidereus Nuncius* as no. 3 in his study of Galileo's library, on the testimony of Viviani's inventory: Favaro 1887 p. 235, and cf. pp. 227–228.

28 Favaro 1992 (1892), pp. 197, 199. In the nineteenth century, Viviani's translation, headed "Auuiso Astronomico", was bound up with Galileo's Draft in Ms. Gal. 48, as fos. 46–57 (fos. 53v–57v blank).

29 In the Stanford fine-paper copy, the word COSMICA is thoroughly eradicated; possibly the slip has been lost. In the Hiroshima copy there is no correction.

cial issue was no longer under his supervision, but he created his own manuscript cancel slip to correct this, present in all the fine-paper copies. Third, he noticed the textual corruption "exitare tamen caepit" on fo. 17v. At this time, he had already sent out some of the fine-paper copies, so the correction was only made in a few copies: partially, in the Munich and Münster copies, and fully only in the Strasbourg copy. It may be that because of his own handwriting, Galileo's eyes did not immediately land on "exitare" as a misspelling. Finally, after all the fine-paper copies had been distributed, Galileo realized that the correction to "semidiametros" on fo. 5r required a corresponding correction on fo. 6v: he entered this in his desk copy.

After Galileo's death, the desk copy of *Sidereus Nuncius* came as noted into the hands of Viviani, and in 1656, when the latter began a translation as part of his plan to produce a folio dual-language edition of Galileo's writings, Latin and Italian, he naturally followed the corrected version. After Viviani's death his nephew and heir, Jacopo Panzanini, cooperated with the editor of the 1718 Florence *Opere* of Galileo, Tommaso Buonaventuri (see chapter V). Directly or indirectly from the desk copy the four authorial corrections in *Sidereus Nuncius* were all incorporated in the 1718 text. They appear also in the 1843 Florence *Opere*, vol. III, but the editorial note in that volume leaves the source of the corrections unclear. In Favaro's Edizione Nazionale the second correction of "semidiametros" fell out by oversight. As for the first correction to "semidiametros", Favaro explicitly stated that he was making his own emendation, unaware that Galileo had corrected the passage repeatedly almost three centuries earlier.[30] Regarding "haesitare tamen coepi" and "Planetae Medicei", Favaro was editorially silent, but here too he presumably thought he was emending, not restoring.

Two more ordinary-paper copies must be mentioned, for they add to the record of Galileo's corrections of *Sidereus Nuncius*. The copy now at the University of Oklahoma was presented by Galileo to the "most illustrious" poet Gabriello Chiabrera of Savona. It appears to be the only copy known with an explicit presentation in Galileo's hand (figure XIV.9). The reason for the presentation is well recorded. As noted earlier in this chapter, in his 19 March letter to Vinta, Galileo was already projecting a second, Italian edition of *Sidereus Nuncius*, and it would be provided with commendatory verses from various poets (which there had been no time to procure in the short production period of *Sidereus Nuncius*): the Tuscan muses would not be reluctant to sing the glories of the most serene Medici House (*credo che le Muse toscane non taceranno in così grande occasione le glorie di questa Ser.ma Casa*).[31] Chiabrera was to be one of those poets, as was Piero de' Bardi, mentioned in chapter XII, and other names are found in Galileo's correspondence. To diverge for a moment from the central topic, it should be noted that separate from these plans for a well-commended second edition of *Sidereus Nuncius* were the verses got up by a friend in Venice, the glass manufacturer and belle-lettrist Girolamo Magagnati: *Medi-*

30 *Opere* III, p. 10 n. 2: "Abbiamo corretto … un errore di penna occorso a Galileo, parlando della distanza della Luna dalla Terra: *diametros*, dato dai due manoscritti autografi e dalla edizione originali, in *semidiametros*."
31 *Opere* X, p. 300 (Ep. 277).

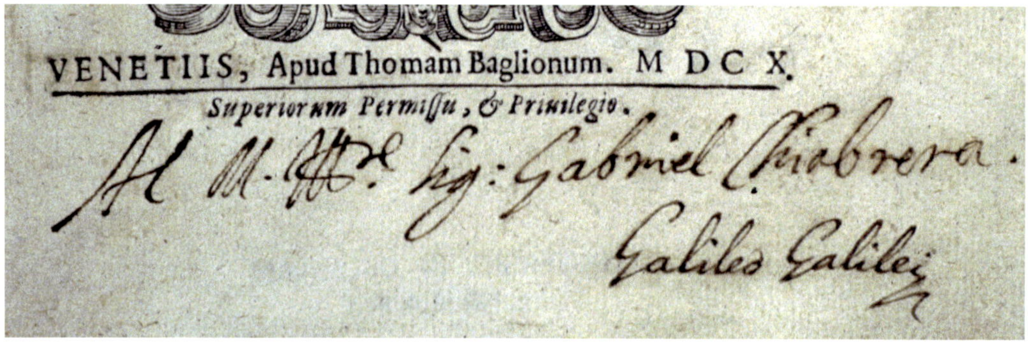

XIV.9: Sidereus Nuncius, A1r, University of Oklahoma.

tazione poetica ... sopra i pianeti medicei, published in Venice by the heirs of Altobello Salicato, a quarto of 6 leaves dedicated to Cosimo II. In May 1610 Galileo personally presented a copy of this work to Cosimo, and Magagnati sent copies also to the Accademia della Crusca, to Jacques Badover, and to Chiabrera.[32] The title-page of Magagnati's poem bears an emblem that Galileo surely approved: the Medici arms, handsomely engraved within a grotesque cartouche, were given a crown whose tines are surmounted by stars: the large star of Jupiter atop the central Florentine lily, flanked by its four small companions.

Galileo's own enthusiasm for a second edition seems to have slowly ebbed in the summer months of 1610. On 10 July Alessandro Sertini reported to Galileo that although Chiabrera had promised to supply verses, they had not yet been sent.[33] For our purposes, the significant point of Galileo's presentation to Chiabrera is that in this copy too he hand-corrected "diametros" to "semidiametros" on fo. 5r, and "MEDICEA" to "MEDICEI" on the last page. As to another ordinary-paper copy, at the University of Arizona, it is unfortunate that it preserves no clues to early ownership, for it almost certainly was given by Galileo to some friend. It contains all three of the major Galilean corrections: "semidiametros" on B1r, "hesitare" for "exitare" on E1v (but without changing "caepit" to "caepi"), and "MEDICEI" on the last page (corrected by a marginal annotation, not a paste-on slip). Thus, Galilean hand corrections are found in the nine known fine-paper copies, and in three ordinary-paper copies that were directly connected to Galileo and therefore lie outside the channel of commercial distribution. The distribution of these corrections may be shown in a table (Yes/No):

32 Drake 1964; the edition is EV17 M39.

33 *Opere* X, p. 399 (Ep. 357*). For a fully documented account of the projected second edition, see Pantin 1993, pp. LIII–LVII.

Copy	B1r semi (diametros)	B2v semi (diametros)	E1v haesitare tamen caepi	G4v MEDICEI
Fine: Florence BNC	Y	N	N	Y
Fine: Hiroshima UL	Y	N	N	Y
Fine: Kansas City LHL	Y	N	N	Y
Fine: Munich BSB	N	N	Partial	Y
Fine: Münster UB	Y	N	Partial	Y
Fine: Naples BN	Y	N	N	Y
Fine: Stanford UL	Y	N	N	Y
Fine: Strasbourg BU	Y	N	Y	Y
Fine: Vienna ÖNB	N	N	N	Y
Ord.: U. Oklahoma (presented to Chiabrera)	Y	N	N	Y
Ord.: U. Arizona	Y	N	Partial	Y
Ord.: Brown U. (Galileo's copy)	Y	Y	Y	Y

The history of textual transmission is filled beyond number with curiosities, puzzles, and seeming paradoxes. In essence, anything that can go wrong in textual transmission somewhere, sometime, somehow has gone wrong. If there is a way to make an error, a human being will find that way. In the history of *Sidereus Nuncius*, the modern chain of false reports of "missing stars" in the Jupiter-moon diagrams (chapter X) provides a sterling example. It is a small irony in the transmission history of Galileo's writings that his own textual corrections to *Sidereus Nuncius*, fully documented in surviving copies, have been so thoroughly submerged. Tommaso Buonaventuri, in the 1718 Florence edition, was perhaps the last editor to realize that at the four places in question he was not emending but reinstating authorial corrections.

Early printed books always contain physical traces of the processes by which they were made. The greater the number of copies of a given edition that are examined, the more reliable the hypotheses about the making of that edition are likely to be; or perhaps it is better to say, the more testable they become. New evidence from one more copy may well challenge or conceivably even overthrow entirely an hypothesis. Our hypotheses of the working practices of early printing shops ought to be framed so clearly that if new evidence should emerge that challenges or contradicts them, there will be little doubt of their deficiency, and of the need for fresh analysis. For example, one of the hypotheses of the preceding analysis is that all fine-paper copies will show frisket bite on D6r. If a fine-paper copy should come to light that does not have frisket bite at this place, the present writer will be compelled to acknowledge that in certain respects his picture of the making of *Sidereus Nuncius* is fundamentally wrong.

So many small details within *Sidereus Nuncius* have been examined, that it will be helpful to review the chief categories that distinguish one copy from another. There are three fundamental strata: the SNML copy, the ordinary-paper copies, and the fine-paper copies. Among the ordinary-paper copies there are two main groups, those lacking and those possessing the paste-on cancel slip "MEDICEA" on B1r. Within the ordinary-paper copies lacking the cancel slip, there is a sub-section: the copies that lack the plates depicting the Moon. The fine-paper copies, and three ordinary-paper copies with direct connection to Galileo, contain certain corrections in his hand. In the typographic setting itself a small number of variations have been identified, such as the two words corrected in the course of printing on A3v. It may easily be that as more copies are studied, and as more complete collations of one copy against another are carried out, more variations will come to light, and these may change, incrementally or perhaps dramatically, our knowledge of the creation of this book that set in motion a radical reconception of the structure of the heavens.

In conclusion, we may spare a grateful thought for the actors of the book: not only its author, who knew fully the gift he had been given, but many actors without names: the makers of paper, of types and woodblocks; the workers of the printing shop; the carriers of mesages between the Veneto and Florence; the bargemen and carters; and finally the mules and oxen whose humble labor moved sheets of *Sidereus Nuncius* across the face of Europe, under the eyes of the boundless sky.

Appendix

Physical Description and Copies Examined

Transcriptions do not reproduce the long s nor the accented "q" character. The abbreviation character representing –que is transcribed as "q3", which it roughly resembles. I have personally examined the copies listed except where brackets after the shelfmark give the name of someone who kindly examined the copy according to a questionnaire. Within the same brackets the word "photographs" means that I examined not the original but a photographic facsimile of whatever nature. The index of "Sidereus Neighbors", additional works bound with copies of Sidereus Nuncius, could not always be based on direct title-page transcriptions, and so there are undoubtedly scattered inconsistencies in the wordings of the titles. The places of publication are regularized to the English vernacular, and dates in roman numerals are transposed to arabic, but in principle the printer and publisher statements are quoted from the title-pages. The descriptions of copies note, where possible, the library shelfmark; the presence or absence of the "MEDICEA" cancel slip on B1r, and the presence of absence of the Moon etchings; the leaf dimensions and the nature of the binding description; former owners; and where applicable, other printed works bound (in a few cases, formerly bound) with the copy. Regrettably, it was not possible to include this information for copy 45.

Galilei, Galileo. Sidereus Nuncius. Venice: Tommaso Baglioni, 1610. 4°. 7½ sheets.

SIDEREVS | NVNCIVS | MAGNA, LONGEQVE ADMIRABILIA | Spectacula pandens, suspiciendaque proponens | vnicuique, præsertim verò | PHILOSOPHIS, atq3 ASTRONOMIS, quæ à | GALILEO GALILEO | PATRITIO FLORENTINO | Patauini Gymnasij Publico Mathematico | PERSPICILLI | Nuper à se reperti beneficio sunt obseruata in LVNÆ FACIE, FIXIS IN- | NVMERIS, LACTEO CIRCVLO, STELLIS NEBVLOSIS, | Apprime verò in | QVATVOR PLANETIS | Circa IOVIS Stellam disparibus interuallis, atque periodis, celeri- | tate mirabili circumuolutis; quos, nemini in hanc vsque | diem cognitos, nouissimè Author depræ- | hendit primus; atque | MEDICEA SIDERA | NVNCVPANDOS DECREVIT. | [Publisher's Device] | VENETIIS, Apud Thomam Baglionum. M D C X. | [rule] | Superiorum Permissu, & Priuilegio.

COLLATION:

A4; B–C4 D6 E–G4: 30 leaves. A2–D4 foliated 2–16, D5–6 unfoliated, E1–G4 foliated 17–28.

CONTENTS:

A1r Title, verso blank
A2r–4r [Dedication:] [headpiece] SERENISSIMO COSMO MEDICES II. MAGNO HÆTRVRIÆ DVCI IIII. [P6]Ræclarum sanè, atque humanitatis plenum … quàm diutissimè fruere. Datum Patauij 4. Idus Martij, M DC X. Celsitudinis tuæ Addictissimus Seruus Galileus Galileus.
A4v [License:] Gli Eccellentissimi Signori Capi … come nel libro Intitolato SYDEREVS NVNCIVS, &c. di D. Galileo Galilei non si troua alcuna cosa contraria alla Santa Fede Cattolica, Prencipi, & buoni costumi … concedono licenza, che possi esser stampato in questa Città. Datum Die primo Martij 1610. D. M. Ant. Valaresso D. Nicolò Bon D. Lunardo Marcello } Capi dell'Ecc. Cons. de'X. ….
B1r–G4v [Text:] [headpiece] ASTRONOMICVS NVNCIVS OBSERVATIONES RECENS HABITAS Noui Perspicilli beneficio in Lunæ facie, Lacteo circulo Stellisq3 nebulosis, innumeris fixis, necnon in quatuor Planetis COSMICA SYDERA nuncupatis, nunquam conspectis adhuc continens, atque declarans. [M8]AGNA æquidem in hac exigua tractatione … Vlterius progredi temporis angustia inhibet; plura de his breuì candidus Lector expectet. FINIS.

PAPER:

Super-Median (SNML proof copy): watermark: LA monogram, cornermark Tp [figures II.4–5]
Super-Median (ordinary-paper issue), uncut quarto leaf dimensions ca. 26×19 cm: three stocks:
(1) watermark: LA monogram, with cornermark Tp, a different stock from that of the SNML proof copy [figures II.6–8], used to print quires B–G
(2) watermark: Anchor in circle [figure II.9], used to print quire A
(3) watermark: GZ monogram [figure II.10], used to print some copies of the half-sheet D3.4, this stock being slightly smaller.
Median (fine-paper issue), uncut quarto leaf dimensions ca. 24×17 cm: watermark: Crown with star surmount [figure II.11].

PRINTING, PRINTING MATERIALS AND LAYOUT:

Typesetting totals approximately 61,000 ens. Four-page formes, inner formes printed first. D3.4 printed from two-page formes, as also the SNML proof copy. Dedication set in italic (20 lines = 113 mm), 29 lines (A2v), 164×101 mm + headline + direction line. Text (and License) set in roman (20 lines = 94 mm), 34 lines (B2r), 160×95 mm + headline + direction line. B1v set to 33 lines. For variations in line width of some pages see chapter X. Additional fonts were used for the title-page and drop-title [figure VI.3], for a two-line initial on B2r [figure VI.6], for the O of the 65 Jupiter-moons diagrams of sheets E–G [figures X.7, 14, 16], and for the small word "fixa" set within five of the diagrams [figure X.18].

Woodcuts:
A1r: publisher's device of Religion in female personification with legend HINC RELIGIO VERA (60 × 80 mm) [figure XII.1]
A2r, B1r: decorative headpiece (18 × 95 mm) [figures IV.1, 2]
A2r: unbordered historiated initial P (33 × 33 mm) [figure IV.1]
B1r: bordered historiated initial M (33 × 34 mm) [figure IV.2]
B3r: Telescope (min. 28 × 165 mm) [figure VI.5]
C4v: Lunar atmosphere (min. 43 × 111 mm) [figure VII.15]
D1v: Geometric calculation of the height of the Moon's mountains (min. 78 × 76 mm) [figure IV.3]
D5v: Stars of Orion (min. 205 × 175 mm) [figure IX.13]
D6r: Stars of the Pleiades (min. 74 × 120 mm) [figure IX.15]
D6v: Nebula of the head of Orion (min. 54 × 45 mm); and Nebula of Praesaepe (min. 87 × 54 mm) [figure IX.16].

Etchings:
Four copperplate etchings of the Moon (approximately 140 × 150 mm), printed in reserved spaces within the letterpress: Moon I, B4r [figure VII.8, detail]; Moon II, C1v [figure VII.12, detail]; Moon III, C2r and C2v [figure VII.9]; Moon IV, C2v [figure VII.9]. In several copies the Moon etchings are incorrectly or imperfectly printed: mispositioned (copy 21), skewed (copy 25), or upside-down (copies 11, 34, 46) (see chapter XIV). The Moon etchings were not supplied in the ordinary-paper copies 4, 13, 14, 15, 16, 18, 23, 38, 42, 44, 61, 74.

Signatures, catchwords, headlines, foliation:
Signed to the fold (B, B2; D, D2, D3; and similarly). Catchwords on every page (but none on A4r–v, C2v, C4v). The C2r catchword "Vnum" connects to C3r, there being no text on C2v; the D5r catchword "PLEIA-" connects to the running title of D6r, there being no text on D5v. The headlines of quire A have foliation only, recto pages. The headlines of quires B–G have recto foliation plus centered running titles across verso-recto openings: OBSERVAT. SIDEREAE || RECENS HABITAE. (C4v, E1v, F1v: OBSERVATIONES SIDEREAE; D6r: PLEIADVM CONSTELLATIO.; C2v, D5v: no headline). Foliation and catchwords are inset from the right margins by about an em-width.

SETTING VARIANTS:

(1) A3v line 1 "vnà": the "n" and "à" are damaged characters in the SNML proof copy and in ordinary-paper copies 6, 22, 35, 59; replaced with undamaged characters in the remaining ordinary-paper and all fine-paper copies.
(2) A3v catchword: "inum" in the SNML proof copy and in ordinary-paper copies 2, 5, 6, 8, 12, 20, 22, 25, 26, 29, 30, 31, 35, 43, 47, 49, 51, 56, 58, 59, 64, 69, 71; corrected to "nium" in the remaining ordinary-paper and all fine-paper copies.

(3) B3r line 8 "bracteas,": the comma is damaged in the SNML proof copy; replaced with undamaged character in the ordinary-paper and fine-paper copies.
(4) C1v line 6 "figutis" is so set in the SNML proof copy and apparently in several ordinary-paper copies (11, 22, 26); corrected to "figuris" in most ordinary-paper and all fine-paper copies.
(5) C3r line 12 "tenebrosæ": the "æ" is damaged in the SNML proof copy; replaced with undamaged character in the ordinary-paper and fine-paper copies.
(6) D6r catchword is "cetum" in the SNML proof copy; corrected to "c tum" in the ordinary-paper and fine-paper copies.
(7) F1r bottom line: the Jupiter-moon diagram has 2 asterisks correctly in the SNML proof copy; 1 asterisk erroneously in ordinary-paper copies 1, 3, 5, 8, 9, 10, 12, 17, 19, 24, 25, 26, 27, 28, 29, 32, 33, 34, 37, 39, 40, 41, 43, 45, 50, 53, 54, 55, 60, 63, 65, 67, 70; and 2 asterisks correctly in the remaining ordinary-paper and all fine-paper copies.
(8) F4v line 1 "minor": the "r" is damaged in the SNML proof copy; replaced with undamaged character in the ordinary-paper and fine-paper copies.
(9) G3r line 2: the Jupiter-moon diagram has a damaged asterisk in the SNML proof copy; replaced with undamaged character in the ordinary-paper and fine-paper copies.
(10) G3v line 12: the Jupiter-moon diagram lacks a small asterisk to the left of the Jupiter O in the SNML proof copy; the missing asterisk is supplied in the ordinary-paper and fine-paper copies.

NOTES:

(1) On the title-page, last line, "Superiorum" is set as "Superiornm". For a selection of other uncorrected setting errors see chapter XIII.
(2) On B1r line 7 "SYDERA" is overlaid with a paste-on cancel "MEDICEA" in the ordinary-paper copies 3, 7, 8, 19, 24 (not present, but glue traces remain), 25, 26, 28, 29, 31, 32, 33, 35, 37, 39, 41, 43, 48, 51, 54, 55, 62, 63, 66, 67, 70, and the fine-paper copies F1, F2, F3, F4, F5, F8, F9 (and perhaps originally also F7, where the word "SYDERA" is heavily scratched out) [figure XIV.4].
(3) On D6r the first two lines of text below the Pleiades woodcut did not print fully at the beginning because of varying slippage of the frisket in many ordinary-paper copies including 3, 4, 6, 13, 14, 18, 19, 22, 24, 25, 26, 27, 28, 29, 33, 35, 38, 39, 41, 42, 43, 44, 49, 54, 55, 60, 61, 65, 67, 74, and in all fine-paper copies [figure III.8]. In copy 68, frisket slip on D4v has blocked the printing of most of the bottom line and of the catchword.
(4) The fine-paper copies and three ordinary-paper copies (nos. 35, 51, 64) contain one or more corrections in Galileo's hand (see chapter XIV).

REFERENCES:

EV17 G-154; Cinti 26

Proof copy

New York, New York, Martayan Lan (cited as SNML). Without B1r "MEDICEA" cancel slip, without Moon etchings (the spaces filled with pen-and-wash drawings from Galileo's hand). 22.2 × 16.2 cm. Mid-17th-century Roman gilt vellum, edges gilt and gauffered. BOUND WITH: Galileo Galilei, Discorso Al Serenissimo Don Cosimo II..... Intorno alle cose, che stanno sù l'Acqua.... Seconda Editione. (Bologna, 1655). – Lodovico delle Colombe. Discorso Apologetico … D'Intorno Al Discorso Del S. Galileo Galilei, Circa le cose, che stanno sù l'Acqua … (Bologna, 1655). – Vincenzo di Gratia. Considerationi … Sopra Il Discorso del Sig. Galileo Galilei Intorno alle cose che stanno sù l'acqua..... (Bologna, 1655). – [Benedetto Castelli]. Risposta Alle Oppositioni Del Sig. Lodovico Delle Colombe E Del Sig. Vincenzo di Gratia, Contro al Trattato Del Sig. Galileo Galilei, Delle cose che stanno su l'Acqua. (Bologna, 1655). FORMER OWNERS: Federico Cesi (stamp); and probably Cassiano Dal Pozzo (binding).

Ordinary-paper copies

1. Amsterdam, Nederlands Scheepvaartmuseum: MM-2247. [Djoeke van Netten]. Without B1r "MEDICEA" cancel slip, with Moon etchings. 22.3 × 16.4 cm. 17th-century Italian (Venetian?) vellum. FORMER OWNERS: an undeciphered early ownership inscription on flyleaf; Dr. Ernst Crone (Hubert J. M. W. Peters, The Crone Library: Books on the Art of Navigation [Nieuwkoop, 1989], no. 124).
2. Ann Arbor, Michigan, University of Michigan, Special Collections: QB 41 .G158si. [Peggy Daub]. Without B1r "MEDICEA" cancel slip, with Moon etchings. 23.2 × 17.2 cm. 19th-century marbled boards, vellum spine, edges plain. Acquired from Gonnelli (Florence), 1927.
3. Augsburg, Staats- und Stadtbibliothek: 4° Math 204 [photographs; Günter Hägele]. With B1r "MEDICEA" cancel slip, with Moon etchings. 22 × 16.3 cm. 19th-century paper boards, edges plain. FORMERLY BOUND WITH: Filippo Ferrari, Nova topographia in Martyrologium Romanum (Venice, 1609). – Giovanni Stefano Ferrerio, S. Eusebij Vercellensis Episc. et Martyris … vita et res gestae (Rome, 1602). – Anselm Boèce de Boodt, Gemmarum & lapidum historia. Hanau, 1609. – Jakob Bornitz, De nummis (Hanau: Marnius, 1608). – Vincenzo Contarini, De frumentaria Romanorum largitione (Venice, 1609), the Sidereus Nuncius bound last in the volume. FORMER OWNER: In the Augsburg City Library by 1633, full contents listed in the order above by Elias Ehinger, Catalogus Bibliothecae Amplissimae Reipublicae Augustanae (Augsburg: Praetorius, 1633, F°), cols. 181–182.
4. Berne, Zentralbibliothek: Bong IV 306:2 [Volker Hartmann]. Without B1r "MEDICEA" cancel slip, without Moon etchings. 22.2 × 16.2 cm. 17th-century vellum. BOUND (2nd) WITH: Tycho Brahe, Epistolarum astronomicarum libri (Uraniborg, 1596). – Johannes Stehn, Pro dignitate philosophiae Euclideae tuenda brevis disquisitio (Paris, 1567). – Johann Kepler, Strena seu De nive sexangula (Frankfurt am Main, 1611). FORMER OWNER: Jacques Bongars

5. Brno, Moravská Zemská Knihovna: ST2-0968.892. [Premysl Bar]. Without B1r "MEDICEA" cancel slip, with Moon etchings. 24 × 17 cm. FORMER OWNER: Pietro Riccardi (bookplate: "Biblioteca Riccardi in Modena S.I. F.17. N.13").
6. Buffalo, New York, Buffalo & Erie County Public Library: Milestones of Science 78. [Amy Pickard]. Without B1r "MEDICEA" cancel slip, with Moon etchings. 23.6 × 17.5 cm. 17th-century Italian vellum, edges plain. FORMER OWNERS: Dr. Cinzio Clemente; Cardinal Cesare Rasponi.
7. Cambridge, Christ's College: Y.3.25. With B1r "MEDICEA" cancel slip, with Moon etchings. 24.2 × 17.6 cm. Contemporary vellum, plain edges. FORMER OWNERS: Galileo Galilei (flyleaf inscription "Authoris dono"); unknown recipient; E. A. A. Hagemeijer (red ink stamp)?; Charles Lesingham Smith (1860-dated bookplate, now missing
8. Cambridge, Trinity College, C.16.77. With B1r "MEDICEA" cancel slip, with Moon etchings. 21.2 × 15.7 cm. 19th-century paste-patterned boards, sheep spine with gilt red leather label; edges red-sprinkled. BOUND WITH (a 19th-century collector's assemblage): Francesco Sizzi, *ΔΙΑΝΟΙΑ* Astronomica (Venice, 1611). – Johann Kepler, Narratio de obseruatis … Iovis satellitibus (Florence, 1611). – Simon Marius, Mundus Iovialis (Nuremberg, 1614). – Galileo Galilei, Istoria e dimostrazioni intorno alle macchie solari (Rome, 1613). – Christoph Scheiner, De maculis solaribus (Rome, 1613). FORMER OWNERS: Pietro Dini (armorial stamp), owner of the Sidereus Nuncius only). Philippus Monh (signature), owner of the Kepler Narratio only.
9. Cambridge, University Library, Adams 5.61.1. Without B1r "MEDICEA" cancel slip, with Moon etchings. 24.4 × 17.5 cm. 17th-century Italian vellum, edges plain, replaced endleaves of c. 1800. FORMER OWNERS: Niccolini family, Florence (monogram stamps of F B in circle, M N in circle); John Couch Adams (bookplate).
10. Cambridge, Massachusetts, Harvard University, Houghton Library: IC6.G1333.610s [copy 1]. Without B1r "MEDICEA" cancel slip, with Moon etchings. 20.5 × 14.1 cm. Early 20th-century vellum (ticket of William Wesley & Son, Booksellers & Publishers, 28 Essex St., Strand, London). FORMER OWNER: Robert Wheeler Willson (pencil signature and acquisition note, February 1910).
11. Cambridge, Massachusetts, Harvard University, Houghton Library: *IC6 G1333.610sa [copy 2]. With B1r "MEDICEA" cancel slip, with Moon etchings. 22.9 × 16.2 cm. 19th-century Italian marbled boards, vellum spine and tips, edges red-sprinkled; old shelfmark "M VIIII. LL.14." FORMER OWNER: David Pingree Wheatland (his gift).
12. Chicago, Illinois, University of Chicago: alcQB41.G196 [Julie Gardner]. Without B1r "MEDICEA" cancel slip, with Moon etchings. 19.5 × 14.5 cm. 20th-century morocco (Sangorski & Sutcliffe). FORMER OWNERS: early marginalia in quires E–G; Herbert M. Evans (purchased by University of Chicago 1960).
13. Cologne, Universitäts- und Stadtbibliothek Köln: N4/61+2 [photographs; Hermann-Josef Eschbach]. Without B1r "MEDICEA" cancel slip, without Moon etchings. 21.7 × 15.7 cm. Modern binding (c.1980), red-stained edges. FORMER OWNERS: Jesuit College, Cologne (former shelfmark GB VIII 157+B); Gymnasial-Bibliothek, Cologne.

14. Copenhagen, Det Kongelige Bibliotek: 8° Astr. 35510. [Torsten Schlichtkrull]. Without B1r "MEDICEA" cancel slip, without Moon etchings (Moon images supplied by hand on paste-in slips, copied from the woodcuts of the 1610 Frankfurt edition). 24.4 × 17.2 cm. 20th-century marbled boards, vellum spine, endleaves preserved from the 17th-century binding. FORMER OWNERS: Petrus Scavenius; King Frederik III (en-bloc purchase of Scavenius' library, 1664/5).
15. Cracow, Biblioteka Jagiellońska: 593922 II. [Joanna Pypłacz]. Without B1r "MEDICEA" cancel slip, without Moon etchings. 22.5 × 17 cm. 17th-century vellum. FORMERLY BOUND WITH: Johann Kepler, Narratio de observatis a se quatuor satellitibus Iovis (Frankfurt, 1611), shelfmark 593923 II.
16. Dayton, Ohio, Stuart Rose [photographs]. Without B1r "MEDICEA" cancel slip, without Moon etchings. 22 × 17 cm. Early paper wrapper. FORMER OWNER: Marquard Gude, Rendsburg; sold Christie's London 8 June 2005, lot 265.
17. Dillingen, Studienbibliothek: VIII 178 [Rüdiger May]. Without B1r "MEDICEA" cancel slip, with Moon etchings. 24.2 × 18 cm. Rebound in 1960s, no record of preceding binding. FORMER OWNER: Jesuit College, Dillingen, the gift of Heinrich von Knöringen, Bishop of Augsburg, 1610 (inscription).
18. Durham, University Library: SB 0255 [Alastair Fraser]. Without B1r "MEDICEA" cancel slip, without Moon etchings. 20.1 × 15.2 cm. 19th-century purple cloth, calf spine, red-stained edges. On endleaf "University Library Durham 1856", circular black ink stamp "Univ. Dunelm.", older shelfmark N.VI.58.
19. Edinburgh, Royal Observatory: Cr.7.145. [Karen Moran]. With B1r "MEDICEA" cancel slip, with Moon etchings. 20.7 cm. 19th-century marbled boards, sheep spine. Acquired 1919–1920.
20. Florence, Biblioteca Nazionale Centrale: Pal. 22.B.5.55. Without B1r "MEDICEA" cancel slip, with Moon etchings. 22.6 × 15.3 cm. 18th-century Italian vellum. FORMER OWNER: Biblioteca Palatina, Florence.
21. Göttingen, Niedersächsische Staats- und Universitätsbibliothek: 4to Astron II, 331 RARA. [photographs; Helmut Rohlfing]. Without B1r "MEDICEA" cancel slip, with Moon etchings. 24.5 × 18 cm. Late 17th/early 18th-century vellum, gilt spine. BOUND WITH: Guidobaldo Dal Monte, Planisphaeriorum universalium theorica (Pesaro, 1579). – Giovanni Battista Porta, Coelestis physiognomiae libri sex (Naples, 1603). FORMER OWNER: Joachim Heinrich von Bülow.
22. Graz, Universitätsbibliothek: I 4099. Without B1r "MEDICEA" cancel slip, with Moon etchings. 22.5 × 17.2 cm. 17th-century German vellum wrapper. FORMER OWNERS: Paul Guldin, S.J. (title-page initials, etc.); Jesuit College, Graz (Collegij societatis JESV Graecij, etc.).
23. Liège, Bibliothèque de l'Université: R.190 C [Carmélia Opsomer]. Without B1r "MEDICEA" cancel slip, without Moon etchings. 24.3 × 17.9 cm. 18th-century gilt brocade paper (made by Johann Lupus Ehinger, Augsburg, fl. 1754–1780), red-sprinkled edges; traces of former sewing on 5 leather thongs. FORMER OWNER: Constantin Le Paige.
24. London, British Library: C.112.c.3. Formerly with B1r "MEDICEA" cancel slip (paste marks remain), with Moon etchings. 20.6 × 15 cm. Early 19th-century olive morocco,

gilt fillets, gilt decorated spine, dentelle-gilt turn ins, marbled endleaves, edges gilt, a British Museum binding. FORMER OWNERS: Henry Gooch (Henricus Gocheus | Christus mihi lucro); probably Sir Hans Sloane (information from John Goldfinch).

25. London, Royal Society: Tracts X2/1. [Rupert Baker]. With B1r "MEDICEA" cancel slip, with Moon etchings. 21.5 × 15.8 cm. Modern buckram. First in a tract volume of nine items, apparently gathered thus in the early nineteenth century, the latest imprint being 1816. FORMER OWNER: Henry Howard, sixth Duke of Norfolk.
26. London, University of London: [H.P.L.] [Galileo] SR. With B1r "MEDICEA" cancel slip, with Moon etchings. 23.5 × 17.5 cm. 19th-century pattern-paper boards, Italian pattern-paper endleaves, edges red-stained; 19th-century shelfmark "L - 9 - 4". FORMER OWNER: Harry Price.
27. Milan, Osservatorio astronomico di Brera: A.2.1.23 [photographs; Agnese Mandrino]. With B1r "MEDICEA" cancel slip, with Moon etchings. 17th-century vellum. BOUND WITH: Galileo Galileo, Discorso ... intorno alle cose che stanno in sù l'acqua (Florence, 1612). – Franc. Sitius, Dianoia astronomica, optica, physica ... (Venice, 1611). – Johann Kepler, Dissertatio cum Nuncio Sidereo (Prague, 1610). – Idem, Dioptrice seu Demonstratio ... praemissae epistolae Galilaei (Augsburg, 1611). FORMER OWNER: Jesuit College, Rome
28. Naples, Biblioteca Nazionale: S.Q. XXVI E 27. [Silvana Acanfora]. With B1r "MEDICEA" cancel slip, with Moon etchings. 27 × 19 cm (uncut). Recent vellum
29. New Haven, Connecticut, Yale University, Beinecke Library: QB41+G33 1610B [copy 1]. With B1r "MEDICEA" cancel slip, with Moon etchings. 26.3 × 18.9 (uncut). 19th-century vellum. FORMER OWNER: Dr. James J. Waring (purchased from Lier & Co., Florence, 1931; donated to Yale, 1951)
30. New Haven, Connecticut, Yale University, Beinecke Library: QB41+G33 1610B [copy 2]. Without B1r "MEDICEA" cancel slip, with Moon etchings. 20.6 × 14.6 cm. BOUND WITH: Johann Kepler, Dissertatio cum Nuncio sidereo (Florence, 1610). – Idem, Narratio de obseruatis ... Iovis satellitibus (Florence, 1611). – Giulio Cesare La Galla, De phoenomenis in orbe lunae ... physica disputatio (Venice, 1612). FORMER OWNERS: J. W. Maxwell Lyte (signature); Alfred T. White (note dated 10 May 1894); Alfred W. Van Sinderen (his gift to Yale)
31. New Haven, Connecticut, Yale University, Yale Medical Historical Library. [no shelfmark]. With B1r "MEDICEA" cancel slip, with Moon etchings. 22.8 × 16.5 cm. 17th-century Italian vellum wrapper. BOUND WITH: Johann Kepler, Dissertatio cum Nuncio sidereo (Florence, 1610). FORMER OWNER: Dr. Harvey Cushing (bookplate, record of purchase from "Taeuber & Weil, München, Aug. 28 1931, M 1500")
32. New York, New York, (former) Dr. Richard Green: sold Christie's New York, 17 June 2008, lot 132. With B1r "MEDICEA" cancel slip, with Moon etchings. 20.4 × 14.8 cm, disbound, formerly in Sammelband: leaves with early foliation 86–115
33. New York, New York, (former) Edward Tufte: sold Christie's New York, 2 December 2010, lot 13. With B1r "MEDICEA" cancel slip, with Moon etchings. 22 × 16 cm. 17th-century Italian limp vellum, title inked on spine, edges plain. BOUND WITH: Marco Antonio de Dominis, De radiis visus et lucis (Venice, 1610). FORMER OWNER: Dr. Haskell Norman.

34. New York, New York, Columbia University Library: Smith 520.1610 G13. Without B1r "MEDICEA" cancel slip, with Moon etchings. 21.2 × 17.1 cm. 19th-century marbled boards, green pebbled cloth spine, edges plain. FORMER OWNER: David Eugene Smith (bookplate and stamp).
35. Norman, Oklahoma, University of Oklahoma Library: Special Collections, 520. G133s. [photographs; Kerry Magruder]. With B1r "MEDICEA" cancel slip, with Moon etchings. 24 × 17 cm. Contemporary vellum. FORMER OWNERS: Galileo Galilei (a correction in his hand and presentation inscription to:); Gabriello Chiabrera; Everett L. DeGolyer (gift to University of Oklahoma).
36. Oxford, Bodleian Library: Antiq.e.I.1610.1 [Cristina Dondi]. Without B1r "MEDICEA" cancel slip, with Moon etchings. 21 × 16.3 cm. Contemporary limp vellum, remains of leather ties. FORMER OWNER: All Souls College (given to the Bodleian Library, 1926).
37. Oxford, Bodleian Library: Rigaud e.148 [Cristina Dondi]. With B1r "MEDICEA" cancel slip, with Moon etchings. 21.2 × 15.2 cm. 17th-century Spanish red morocco gilt, supralibros of Felipe Nuñez de Guzman, Duke of Medina de las Torres. BOUND (2nd) WITH: Galileo Galilei, Il saggiatore (Rome, 1623). – Giulio Cesare La Galla, De phoenomenis in orbe lunae … physica disputatio (Venice, 1612). – Francesco Sizzi, *ΔΙΑΝΟΙΑ* Astronomica (Venice, 1611). – John Wedderburn, Quatuor problematum … confutatio (Padua, 1610). – Mario Guiducci, Lettera…al Tarquinio Galluzzi (Florence, 1620). FORMER OWNERS: Duke of Medina de la Torres (as above); Stephen Peter Rigaud; Radcliffe Observatory (stamp)
38. Oxford, Corpus Christi College, Δ.22.1(6) [Cristina Dondi]. Without B1r "MEDICEA" cancel slip, without Moon etchings (Moon images supplied by hand). 20.3 × 15.2 cm. 19th-century sheep spine, buckram boards, yellow-stained edges. BOUND (last) WITH: Leonard Digges, A prognostication euerlasting. 1556 (London, 1556). –John Feild, Ephemerides trium annorum 58. 59. et 60 (London, 1558). – Petrus Apianus, Cosmographia (Cologne, 1564). – Francesco Sizzi, *ΔΙΑΝΟΙΑ* Astronomica (Venice, 1611). – Giulio Cesare La Galla, De phoenomenis in orbe lunae (Venice, 1612). FORMER OWNER: Brian Twyne (the first three works in the Sam melband having been inherited from his father, Thomas Twyne), his bequest to Corpus Christi
39. Padua, Biblioteca Universitaria: 99 b 67 [Carla Lestani]. With B1r "MEDICEA" cancel slip, with Moon etchings. 23.7 × 18.2 cm. Early 17th-century Italian boards. Former owner: Natio Germanica of the University (purchase inscription of 1626: Sumptibus Nationis Germanicae Juridicae facultatis …)
40. Palo Alto, California, John Warnock [facsimile: Octavo CD-ROM]. Without B1r "MEDICEA" cancel slip, with Moon etchings. 17th-century French red morocco gilt, marbled endleaves, apparently a remboîtage
41. Paris, Bibliothèque de l'Arsenal: S 4° 4607. With B1r "MEDICEA" cancel slip, with Moon etchings. 19.7 × 15.5 cm. 17th-century French citron morocco gilt, edges gilt, De Thou supralibros. BOUND (4th) WITH: Heliodorus. Capita opticorum ([Hamburg], 1610). – Johann Kepler. Narratio de observatis … Iovis satellitibus (Frankfurt, 1611). – Idem, Dissertatio cum Sidereo nuncio (Prague, 1610). – Guilielmus Rech-

perger, Astrolabium eiusque usus … explicatus (Augsburg, 1611). – Willebrord Snell, Apollonius Batavus (Leiden, 1608). FORMER OWNER: Jacques-Auguste de Thou (supralibros, corrections in his hand)

42. Paris, Bibliothèque nationale de France: Rés. p.V.743, earlier shelfmark 7700(5). Without B1r "MEDICEA" cancel slip, without Moon etchings. 19.7 × 13.4 cm. Modern vellum, traces of red-sprinkled edges from earlier binding; bifolium D3.4 in photofacsimile. FORMERLY BOUND WITH (broken up in 1975): Galileo Galilei, Discorso … intorno alle cose che stanno in sù l'acqua … seconda editione (Florence, 1612). – Idem, Della scienza mecanica (Ravenna, 1649). – Giovanni Bardi, Eorum quae vehuntur in aquis experimenta (Rome, 1614). – Mario Guiducci, Lettera al M. R. P. Tarquinio Galluzzi (Florence, 1620). – [Flaminio Papazzoni], Considerazioni … intorno alle cose che stanno in su l'acqua (Pisa, 1612). – Giovanni Battista Stelluti, Scandaglio sopra la libra astronomica et filosofica di Lotario Sarsi (Terni, 1622). FORMER OWNER: Cardinal Mazarin
43. Paris, Institut de France: 4 M 713 C. With B1r "MEDICEA" cancel slip, with Moon etchings. 22.8 × 17 cm. 19th-century French marbled boards, sheep spine, red spine label, marbled endleaves, preserving an earlier endleaf, probably from a 17th-century English binding. FORMER OWNER: Sir George Shuckburgh, Bart. (purchase inscription, 1779)
44. Paris, Observatoire de Paris: Inv. 20174 (1). Without B1r "MEDICEA" cancel slip, without Moon etchings. 20.7 × 15 cm. Late 18th-century polished sheep, gilt supralibros (DÉPÔT GÉNÉRALE DES CARTES ET PLANS MARITIMES). BOUND WITH: see next. FORMER OWNER: Jean Dominique Cassini, with his extensive annotations in quires E–G including numbered identifications of the Jupiter moons
45. Paris, Observatoire de Paris: Inv. 20174 (1bis). Without B1r "MEDICEA" cancel slip, with Moon etchings. 20.7 × 15 cm. Late 18th-century polished sheep (see preceding). BOUND WITH: Sidereus Nuncius (preceding). – Galileo Galilei, Istoria e dimostrazioni intorno alle macchie solari (Rome, 1613, imperfect). – Idem, Istoria e dimostrazioni intorno alle macchie solari (Bologna, 1655). FORMER OWNER: Joseph Nicolai Delisle (autograph notes on interleaves, partly incorporating Cassini's notes in preceding copy)
46. Paris, Observatoire de Paris: Inv. 20276 (1). Without B1r "MEDICEA" cancel slip, with Moon etchings. 20.8 × 15.4 cm. 17th-century French vellum, remains of 2 leather ties, edges plain. BOUND WITH: Marco Antonio de Dominis, De radiis visis et lucis (Venice, 1611). – Giulio Cesare La Gallla, De phoenomenis in orbe lunae … physica disputatio (Venice, 1612)
47. Parma, Biblioteca Palatina: Misc. erud. 4° 247/4. [Andrea De Pasquale]. Without B1r "MEDICEA" cancel slip, with Moon etchings. 21.8 × 15.8 cm. Later vellum with ties, marbled-paper endleaves, sprinkled edges. BOUND WITH: three (or more?) unspecified works "of a scientific nature".
48. Pasadena, California, California Institute of Technology: QB41 415 1610a. [Elisa Piccio]. Without B1r "MEDICEA" cancel slip, with Moon etchings. 20.7 × 15.9 cm. Early 20th-century morocco, supralibros of Rocco di Torrepadula, edges gilt. FORMER OWNER: Don Giampaolo Rocco di Torrepadula (supralibros; Cinti 26)

49. Pisa, Biblioteca Universitaria, S.R. 7.50. [photocopy]. Without B1r "MEDICEA" cancel slip, with Moon etchings. FORMER OWNER: Giuseppe Piazzini
50. Princeton, New Jersey, Institute for Advanced Studies, Lessing J. Rosenwald History of Science Collection. Without B1r "MEDICEA" cancel slip, with Moon etchings. 20.1 × 15 cm. 17th-century vellum, parcel-gilt arabesque centerpiece including MH monogram, edges sprinkled red & blue. BOUND WITH: Johannes Kepler, Narratio de observatis ... Iovis satellitibus (Frankfurt, 1611). – Francesco Sizzi, *ΔΙΑΝΟΙΑ* Astronomica (Venice, 1611). – Christoph Scheiner, De maculis solaribus ... ad Marcum Velserum (Augsburg, 1612). – Christoph Scheiner, Tres epistolae de maculis solaribus (Augsburg, 1612). – Simon Marius, Mundus Iovialis (Nuremberg, 1614). FORMER OWNERS: Chorinsky family of Schloss Wessely a.d. March, including Ignaz Dominic Chorinsky von Ledske, 1729–1790 (supralibros, bookplate), auctioned Hamburg, 17 June 1930; Herbert M. Evans
51. Providence, Rhode Island, Brown University, John Hay Library: QB41 G.48x 1610. With B1r "MEDICEA" cancel slip, with Moon etchings. 20.9 × 14.8 cm. 20th-century vellum, plain edges. FORMER OWNERS: Galileo Galilei (corrections in his hand); Vincenzo Viviani (signature); Albert E. Lownes (bookplate; purchased from John F. Fleming, who acquired at Sotheby's 30 May 1960, lot 110)
52. Ravenna, Biblioteca Classense: 51.4.L.2 [Claudia Giuliani]. Without B1r "MEDICEA" cancel slip, with Moon etchings. 22.4 × 16.5 cm. 18th-century limp vellum. Former owners: Giovan Battista Capponi; Camillo Morigia
53. Rome, Biblioteca Universitaria Alessandrina: XIV.f.4.7. Without B1r "MEDICEA" cancel slip, with Moon etchings. 21.2 × 15.8 cm. Modern wrapper, disbound from earlier binding with red-spotted edges, stamped on t–p. as "7" and with an older ink "7" on title-page, apparently from its position in a now-dispersed Sammelband
54. Rome, Biblioteca Angelica: Rari I.5.15. With B1r "MEDICEA" cancel slip, with Moon etchings. 22.6 × 16.3 cm. Limp vellum, ca. 1700, edges blue-stained
55. Rome, Biblioteca Nazionale Centrale: 69.4.C.118. With B1r "MEDICEA" cancel slip, with Moon etchings. 20.8 × 15.3 cm. 17th-century Italian limp vellum, edges stained light yellow-brown, yellow silk endbands. BOUND WITH: Johann Kepler, Dissertatio cum Nuncio sidereo (Florence, 1610). FORMER OWNERS: Alfonso Borelli (Ex lib. Alph. Borellj); Piarist School, Rome (Biblioth. Schol: Piarum Sancti Pantal.")
56. Rome, Biblioteca Nazionale Centrale: 34.2.I.18. Without B1r "MEDICEA" cancel slip, with Moon etchings. 17.7 × 14.2 cm. Early 18th-century vellum over pasteboard, blue-sprinkled edges. BOUND WITH: Various religious tracts, including Delphinatus, et Calabria Ulterior sanctificata (Rome, 1703); Collegii Romani obsequia Clementi Undecimo ... Exhibita Anno M D CC III. (Rome, 1703), and others printed Torino, 1698; Rome, 1696; Rome, 1699
57. Rome, Osservatorio Astronomico di Roma: VII.D.7.6. Without B1r "MEDICEA" cancel slip, with Moon etchings. 21.9 × 15.5 cm. 19th-century marbled paper, vellum spine and tips, edges plain, smooth blue-violet paper endleaves. FORMER OWNER: A. Wolynski (stamp; also of Museo Copernicano, Roma)
58. San Marino, California, Huntington Library: HEH 487000:71 (copy 1). [Stephen Tabor]. Without B1r "MEDICEA" cancel slip, with Moon etchings. 23.4 × 16 cm. 17th-

century (?) limp vellum. FORMER OWNERS: Giorgio Pernicelli, ex-dono Bartolommeo Pernicelli, 1628 (inscription); Jesuit College, Reggio Emilia (inscription); George E. Hale; Mount Wilson Observatory

59. San Marino, California, Huntington Library: HEH 701395 (copy 2: Burndy Library). [Stephen Tabor]. Without B1r "MEDICEA" cancel slip, with Moon etchings. 23.9 × 17.3 cm. 19th-century(?) limp vellum, earlier blue paper wrappers bound in, blue-stained edges. FORMER OWNER: Vito Volterra (stamps of the Biblioteca Nazionale Centrale, Roma, applied when that collection was deposited); Burndy Library (donated to Huntington Library, 2008)
60. Schenectady, New York, Dudley Observatory QB41.G15, housed at Schaffer Library, Union College. [Marlaine J. DesChamps]. Without B1r "MEDICEA" cancel slip (but "Medicea" supplied by hand), with Moon etchings. 20.5 × 15 cm. Plain paper boards, edges plain; printed ticket (ca. 1850s?) of B. Westermann & Co., 290 Broadway, New York.
61. St Andrews, University Library: For QB41.G2S4 [photocopy; Daryl Green]. Without B1r "MEDICEA" cancel slip, without Moon etchings. 24.1 × 18.6 cm. Marble boards, vellum spine (1965), preserving an earlier wrapper of engraved plates relating to the optics of telescopes. FORMER OWNERS: Caspar Marschalck, parochus of Leinach(?), 1646 (deleted inscription); Benedictines of Neustadt am Main; Williams & Norgate (booksellers, London and Edinburgh: ticket); James David Forbes (gift of his son George Forbes, F.R.S., to St Andrews, 1929)
62. Stockholm, Kungliga Biblioteket: RAR 106A [Wolfgang Undorf]. With B1r "MEDICEA" cancel slip, with Moon etchings. 22 × 16.4 cm. 19th-century red-brown marbled boards, leather spine & tips, gilt-lettered on spine "Galilaei Siderus Nuncius.", edges red-sprinkled, earlier shelfmark C.20.7
63. Stockholm, Universitetsbibliotek: MAG MN R 85. [Clas-Ove Strandberg]. With B1r "MEDICEA" cancel slip, with Moon etchings. 21.9 × 16.7 cm. Contemporary vellum, traces of green silk ties, edges plain, title written on spine. FORMER OWNERS: Jacob Concius (signature and engraved bookplate M. Jacobus Concius Archipresb: Memelen: 1705); Jacob von Eggers; Royal Swedish Academy of Sciences (after 1768, before 1783; deposited at University Library, 1984). Early annotations and corrections, probably preceding Concius' ownership
64. Tucson, Arizona, University of Arizona: Special Collections (Vault) QB41 .G17 1610. [Verónica Reyes-Escudero]. Without B1r "MEDICEA" cancel slip, with Moon etchings. 22.5 × 17.2 cm. Early 19th-century green-stained sheep, gilt turn-ins, sprinkled edges, marbled endleaves. FORMER OWNER: Galileo Galilei (corrections in his hand). Acquired by University 1977–1978
65. Turin, Biblioteca Nazionale Universitaria: Q.V. 144 [Lucia Frattarolo]. Without B1r "MEDICEA" cancel slip, with Moon etchings. 23 × 16.5 cm. Contemporary limp vellum, edges plain. FORMER OWNERS: Dukes of Savoy
66. Uppsala, Universitet, Astronomiska Observatoriet: no shelfmark. [Laila Österlund]. With B1r "MEDICEA" cancel slip, with Moon etchings. 22 × 16 cm. 17th-century limp vellum, red-brown sprinkled edges. BOUND WITH: Johann Kepler, Dissertatio cum Nuncio Sidereo (Florence, 1610). – Franc. Sitius, Dianoia astronomica, optica,

physica … (Venice, 1611). FORMER OWNER: Anders Celsius (probable owner); Olof Hjorter

67. Venice, Biblioteca Nazionale Marciana, Misc. 119.10 [Horst Bredekamp]. Formerly with B1r "MEDICEA" cancel slip (paste stain remains), with Moon etchings. 26.5–27 × 19.3–19.4 cm (uncut). FORMER OWNER: Marco Contarini
68. Vicenza, (late) Giancarlo Beltrame. [Andrea Ditti]. Without B1r "MEDICEA" cancel slip, with Moon etchings. 21.9 × 16.7 cm. Old rose-paper boards with ties. Acquired 1984
69. Virginia, Private Collection. Without B1r "MEDICEA" cancel slip, with Moon etchings. 22.4 × 16.7 cm. Old vellum (re-used: bound c. 1950), edges plain. Former owner: Robert B. Honeyman, sold Sotheby's London, 5 November 1979, lot 1938, and again Christie's New York, 8 June 1990, lot 288
70. Washington, D.C., Smithsonian Institution, Dibner Library of the History of Science and Technology: QB41.G15 SCDIRB. [Kirsten van der Veen]. With B1r "MEDICEA" cancel slip, with Moon etchings. 22.8 × 17 cm. 17th-century vellum, resewn with nrenewed endleaves. FORMER OWNERS: Herbert M. Evans; Bern Dibner (donated to the Smithsonian Institution, 1974)
71. Washington, D.C.: Library of Congress. Without B1r "MEDICEA" cancel slip, with Moon etchings. 26.3 × 19.4 cm (uncut). 17th-century stiff gray pasteboard wrapper with flap. Former owner: Domenico Fratei (?), Verona, 17th century (Ex libris dominici Fratei Veron …)
72. Wellesley, Massachusetts, Wellesley College: Acc. no. 110606. Without B1r "MEDICEA" cancel slip, with Moon etchings. 25 × 17.8 cm, in original pasteboard wrapper, sewn on 3 thongs laced into boards. FORMER OWNERS: RB (18th-century Italian monogram stamp); Louis M. Derr (purchased by Wellesley from his widow, 1924)
73. Williamstown, Massachusetts, Prof. Jay M. Pasachoff. [Wayne Hammond]. Without B1r "MEDICEA" cancel slip, with Moon etchings. Contemporary vellum, front pastedown printed waste, perhaps of an incunable. D5 supplied from a copy of the fine-paper issue. BOUND WITH: Francesco Sizzi, *ΔIANOIA* Astronomica (Venice, 1611). – Marco Antonio de Dominis, De radiis visus et lucis (Venice, 1611). – Giov. Ant. Magini, Breve instruttione sopra l'apparenze, et mirabili effetti dello Specchio concavo sferico (Bologna, 1611). Formerly Sotheby's London 19 July 1984, lot 253
74. Zurich, Eidgenössische Technische Hochschule (ETH), Rar. 4342:1 [photographs; Franziska Geisser]. Without B1r "MEDICEA" cancel slip, without Moon etchings. 20 × 15.5 cm. 17th-century vellum with renewed ties, blue-stained edges. BOUND WITH: Johann Kepler, Strena seu de niue sexangula (Frankfurt, 1611). – Marco Antonio de Dominis, De radiis visus et lucis (Venice, 1610). – Francesco Sizzi, *ΔIANOIA* Astronomica (Venice, 1611). – Johann Kepler, Phaenomenon singulare seu Mercurius in Sole (Leipzig, 1609). –. Idem, Dissertatio cum Nuncio sidereo (Prague, 1610). – Idem, Ad rerum coelestium amatores universos … De Solis deliquio (Prague, 1605). FORMER OWNERS: Winandus Reuthel (?), vicarius of St. Gervinus, Cologne, 1658; Matthias Holtzenius (?), Ph. and M.D., purchased Cologne, 1680.

Fine paper copies

F1. Florence, Biblioteca Nazionale Centrale: Postillato 110. With B1r "MEDICEA" cancel slip, with Moon etchings. 22.4 × 16.1 cm. 17th-century Italian limp vellum, edges plain, endbands of orange and green silk over rawhide cores. With corrections in Galileo's autograph. FORMER OWNER: Extensive early marginal annotations on fos. 6r, 6v, 7r, 7v, 8r, 8v, 10v, 11r, 11v, 12r, 12v, 13r, 13v, 14r, 14v, 15r, 15v, 16r.

F2. Hiroshima University of Economics: no shelfmark [photographs; Hideharu Nishikawa]. With B1r "MEDICEA" cancel slip, with Moon etchings. 21.5 × 15.5 cm. 17th-century Italian vellum. With corrections in Galileo's autograph. BOUND WITH: John Wedderburn, Quatuor problematum quae Martinus Horky contra Nuntium sidereum de quatuor planetis novis distputanda proposuit, confutatio (Padua, 1610). FORMER OWNER: Niccolini family, Florence (stamp: P. G. N.)

F3. Kansas City, Missouri, Linda Hall Library of Science, Engineering & Technology: QB41.G15 1610a. With B1r "MEDICEA" cancel slip, with Moon etchings. 22.1 × 16.5 cm. 17th-century Italian limp vellum, edges plain. With corrections in Galileo's autograph. FORMER OWNERS: Benedictine Convent, Mantua? (inscription, "Ct. [?] S. Bened. Mant.ni"); A. Cristofori, Mantua; British Rail Pension Fund.

F4. Munich, Bayerische Staatsbibliothek: Res. 4° Astr. p. 313 [photographs; Bettina Wagner]. With B1r "MEDICEA" cancel slip, with Moon etchings. 21.5 × 16.2 cm. 17th-century plain vellum (Munich, presumably), two fore-edge slits for silk ribbon ties. With corrections in Galileo's autograph. FORMER OWNER: Bibliotheca electoralis, Munich (bookplates).

F5. Münster in Westfalen, Universitäts- und Landesbibliothek: 4° M4 278 So. Mag. [photographs; Horst Bredekamp]. With B1r "MEDICEA" cancel slip, with Moon etchings. 21.3 × 16.4 cm. 17th-century German vellum over pasteboard, decorative border in blind, red-stained edges, remains of two fore-edge leather ties. With corrections in Galileo's autograph. BOUND WITH: Philippus Apianus, De utilitate trientis, instrumenti astronomici novi (Tübingen, 1586). FORMER OWNER: Sebastianus Menge, Premonstratensian of Wedinghausen.

F6. Naples, Biblioteca Nazionale: S.Q. XXXI B 8. [Silvana Acanfora]. Without B1r "MEDICEA" cancel slip; with Moon etchings. 21.8 × 16 cm (D5 folded in to preserve the Orion woodcut). With corrections in Galileo's autograph. FORMER OWNER: Dominican convent (Santa Catarina a Formiello), Naples.

F7. Stanford, California, Stanford University: BARCH QB41 G148 1610. [photographs; John Mustain]. Without B1r "MEDICEA" cancel slip (perhaps formerly present: the word COSMICA eradicated), with Moon etchings. 21.3 × 16 cm. 17th-century Italian vellum gilt with arabesque interlace centerpiece. With corrections in Galileo's autograph. FORMER OWNERS: Marchese Giacomo Filippo Durazzo; Franklin Institute, Philadelphia; Samuel and Cecile Barchas.

F8. Strasbourg, Université Louis Pasteur: H 109.498 [photographs]. With B1r "MEDICEA" cancel slip, with Moon etchings. ca. 20.4 × 19.5 cm. With corrections in Galileo's autograph. FORMER OWNER: Faculté de Médécine, Strasbourg (stamp).

F9. Vienna, Österreichische Nationalbibliothek: 72.J.106. With B1r "MEDICEA" cancel slip, with Moon etchings. 20.9 × 15.2 cm. 19th-century marbled boards, sprinkled edges. With one correction in Galileo's autograph.

Copies recorded but not examined:

Seven additional copies may be briefly noted:

(1) Property of an American private collector, with whom I have had no contact; leaf dimensions 22 × 16.5 cm, the binding a remboîtage of 17th-century calf; with the Moon etchings but lacking the "MEDICEA" cancel slip on B1r.

(2) The former Prince Boncompagni copy, auctioned in Rome, 1898, Boncompagni sale (Catalogo della Biblioteca Boncompagni ...), pt. IV, lot 912. BOUND WITH Francesco Sizzi, *ΔIANOIA* Astronomica (Venice, 1611). – Johann Kepler. Narratio de observatis ... Iovis satellitibus (Frankfurt, 1611). As no so-far recorded copy above contains just these three editions, the Boncompagni copy may be lost; or may survive, as yet unrecorded; or its volume may have been broken up, with its Sidereus Nuncius one of those recorded above but unrecognized as such.

(3) The former Robert S. Dunham copy, sold Sotheby's New York, 11 December 1993, lot 679, 22.9 × 15.9 cm, modern vellum; with the Moon etchings but lacking the "MEDICEA" cancel slip on B1r; now belonging to an American private collector, with whom I have had no contact.

(4) The former Earls of Macclesfield copy, sold Sotheby's London 4 November 2004, lot 830; with the Moon etchings, no information on the "MEDICEA" cancel slip, much cut down in a late Sammelband with five other works dated between 1650 and 1710.

(5) A copy sold Sotheby's London, 19 February 1979, lot 107: in contemporary vellum, with the Moon etchings, no information on the "MEDICEA" cancel slip (conceivably identical with a copy already listed above).

(6) Rovigo, Accademia dei Concordi, shelfmark not known.

(7) Rovereto, Biblioteca Rosminiana, shelfmark A.1.4.

Two copies cited by Grassi 1989, p. 272, are ghosts: Bologna, Osservatorio Astronomico (not listed in the online catalogue of its library), and Hamburg, Hamburger Sternwarte (Anke Vollersen has verified that there is no copy there). Bruni & Evans 2316 cite a copy at Emmanuel College Cambridge, which possesses rather a copy of the 1610 Frankfurt am Main octavo reprint.

Former owners

B R / J B R (?), 18th-century Italian crowned monogram stamp: 72

Adams, John Couch, 1819–1892, astronomer, Cambridge, co-discoverer of Neptune: 9

Barchas, Stanley & Cecile, American collectors of history of science, their library now at Stanford University: F7

Beltrame, Giancarlo, d. 2011, Vicenza, collector of history of science: 68

Dibner, Bern, 1897–1988, electrical engineer, collector of history of science, founder of the Burndy Library, divided between the Smithsonian Institution, Washington, D.C., and the Huntington Library, San Marino, Cal.: 70

DILLINGEN, Jesuit College, gift of Heinrich von Knöringen, Bishop of Augsburg, q.v.: 17

Dini, Pietro, c. 1550–1625, correspondent of Galileo, Archbishop of Fermo 1621 (armorial stamp): 8

Dunham, Robert S., d. 1991, New York patent lawyer: unexamined copy sold Sotheby's New York, 11 December 1993

Durazzo, Giacomo Filippo, Marchese di Gabiano, 1729–1812 (bookplate): F7

Eggers, Jacob von, 1704–1773, member of Royal Swedish Academy of Sciences: 63

Evans, Herbert McLean, 1882–1971, M.D., American medical researcher, collector of history of science: 12, 50, 70

FLORENCE, Bibliotheca Palatina: 20

Forbes, James David, 1809–1868, Scottish physicist and geologist, professor at Edinburgh and St Andrews (his science library presented to St Andrews in 1929 by his son George Forbes, F.R.S.): 61

Fratei (?), Domenico, of Verona, 17th century: 71

Frederik III, 1609–1670, King of Denmark: 14

Galilei, Galileo, 1564–1642: the SNML proof copy (inscribed by him), 7 (presented to unknown recipient), 35 (presented to Gabriello Chiabrera), 51 (with autograph corrections and emendations), 64 (with autograph corrections); and, probably, all the fine-paper copies

Gooch, Henry, B.A. Cambridge 1600/1 (M.A. 1604, B.D. 1620, D.D. 1624), Fellow of Trinity College, traveller to Persia, former owner of several medieval manuscripts now at Emmanuel College Cambridge and Cambridge University Library: 24

GRAZ, Jesuit College: 22

Green, Richard, M.D., American physician and collector of history of science, his books sold Christie's New York, 17 June 2008: 32

Gude, Marquard, 1635–1689, of Schleswig; classical scholar and antiquary, librarian to Herzog Christian Albert von Holstein-Gottorp: 16

Guldin, Paul, SJ, 1577–1643, mathematician, Graz, correspondent of Galileo and Kepler (inscription): 22

Hagemeijer, E. A. A. (a red ink stamp, so identified). Also with this volume is a loose bookplate "Ex libris Joannes [sic] Theodorus Hagemeijer et ex libris sui filii Adrianus [sic] Hagemeijer", date uncertain. As the volume was given to Christ's College Cambridge in the nineteenth century by Charles Lesingham Smith, q.v., it is doubtful whether the Hagemeijer ownership is authentic: 7

Hale, George Ellery, 1868–1938, American astronomer, founder of Mount Wilson Observatory: 58

Hjorter, Olof, 1696–1750, Observator regius, Uppsala University, brother-in-law of Anders Celsius: 66

Holtzenius (?), Matthias, Philosophiae et Medicinae Doctor, 17th century (emptus Coloniae 1680): 74

Sidereus neighbors: works bound or formerly bound with copies of sidereus Nuncius

Apianus, Philipppus (1531–1589). De utilitate trientis, instrumenti astronomici novi, libellus. Tübingen: n.pr., 1586. 4°: F5

Bardi, Giovanni (1534–1612). Eorum quae vehuntur in aquis experimenta a Ioanne Bardio Florentino ad Archimedis trutinam examinata. Rome: Ex typographiae Bartholomaei Zannetti, 1614. 4°: 42

Boodt, Anselm Boèce de (1550–1632). Gemmarum & lapidum historia ... opus principibus, medicis, chymicis, physicis, ac liberalioribus ingeniis vtilissimum. Hanau: Typis Wechelianis apud C. Marnium & heredes J. Aubrii, 1609. 4°: 3

Bornitz, Jakob (ca. 1560–1625). De Nummis in Repub. percutiendis & conservandis Libri Duo. Ex Systemate politico deprompti. Hanau: Typis Wechelianis, apud Claudium Marnium & heredes Ioannis Aubrii, 1608. 4°: 3

Castelli, Benedetto (1578–1643). Risposta Alle Oppositioni Del Sig. Lodovico Delle Colombe E Del Sig. Vincenzo di Gratia, Contro al Trattato Del Sig. Galileo Galilei, Delle cose che stanno su l'Acqua. Bologna: Per gli HH. del Dozza, 1655. 4°: SNML proof copy

Colombe, Lodovico delle (1565–1616). Discorso Apologetico di Lodovico Delle Colombe. D'Intorno Al Discorso Del S. Galileo Galilei, Circa le cose, che stanno sù l'Acqua ... Bologna: Per gli HH. del Dozza, 1655. 4°: SNML proof copy

Contarini, Vincenzo (1570–1619). De frumentaria Romanorum largitione liber in quo ea praecipue, quae sunt a Iusto Lipsio cum in electis, tum in admirandis de eadem prodita, examinantur. Eiusdem De militari Romanorum stipendio commentarius. Venice: apud Nicolaum Polum, 1609. 4°: 3

Dal Monte, Guidobaldo, Marchese (1545–1607). Planisphaeriorum uniuersalium theorica. Pesaro: Apud Hieronymum Concordiam, 1579. 4°: 21

Della Porta, Giambattista (1535–1615). Coelestis physiognomiae libri 6. Naples: Subtilis, 1603. 4°: 21

De Dominis, Marco Antonio (1560–1624). De radiis visus et lucis in vitris perspectivis et iride tractatus ... per Joannem Bartolum in lucem editus. In quo inter alia ostenditur ratio Instrumenti cuiusdam ad clarè videndum, quae sunt valde remota excogitati. Venice: Apud Thomam Baglionum, 1611. 4°: 33, 46, 73, 74

Ferrari, Filippo (1551–1626). Nova topographia in Martyrologium Romanum. Venice: apud Bernardum Iuntam, Io. Baptistam Ciottum, & socios, 1609. 4°: 3

Ferrerio, Giovanni Stefano (1568–1610). S. Eusebij Vercellensis Episc. et Martyris ... vita et res gestae. Rome: Apud Aloysium Zannettum, 1602. 4°: 3

Galilei, Galileo (1564–1642). Il saggiatore nel quale con bilancia esquisita e giusta si ponderano le cose contenute nella Libra astronomica e filosofica di Lotario Sarsi Sigensano scritto in forma di lettera all' Illmo. et Reuermo. Monsre. D. Virginio Cesarini. Rome: Giacomo Mascardi, 1623. 4°: 37

Galilei, Galileo (1564–1642). Istoria e dimostrazioni intorno alle macchie solari ... Rome: Appresso Giacomo Mascardi, 1613. 4°: 8, 45

Galilei, Galileo (1564–1642). Discorso al serenissimo don Cosimo II. Gran Duca di Toscana intorno alle cose, che stanno in sù l'acqua,ò che in quella si muouono. Florence: Appresso Cosimo Giunti, 1612. 4°: 27

Galilei, Galileo (1564–1642). Discorso al serenissimo don Cosimo II. Gran Duca di Toscana intorno alle cose, che stanno in sù l'acqua,ò che in quella si muouono, seconda editione. Florence: Appresso Cosimo Giunti, 1612. 4°: 42

Galilei, Galileo (1564–1642). Discorso al serenissimo don Cosimo II. Gran Duca di Toscana intorno alle cose, che stanno in sù l'acqua,ò che in quella si muouono. Bologna: Per gli HH. del Dozza, 1655. 4°: SNML proof copy

Galilei, Galileo (1564–1642). Della scienza mecanica e delle utilità che si traggono da gl' istromenti di quella, opera cavata da manoscritti dell' eccelentissimo matematico Galileo Galilei, dal cavaliere Luca Danesi. Ravenna: Appresso gli Stamp. Camerali, 1649. 4°: 42

Gratia, Vincenzo di (fl. c. 1600–1630). Considerationi ... Sopra Il Discorso del Sig. Galileo Galilei Intorno alle cose che stanno sù l'acqua.... Bologna: Per gli HH. del Dozza, 1655. 4°: SNML proof copy

Guiducci, Mario (1585–1646). Lettera al M. R. P. Tarquinio Galluzzi ... Nella quale si giustifica dell'imputazion dategli da Lottario Sarsi Sigensano nella Libra astronomica, e Filosofica. Florence: Nella Stamperia di Zanobi Pignoni, 1620. 4°: 37, 42

Heliodorus of Larissa (3rd century A.D.?). Capita opticorum. Ad Hetrusci codicis fidem graece et latina edita & recensita. Ex bibliotheca Fr. Lindenbrogi. [Hamburg:] In librario Heringiano, 1610. 4°: 41

Kepler, Johann (1571–1630). Ad rerum coelestium amatores universos ... De Solis deliquio, quod hoc anno 1605. mense Octobri contigit, epistola. Prague: E Typographîo Schumaniano [1605]. 4°: 74

Kepler, Johann (1571–1630). Dioptrice seu Demonstratio eorum quæ visui & visibilibus propter Conspicilla non ita pridem inventa accidunt. Præmissa Epistolæ Galilæi de ijs, quæ post editionem Nuncij siderij ope Perspicilli, nova & admiranda in cœlo deprehensa sunt. Augsburg: typis Davidis Franci, 1611. 4°: 27

Kepler, Johann (1571–1630). Dissertatio cum Sidereo nuncio nuper ad mortales misso à Galilaeo Galilaeo Mathematico Patavino. Prague: Typis Danielis Sedesani, 1610. 4°: 27, 41, 74

Kepler, Johann (1571–1630). Dissertatio cum Sidereo nuncio nuper ad mortales misso à Galilaeo Galilaeo Mathematico Patavino. Florence: Apud Io. Antonium Caneum, 1610. 4°: 30, 31, 55, 66

Kepler, Johann (1571–1630). Narratio de observatis a se quatuor Iouis satellitibus erronibus, quos Galilaeus Galilaeus mathematicus Florentinus iure inuentionis Medicaea sidera nuncupauit. Cum adiuncta dissertatione de Nuncio sidereo nuper ad mortales misso. Frankfurt am Main: Sumptibus Zachariae Palthenii, 1611, 4°: 41, 50, former Boncompagni copy

Kepler, Johann (1571–1630). Narratio de observatis a se quatuor Iouis satellitibus erronibus, quos Galilaeus Galilaeus mathematicus Florentinus iure inuentionis Medicaea sidera nuncupauit. Cum adiuncta dissertatione de Nuncio sidereo nuper ad mortales misso. Florence: Apud Cosmam Iunctam, 1611. 4°: 30

Kepler, Johann (1571–1630). Phaenomenon singulare seu Mercurius in Sole. Cum digressione de causis, cur Dionysius Abbas Christanos minus iusto à Nativitate Christi Domini numerare docuerit: De capite & anni Ecclesiastici. Leipzig: Impensis Thomae Schureri Bibliopolae, 1609. 4°: 74

Kepler, Johann (1571–1630). Strena Seu De Niue Sexangula. Frankfurt am Main: apud Godefridum Tampach, 1611. 4°: 74

La Galla, Giulio Cesare (1576–1624). De phoenomenis in orbe lunae noui telescopij usu a d. Galileo Galileo nunc iterum suscitatis physica disputatio. Venice: Apud Thomam Baglionum, 1612. 4°: 30, 37, 38, 45

Magini, Giovanni Antonio (1555–1617). Breve instruttione sopra l'apparenze, et mirabili effetti dello specchio concauo sferico. Bologna: Presso Gio. Battista Bellagamba, 1611. 4°: 73

Marius (Mayr), Simon, Guntzenhusanus (1573–1624). Mundus Iovialis anno M.DC.IX. detectus ope perspicilli Belgici, Hoc est, quatuor Jovialium planetarum, cum theoria, tum tabulæ, propriis observationibus maxime fundatæ, ex quibus situs illorum ad Iovem, ad quodvis tempus datum promptissimè & facilimè supputari potest. Nuremberg: Sumptibus & Typis Iohannis Lauri, 1614, 4°: 8, 50

[Pannocchieschi d'Elci, Arturo, ca. 1564–1614]. Accademico Incognito. Considerazioni sopra il discorso del sig. Galileo Galilei intorno alle cose che stanno in sù l'acqua … fatte a difesa, e dichiarazione dell'opinione d'Aristotile da Accademico incognito. Pisa: Appresso Gio. Batista Boschetti, & Giouanni Fontani, 1612. 4° (the dedication to Maria Magdalena, Archduchess of Tuscany, signed, as translator, by "Arturo Pannochieschi de'Conti d'Elci Pr."[1]): 42

Rechperger, Guilielmus (fl. 1607–1629). Astrolabium eiusque usus tam astronomicus quam geometricus, breuibus dilucidis ac facilibus regulis explicatus. Augsburg: Typis Davidis Franci, 1611. 4°: 41

1 This work has been diversely attributed: see Camerota 2001 for arguments in favor of Panocchieschi, provveditore generale of the University of Pisa.

[Scheiner, Christoph, S.J. (1575–1650)]. Apelles … vel si mavis Ulysses. De maculis solarib. et stellis circa Iouem errantibus, accuratior disquisitio, ad Marcum Velserum … Perscripta. Augsburg: Ad insigne pinus, Idib. Septembr. 1612, 4°: 50

[Scheiner, Christoph, S.J. (1575–1650)]. Tres epistolae de maculis solaribus scriptae ad Marcum Velserum … cum obseruationum iconismis. Augsburg: Ad insigne pinus, Non. Ian. 1612, 4°: 50

Sitius (Sizzi), Franciscus (d.1618). *ΔΙΑΝΟΙΑ* astronomica, optica, physica, Qua Syderei Nuncij rumor de Quatuor Planetis à Galilaeo Galilaeo Mathematico Celeberrimo recens perspicilli cuiusdam ope conspectis, vanus redditur. Venice: Apud Petrum Mariam Bertanum, 1611, 4°: 8, 27, 37, 38, 50, 66, 73, 74, former Boncompagni copy

Snellius, Wilebrordus (1580–1626). Apollonius Batavus, seu exsuscitata Apollonii Pergaei *περι ΔΙΟΡΙΣΜΕΝΗΣ ΤΟΜΗΣ* geometria. Leiden: Excudebat Iohannes a Dorp, Prostant apud Iohannem Maire, 1608. 4°: 41

Stelluti, Giovanni Battista (1577–1643). Scandaglio sopra la libra astronomica et filosofica di Lotario Sarsi Nella controuersia delle Comete e particolarmente delle tre vltimamente vedute l'Anno 1618. Terni: Apresso Tomasso Guerrieri, 1622. 4°: 42

Wodderbornius, Joannes, Scottobritannus (John Wedderburn, 1583–1651). Quatuor problematum quae Martinus Horky contra Nuntium sidereum de quatuor planetis novis distputanda proposuit, confutatio. Padua: Ex typographia Petri Marinelli, 1610. 4°: 37, F2

Acknowledgements

I have received help and generosity at every turn while studying *Sidereus Nuncius*. For continuous moral support as well as good advice and information I am deeply grateful to Richard Lan, a phone call from whom set the investigations underway; to Horst Bredekamp, whose generosity of knowledge and spirit has been beyond measure; and to the dedicatee of this work. These constitute the "without whom ..." triad. David L. Vander Meulen's contribution is only partially reflected in our joint authorship of chapter XI. There is no one better to consult on matters of bibliographical analysis, and when I first sent him a small inquiry, I felt as clever as Tom Sawyer whitewashing the fence, knowing well what would follow. A brilliant result by Irene Brückle and Manfred Mayer led me to a complete re-examination of the paper of the Martayan Lan copy, and Wolfgang Undorf's notes on a damaged type I had overlooked similarly was the catalyst of a new analysis. Nicholas Wilding graciously shared with me his important recent researches into *Sidereus Nuncius*. I owe special thanks to an anonymous American private collector who made it possible for me to carry out studies in Florence, Rome, and Paris. Without the opportunity to study the originals of Galileo's drafts in Florence, a central theme of this work could never have been satisfactory. For the design and production of the book I am deeply indebted to Petra Florath, Glenn Vincent Kraft, and Alexis Ruccius. For permission to reproduce images of *Sidereus Nuncius* and of its related autographs, special thanks go to the Biblioteca Nazionale Centrale, Florence; the Institute for Advanced Study, Princeton, N.J.; the Linda Hall Library of Science, Engineering and Technology, Kansas City, Mo.; the Special Collections Library, University of Michigan, Ann Arbor, Mich.; and the Houghton Library, Harvard University, Cambridge, Mass., as well as to the firm of Martayan Lan, New York City. The footnotes record many intellectual obligations, but are inadequate in expressing how much I have learned from Isabelle Pantin, whose annotated editions of *Sidereus Nuncius* and of Kepler's responses are the foundation stones of *Sidereus Nuncius* studies.

Many others have provided invaluable assistance, both at libraries I visited and those to which I sent inquiries. Among these sharp-eyed helpers are, with apologies to anyone whose contributions I may have overlooked: Djoeke van Netten (Amsterdam); Peggy Daub (Ann Arbor, Mich.); Günter Hägele (Augsburg); Volker Hartmann and Rieke Wolper (Berne); Ben Weiss (Boston); Premysl Bar (Brno); Amy Pickard (Buffalo, N.Y.); David McKitterick and Jill Whitelock (Cambridge); Owen Gingerich, Hope Mayo and William P. Stoneman (Cambridge, Mass.); Edward Tufte (Cheshire, Conn.); Julie Gardner (Chicago);

Hermann-Josef Eschbach (Cologne); Torsten Schlichtkrull (Copenhagen); Joanna Pypłacz (Cracow); Rüdiger May (Dillingen); Alistair Fraser (Durham); Brian Hillyard and Karen Moran (Edinburgh); Neil Harris, Piero Scapecchi and Isabella Truci (Florence); Helmut Rohlfing (Göttingen); Manfred Mayer (Graz); Hideharu Nishikawa (Hiroshima); Katherine Reagan and David W. Corson (Ithaca, N.Y.); Bruce Bradley and Cindy Rogers (Kansas City, Mo.); Carmélia Opsomer (Liège); Rupert Baker, Stephen Parkin, and especially Cristina Dondi, who helped in a multitude of ways (London); Agnese Mandrino (Milan); Bettina Wagner (Munich); Silvana Acanfora (Naples); Robert Babcock (formerly New Haven, Conn.); Gerald W. Cloud and Jonathan A. Hill (New York City); Kerry Magruder (Norman, Okla.); Tiziana Bascelli, Carla Lestani and William Shea (Padua); John E. Mustain (Palo Alto, Cal.); Laurence Bobis, Isabelle de Conihaut, Dominique Coq, Martine Lefèvre, Nicolas Petit and Mireille Pastoureau (Paris); Andrea De Pasquale (formerly Parma); Shelley Irwin and Elisa Piccio (Pasadena, Cal.); Roel Muñoz, AnnaLee Pauls and Marcia Tucker (Princeton, N.J.); William Monroe (Providence, R.I.); Claudia Giuliani (Ravenna); Marinella Calisi, Fiammetta Terlizzi and especially Marina Venier, whose help went far beyond the call of duty (Rome); Daniel Lewis and Stephen Tabor (San Marino, Cal.); Marlaine J. DesChamps (Schenectady, N.Y.); Daryl Green (St Andrews); Clas-Ove Strandberg and Wolfgang Undorf (Stockholm); Verónica Reyes (Tucson, Ariz.); Lucia Frattarola (Turin); Laila Österlund (Uppsala); Albert van Helden (Utrecht); Andrea Ditti (Vicenza); Konstanze Mittendorfer (Vienna); Julia Blakely, Daniel de Simone and Kirsten van der Veen (Washington, D.C.); Ruth Rogers (Wellesley, Mass.); Wayne Hammond and Jay M. Pasachoff (Williamstown, Mass.); Franziska Geisser (Zurich)

Works Cited

Abbreviations

BNCF	Florence, Biblioteca Nazionale Centrale
DBI	*Dizionario biografico degli Italiani* (Rome, 1960–)
EV17	see Griffante
Opere	Galileo Galilei, *Opere*

Ahmanson-Murphy (2001)
The Aldine Press, Catalogue of the Ahmanson-Murphy Collection of Books by or relating to the Press in the Library of the University of California, Los Angeles incorporating Works recorded Elsewhere (Berkeley and Los Angeles)

Barberi, Francesco (1969)
Il Frontespizio nel libro italiano del Quattrocento e del Cinquecento (2 vols., Milan)

Barbieri, Edoardo (2005)
"Una prassi correttoria della tipografia manuale: il cartiglio incollato," *La Bibliofilía* 107, pp. 115–142

Bascelli, Tiziana (2009), *Galileo Galilei, Sidereus Nuncius ovvero Avviso Sidereo, Traduzione di Tiziana Bascelli, Introduzione e note di William Shea e Tiziana Bascelli* (Venice)

Battistini, Andrea (1993), Galileo Galilei, *Sidereus Nuncius*, a cura di Andrea Battistini, traduzione di Maria Timpanaro Cardini, con teste a fronte (Venice)

Biagioli, Mario (1993), *Galileo, Courtier: The Practice of Science in the Culture of Absolutism* (Chicago)

Biagioli, Mario (2006), *Galileo's Instruments of Credit: Telescopes, Images, Secrecy* (Chicago)

Boghardt, Martin (1990), *Archäologie des gedruckten Buches*, ed. Paul Needham and Julie Boghardt (Wiesbaden, 2008), ch. 1, "Der Buchdruck und das Prinzip des typographischen Kreislaufs: Modell einer Erfindung" (first published 1990)

Bonelli, Maria Luisa (1972), "L'ultimo discepolo: Vincenzio Viviani," in *Saggi su Galileo Galileo*, ed. Carlo Maccagni (Florence), II, pp. 656–688

Bowers, Fredson (1941), "The Headline in Early Books," *English Institute Annual 1941* (New York, 1942), pp. 185–205; reprinted in his *Essays in Bibliography, Text, and Editing* (Charlottesville, 1975), pp. 199–211

Bredekamp, Horst (2007), *Galilei der Künstler* (Berlin)

Breman, Paul (1992), *Books on Military Architecture Printed in Venice: An Annotated Catalogue* ('t Goy-Houten: Hes & De Graaf)

Briquet, Charles-Moïse, *Les Filigranes: Dictionnaire historique des marques du papier dès leur apparition vers 1282 jusqu'en 1600.* 4 volumes (Geneva, 1907; reprint, ed. Allan H. Stevenson, Amsterdam, 1968)

Bruni & Evans: Roberto L. Bruni, D. Wyn Evans, *Italian 17th-century Books in Cambridge Libraries: A Short-title Catalogue* (Florence, 1997)

Bucciantini, Massimo (1998), "Celebration and Conservation: the Galilean Collection of the National Library of Florence," in *Archives of the Scientific Revolution: The Formation and Exchange of Ideas in Seventeenth-Century Europe*, ed. Michael Hunter (Boydell Press, 1998), pp. 21–34

Bühler, Curt (1949), "The Headlines of William de Machlinia's Year-Book, 37 Henry VI," *Studies in Bibliography* I, pp. 125–132

Camerota, Michele (1992), *Gli scritti De motu antiquiora di Galileo Galilei: il Ms. Gal. 71. Un'analisi storico-critica* (Cagliari)

Camerota, Michele (2001), "Galileo Galilei und der Accademico Incognito," in *Der ungebändigte Galilei*, Beiträge zu einem Symposion, ed. Michael Segre and Eberhard Knobloch (Stuttgart), pp. 35–48

Chaberlot, Frédéric (2003), *La Voie lactée: Histoire des conceptions et des modéles de notre Galaxie des temps anciens aux années 1930* (Paris)

Cinti, Dino (1957), *Biblioteca Galileiana raccolta dal Principe Giampaolo Rocco di Torrepadula* (Florence)

Culture et Civilisation (1967), *Sidereus Nuncius magna, longeque admirabilia spectacula pandens* ... (Brussels: Culture et Civilisation, 1967, "impression anastaltique" [sic])

Derolez, Albert (1984), *Codicologie des manuscrits en écriture humanistique sur parchemin.* 2 vols. (Turnhout)

Derolez, Albert (2003), *The Palaeography of Gothic Manuscript Books* (Cambridge)

Dibner, Bern (1980), *Heralds of Science, as Represented by Two Hundred Epochal Books and Pamphlets in the Dibner Library, Smithsonian Institution* (New York)

Dietz, Alexander (1910), *Frankfurter Handelsgeschichte* (4 volumes, Frankfurt/Main, 1910–1925)

Domus Galilaeana (1964), *Il Sidereus Nuncius di Galileo Galilei.* Riproduzione Anastatica dell'editio princeps, a cura del Comitato Nazionale per le Manifestazioni Celebrative del IV Centenario della Nascità di Galileo (Pisa: Domus Galilaeana, 1964)

Drake, Stillman (1962), "An Unpublished Fragment Relating to the Telescope and the Medicean Stars," *Physis* 4, pp. 342–344

Drake, Stillman (1964), "Galileo and Girolamo Magagnati," *Physis* 6, pp. 269–286

Drake, Stillman (1972), "On the Probable Order of Galileo's Notes on Motion," *Physis* 14, pp. 55–56

Drake, Stillman (1976), "Galileo's First Telescopic Observations," *Journal for the History of Astronomy* 7, pp. 153–168

Drake, Stillman (1978), *Galileo at Work: His Scientific Biography* (Chicago)

Drake, Stillman (1978b), *Galileo Galilei, Operations of the Geometric and Military Compass 1606*. Translated, with an Introduction and Commentary by Stillman Drake (Washington, D.C.)

Drake, Stillman (1979), *Galileo's Notes on Motion, Arranged in Probable Order of Composition and Presented in Reduced Facsimile* (Florence: Supplemento agli Annali dell'Istituto e Museo di Storia della Scienza, Anno 1979 – Fascicolo 2)

Drake, Stillman (1983), *Telescopes, Tides, and Tactics: A Galilean Dialogue about the Starry Messenger and Systems of the World* (Chicago)

Dressler, Fridolin (1972), *Die Exlibris der Bayerischen Hof- und Staatsbibliothek, 17. bis 20. Jahrhundert* (Wiesbaden, 1972)

[Evans, Herbert M.] (1934), *Exhibition of First Editions of Epochal Achievements in the History of Science* (Berkeley, June 1934)

Favaro, Antonio (1886), *Documenti inediti per la storia dei manoscritti galileiani nella Biblioteca Nazionale di Firenze* (Rome, 1886, 192 pp.), offprinted from *Bullettino di bibliografia e di storia delle scienze matematiche e fisiche*, 18 (1885), pp. 1–112, pp. 151–230.

Favaro, Antonio (1887), "La Libreria di Galileo Galilei descritta ed illustrata," *Bullettino di bibliografia e di storia delle scienze matematiche e fisiche* 19 (1887), pp. 219–293

Favaro, Antonio (1889), "Intorno alla licenza di stampa del Sidereus Nuncius di Galileo Galilei," *Rivista delle biblioteche* 2 (1889) pp. 98–103

Favaro, Antonio (1992), *Scampoli Galileiani*, ed. Lucia Rossetti & Maria Laura Soppelsa, 2 vols. (Trieste, 1992). Originally published as contributions to the *Atti e memorie* of the Accademia di scienze, lettere et arti in Padua, 1885 and after.

Foxon, David (1970), "The Varieties of Early Proof: Cartwright's Royal Slave, 1639, 1640," *The Library* 5th Series 25, pp. 151–154

Galilei, Galileo, *Le opere* … [ed. Antonio Favaro], Edizione Nazionale, 20 volumes (Florence, 1890–1909). Vol. III, Parte Seconda (Florence, 1931) [cited as: *Opere*]

Gaskell, Philip (1972), *A New Introduction to Bibliography* (New York and Oxford)

Gingerich, Owen (1975), "Dissertatio cum Professore Righini et Sidereo Nuncio," in *Reason, Experiment and Mysticism in the Scientific Revolution*, ed. Maria Luisa Bonelli and William R. Shea (New York), pp. 77–88

Gingerich, Owen & Albert Van Helden (2003), "From Occhiale to Printed Page: The Making of Galileo's Sidereus Nuncius," *Journal for the History of Astronomy* 34, pp. 251–267

Gingerich, Owen (2009), "The Curious Case of the M–L Sidereus Nuncius," *Galilaeana* 6, pp. 141–165

Grendler, Paul F. (1977), *The Roman Inquisition and the Venetian Press, 1540–1605* (Princeton)

Gingerich, Owen (1978), "Books for Sarpi: The Smuggling of Prohibited Books into Venice during the Interdict of 1606–1607," in *Essays Presented to Myron P. Gilmore*, ed. Sergio Bertelli and Gloria Ramakus, 2 vols. (Florence), II, pp. 105–114

Griffante, Caterina (2003–2006), *Edizioni veneziane del Seicento: censimento*, 2 vols. (Milan) [cited as EV17]

Hellinga, Lotte (1986), "Proof-reading in 1459: the Munich copy of Guillelmus Duranti, Rationale," in *Ars Impressoria ... Festgabe für Severin Corsten* (Munich), pp. 183–202

Hinman, Charlton (1963), *The Printing and Proof-reading of the First Folio of Shakespeare*, 2 vols. (Oxford)

Horblit, Harrison D. (1964), *One Hundred Books Famous in Science* (New York)

Lambert, Ladina Bezzola (2002), *Imagining the Unimaginable: The Poetics of Early Modern Astronomy* (Amsterdam and New York)

Mallinckrodt, Bernard von (1640), *De ortu ac progressu artis typographicae dissertatio historica* … (Cologne, 1640).

Marabini, Stefano, Lucio Donati & Gian Battista Vai (2003), "Ulisse Aldrovandi's printing contract, 1594," in *Four Centuries of the Word Geology, Ulisse Aldrovandi 1603 in Bologna*, ed. Gian Battista Vai & William Cavazza (Bologna), pp. 113–125

McKenzie, Donald F. (1973), "Eight Quarto Proof Sheets of 1594 Set by Formes: A frutefull commentarie," *The Library* 5th Series 28, pp. 1–13

McKerrow, R. B. (1935), review of Percy Simpson, *Proof-reading in the Sixteenth Seventeenth and Eighteenth Centuries*, in *The Library* 4th Series 16, pp. 347–352.

Meeus, Jean (1964), "Galileo's First Records of Jupiter's Satellites," *Sky and Telescope* 27 no. 2 (February 1964), pp. 105–106

Mirto, Alfonso (2001), "I rapporti epistolari tra Cassiano dal Pozzo e Carlo Roberto Dati," *Nouvelles de la République des Lettres* 2001 no. 2, pp. 7–102

Mortimer, Ruth (1964, 1974), Harvard College Library Department of Printing and Graphic Arts, *Catalogue of Books and Manuscripts, Part I: French 16th Century Books*, 2 volumes (Cambridge, Mass.); ––, *Part II: Italian 16th Century Books*, 2 volumes (Cambridge, Mass.)

Motta, Emilio. (1893), "Demetrio Calcondila, editore," *Archivio storico lombardo* 20, pp. 143–166

Moxon, Joseph (1683, 1962), *Mechanick Exercises on the Whole Art of Printing (1683–1684)* … edited by Herbert Davis & Harry Carter. Second Edition (Oxford, 1962)

Needham, Paul (1986), "The Cambridge Proof Sheets of Mentelin's Latin Bible," *Transactions of the Cambridge Bibliographical Society* IX, pp. 1–35

Needham, Paul (1986b), "A Gutenberg Bible used as Printer's Copy," ibid., pp. 36–75

Needham, Paul (1994), "Res papirea: Sizes and Formats of the Late Medieval Book," in *Rationalisierung der Buchherstellung im Mittelalter und in der frühen Neuzeit*, ed. Peter Rück and Martin Boghardt (Marburg an der Lahn: elementa diplomatica, 2), pp. 123–145

Needham, Paul (1994b), "Aldus Manutius's Paper Stocks: The Evidence of Two Uncut Books," *Princeton University Library Chronicle* 55, pp. 287–307

Needham, Paul (2000), "Concepts of Paper Study," in *Puzzles in Paper: Concepts in Historical Watermarks*, ed. Daniel W. Mosser, Michael Saffle & Ernest W. Sullivan II (London), pp. 1–36

Needham, Paul (2004), "The Canterbury Tales and the Rosary: A Mirror of Caxton's Devotions?" in *The Medieval Book and a Modern Collector: Essays in Honour of*

Toshiyuki Takamiya, ed. Takami Matsuda, Richard A. Linenthal, and John Scahill (D. S. Brewer & Yushodo Press), pp. 313–356

Pantin, Isabelle (1992), Ed., Tr., Comm.: Galileo Galilei, *Sidereus Nuncius Le messager céleste* (Paris)

Pantin, Isabelle (1993), Ed., Tr., Comm.: Johannes Kepler, *Dissertatio cum nuncio sidereo ... Narratio de observatis Jovis satellitibus* (Paris)

Pesendorfer, Franz (1987), *Il governo di famiglia in Toscana. Le memorie del granduca Leopoldo II di Lorena* (Florence)

Petrucciani, Alberto (1988), *Gli Incunaboli della Biblioteca Durazzo* (Genoa)

Pighi, Giovanni Battista (1972), "Il latino di Galileo Galilei," in *Saggi su Galileo Galileo*, ed. Carlo Maccagni (Florence), II, pp. 541–550

Pollard, Alfred W. (1891), *Last Words on the History of the Title-Page with Notes on Some Colophons and Twenty-Seven Facsimiles of Title-Pages* (London, 1891)

Poppi, Antonino (1998), "Una implicita ritrattazione di Antonio Favaro sulla licenza di stampa del Sidereus Nuncius," *Atti e memorie dell'Accademia Patavina di scienze lettere ed arti*, vol. 110 pt. 3 (1998), pp. 99–105

Povey, Kenneth (1965), "Working to Rule, 1600–1800: A Study of Pressmen's Practice," *The Library* 5th Series 20 (1965), pp. 13–54

Printing and the Mind of Man (1967): *Printing and the Mind of Man: a Descriptive Catalogue Illustrating the Impact of Print on the Evolution of Western Civilization during Five Centuries*, ed. John Carter and Percy H. Muir (London and New York)

Procissi, Angiolo (1959), *La Collezione Galileiana della Biblioteca Nazionale di Firenze*, Volume I (Rome, 1959)

Reeves, Eileen (1997), *Painting the Heavens: Art and Science in the Age of Galileo* (Princeton)

Reeves, Eileen (2008), *Galileo's Glassworks: The Telescope and the Mirror* (Cambridge, Mass.)

Righini, Guglielmo (1975), "New Light of Galileo's Lunar Observations," in *Reason, Experiment and Mysticism in the Scientific Revolution*, ed. Maria Luisa Bonelli and William R. Shea (New York), pp. 59–76

Righini, Guglielmo (1976), "L'Oroscopo Galileiano di Cosimo de' Medici," *Annali dell'Istituto e Museo di Storia della Scienza di Firenze* 1, pp. 29–36

Rosen, Edward (1952), "Galileo on the Distance between the Earth and the Moon," *Isis* 43, pp. 344–348

Sarpi, Paolo (1931), *Lettere ai Protestanti*, ed. Manlio Duilio Busnelli, 2 volumes (Bari)

Sarpi, Paolo (1961), *Lettere ai Gallicani*, ed. Boris Ulianich (Wiesbaden)

Sayce, R. A. (1979), "Compositorial Practices and the Localization of Printed Books 1530–1800," *The Library* 5th Series 21, pp. 1–45; reprinted with additions: Oxford Bibliographical Society 1979, Occasional Publications 13

Sharratt, Michael (1994), *Galileo: Decisive Innovator* (Cambridge)

Shea, William R. (2009), *Galileo's Sidereus Nuncius or A Sidereal Message. Translated from the Latin by William R. Shea, Introduction and Notes by William R. Shea and Tiziana Bascelli* (Sagamore Beach, Mass.)

Simpson, Percy (1935), *Proof-reading in the Sixteenth, Seventeenth and Eighteenth Centuries* (Oxford)

Smith, Logan Pearsall (1907), *The Life and Letters of Sir Henry Wotton*, 2 volumes (London)

Smith, Margaret M. (1988), "Printed foliation: forerunner to printed page-numbers?" *Gutenberg-Jahrbuch 1988*, pp. 54–70

Smith, Margaret M. (2000), *The Title-page: Its Early Development 1460–1510* (London)

Sorbelli, Albano (1925), "Di un particolare forma de 'richiamo' in antichi incunabuli," in *Gutenberg Festschrift zur Feier des 25iaehrigen bestehens des Gutenbergmuseums in Mainz*, ed. Aloys Ruppel (Mainz), pp. 325–330

Swerdlow, Noel (1998), "Galileo's Discoveries with the Telescope and their Evidence for the Copernican Theory," *The Cambridge Companion to Galileo*, ed. Peter Machamer (Cambridge), pp. 244–270

Tanselle, G. Thomas (2002), *Introduction to Bibliography: Seminar Syllabus* (Charlottesville, Va.)

Van Helden, Albert (1977), *The Invention of the Telescope* (Philadelphia: *Transactions of the American Philosophical Society* new series 67 no. 4)

Van Helden, Albert (1989), Ed., Tr., Comm.: Galileo Galilei, *Sidereus Nuncius or The Sidereal Messenger, translated with introduction, conclusion, and notes* (Chicago)

Vergara Caffarelli, Roberto (2004), *Galileo e Pisa* (Pisa)

Vianini Tolomei, Claudio (1991), *Legatura romana barocca 1565–1700*, ed. Claudio Vianini Tolomei (Rome)

Weiss, Adrian (2007), "Casting Compositors, Foul Cases, and Skeletons: Printing in Middleton's Age," in: *Thomas Middleton and Early Modern Textual Culture: Companion to the Collected Works of Thomas Middleton*, ed. Gary Taylor et al. (Oxford), pp. 195–225

Whitaker, Ewan (1978), "Galileo's Lunar Observations and the Dating of the Composition of Sidereus Nuncius," *Journal for the History of Astronomy* 9, pp. 155–169

Wilson, Nigel (1995), *Mediaeval Greek Bookhands, Examples Selected from Greek Manuscripts in Oxford Libraries* (Cambridge, Mass., Mediaeval Academy of America, 1972–1973; reprinted 1995)

Woodward, David (1996), *Catalogue of Watermarks in Italian Printed Maps ca 1540–1600* (Chicago)

Wootton, David (2009), "New Light on the Composition and Publication of the Sidereus Nuncius," *Galilaeana* 6, pp. 123–140

Index

The Index records names except for Galileo, a constant presence; and places. A number of technical terms and features are brought under "Paper" and "Printing", and cited early editions, except for the first edition of *Sidereus Nuncius* itself, are brought under "Printed editions".